FUNCTIONAL ANALYSIS

FUNCTIONAL ANALYSIS

Third Revised Edition

WALTER RUDIN
Professor of Mathematics
University of Wisconsin

MedTech Science Press
A Division of Scientific International
Engaging Sciences—Developing Minds!

FUNCTIONAL ANALYSIS

MedTech Science Press
A Division of
Scientific International Pvt. Ltd.

Third Revised Edition: 2024

ISBN: 978-93-87465-86-2

Published by: Rajan Jain, Scientific International (Pvt.) Ltd.

Registered Office: • **New Delhi:** 4222/1, Ansari Road, Daryaganj, New Delhi-110002

Branch Office: • **Bengaluru:** House No. 1665, Ground Floor, 18th Main, BSK 2nd Stage, Bengaluru-560070

CONTENTS

PREFACE TO THE NEW EDITION

Functional analysis delves into topological-algebraic structures and explores methodologies for applying this knowledge to analytical problems. An effective introductory text on this subject should cover axiomatics, delve into specific topics, and present intriguing applications to other mathematical branches. I believe the present book adheres to these criteria.

Given the expansive and rapidly evolving nature of the subject, certain areas had to be selected for inclusion while others were regrettably left out, acknowledging that experts may miss their favorite topics. The intention was not to create an encyclopedic treatise but a book that sparks further exploration.

Esoteric topics in the general theory of topological vector spaces, such as uniform spaces, Moore-Smith convergence, nets, filters, bornological spaces, and barreled ones, have been omitted to maintain a manageable size. The focus remains on the interplay between abstract concepts and concrete applications, emphasizing the subject's practical and captivating aspects.

Part I of the book provides comprehensive material for most concrete problem applications. This intentional exclusion of certain esoteric topics allows for a balance between abstraction and practicality. The selected material features a presentation of the general theory without assuming local convexity, deriving basic properties of compact operators from duality theory in Banach spaces, incorporating the Krein-Milman theorem, and exploring the theory of distributions and Fourier transforms in detail.

I assume readers' familiarity with the theory of measure and Lebesgue integration, basic properties of holomorphic functions, and the corresponding topological background. Appendix A briefly presents some additional topological facts, requiring minimal algebraic background beyond understanding homomorphisms.

Historical references are compiled in Appendix B, referencing original sources and additional resources. Applications are concentrated in Chapters 5, 8, and 9, as well as in Chapter 11 and over 250 exercises, with many exercises accompanied by hints. The interdependence of chapters is illustrated in the provided diagram.

While suggestions were made to introduce applications earlier in the text, Chapter 5 starts with background indications for each application, facilitating study at an earlier stage if desired. In this edition, some material from Chapter 10 has been removed, and new content, such as the mean ergodic theorem of von Neumann, the Hille-Yosida theorem, fixed point theorems, Bonsall's application of the closed range theorem, and Lomonosov's invariant subspace theorem, has been added. Sections have been revised for clarity, and certain proofs have been shortened and simplified.

These changes stem from valuable suggestions from friends and colleagues, with special acknowledgment to Justin Peters, Ralph Raimi, and the Russian translator of the first edition. My gratitude to all who have contributed to the improvement of this work.

FUNCTIONAL ANALYSIS

PART I

GENERAL THEORY

CHAPTER 1

TOPOLOGICAL VECTOR SPACES

Introduction

1.1 Many problems that analysts study are not primarily concerned with a single object such as a function, a measure, or an operator, but they deal instead with large classes of such objects. Most of the interesting classes that occur in this way turn out to be vector spaces, either with real scalars or with complex ones. Since limit processes play a role in every analytic problem (explicitly or implicitly), it should be no surprise that these vector spaces are supplied with metrics, or at least with topologies, that bear some natural relation to the objects of which the spaces are made up. The simplest and most important way of doing this is to introduce a *norm*. The resulting structure (defined below) is called a normed vector space, or a normed linear space, or simply a *normed space*.

Throughout this book, the term *vector space* will refer to a vector space over the complex field $\mathcal{C}$ or over the real field R. For the sake of completeness, detailed definitions are given in Section 1.4.

1.2 Normed spaces A vector space X is said to be a *normed space* if to every $x \in X$ there is associated a nonnegative real number $\|x\|$, called the *norm* of x, in such a way that

(*a*) $\|x + y\| \le \|x\| + \|y\|$ for all x and y in X,
(*b*) $\|\alpha x\| = |\alpha| \, \|x\|$ if $x \in X$ and α is a scalar,
(*c*) $\|x\| > 0$ if $x \neq 0$.

The word "norm" is also used to denote the *function* that maps x to $\|x\|$.

Every normed space may be regarded as a metric space, in which the distance $d(x, y)$ between x and y is $\|x - y\|$. The relevant properties of d are

(*i*) $0 \le d(x, y) < \infty$ for all x and y,
(*ii*) $d(x, y) = 0$ if and only if $x = y$,
(*iii*) $d(x, y) = d(y, x)$ for all x and y,
(*iv*) $d(x, z) \le d(x, y) + d(y, z)$ for all x, y, z.

In any metric space, the *open ball* with center at x and radius r is the set

$$B_r(x) = \{y: d(x, y) < r\}.$$

In particular, if X is a normed space, the sets

$$B_1(0) = \{x: \|x\| < 1\} \qquad \text{and} \qquad \bar{B}_1(0) = \{x: \|x\| \le 1\}$$

are the *open unit ball* and the *closed unit ball* of X, respectively.

By declaring a subset of a metric space to be open if and only if it is a (possibly empty) union of open balls, a *topology* is obtained. (See Section 1.5.) It is quite easy to verify that the vector space operations (addition and scalar multiplication) are continuous in this topology, if the metric is derived from a norm, as above.

A *Banach space* is a normed space which is *complete* in the metric defined by its norm; this means that every Cauchy sequence is required to converge.

1.3 Many of the best-known function spaces are Banach spaces. Let us mention just a few types: spaces of continuous functions on compact spaces; the familiar L^p-spaces that occur in integration theory; Hilbert spaces — the closest relatives of euclidean spaces; certain spaces of differentiable functions; spaces of continuous linear mappings from one Banach space into another; Banach algebras. All of these will occur later on in the text.

But there are also many important spaces that do not fit into this framework. Here are some examples:

(*a*) $C(\Omega)$, the space of all continuous complex functions on some open set Ω in a euclidean space R^n.

(*b*) $H(\Omega)$, the space of all holomorphic functions in some open set Ω in the complex plane.

(*c*) C_K^∞, the space of all infinitely differentiable complex functions on R^n that vanish outside some fixed compact set K with nonempty interior.

(*d*) The test function spaces used in the theory of distributions, and the distributions themselves.

These spaces carry natural topologies that cannot be induced by norms, as we shall see later. They, as well as the normed spaces, are examples of *topological vector spaces*, a concept that pervades all of functional analysis.

After this brief attempt at motivation, here are the detailed definitions, followed (in Section 1.9) by a preview of some of the results of Chapter 1.

1.4 Vector spaces The letters R and $\mathcal{C}$ will always denote the field of real numbers and the field of complex numbers, respectively. For the moment, let Φ stand for either R or $\mathcal{C}$. A *scalar* is a member of the *scalar field* Φ. A *vector space over* Φ is a set X, whose elements are called vectors, and in which two operations, *addition* and *scalar multiplication*, are defined, with the following familiar algebraic properties:

(*a*) To every pair of vectors x and y corresponds a vector $x + y$, in such a way that

$$x + y = y + x \qquad \text{and} \qquad x + (y + z) = (x + y) + z;$$

X contains a unique vector 0 (the *zero vector* or *origin* of X) such that $x + 0 = x$ for every $x \in X$; and to each $x \in X$ corresponds a unique vector $-x$ such that $x + (-x) = 0$.

(*b*) To every pair (α, x) with $\alpha \in \Phi$ and $x \in X$ corresponds a vector αx, in such a way that

$$1x = x, \qquad \alpha(\beta x) = (\alpha\beta)x,$$

and such that the two distributive laws

$$\alpha(x + y) = \alpha x + \alpha y, \qquad (\alpha + \beta)x = \alpha x + \beta x$$

hold.

The symbol 0 will of course also be used for the zero element of the scalar field.

A *real vector space* is one for which $\Phi = R$; a *complex vector space* is one for which $\Phi = \mathcal{C}$. Any statement about vector spaces in which the scalar field is not explicitly mentioned is to be understood to apply to both of these cases.

If X is a vector space, $A \subset X$, $B \subset X$, $x \in X$, and $\lambda \in \Phi$, the following notations will be used:

$$x + A = \{x + a : a \in A\},$$
$$x - A = \{x - a : a \in A\},$$
$$A + B = \{a + b : a \in A, b \in B\},$$
$$\lambda A = \{\lambda a : a \in A\}.$$

In particular (taking $\lambda = -1$), $-A$ denotes the set of all additive inverses of members of A.

A word of warning: With these conventions, it may happen that $2A \neq A + A$ (Exercise 1).

A set $Y \subset X$ is called a *subspace* of X if Y is itself a vector space (with respect to the same operations, of course). One checks easily that this happens if and only if $0 \in Y$ and

$$\alpha Y + \beta Y \subset Y$$

for all scalars α and β.

A set $C \subset X$ is said to be *convex* if

$$tC + (1 - t)C \subset C \qquad (0 \le t \le 1).$$

In other words, it is required that C should contain $tx + (1 - t)y$ if $x \in C$, $y \in C$, and $0 \le t \le 1$.

A set $B \subset X$ is said to be *balanced* if $\alpha B \subset B$ for every $\alpha \in \Phi$ with $|\alpha| \le 1$.

A vector space X has *dimension n* ($\dim X = n$) if X has a *basis* $\{u_1, \ldots, u_n\}$. This means that every $x \in X$ has a unique representation of the form

$$x = \alpha_1 u_1 + \cdots + \alpha_n u_n \qquad (\alpha_i \in \Phi).$$

If $\dim X = n$ for some n, X is said to have *finite dimension.* If $X = \{0\}$, then $\dim X = 0$.

Example. If $X = \mathbb{C}$ (a one-dimensional vector space over the scalar field $\mathbb{C}$), the balanced sets are $\mathbb{C}$, the empty set $\emptyset$, and every circular disc (open or closed) centered at 0. If $X = R^2$ (a two-dimensional vector space over the scalar field R), there are many more balanced sets; any line segment with midpoint at (0, 0) will do. The point is that, in spite of the well-known and obvious identification of $\mathbb{C}$ with R^2, these two are entirely different as far as their vector space structure is concerned.

1.5 Topological spaces A *topological space* is a set S in which a collection τ of subsets (called *open sets*) has been specified, with the following properties: S is open, $\varnothing$ is open, the intersection of any two open sets is open, and the union of every collection of open sets is open. Such a collection τ is called a *topology on S*. When clarity seems to demand it, the topological space corresponding to the topology τ will be written (S, τ) rather than S.

Here is some of the standard vocabulary that will be used, if S and τ are as above.

A set $E \subset S$ is *closed* if and only if its complement is open. The *closure* $\bar{E}$ of E is the intersection of all closed sets that contain $\bar{E}$. The *interior* E° of E is the union of all open sets that are subsets of E. A *neighborhood* of a point $p \in S$ is any open set that contains p. (S, τ) is a *Hausdorff space*, and τ is a *Hausdorff topology*, if distinct points of S have disjoint neighborhoods. A set $K \subset S$ is *compact* if every open cover of K has a finite subcover. A collection $\tau' \subset \tau$ is a *base* for τ if every member of τ (that is, every open set) is a union of members of τ'. A collection γ of neighborhoods of a point $p \in S$ is a *local base at p* if every neighborhood of p contains a member of γ.

If $E \subset S$ and if σ is the collection of all intersections $E \cap V$, with $V \in \tau$, then σ is a topology on E, as is easily verified; we call this the topology that E *inherits* from S.

If a topology τ is induced by a metric d (see Section 1.2) we say that d and τ are *compatible* with each other.

A sequence $\{x_n\}$ in a Hausdorff space X *converges* to a point $x \in X$ (or $\lim_{n\to\infty} x_n = x$) if every neighborhood of x contains all but finitely many of the points x_n.

1.6 Topological vector spaces Suppose τ is a topology on a vector space X such that

(*a*) *every point of X is a closed set, and*

(*b*) *the vector space operations are continuous with respect to* τ.

Under these conditions, τ is said to be a *vector topology* on X, and X is a *topological vector space*.

Here is a more precise way of stating (*a*): For every $x \in X$, the set $\{x\}$ which has x as its only member is a closed set.

In many texts, (*a*) is omitted from the definition of a topological vector space. Since (*a*) is satisfied in almost every application, and since most theorems of interest require (*a*) in their hypotheses, it seems best to include it in the axioms. [Theorem 1.12 will show that (*a*) and (*b*) together imply that τ is a Hausdorff topology.]

To say that addition is *continuous* means, by definition, that the mapping

$$(x, y) \to x + y$$

of the cartesian product $X \times X$ into X is continuous: If $x_i \in X$ for $i = 1, 2$, and if V is a neighborhood of $x_1 + x_2$, there should exist neighborhoods V_i of x_i such that

$$V_1 + V_2 \subset V.$$

Similarly, the assumption that scalar multiplication is continuous means that the mapping

$$(\alpha, x) \to \alpha x$$

of $\Phi \times X$ into X is continuous: If $x \in X$, α is a scalar, and V is a neighborhood of αx, then for some $r > 0$ and some neighborhood W of x we have $\beta W \subset V$ whenever $|\beta - \alpha| < r$.

A subset E of a topological vector space is said to be *bounded* if to every neighborhood V of 0 in X corresponds a number $s > 0$ such that $E \subset tV$ for every $t > s$.

1.7 Invariance Let X be a topological vector space. Associate to each $a \in X$ and to each scalar $\lambda \neq 0$ the *translation operator* T_a and the *multiplication operator* M_λ, by the formulas

$$T_a(x) = a + x, \qquad M_\lambda(x) = \lambda x \qquad (x \in X).$$

The following simple proposition is very important:

Proposition. *T_a and M_λ are homeomorphisms of X onto X.*

PROOF. The vector space axioms alone imply that T_a and M_λ are one-to-one, that they map X onto X, and that their inverses are T_{-a} and $M_{1/\lambda}$, respectively. The assumed continuity of the vector space operations implies that these four mappings are continuous. Hence each of them is a homeomorphism (a continuous mapping whose inverse is also continuous). ////

One consequence of this proposition is that every vector topology τ is *translation-invariant* (or simply *invariant*, for brevity): A set $E \subset X$ is open if and only if each of its translates $a + E$ is open. Thus τ is completely determined by any local base.

In the vector space context, the term *local base* will always mean a local base at 0. A local base of a topological vector space X is thus a collection $\mathscr{B}$ of neighborhoods of 0 such that every neighborhood of 0 contains a member of $\mathscr{B}$. The open sets of X are then precisely those that are unions of translates of members of $\mathscr{B}$.

A metric d on a vector space X will be called *invariant* if

$$d(x + z, y + z) = d(x, y)$$

for all x, y, z in X.

1.8 Types of topological vector spaces In the following definitions, X always denotes a topological vector space, with topology τ.

(*a*) X is *locally convex* if there is a local base $\mathscr{B}$ whose members are convex.
(*b*) X is *locally bounded* if 0 has a bounded neighborhood.
(*c*) X is *locally compact* if 0 has a neighborhood whose closure is compact.
(*d*) X is *metrizable* if τ is compatible with some metric d.
(*e*) X is an *F-space* if its topology τ is induced by a complete invariant metric d. (Compare Section 1.25.)
(*f*) X is a *Fréchet space* if X is a locally convex F-space.
(*g*) X is *normable* if a norm exists on X such that the metric induced by the norm is compatible with τ.
(*h*) *Normed spaces* and *Banach spaces* have already been defined (Section 1.2).
(*i*) X has the *Heine-Borel property* if every closed and bounded subset of X is compact.

The terminology of (*e*) and (*f*) is not universally agreed upon: In some texts, local convexity is omitted from the definition of a Fréchet space, whereas others use F-space to describe what we have called Fréchet space.

1.9 Here is a list of some relations between these properties of a topological vector space X.

(*a*) If X is locally bounded, then X has a countable local base [part (*c*) of Theorem 1.15].
(*b*) X is metrizable if and only if X has a countable local base (Theorem 1.24).
(*c*) X is normable if and only if X is locally convex and locally bounded (Theorem 1.39).
(*d*) X has finite dimension if and only if X is locally compact (Theorems 1.21, 1.22).
(*e*) If a locally bounded space X has the Heine-Borel property, then X has finite dimension (Theorem 1.23).

The spaces $H(\Omega)$ and C_K^∞ mentioned in Section 1.3 are infinite-dimensional Fréchet spaces with the Heine-Borel property (Sections 1.45, 1.46). They are therefore not locally bounded, hence not normable; they also show that the converse of (*a*) is false.

On the other hand, there exist locally bounded *F*-spaces that are not locally convex (Section 1.47).

Separation Properties

1.10 Theorem *Suppose K and C are subsets of a topological vector space X, K is compact, C is closed, and $K \cap C = \varnothing$. Then 0 has a neighborhood V such that*

$$(K + V) \cap (C + V) = \varnothing.$$

Note that $K + V$ is a union of translates $x + V$ of V $(x \in K)$. Thus $K + V$ is an open set that contains K. The theorem thus implies the existence of disjoint open sets that contain K and C, respectively.

PROOF. We begin with the following proposition, which will be useful in other contexts as well:

If W is a neighborhood of 0 in X, then there is a neighborhood U of 0 which is symmetric (in the sense that $U = -U$) and which satisfies $U + U \subset W$.

To see this, note that $0 + 0 = 0$, that addition is continuous, and that 0 therefore has neighborhoods V_1, V_2 such that $V_1 + V_2 \subset W$. If

$$U = V_1 \cap V_2 \cap (-V_1) \cap (-V_2),$$

then U has the required properties.

The proposition can now be applied to U in place of W and yields a new symmetric neighborhood U of 0 such that

$$U + U + U + U \subset W.$$

It is clear how this can be continued.

If $K = \varnothing$, then $K + V = \varnothing$, and the conclusion of the theorem is obvious. We therefore assume that $K \neq \varnothing$, and consider a point $x \in K$. Since C is closed, since x is not in C, and since the topology of X is invariant under translations, the preceding proposition shows that 0 has a symmetric neighborhood V_x such that $x + V_x + V_x + V_x$ does not intersect C; the symmetry of V_x shows then that

$$(x + V_x + V_x) \cap (C + V_x) = \varnothing. \tag{1}$$

Since K is compact, there are finitely many points $x_1, \ldots, x_n$ in K such that

$$K \subset (x_1 + V_{x_1}) \cup \cdots \cup (x_n + V_{x_n}).$$

Put $V = V_{x_1} \cap \cdots \cap V_{x_n}$. Then

$$K + V \subset \bigcup_{i=1}^{n} (x_i + V_{x_i} + V) \subset \bigcup_{i=1}^{n} (x_i + V_{x_i} + V_{x_i}),$$

and no term in this last union intersects $C + V$, by (1). This completes the proof. ////

Since $C + V$ is open, it is even true that the *closure* of $K + V$ does not intersect $C + V$; in particular, the closure of $K + V$ does not intersect C. The following special case of this, obtained by taking $K = \{0\}$, is of considerable interest.

1.11 Theorem *If $\mathscr{B}$ is a local base for a topological vector space X, then every member of $\mathscr{B}$ contains the closure of some member of $\mathscr{B}$.*

So far we have not used the assumption that every point of X is a closed set. We now use it and apply Theorem 1.10 to a pair of distinct points in place of K and C. The conclusion is that these points have disjoint neighborhoods. In other words, the Hausdorff separation axiom holds:

1.12 Theorem *Every topological vector space is a Hausdorff space.*

We now derive some simple properties of closures and interiors in a topological vector space. See Section 1.5 for the notations $\bar{E}$ and E°. Observe that a point p belongs to $\bar{E}$ if and only if every neighborhood of p intersects E.

1.13 Theorem *Let X be a topological vector space.*

(*a*) *If $A \subset X$ then $\bar{A} = \bigcap (A + V)$, where V runs through all neighborhoods of 0.*

(*b*) *If $A \subset X$ and $B \subset X$, then $\bar{A} + \bar{B} \subset \overline{A + B}$.*

(*c*) *If Y is a subspace of X, so is $\bar{Y}$.*

(*d*) *If C is a convex subset of X, so are $\bar{C}$ and C°.*

(*e*) *If B is a balanced subset of X, so is $\bar{B}$; if also $0 \in B^\circ$ then B° is balanced.*

(*f*) *If E is a bounded subset of X, so is $\bar{E}$.*

PROOF. (*a*) $x \in \bar{A}$ if and only if $(x + V) \cap A \neq \varnothing$ for every neighborhood V of 0, and this happens if and only if $x \in A - V$ for every such V. Since $-V$ is a neighborhood of 0 if and only if V is one, the proof is complete.

(*b*) Take $a \in \bar{A}$, $b \in \bar{B}$; let W be a neighborhood of $a + b$. There are neighborhoods W_1 and W_2 of a and b such that $W_1 + W_2 \subset W$. There exist $x \in A \cap W_1$ and $y \in B \cap W_2$, since $a \in \bar{A}$ and $b \in \bar{B}$. Then $x + y$ lies in $(A + B) \cap W$, so that this intersection is not empty. Consequently, $a + b \in \overline{A + B}$.

(*c*) Suppose α and β are scalars. By the proposition in Section 1.7, $\alpha\bar{Y} = \overline{\alpha Y}$ if $\alpha \neq 0$; if $\alpha = 0$, these two sets are obviously equal. Hence it follows from (*b*) that

$$\alpha\bar{Y} + \beta\bar{Y} = \overline{\alpha Y} + \overline{\beta Y} \subset \overline{\alpha Y + \beta Y} \subset \bar{Y};$$

the assumption that Y is a subspace was used in the last inclusion.

The proofs that convex sets have convex closures and that balanced sets have balanced closures are so similar to this proof of (*c*) that we shall omit them from (*d*) and (*e*).

(*d*) Since $C^\circ \subset C$ and C is convex, we have

$$tC^\circ + (1 - t)C^\circ \subset C$$

if $0 < t < 1$. The two sets on the left are open; hence so is their sum. Since every open subset of C is a subset of C°, it follows that C° is convex.

(*e*) If $0 < |\alpha| \leq 1$, then $\alpha B^\circ = (\alpha B)^\circ$, since $x \to \alpha x$ is a homeomorphism. Hence $\alpha B^\circ \subset \alpha B \subset B$, since B is balanced. But αB° is open. So $\alpha B^\circ \subset B^\circ$. If B° contains the origin, then $\alpha B^\circ \subset B^\circ$ even for $\alpha = 0$.

(*f*) Let V be a neighborhood of 0. By Theorem 1.11, $\bar{W} \subset V$ for some neighborhood W of 0. Since E is bounded, $E \subset tW$ for all sufficiently large t. For these t, we have $\bar{E} \subset t\bar{W} \subset tV$. ////

1.14 Theorem *In a topological vector space X,*

(*a*) *every neighborhood of* 0 *contains a balanced neighborhood of* 0, *and*

(*b*) *every convex neighborhood of* 0 *contains a balanced convex neighborhood of* 0.

PROOF. (*a*) Suppose U is a neighborhood of 0 in X. Since scalar multiplication is continuous, there is a $\delta > 0$ and there is a neighborhood V of 0 in X such that $\alpha V \subset U$ whenever $|\alpha| < \delta$. Let W be the union of all these sets αV. Then W is a neighborhood of 0, W is balanced, and $W \subset U$.

(*b*) Suppose U is a convex neighborhood of 0 in X. Let $A = \bigcap \alpha U$, where α ranges over the scalars of absolute value 1. Choose W as in part (*a*). Since W is balanced, $\alpha^{-1}W = W$ when $|\alpha| = 1$; hence $W \subset \alpha U$. Thus $W \subset A$, which implies that the interior A° of A is a neighborhood of 0. Clearly $A^\circ \subset U$. Being an intersection of convex sets, A is convex; hence so is A°. To prove that A° is a neighborhood with the desired properties, we have to show that A° is balanced; for this it suffices to prove that A is balanced. Choose r and β so that $0 \le r \le 1$, $|\beta| = 1$. Then

$$r\beta A = \bigcap_{|\alpha|=1} r\beta\alpha U = \bigcap_{|\alpha|=1} r\alpha U.$$

Since αU is a convex set that contains 0, we have $r\alpha U \subset \alpha U$. Thus $r\beta A \subset A$, which completes the proof. ////

Theorem 1.14 can be restated in terms of local bases. Let us say that a local base $\mathscr{B}$ is *balanced* if its members are balanced sets, and let us call $\mathscr{B}$ *convex* if its members are convex sets.

Corollary

(*a*) *Every topological vector space has a balanced local base.*
(*b*) *Every locally convex space has a balanced convex local base.*

Recall also that Theorem 1.11 holds for each of these local bases.

1.15 Theorem *Suppose V is a neighborhood of* 0 *in a topological vector space X.*

(*a*) *If $0 < r_1 < r_2 < \cdots$ and $r_n \to \infty$ as $n \to \infty$, then*

$$X = \bigcup_{n=1}^{\infty} r_n V.$$

(*b*) *Every compact subset K of X is bounded.*
(*c*) *If $\delta_1 > \delta_2 > \cdots$ and $\delta_n \to 0$ as $n \to \infty$, and if V is bounded, then the collection*

$$\{\delta_n V : n = 1, 2, 3, \ldots\}$$

is a local base for X.

PROOF. (*a*) Fix $x \in X$. Since $\alpha \to \alpha x$ is a continuous mapping of the scalar field into X, the set of all α with $\alpha x \in V$ is open, contains 0, hence contains $1/r_n$ for all large n. Thus $(1/r_n)x \in V$, or $x \in r_n V$, for large n.

(*b*) Let W be a balanced neighborhood of 0 such that $W \subset V$. By (*a*),

$$K \subset \bigcup_{n=1}^{\infty} nW.$$

Since K is compact, there are integers $n_1 < \cdots < n_s$ such that

$$K \subset n_1 W \cup \cdots \cup n_s W = n_s W.$$

The equality holds because W is balanced. If $t > n_s$, it follows that $K \subset tW \subset tV$.

(*c*) Let U be a neighborhood of 0 in X. If V is bounded, there exists $s > 0$ such that $V \subset tU$ for all $t > s$. If n is so large that $s\delta_n < 1$, it follows that $V \subset (1/\delta_n)U$. Hence U actually contains all but finitely many of the sets $\delta_n V$. ////

Linear Mappings

1.16 Definitions When X and Y are sets, the symbol

$$f: X \to Y$$

will mean that f is a mapping of X into Y. If $A \subset X$ and $B \subset Y$, the *image* $f(A)$ of A and the *inverse image* or *preimage* $f^{-1}(B)$ of B are defined by

$$f(A) = \{f(x): x \in A\}, \qquad f^{-1}(B) = \{x: f(x) \in B\}.$$

Suppose now that X and Y are vector spaces *over the same scalar field*. A mapping $\Lambda: X \to Y$ is said to be *linear* if

$$\Lambda(\alpha x + \beta y) = \alpha\Lambda x + \beta\Lambda y$$

for all x and y in X and all scalars α and β. Note that one often writes Λx, rather than $\Lambda(x)$, when Λ is linear.

Linear mappings of X into its scalar field are called *linear functionals*.

For example, the multiplication operators M_α of Section 1.7 are linear, but the translation operators T_a are not, except when $a = 0$.

Here are some properties of linear mappings $\Lambda: X \to Y$ whose proofs are so easy that we omit them; it is assumed that $A \subset X$ and $B \subset Y$:

(*a*) $\Lambda 0 = 0$.

(*b*) If A is a subspace (or a convex set, or a balanced set) the same is true of $\Lambda(A)$.

(*c*) If B is a subspace (or a convex set, or a balanced set) the same is true of $\Lambda^{-1}(B)$.

(*d*) In particular, the set

$$\Lambda^{-1}(\{0\}) = \{x \in X : \Lambda x = 0\} = \mathscr{N}(\Lambda)$$

is a subspace of X, called the *null space* of Λ.

We now turn to continuity properties of linear mappings.

1.17 Theorem *Let X and Y be topological vector spaces. If $\Lambda: X \to Y$ is linear and continuous at 0, then Λ is continuous. In fact, Λ is uniformly continuous, in the following sense: To each neighborhood W of 0 in Y corresponds a neighborhood V of 0 in X such that*

$$y - x \in V \text{ implies } \Lambda y - \Lambda x \in W.$$

PROOF. Once W is chosen, the continuity of Λ at 0 shows that $\Lambda V \subset W$ for some neighborhood V of 0. If now $y - x \in V$, the linearity of Λ shows that $\Lambda y - \Lambda x = \Lambda(y - x) \in W$. Thus Λ maps the neighborhood $x + V$ of x into the preassigned neighborhood $\Lambda x + W$ of Λx, which says that Λ is continuous at x. ////

1.18 Theorem *Let Λ be a linear functional on a topological vector space X. Assume $\Lambda x \neq 0$ for some $x \in X$. Then each of the following four properties implies the other three:*

(*a*) *Λ is continuous.*

(*b*) *The null space $\mathscr{N}(\Lambda)$ is closed.*

(*c*) *$\mathscr{N}(\Lambda)$ is not dense in X.*

(*d*) *Λ is bounded in some neighborhood V of 0.*

PROOF. Since $\mathscr{N}(\Lambda) = \Lambda^{-1}(\{0\})$ and $\{0\}$ is a closed subset of the scalar field Φ, (*a*) implies (*b*). By hypothesis, $\mathscr{N}(\Lambda) \neq X$. Hence (*b*) implies (*c*).

Assume (*c*) holds; i.e., assume that the complement of $\mathscr{N}(\Lambda)$ has nonempty interior. By Theorem 1.14,

$$(1) \qquad (x + V) \cap \mathscr{N}(\Lambda) = \varnothing$$

for some $x \in X$ and some balanced neighborhood V of 0. Then ΛV is a balanced subset of the field Φ. Thus either ΛV is bounded, in which case (*d*) holds, or $\Lambda V = \Phi$. In the latter case, there exists $y \in V$ such that $\Lambda y = -\Lambda x$, and so $x + y \in \mathscr{N}(\Lambda)$, in contradiction to (1). Thus (*c*) implies (*d*).

Finally, if (*d*) holds then $|\Lambda x| < M$ for all x in V and for some $M < \infty$. If $r > 0$ and if $W = (r/M)V$, then $|\Lambda x| < r$ for every x in W. Hence Λ is continuous at the origin. By Theorem 1.17, this implies (*a*). ////

Finite-Dimensional Spaces

1.19 Among the simplest Banach spaces are R^n and $\mathcal{C}^n$, the standard n-dimensional vector spaces over R and $\mathcal{C}$, respectively, normed by means of the usual euclidean metric: If, for example,

$$z = (z_1, \ldots, z_n) \qquad (z_i \in \mathcal{C})$$

is a vector in $\mathcal{C}^n$, then

$$\|z\| = (|z_1|^2 + \cdots + |z_n|^2)^{1/2}.$$

Other norms can be defined on $\mathcal{C}^n$. For example,

$$\|z\| = |z_1| + \cdots + |z_n| \qquad \text{or} \qquad \|z\| = \max\,(|z_i| : 1 \le i \le n).$$

These norms correspond, of course, to different metrics on $\mathcal{C}^n$ (when $n > 1$) but one can see very easily that they all induce the same topology on $\mathcal{C}^n$. Actually, more is true.

If X is a topological vector space over $\mathcal{C}$, and $\dim X = n$, then every basis of X induces an isomorphism of X onto $\mathcal{C}^n$. Theorem 1.21 will prove that this *isomorphism must be a homeomorphism.* In other words, this says that *the topology of $\mathcal{C}^n$ is the only vector topology that an n-dimensional complex topological vector space can have.*

We shall also see that finite-dimensional subspaces are always closed and that no infinite-dimensional topological vector space is locally compact.

Everything in the preceding discussion remains true with real scalars in place of complex ones.

1.20 Lemma *If X is a complex topological vector space and $f: \mathcal{C}^n \to X$ is linear, then f is continuous.*

PROOF. Let $\{e_1, \ldots, e_n\}$ be the standard basis of $\mathcal{C}^n$: The kth coordinate of e_k is 1, the others are 0. Put $u_k = f(e_k)$, for $k = 1, \ldots, n$. Then $f(z) = z_1 u_1 + \cdots + z_n u_n$ for every $z = (z_1, \ldots, z_n)$ in $\mathcal{C}^n$. Every z_k is a continuous function of z. The continuity of f is therefore an immediate consequence of the fact that addition and scalar multiplication are continuous in X. ////

1.21 Theorem *If n is a positive integer and Y is an n-dimensional subspace of a complex topological vector space X, then*

(*a*) *every isomorphism of $\mathcal{C}^n$ onto Y is a homeomorphism, and*

(*b*) *Y is closed.*

PROOF. Let S be the sphere which bounds the open unit ball B of $\mathbb{C}^n$. Thus $z \in S$ if and only if $\Sigma|z_i|^2 = 1$, and $z \in B$ if and only if $\Sigma|z_i|^2 < 1$.

Suppose $f: \mathbb{C}^n \to Y$ is an isomorphism. This means that f is linear, one-to-one, and $f(\mathbb{C}^n) = Y$. Put $K = f(S)$. Since f is continuous (Lemma 1.20), K is compact. Since $f(0) = 0$ and f is one-to-one, $0 \notin K$, and therefore there is a balanced neighborhood V of 0 in X which does not intersect K. The set

$$E = f^{-1}(V) = f^{-1}(V \cap Y)$$

is therefore disjoint from S. Since f is linear, E is balanced, and hence connected. Thus $E \subset B$, because $0 \in E$, and this implies that the linear map f^{-1} takes $V \cap Y$ into B. Since f^{-1} is an n-tuple of linear functionals on Y, the implication $(d) \to (a)$ in Theorem 1.18 shows that f^{-1} is continuous. Thus f is a homeomorphism.

To prove (b), choose $p \in \bar{Y}$, and let f and V be as above. For some $t > 0$, $p \in tV$, so that p lies in the closure of

$$Y \cap (tV) \subset f(tB) \subset f(t\bar{B}).$$

Being compact, $f(t\bar{B})$ is closed in X. Hence $p \in f(t\bar{B}) \subset Y$, and this proves that $\bar{Y} = Y$. ////

1.22 Theorem *Every locally compact topological vector space X has finite dimension.*

PROOF. The origin of X has a neighborhood V whose closure is compact. By Theorem 1.15, V is bounded, and the sets $2^{-n}V$ $(n = 1, 2, 3, \ldots)$ form a local base for X.

The compactness of $\bar{V}$ shows that there exist $x_1, \ldots, x_m$ in X such that

$$\bar{V} \subset (x_1 + \tfrac{1}{2}V) \cup \cdots \cup (x_m + \tfrac{1}{2}V).$$

Let Y be the vector space spanned by $x_1, \ldots, x_m$. Then $\dim Y \le m$. By Theorem 1.21, Y is a *closed* subspace of X.

Since $V \subset Y + \tfrac{1}{2}V$ and since $\lambda Y = Y$ for every scalar $\lambda \ne 0$, it follows that

$$\tfrac{1}{2}V \subset Y + \tfrac{1}{4}V$$

so that

$$V \subset Y + \tfrac{1}{2}V \subset Y + Y + \tfrac{1}{4}V = Y + \tfrac{1}{4}V.$$

If we continue in this way, we see that

$$V \subset \bigcap_{n=1}^{\infty} (Y + 2^{-n}V).$$

Since $\{2^{-n}V\}$ is a local base, it now follows from (*a*) of Theorem 1.13 that $V \subset \bar{Y}$. But $\bar{Y} = Y$. Thus $V \subset Y$, which implies that $kV \subset Y$ for $k = 1, 2, 3, \ldots$. Hence $Y = X$, by (*a*) of Theorem 1.15, and consequently $\dim X \leq m$. ////

1.23 Theorem *If X is a locally bounded topological vector space with the Heine-Borel property, then X has finite dimension.*

PROOF. By assumption, the origin of X has a bounded neighborhood V. Statement (*f*) of Theorem 1.13 shows that $\bar{V}$ is also bounded. Thus $\bar{V}$ is compact, by the Heine-Borel property. This says that X is locally compact, hence finite-dimensional, by Theorem 1.22.

Metrization

We recall that a topology τ on a set X is said to be *metrizable* if there is a metric d on X which is compatible with τ. In that case, the balls with radius $1/n$ eentered at x form a local base at x. This gives a necessary condition for metrizability which, for topological vector spaces, turns out to be also sufficient.

1.24 Theorem *If X is a topological vector space with a countable local base, then there is a metric d on X such that*

(*a*) *d is compatible with the topology of X,*
(b) *the open balls centered at 0 are balanced, and*
(c) *d is invariant: $d(x + z, y + z) = d(x, y)$ for $x, y, z \in X$.*

If, in addition, X is locally convex, then d can be chosen so as to satisfy (*a*), (*b*), (*c*), *and also*

(*d*) *all open balls are convex.*

PROOF. By Theorem 1.14, X has a balanced local base $\{V_n\}$ such that

$$V_{n+1} + V_{n+1} + V_{n+1} + V_{n+1} \subset V_n \qquad (n + 1, 2, 3, \ldots); \tag{1}$$

when X is locally convex, this local base can be chosen so that each V_n is also convex.

Let D be the set of all rational numbers r of the form

(2) $$r = \sum_{n=1}^{\infty} c_n(r)2^{-n},$$

where each of the "digits" $c_i(r)$ is 0 or 1 and only finitely many are 1. Thus each $r \in D$ satisfies the inequalities $0 \leq r < 1$.

Put $A(r) = X$ if $r \geq 1$; for any $r \in D$, define

(3) $$A(r) = c_1(r)V_1 + c_2(r)V_2 + c_3(r)V_3 + \cdots.$$

Note that each of these sums is actually finite. Define

(4) $$f(x) = \inf \{r: x \in A(r)\} \qquad (x \in X)$$

and

(5) $$d(x, y) = f(x - y) \qquad (x \in X, y \in X).$$

The proof that this d has the desired properties depends on the inclusions

(6) $$A(r) + A(s) \subset A(r + s) \qquad (r \in D, s \in D).$$

Before proving (6), let us see how the theorem follows from it. Since every $A(s)$ contains 0, (6) imples

(7) $$A(r) \subset A(r) + A(t - r) \subset A(t) \quad \text{if} \quad r < t.$$

Thus $\{A(r)\}$ is totally ordered by set inclusion. We claim that

(8) $$f(x + y) \leq f(x) + f(y) \qquad (x \in X, y \in X).$$

In the proof of (8) we may, of course, assume that the right side is <1. Fix $\varepsilon > 0$. There exist r and s in D such that

$$f(x) < r, \qquad f(y) < s, \qquad r + s < f(x) + f(y) + \varepsilon.$$

Thus $x \in A(r)$, $y \in A(s)$, and (6) implies $x + y \in A(r + s)$. Now (8) follows, because

$$f(x + y) \leq r + s < f(x) + f(y) + \varepsilon,$$

and ε was arbitrary.

Since each $A(r)$ is balanced, $f(x) = f(-x)$. It is obvious that $f(0) = 0$. If $x \neq 0$, then $x \notin V_n = A(2^{-n})$ for some n, and so $f(x) \geq 2^{-n} > 0$.

These properties of f show that (5) defines a translation-invariant metric d on X. The open balls centered at 0 are the open sets

(9) $$B_\delta(0) = \{x: f(x) < \delta\} = \bigcup_{r<\delta} A(r).$$

If $\delta < 2^{-n}$, then $B_\delta(0) \subset V_n$. Hence $\{B_\delta(0)\}$ is a local base for the topology of X. This proves (*a*). Since each $A(r)$ is balanced, so is each $B_\delta(0)$.

If each V_n is convex, so is each $A(r)$, and (9) implies that the same is true of each $B_\delta(0)$, hence also of each translate of $B_\delta(0)$.

We turn to the proof of (6). If $r + s \geq 1$, then $A(r + s) = X$ and (6) is obvious. We may therefore assume that $r + s < 1$, and we will use the following simple proposition about addition in the binary system of notation:

If r, s, and $r + s$ are in D and $c_n(r) + c_n(s) \neq c_n(r + s)$ for some n, then at the smallest n where this happens we have $c_n(r) = c_n(s) = 0$, $c_n(r + s) = 1$.

Put $\alpha_n = c_n(r)$, $\beta_n = c_n(s)$, $\gamma_n = c_n(r + s)$. If $\alpha_n + \beta_n = \gamma_n$ for all n then (3) shows that $A(r) + A(s) = A(r + s)$. In the other case, let N be the smallest integer for which $\alpha_N + \beta_N \neq \gamma_N$. Then, as mentioned above, $\alpha_N = \beta_N = 0$, $\gamma_N = 1$. Hence

$$A(r) \subset \alpha_1 V_1 + \cdots + \alpha_{N-1} V_{N-1} + V_{N+1} + V_{N+2} + \cdots$$
$$\subset \alpha_1 V_1 + \cdots + \alpha_{N-1} V_{N-1} + V_{N+1} + V_{N+1}.$$

Likewise

$$A(s) \subset \beta_1 V_1 + \cdots + \beta_{N-1} V_{N-1} + V_{N+1} + V_{N+1}.$$

Since $\alpha_n + \beta_n = \gamma_n$ for all $n < N$, (1) now leads to

$$A(r) + A(s) \subset \gamma_1 V_1 + \cdots + \gamma_{N-1} V_{n-1} + V_N \subset A(r + s)$$

because $\gamma_N = 1$. ////

1.25 Cauchy sequences (*a*) Suppose d is a metric on a set X. A sequence $\{x_n\}$ in X is a *Cauchy sequence* if to every $\varepsilon > 0$ there corresponds an integer N such that $d(x_m, x_n) < \varepsilon$ whenever $m > N$ and $n > N$. If every Cauchy sequence in X converges to a point of X, then d is said to be a *complete* metric on X.

(*b*) Let τ be the topology of a topological vector space X. The notion of Cauchy sequence can be defined in this setting without reference to any metric: Fix a local base $\mathscr{B}$ for τ. A sequence $\{x_n\}$ in X is then said to be a *Cauchy sequence* if to every $V \in \mathscr{B}$ corresponds an N such that $x_n - x_m \in V$ if $n > N$ and $m > N$.

It is clear that different local bases for the same τ give rise to the same class of Cauchy sequences.

(*c*) Suppose now that X is a topological vector space whose topology τ is compatible with an *invariant* metric d. Let us temporarily use the terms d-Cauchy sequence and τ-Cauchy sequence for the concepts defined in (*a*) and (*b*), respectively. Since

$$d(x_n, x_m) = d(x_n - x_m, 0),$$

and since the d-balls centered at the origin form a local base for τ, we conclude:

A sequence $\{x_n\}$ in X is a d-Cauchy sequence if and only if it is a τ-Cauchy sequence.

Consequently, any two invariant metrics on X that are compatible with τ have the same Cauchy sequences. They clearly also have the same convergent sequences (namely, the τ-convergent ones). These remarks prove the following fact:

If d_1 and d_2 are invariant metrics on a vector space X which induce the same topology on X, then

(a) d_1 and d_2 have the same Cauchy sequences, and
(b) d_1 is complete if and only if d_2 is complete.

Invariance is needed in the hypothesis (Exercise 12).
The following " dilation principle " will be used several times.

1.26 Theorem *Suppose that (X, d_1) and (Y, d_2) are metric spaces, and (X, d_1) is complete. If E is a closed set in X, $f: E \to Y$ is continuous, and*

$$d_2(f(x'), f(x'')) \geq d_1(x', x'')$$

for all $x', x'' \in E$, then $f(E)$ is closed.

PROOF. Pick $y \in \overline{f(E)}$. There exist points $x_n \in E$ so that $y = \lim f(x_n)$. Thus $\{f(x_n)\}$ is Cauchy in Y. Our hypothesis implies therefore that $\{x_n\}$ is Cauchy in X. Being a closed subset of a complete metric space, E is complete; hence there exists $x = \lim x_n$ in E. Since f is continuous,

$$f(x) = \lim f(x_n) = y.$$

Thus $y \in f(E)$. ////

1.27 Theorem *Suppose Y is a subspace of a topological vector space X, and Y is an F-space (in the topology inherited from X). Then Y is a closed subspace of X.*

PROOF. Choose an invariant metric d on Y, compatible with its topology. Let

$$B_{1/n} = \left\{y \in Y: d(y, 0) < \frac{1}{n}\right\},$$

let U_n be a neighborhood of 0 in X such that $Y \cap U_n = B_{1/n}$, and choose symmetric neighborhoods V_n of 0 in X such that $V_n + V_n \subset U_n$ and $V_{n+1} \subset V_n$.

Suppose $x \in \bar{Y}$, and define

$$E_n = Y \cap (x + V_n) \qquad (n = 1, 2, 3, \ldots).$$

If $y_1 \in E_n$ and $y_2 \in E_n$, then $y_1 - y_2$ lies in Y and also in $V_n + V_n \subset U_n$, hence in $B_{1/n}$. The diameters of the sets E_n therefore tend to 0. Since each E_n is nonempty and since Y is complete, it follows that the Y-closures of the sets E_n have exactly one point y_0 in common.

Let W be a neighborhood of 0 in X, and define

$$F_n = Y \cap (x + W \cap V_n).$$

The preceding argument shows that the Y-closures of the sets F_n have one common point y_W. But $F_n \subset E_n$. Hence $y_W = y_0$. Since $F_n \subset x + W$, it follows that y_0 lies in the X-closure of $x + W$, for every W. This implies $y_0 = x$. Thus $x \in Y$. This proves that $\bar{Y} = Y$. ////

The following simple facts are sometimes useful.

1.28 Theorem

(a) *If d is a translation-invariant metric on a vector space X then*

$$d(nx, 0) \le nd(x, 0)$$

for every $x \in X$ and for $n = 1, 2, 3, \ldots$.

(b) *If $\{x_n\}$ is a sequence in a metrizable topological vector space X and if $x_n \to 0$ as $n \to \infty$, then there are positive scalars γ_n such that $\gamma_n \to \infty$ and $\gamma_n x_n \to 0$.*

PROOF. Statement (a) follows from

$$d(nx, 0) \le \sum_{k=1}^{n} d(kx, (k-1)x) = nd(x, 0).$$

To prove (b), let d be a metric as in (a), compatible with the topology of X. Since $d(x_n, 0) \to 0$, there is an increasing sequence of positive integers n_k such that $d(x_n, 0) < k^{-2}$ if $n \ge n_k$. Put $\gamma_n = 1$ if $n < n_1$; put $\gamma_n = k$ if $n_k \le n < n_{k+1}$. For such n,

$$d(\gamma_n x_n, 0) = d(kx_n, 0) \le kd(x_n, 0) < k^{-1}.$$

Hence $\gamma_n x_n \to 0$ as $n \to \infty$. ////

Boundedness and Continuity

1.29 Bounded sets The notion of a *bounded subset of a topological vector space* X was defined in Section 1.6 and has been encountered several times since then. When X is metrizable, there is a possibility of misunderstanding, since another very familiar notion of boundedness exists in metric spaces.

If d is a metric on a set X, a set $E \subset X$ is said to be d-bounded if there is a number $M < \infty$ such that $d(z, y) \leq M$ for all x and y in E.

If X is a topological vector space with a compatible metric d, the bounded sets and the d-bounded ones need not be the same, even if d is invariant. For instance, if d is a metric such as the one constructed in Theorem 1.24, then X itself is d-bounded (with $M = 1$) but, as we shall see presently, X cannot be bounded, unless $X = \{0\}$. If X is a normed space and d is the metric induced by the norm, then the two notions of boundedness coincide; but if d is replaced by $d_1 = d/(1 + d)$ (an invariant metric which induces the same topology) they do not.

Whenever bounded subsets of a topological vector space are discussed, *it will be understood that the definition is as in Section* 1.6: A set E is bounded if, for every neighborhood V of 0, we have $E \subset tV$ for all sufficiently large t.

We already saw (Theorem 1.15) that *compact sets are bounded.* To see another type of example, let us prove that *Cauchy sequences are bounded* (hence *convergent sequences are bounded*): If $\{x_n\}$ is a Cauchy sequence in X, and V and W are balanced neighborhoods of 0 with $V + V \subset W$, then [part (*b*) of Section 1.25] there exists N such that $x_n \in x_N + V$ for all $n \geq N$. Take $s > 1$ so that $x_N \in sV$. Then

$$x_n \in sV + V \subset sV + sV \subset sW \qquad (n \geq N).$$

Hence $x_n \in tW$ for all $n \geq 1$, if t is sufficiently large.

Also, closures of bounded sets are bounded (Theorem 1.13).

On the other hand, if $x \neq 0$ and $E = \{nx\colon n = 1, 2, 3, \ldots\}$, then E is not bounded, because there is a neighborhood V of 0 that does not contain x; hence nx is not in nV; it follows that no nV contains E.

Consequently, *no subspace of X (other than $\{0\}$) can be bounded.*

The next theorem characterizes boundedness in terms of sequences.

1.30 Theorem *The following two properties of a set E in a topological vector space are equivalent:*

(*a*) *E is bounded.*

(*b*) *If $\{x_n\}$ is a sequence in E and $\{\alpha_n\}$ is a sequence of scalars such that $\alpha_n \to 0$ as $n \to \infty$, then $\alpha_n x_n \to 0$ as $n \to \infty$.*

PROOF. Suppose E is bounded. Let V be a balanced neighborhood of 0 in X. Then $E \subset tV$ for some t. If $x_n \in E$ and $\alpha_n \to 0$, there exists N such that $|\alpha_n| t < 1$ if $n > N$. Since $t^{-1}E \subset V$ and V is balanced, $\alpha_n x_n \in V$ for all $n > N$. Thus $\alpha_n x_n \to 0$.

Conversely, if E is not bounded, there is a neighborhood V of 0 and a sequence $r_n \to \infty$ such that no $r_n V$ contains E. Choose $x_n \in E$ such that $x_n \notin r_n V$. Then no $r_n^{-1} x_n$ is in V, so that $\{r_n^{-1} x_n\}$ does not converge to 0. ////

1.31 Bounded linear transformations Suppose X and Y are topological vectc. spaces and $\Lambda: X \to Y$ is linear. Λ is said to be *bounded* if Λ maps bounded sets into bounded sets, i.e., if $\Lambda(E)$ is a bounded subset of Y for every bounded set $E \subset X$.

This definition conflicts with the usual notion of a bounded function as being one whose range is a bounded set. In that sense, no linear function (other than 0) could ever be bounded. Thus when bounded linear mappings (or transformations) are discussed, it is to be understood that the definition is in terms of bounded sets, as above.

1.32 Theorem *Suppose X and Y are topological vector spaces and $\Lambda: X \to Y$ is linear. Among the following four properties of Λ, the implications*

$$(a) \to (b) \to (c)$$

hold. If X is metrizable, then also

$$(c) \to (d) \to (a),$$

so that all four properties are equivalent.

(*a*) *Λ is continuous.*
(*b*) *Λ is bounded.*
(*c*) *If $x_n \to 0$ then $\{\Lambda x_n : n = 1, 2, 3, \ldots\}$ is bounded.*
(*d*) *If $x_n \to 0$ then $\Lambda x_n \to 0$.*

Exercise 13 contains an example in which (*b*) holds but (*a*) does not.

PROOF. Assume (*a*), let E be a bounded set in X, and let W be a neighborhood of 0 in Y. Since Λ is continuous (and $\Lambda 0 = 0$) there is a neighborhood V of 0 in X such that $\Lambda(V) \subset W$. Since E is bounded,

$E \subset tV$ for all large t, so that

$$\Lambda(E) \subset \Lambda(tV) = t\Lambda(V) \subset tW.$$

This shows that $\Lambda(E)$ is a bounded set in Y.

Thus $(a) \to (b)$. Since convergent sequences are bounded, $(b) \to (c)$.

Assume now that X is metrizable, that Λ satisfies (c), and that $x_n \to 0$. By Theorem 1.28, there are positive scalars $\gamma_n \to \infty$ such that $\gamma_n x_n \to 0$. Hence $\{\Lambda(\gamma_n x_n)\}$ is a bounded set in Y, and now Theorem 1.30 implies that

$$\Lambda x_n = \gamma_n^{-1}\Lambda(\gamma_n x_n) \to 0 \qquad \text{as} \quad n \to \infty.$$

Finally, assume that (a) fails. Then there is a neighborhood W of 0 in Y such that $\Lambda^{-1}(W)$ contains no neighborhood of 0 in X. If X has a countable local base, there is therefore a sequence $\{x_n\}$ in X so that $x_n \to 0$ but $\Lambda x_n \notin W$. Thus (d) fails. ////

Seminorms and Local Convexity

1.33 Definitions A *seminorm* on a vector space X is a real-valued function p on X such that

(*a*) $p(x + y) \le p(x) + p(y)$ and

(*b*) $p(\alpha x) = |\alpha|\, p(x)$

for all x and y in X and all scalars α.

Property (a) is called *subadditivity*. Theorem 1.34 will show that a seminorm p is a norm if it satisfies

(*c*) $p(x) \ne 0$ if $x \ne 0$.

A family $\mathscr{P}$ of seminorms on X is said to be *separating* if to each $x \ne 0$ corresponds at least one $p \in \mathscr{P}$ with $p(x) \ne 0$.

Next, consider a convex set $A \subset X$ which is *absorbing*, in the sense that every $x \in X$ lies in tA for some $t = t(x) > 0$. [For example, (a) of Theorem 1.15 implies that every neighborhood of 0 in a topological vector space is absorbing. Every absorbing set obviously contains 0.] The *Minkowski functional* μ_A of A is defined by

$$\mu_A(x) = \inf\{t > 0: t^{-1}x \in A\} \qquad (x \in X).$$

Note that $\mu_A(x) < \infty$ for all $x \in X$, since A is absorbing. The seminorms on X will turn out to be precisely the Minkowski functionals of *balanced* convex absorbing sets.

Seminorms are closely related to local convexity, in two ways: In every locally convex space there exists a separating family of *continuous* seminorms. Conversely, if $\mathscr{P}$ is a separating family of seminorms on a vector space X, then $\mathscr{P}$ can be used to define a locally convex topology on X with the property that every $p \in \mathscr{P}$ is continuous. This is a frequently used method of introducing a topology. The details are contained in Theorems 1.36 and 1.37.

1.34 Theorem *Suppose p is a seminorm on a vector space X. Then*

(*a*) $p(0) = 0$.
(*b*) $|p(x) - p(y)| \leq p(x - y)$.
(*c*) $p(x) \geq 0$.
(*d*) $\{x: p(x) = 0\}$ *is a subspace of* X.
(*e*) *The set* $B = \{x: p(x) < 1\}$ *is convex, balanced, absorbing, and* $p = \mu_B$.

PROOF. Statement (*a*) follows from $p(\alpha x) = |\alpha| p(x)$, with $\alpha = 0$. The subadditivity of p shows that

$$p(x) = p(x - y + y) \leq p(x - y) + p(y)$$

so that $p(x) - p(y) \leq p(x - y)$. This also holds with x and y interchanged. Since $p(x - y) = p(y - x)$, (*b*) follows. With $y = 0$, (*b*) implies (*c*). If $p(x) = p(y) = 0$ and α, β are scalars, (*c*) implies

$$0 \leq p(\alpha x + \beta y) \leq |\alpha| p(x) + |\beta| p(y) = 0.$$

This proves (*d*).

As to (*e*), it is clear that B is balanced. If $x \in B$, $y \in B$, and $0 < t < 1$, then

$$p(tx + (1 - t)y) \leq tp(x) + (1 - t)p(y) < 1.$$

Thus B is convex. If $x \in X$ and $s > p(x)$ then $p(s^{-1}x) = s^{-1}p(x) < 1$. This shows that B is absorbing and also that $\mu_B(x) \leq s$. Hence $\mu_B \leq p$. But if $0 < t \leq p(x)$ then $p(t^{-1}x) \geq 1$, and so $t^{-1}x$ is not in B. This implies $p(x) \leq \mu_B(x)$ and completes the proof. ////

1.35 Theorem *Suppose A is a convex absorbing set in a vector space X. Then*

(*a*) $\mu_A(x + y) \leq \mu_A(x) + \mu_A(y)$.
(*b*) $\mu_A(tx) = t\mu_A(x)$ *if* $t \geq 0$.

(c) *μ_A is a seminorm if A is balanced.*

(d) *If $B = \{x: \mu_A(x) < 1\}$ and $C = \{x: \mu_A(x) \le 1\}$, then $B \subset A \subset C$ and $\mu_B = \mu_A = \mu_C$.*

PROOF. If $t = \mu_A(x) + \varepsilon$ and $s = \mu_A(y) + \varepsilon$, for some $\varepsilon > 0$, then x/t and y/s are in A; hence so is their convex combination

$$\frac{x+y}{s+t} = \frac{t}{s+t}\cdot\frac{x}{t} + \frac{s}{s+t}\cdot\frac{y}{s}.$$

This shows that $\mu_A(x+y) \le s + t = \mu_A(x) + \mu_A(y) + 2\varepsilon$, and (*a*) is proved.

Property (*b*) is clear, and (*c*) follows from (*a*) and (*b*).

When we turn to (*d*), the inclusions $B \subset A \subset C$ show that $\mu_C \le \mu_A \le \mu_B$. To prove equality, fix $x \in X$, and choose s, t so that $\mu_C(x) < s < t$. Then $x/s \in C$, $\mu_A(x/s) \le 1$, $\mu_A(x/t) \le s/t < 1$; hence $x/t \in B$, so that $\mu_B(x) \le t$. This holds for every $t > \mu_C(x)$. Hence $\mu_B(x) \le \mu_C(x)$. ////

1.36 Theorem *Suppose $\mathscr{B}$ is a convex balanced local base in a topological vector space X. Associate to every $V \in \mathscr{B}$ its Minkowski functional μ_V. Then*

(*a*) *$V = \{x \in X: \mu_V(x) < 1\}$, for every $V \in \mathscr{B}$, and*

(*b*) *$\{\mu_V: V \in \mathscr{B}\}$ is a separating family of continuous seminorms on X.*

PROOF. If $x \in V$, then $x/t \in V$ for some $t < 1$, because V is open; hence $\mu_V(x) < 1$. If $x \notin V$, then $x/t \in V$ implies $t \ge 1$, because V is balanced; hence $\mu_V(x) \ge 1$. This proves (*a*).

Theorem 1.35 shows that each μ_V is a seminorm. If $r > 0$, it follows from (*a*) and Theorem 1.34 that

$$|\mu_V(x) - \mu_V(y)| \le \mu_V(x - y) < r$$

if $x - y \in rV$. Hence μ_V is continuous. If $x \in X$ and $x \ne 0$, then $x \notin V$ for some $V \in \mathscr{B}$. For this V, $\mu_V(x) \ge 1$. Thus $\{\mu_V\}$ is separating. ////

1.37 Theorem *Suppose $\mathscr{P}$ is a separating family of seminorms on a vector space X. Associate to each $p \in \mathscr{P}$ and to each positive integer n the set*

$$V(p, n) = \left\{x: p(x) < \frac{1}{n}\right\}.$$

Let $\mathscr{B}$ be the collection of all finite intersections of the sets $V(p, n)$. Then $\mathscr{B}$ is a convex balanced local base for a topology τ on X, which turns X into a locally convex space such that

(a) *every $p \in \mathscr{P}$ is continuous, and*

(b) *a set $E \subset X$ is bounded if and only if every $p \in \mathscr{P}$ is bounded on E.*

PROOF. Declare a set $A \subset X$ to be open if and only if A is a (possibly empty) union of translates of members of $\mathscr{B}$. This clearly defines a translation-invariant topology τ on X; each member of $\mathscr{B}$ is convex and balanced, and $\mathscr{B}$ is a local base for τ.

Suppose $x \in X$, $x \neq 0$. Then $p(x) > 0$ for some $p \in \mathscr{P}$. Since x is not in $V(p, n)$ if $np(x) > 1$, we see that 0 is not in the neighborhood $x - V(p, n)$ of x, so that x is not in the closure of $\{0\}$. Thus $\{0\}$ is a closed set, and since τ is translation-invariant, every point of X is a closed set.

Next we show that addition and scalar multiplication are continuous. Let U be a neighborhood of 0 in X. Then

$$U \supset V(p_1, n_1) \cap \cdots \cap V(p_m, n_m) \tag{1}$$

for some $p_1, \ldots, p_m \in \mathscr{P}$ and some positive integers $n_1, \ldots, n_m$. Put

$$V = V(p_1, 2n_1) \cap \cdots \cap V(p_m, 2n_m). \tag{2}$$

Since every $p \in \mathscr{P}$ is subadditive, $V + V \subset U$. This proves that addition is continuous.

Suppose now that $x \in X$, α is a scalar, and U and V are as above. Then $x \in sV$ for some $s > 0$. Put $t = s/(1 + |\alpha| s)$. If $y \in x + tV$ and $|\beta - \alpha| < 1/s$, then

$$\beta y - \alpha x = \beta(y - x) + (\beta - \alpha)x$$

which lies in

$$|\beta| tV + |\beta - \alpha| sV \subset V + V \subset U$$

since $|\beta| t \leq 1$ and V is balanced. This proves that scalar multiplication is continuous.

Thus X is a locally convex space. The definition of $V(p, n)$ shows that every $p \in \mathscr{P}$ is continuous at 0. Hence p is continuous on X, by (b) of Theorem 1.34.

Finally, suppose $E \subset X$ is bounded. Fix $p \in \mathscr{P}$. Since $V(p, 1)$ is a neighborhood of 0, $E \subset kV(p, 1)$ for some $k < \infty$. Hence $p(x) < k$ for every $x \in E$. It follows that every $p \in \mathscr{P}$ is bounded on E.

Conversely, suppose E satisfies this condition, U is a neighborhood of 0, and (1) holds. There are numbers $M_i < \infty$ such that $p_i < M_i$ on E $(1 \leq i \leq m)$. If $n > M_i n_i$ for $1 \leq i \leq m$, it follows that $E \subset nU$, so that E is bounded. ////

1.38 Remarks (a) It was necessary to take finite intersections of the sets $V(p, n)$ in Theorem 1.37; the sets $V(p, n)$ themselves need not form a local base. (They do form what is usually called a *subbase* for the constructed topology.) To see an example of this, take $X = R^2$, and let $\mathscr{P}$ consist of the seminorms p_1 and p_2 defined by $p_i(x) = |x_i|$; here $x = (x_1, x_2)$. Exercise 8 develops this comment further.

(*b*) Theorems 1.36 and 1.37 raise a natural problem: If $\mathscr{B}$ is a convex balanced local base for the topology τ of a locally convex space X, then $\mathscr{B}$ generates a separating family $\mathscr{P}$ of continuous seminorms on X, as in Theorem 1.36. This $\mathscr{P}$ in turn induces a topology τ_1 on X, by the process described in Theorem 1.37. Is $\tau = \tau_1$?

The answer is affirmative. To see this, note that every $p \in \mathscr{P}$ is τ-continuous, so that the sets $V(p, n)$ of Theorem 1.37 are in τ. Hence $\tau_1 \subset \tau$. Conversely, if $W \in \mathscr{B}$ and $p = \mu_W$, then

$$W = \{x\colon \mu_W(x) < 1\} = V(p, 1).$$

Thus $W \in \tau_1$ for every $W \in \mathscr{B}$; this implies that $\tau \subset \tau_1$.

(*c*) If $\mathscr{P} = \{p_i\colon i = 1, 2, 3, \ldots\}$ is a countable separating family of seminorms on X, Theorem 1.37 shows that $\mathscr{P}$ induces a topology τ with a countable local base. By Theorem 1.24, τ is metrizable. In the present situation, a compatible translation-invariant metric can be defined directly in terms of $\{p_i\}$ by setting

$$(1) \qquad d(x, y) = \max_i \frac{c_i p_i(x - y)}{1 + p_i(x - y)},$$

where $\{c_i\}$ is some fixed sequence of positive numbers which converges to 0 as $i \to \infty$.

It is easy to verify that d is a metric on X.

We claim that the balls

$$(2) \qquad B_r = \{x\colon d(0, x) < r\} \qquad (0 < r < \infty)$$

form a convex balanced local base for τ.

Fix r. If $c_i \le r$ (which holds for all but finitely many i, since $c_i \to 0$), then $c_i p_i/(1 + p_i) < r$. Hence B_r is the intersection of *finitely* many sets of the form

$$(3) \qquad \left\{x\colon p_i(x) < \frac{r}{c_i - r}\right\},$$

namely those for which $c_i > r$. These sets are open, since each p_i is continuous (Theorem 1.37). Thus B_r is open, and, by Theorem 1.34, is also convex and balanced.

Next, let W be a neighborhood of 0 in X. The definition of τ shows that W contains the intersection of appropriately chosen sets

$$(4) \qquad V(p_i, \delta_i) = \{x\colon p_i(x) < \delta_i < 1\} \qquad (1 \le i \le k).$$

If $2r < \min\{c_1\delta_1, \ldots, c_k\delta_k\}$ and $x \in B_r$, then

$$\frac{c_i p_i(x)}{1 + p_i(x)} < r < \frac{c_i \delta_i}{2} \qquad (1 \le i \le k), \tag{5}$$

which implies $p_i(x) < \delta_i$. Thus $B_r \subset W$.

This proves our claim and also shows that d is compatible with τ.

1.39 Theorem *A topological vector space X is normable if and only if its origin has a convex bounded neighborhood.*

PROOF. If X is normable, and if $\|\cdot\|$ is a norm that is compatible with the topology of X, then the open unit ball $\{x: \|x\| < 1\}$ is convex and bounded.

For the converse, assume V is a convex bounded neighborhood of 0. By Theorem 1.14, V contains a convex balanced neighborhood U of 0; of course, U is also bounded. Define

$$\|x\| = \mu(x) \qquad (x \in X) \tag{1}$$

where μ is the Minkowski functional of U.

By (*c*) of Theorem 1.15, the sets rU $(r > 0)$ form a local base for the topology of X. If $x \neq 0$, then $x \notin rU$ for some $r > 0$; hence $\|x\| \ge r$. It now follows from Theorem 1.35 that (1) defines a norm. The definition of the Minkowski functional, together with the fact that U is open, implies that

$$\{x: \|x\| < r\} = rU \tag{2}$$

for every $r > 0$. The norm topology coincides therefore with the given one. ////

Quotient Spaces

1.40 Definitions Let N be a subspace of a vector space X. For every $x \in X$, let $\pi(x)$ be the coset of N that contains x; thus

$$\pi(x) = x + N.$$

These cosets are the elements of a vector space X/N, called the *quotient space of X modulo N*, in which addition and scalar multiplication are defined by

$$\pi(x) + \pi(y) = \pi(x + y), \qquad \alpha\pi(x) = \pi(\alpha x). \tag{1}$$

[Note that now $\alpha\pi(x) = N$ when $\alpha = 0$. This differs from the usual notation, as introduced in Section 1.4.] Since N is a vector space, the operations (1) are well defined. This means that if $\pi(x) = \pi(x')$ (that is, $x' - x \in N$) and

$\pi(y) = \pi(y')$ then

$$(2) \qquad \pi(x) + \pi(y) = \pi(x') + \pi(y'), \qquad \alpha\pi(x') = \alpha\pi(x).$$

The origin of X/N is $\pi(0) = N$. By (1), π is a linear mapping of X onto X/N with N as its null space; π is often called the *quotient map* of X onto X/N.

Suppose now that τ is a vector topology on X and that N is a *closed* subspace of X. Let τ_N be the collection of all sets $E \subset X/N$ for which $\pi^{-1}(E) \in \tau$. Then τ_N turns out to be a topology on X/N, called the *quotient topology.* Some of its properties are listed in the next theorem. Recall that an *open mapping* is one that maps open sets to open sets.

1.41 Theorem *Let N be a closed subspace of a topological vector space X. Let τ be the topology of X and define τ_N as above.*

(*a*) *τ_N is a vector topology on X/N; the quotient map $\pi: X \to X/N$ is linear, continuous, and open.*
(*b*) *If $\mathscr{B}$ is a local base for τ, then the collection of all sets $\pi(V)$ with $V \in \mathscr{B}$ is a local base for τ_N.*
(*c*) *Each of the following properties of X is inherited by X/N: local convexity, local boundedness, metrizability, normability.*
(*d*) *If X is an F-space, or a Fréchet space, or a Banach space, so is X/N.*

PROOF. Since $\pi^{-1}(A \cap B) = \pi^{-1}(A) \cap \pi^{-1}(B)$ and

$$\pi^{-1}(\bigcup E_\lambda) = \bigcup \pi^{-1}(E_\lambda),$$

τ_N is a topology. A set $F \subset X/N$ is τ_N-closed if and only if $\pi^{-1}(F)$ is τ-closed. In particular, every point of X/N is closed, since

$$\pi^{-1}(\pi(x)) = N + x$$

and N was assumed to be closed.

The continuity of π follows directly from the definition of τ_N. Next, suppose $V \in \tau$. Since

$$\pi^{-1}(\pi(V)) = N + V$$

and $N + V \in \tau$, it follows that $\pi(V) \in \tau_N$. Thus π is an open mapping.

If now W is a neighborhood of 0 in X/N, there is a neighborhood V of 0 in X such that

$$V + V \subset \pi^{-1}(W).$$

Hence $\pi(V) + \pi(V) \subset W$. Since π is open, $\pi(V)$ is a neighborhood of 0 in X/N. Addition is therefore continuous in X/N.

The continuity of scalar multiplication in X/N is proved in the same manner. This establishes (*a*).

It is clear that (*a*) implies (*b*). With the aid of Theorems 1.32, 1.24, and 1.39, it is just as easy to see that (*b*) implies (*c*).

Suppose next that d is an invariant metric on X, compatible with τ. Define ρ by

$$\rho(\pi(x), \pi(y)) = \inf \{d(x - y, z): z \in N\}.$$

This may be interpreted as the distance from $x - y$ to N. We omit the verifications that are now needed to show that ρ is well defined and that it is an invariant metric on X/N. Since

$$\pi(\{x: d(x, 0) < r\}) = \{u: \rho(u, 0) < r\},$$

it follows from (*b*) that ρ is compatible with τ_N.

If X is normed, this definition of ρ specializes to yield what is usually called the *quotient norm* of X/N:

$$\|\pi(x)\| = \inf \{\|x - z\|: z \in N\}.$$

To prove (*d*) we have to show that ρ is a complete metric whenever d is complete.

Suppose $\{u_n\}$ is a Cauchy sequence in X/N, relative to ρ. There is a subsequence $\{u_{n_i}\}$ with $\rho(u_{n_i}, u_{n_{i+1}}) < 2^{-i}$. One can then inductively choose $x_i \in X$ such that $\pi(x_i) = u_{n_i}$ and $d(x_i, x_{i+1}) < 2^{-i}$. If d is complete, the Cauchy sequence $\{x_i\}$ converges to some $x \in X$. The continuity of π implies that $u_{n_i} \to \pi(x)$ as $i \to \infty$. But if a Cauchy sequence has a convergent subsequence then the full sequence must converge. Hence ρ is complete, and so is the proof of Theorem 1.41. ////

Here is an easy application of these concepts:

1.42 Theorem *Suppose N and F are subspaces of a topological vector space X, N is closed, and F has finite dimension. Then $N + F$ is closed.*

PROOF. Let π be the quotient map of X onto X/N, and give X/N its quotient topology. Then $\pi(F)$ is a finite-dimensional subspace of X/N; since X/N is a topological vector space, Theorem 1.21 implies that $\pi(F)$ is closed in X/N. Since $N + F = \pi^{-1}(\pi(F))$ and π is continuous, we conclude that $N + F$ is closed. (Compare Exercise 20.) ////

1.43 Seminorms and quotient spaces Suppose p is a seminorm on a vector space X and

$$N = \{x: p(x) = 0\}.$$

Then N is a subspace of X (Theorem 1.34). Let π be the quotient map of X onto X/N, and define

$$\tilde{p}(\pi(x)) = p(x).$$

If $\pi(x) = \pi(y)$, then $p(x - y) = 0$, and since

$$|p(x) - p(y)| \le p(x - y)$$

it follows that $\tilde{p}(\pi(x)) = \tilde{p}(\pi(y))$. Thus $\tilde{p}$ is well defined on X/N, and it is now easy to verify that $\tilde{p}$ is a *norm* on X/N.

Here is a familiar example of this. Fix r, $1 \le r < \infty$; let L^r be the space of all Lebesgue measurable functions on $[0, 1]$ for which

$$p(f) = \|f\|_r = \left\{\int_0^1 |f(t)|^r \, dt\right\}^{1/r} < \infty.$$

This defines a seminorm on L^r, not a norm, since $\|f\|_r = 0$ whenever $f = 0$ almost everywhere. Let N be the set of these "null functions." Then L^r/N is the Banach space that is usually called L^r. The norm of L^r is obtained by the above passage from p to $\tilde{p}$.

Examples

1.44 The spaces $C(\Omega)$ If Ω is a nonempty open set in some euclidean space, then Ω is the union of countably many compact sets $K_n \neq \varnothing$ which can be chosen so that K_n lies in the interior of K_{n+1} $(n = 1, 2, 3, \ldots)$. $C(\Omega)$ is the vector space of all complex-valued continuous functions on Ω, topologized by the separating family of seminorms

(1) $$p_n(f) = \sup\{|f(x)| : x \in K_n\},$$

in accordance with Theorem 1.37. Since $p_1 \le p_2 \le \cdots$, the sets

(2) $$V_n = \left\{f \in C(\Omega) : p_n(f) < \frac{1}{n}\right\} \qquad (n = 1, 2, 3, \ldots)$$

form a convex local base for $C(\Omega)$. According to remark (*c*) of Section 1.38, the topology of $C(\Omega)$ is compatible with the metric

(3) $$d(f, g) = \max_n \frac{2^{-n} p_n(f - g)}{1 + p_n(f - g)}.$$

If $\{f_i\}$ is a Cauchy sequence relative to this metric, then $p_n(f_i - f_j) \to 0$ for every n, as $i, j \to \infty$, so that $\{f_i\}$ converges uniformly on K_n, to a function $f \in C(\Omega)$. An easy computation then shows $d(f, f_i) \to 0$. Thus d is a complete metric. We have now proved *that $C(\Omega)$ is a Fréchet space.*

By (*b*) of Theorem 1.37, a set $E \subset C(\Omega)$ is bounded if and only if there are numbers $M_n < \infty$ such that $p_n(f) \le M_n$ for all $f \in E$; explicitly,

(4) $$|f(x)| \le M_n \qquad \text{if } f \in E \text{ and } x \in K_n.$$

Since every V_n contains an f for which $p_{n+1}(f)$ is as large as we please, it follows that no V_n is bounded. *Thus $C(\Omega)$ is not locally bounded, hence is not normable.*

1.45 The spaces $H(\Omega)$ Let Ω now be a nonempty open subset of the complex plane, define $C(\Omega)$ as in Section 1.44, and let $H(\Omega)$ be the subspace of $C(\Omega)$ that consists of the holomorphic functions in Ω. Since sequences of holomorphic functions that converge uniformly on compact sets have holomorphic limits, $H(\Omega)$ is a closed subspace of $C(\Omega)$. *Hence $H(\Omega)$ is a Fréchet space.*

We shall now prove that $H(\Omega)$ *has the Heine-Borel property.* It will then follow from Theorem 1.23 that $H(\Omega)$ *is not locally bounded, hence is not normable.*

Let E be a closed and bounded subset of $H(\Omega)$. Then E satisfies inequalities such as (4) of Section 1.44. Montel's classical theorem about normal families (Th. 14.6 of [23][1]) implies therefore that every sequence $\{f_i\} \subset E$ has a subsequence that converges uniformly on compact subsets of Ω [hence in the topology of $H(\Omega)$] to some $f \in H(\Omega)$. Since E is closed, $f \in E$. This proves that E is compact.

1.46 The spaces $C^\infty(\Omega)$ and $\mathscr{D}_K$ We begin this section by introducing some terminology that will be used in our later work with distributions.

In any discussion of functions of n variables, the term *multi-index* denotes an ordered n-tuple

$$\alpha = (\alpha_1, \ldots, \alpha_n) \tag{1}$$

of nonnegative integers α_i. With each multi-index α is associated the differential operator

$$D^\alpha = \left(\frac{\partial}{\partial x_1}\right)^{\alpha_1} \cdots \left(\frac{\partial}{\partial x_n}\right)^{\alpha_n} \tag{2}$$

whose *order* is

$$|\alpha| = \alpha_1 + \cdots + \alpha_n. \tag{3}$$

If $|\alpha| = 0$, $D^\alpha f = f$.

A complex function f defined in some nonempty open set $\Omega \subset R^n$ is said to belong to $C^\infty(\Omega)$ if $D^\alpha f \in C(\Omega)$ for every multi-index α.

[1] Numbers in brackets refer to sources listed in the bibliography.

The *support* of a complex function f (on any topological space) is the closure of $\{x: f(x) \neq 0\}$.

If K is a compact set in R^n, then $\mathscr{D}_K$ denotes the space of all $f \in C^\infty(R^n)$ whose support lies in K. (The letter $\mathscr{D}$ has been used for these spaces ever since Schwartz published his work on distributions.) If $K \subset \Omega$, then $\mathscr{D}_K$ may be identified with a subspace of $C^\infty(\Omega)$.

We now define a topology on $C^\infty(\Omega)$ which makes $C^\infty(\Omega)$ into a Fréchet space with the Heine-Borel property, such that $\mathscr{D}_K$ is a closed subspace of $C^\infty(\Omega)$ whenever $K \subset \Omega$.

To do this, choose compact sets K_i $(i = 1, 2, 3, \ldots)$ such that K_i lies in the interior of K_{i+1} and $\Omega = \bigcup K_i$. Define seminorms p_N on $C^\infty(\Omega)$, $N = 1, 2, 3, \ldots$, by setting

(4) $$p_N(f) = \max \{|D^\alpha f(x)|: x \in K_N, |\alpha| \leq N\}.$$

They define a metrizable locally convex topology on $C^\infty(\Omega)$; see Theorem 1.37 and remark (c) of Section 1.38. For each $x \in \Omega$, the functional $f \to f(x)$ is continuous in this topology. Since $\mathscr{D}_K$ is the intersection of the null spaces of these functionals, as x ranges over the complement of K, it follows that $\mathscr{D}_K$ is closed in $C^\infty(\Omega)$.

A local base is given by the sets

(5) $$V_N = \left\{f \in C^\infty(\Omega): p_N(f) < \frac{1}{N}\right\} \qquad (N = 1, 2, 3, \ldots).$$

If $\{f_i\}$ is a Cauchy sequence in $C^\infty(\Omega)$ (see Section 1.25) and if N is fixed, then $f_i - f_j \in V_N$ if i and j are sufficiently large. Thus $|D^\alpha f_i - D^\alpha f_j| < 1/N$ on K_N, if $|\alpha| \leq N$. It follows that each $D^\alpha f_i$ converges (uniformly on compact subsets of Ω) to a function g_α. In particular, $f_i(x) \to g_0(x)$. It is now evident that $g_0 \in C^\infty(\Omega)$, that $g_\alpha = D^\alpha g_0$, and that $f_i \to g$ in the topology of $C^\infty(\Omega)$.

Thus $C^\infty(\Omega)$ is a Fréchet space. The same is true of each of its closed subspaces $\mathscr{D}_K$.

Suppose next that $E \subset C^\infty(\Omega)$ is closed and bounded. By Theorem 1.37, the boundedness of E is equivalent to the existence of numbers $M_N < \infty$ such that $p_N(f) \leq M_N$ for $N = 1, 2, 3, \ldots$ and for all $f \in E$. The inequalities $|D^\alpha f| \leq M_N$, valid on K_N when $|\alpha| \leq N$, imply the equicontinuity of $\{D^\beta f: f \in E\}$ on K_{N-1}, if $|\beta| \leq N - 1$. It now follows from Ascoli's theorem (proved in Appendix A) and Cantor's diagonal process that every sequence in E contains a subsequence $\{f_i\}$ for which $\{D^\beta f_i\}$ converges, uniformly on compact subsets of Ω, for each multi-index β. Hence $\{f_i\}$ converges in the topology of $C^\infty(\Omega)$. This proves that E is compact.

Hence $C^\infty(\Omega)$ has the Heine-Borel property. It follows from Theorem 1.23 that $C^\infty(\Omega)$ *is not locally bounded, hence not normable.* The same conclusion holds for $\mathscr{D}_K$ whenever K has nonempty interior (otherwise $\mathscr{D}_K = \{0\}$), because $\dim \mathscr{D}_K = \infty$ in that case. This last statement is a consequence of the following proposition:

If B_1 and B_2 are concentric closed balls in R^n, with B_1 in the interior of B_2, then there exists $\phi \in C^\infty(R^n)$ such that $\phi(x) = 1$ for every $x \in B_1$, $\phi(x) = 0$ for every x outside B_2, and $0 \le \phi \le 1$ on R^n.

To find such a ϕ, we construct $g \in C^\infty(R^1)$ such that $g(x) = 0$ for $x < a$, $g(x) = 1$ for $x > b$ (where $0 < a < b < \infty$ are preassigned) and put

(6) $$\phi(x_1, \ldots, x_n) = 1 - g(x_1^2 + \cdots + x_n^2).$$

The following construction of g has the advantage that suitable choices of $\{\delta_i\}$ can lead to functions with other desired properties.

Suppose $0 < a < b < \infty$. Choose positive numbers $\delta_0, \delta_1, \delta_2, \ldots$, with $\Sigma\delta_i = b - a$; put

(7) $$m_n = \frac{2^n}{\delta_1 \cdots \delta_n} \qquad (n = 1, 2, 3, \ldots);$$

let f_0 be a continuous monotonic function such that $f_0(x) = 0$ when $x < a$, $f_0(x) = 1$ when $x > a + \delta_0$; and define

(8) $$f_n(x) = \frac{1}{\delta_n} \int_{x-\delta_n}^{x} f_{n-1}(t)\, dt \qquad (n = 1, 2, 3, \ldots).$$

Differentiation of this integral shows, by induction, that f_n has n continuous derivatives and that $|D^n f_n| \le m_n$. If $n > r$, then

(9) $$D^r f_n(x) = \frac{1}{\delta_n} \int_0^{\delta_n} (D^r f_{n-1})(x - t)\, dt,$$

so that

(10) $$|D^r f_n| \le m_r \qquad (n \ge r),$$

again by induction on n. The mean value theorem, applied to (9), shows that

(11) $$|D^r f_n - D^r f_{n-1}| \le m_{r+1}\, \delta_n \qquad (n \ge r + 2).$$

Since $\Sigma\delta_n < \infty$, each $\{D^r f_n\}$ converges, uniformly on $(-\infty, \infty)$, as $n \to \infty$. Hence $\{f_n\}$ converges to a function g, with $|D^r g| \le m_r$ for $r = 1, 2, 3, \ldots$, such that $g(x) = 0$ for $x < a$ and $g(x) = 1$ for $x > b$.

1.47 The spaces L^p with $0 < p < 1$ Consider a fixed p in this range. The elements of L^p are those Lebesgue measurable functions f on $[0, 1]$ for which

(1) $$\Delta(f) = \int_0^1 |f(t)|^p\, dt < \infty,$$

with the usual identification of functions that coincide almost everywhere. Since $0 < p < 1$, the inequality

(2) $$(a + b)^p \le a^p + b^p$$

holds when $a \ge 0$ and $b \ge 0$. This gives

(3) $$\Delta(f + g) \le \Delta(f) + \Delta(g),$$

so that

(4) $$d(f, g) = \Delta(f - g)$$

defines an *invariant metric* on L^p. That this d is *complete* is proved in the same way as in the familiar case $p \ge 1$. The balls

(5) $$B_r = \{f \in L^p: \Delta(f) < r\}$$

form a local base for the topology of L^p. Since $B_1 = r^{-1/p}B_r$, for all $r > 0$, B_1 is bounded.

Thus L^p is a locally bounded F-space.

We claim that *L^p contains no convex open sets, other than $\varnothing$ and L^p.*

To prove this, suppose $V \ne \varnothing$ is open and convex in L^p. Assume $0 \in V$, without loss of generality. Then $V \supset B_r$, for some $r > 0$. Pick $f \in L^p$. Since $p < 1$, there is a positive integer n such that $n^{p-1}\,\Delta(f) < r$. By the continuity of the indefinite integral of $|f|^p$, there are points

$$0 = x_0 < x_1 < \cdots < x_n = 1$$

such that

(6) $$\int_{x_{i-1}}^{x_i} |f(t)|^p\,dt = n^{-1}\,\Delta(f) \qquad (1 \le i \le n).$$

Define $g_i(t) = nf(t)$ if $x_{i-1} < t \le x_i$, $g_i(t) = 0$ otherwise. Then $g_i \in V$, since (6) shows

(7) $$\Delta(g_i) = n^{p-1}\,\Delta(f) < r \qquad (1 \le i \le n)$$

and $V \supset B_r$. Since V is convex and

(8) $$f = \frac{1}{n}(g_1 + \cdots + g_n),$$

it follows that $f \in V$. Hence $V = L^p$.

This lack of convex open sets has a curious consequence.

Suppose $\Lambda: L^p \to Y$ is a continuous linear mapping of L^p into some locally convex space Y. Let $\mathscr{B}$ be a convex local base for Y. If $W \in \mathscr{B}$, then $\Lambda^{-1}(W)$ is convex, open, not empty. Hence $\Lambda^{-1}(W) = L^p$. Consequently, $\Lambda(L^p) \subset W$ for every $W \in \mathscr{B}$. We conclude that $\Lambda f = 0$ for every $f \in L^p$.

Thus 0 is the only continuous linear mapping of L^p into any locally convex space Y, if $0 < p < 1$. In particular, 0 is the only continuous linear functional on these L^p-spaces.

This is, of course, in violent contrast to the familiar case $p \ge 1$.

Exercises

1. Suppose X is a vector space. All sets mentioned below are understood to be subsets of X. Prove the following statements from the axioms as given in Section 1.4. (Some of these are tacitly used in the text.)
   (*a*) If $x \in X$ and $y \in X$ there is a unique $z \in X$ such that $x + z = y$.
   (*b*) $0x = 0 = \alpha 0$ if $x \in X$ and α is a scalar.
   (*c*) $2A \subset A + A$; it may happen that $2A \neq A + A$.
   (*d*) A is convex if and only if $(s + t)A = sA + tA$ for all positive scalars s and t.
   (*e*) Every union (and intersection) of balanced sets is balanced.
   (*f*) Every intersection of convex sets is convex.
   (*g*) If Γ is a collection of convex sets that is totally ordered by set inclusion, then the union of all members of Γ is convex.
   (*h*) If A and B are convex, so is $A + B$.
   (*i*) If A and B are balanced, so is $A + B$.
   (*j*) Show that parts (*f*), (*g*), and (*h*) hold with subspaces in place of convex sets.
2. The *convex hull* of a set A in a vector space X is the set of all *convex combinations* of members of A, that is, the set of all sums

$$t_1 x_1 + \cdots + t_n x_n$$

   in which $x_i \in A$, $t_i \geq 0$, $\sum t_i = 1$; n is arbitrary. Prove that the convex hull of A is convex and that it is the intersection of all convex sets that contain A.
3. Let X be a topological vector space. All sets mentioned below are understood to be the subsets of X. Prove the following statements.
   (*a*) The convex hull of every open set is open.
   (*b*) If X is locally convex then the convex hull of every bounded set is bounded. (This is false without local convexity; see Section 1.47.)
   (*c*) If A and B are bounded, so is $A + B$.
   (*d*) If A and B are compact, so is $A + B$.
   (*e*) If A is compact and B is closed, then $A + B$ is closed.
   (*f*) The sum of two closed sets may fail to be closed. [The inclusion in (*b*) of Theorem 1.13 may therefore be strict.]
4. Let $B = \{(z_1, z_2) \in \mathcal{C}^2 : |z_1| \leq |z_2|\}$. Show that B is balanced but that its interior is not. [Compare with (*e*) of Theorem 1.13.]
5. Consider the definition of "bounded set" given in Section 1.6. Would the content of this definition be altered if it were required merely that to every neighborhood V of 0 corresponds *some* $t > 0$ such that $E \subset tV$?
6. Prove that a set E in a topological vector space is bounded if and only if every countable subset of E is bounded.
7. Let X be the vector space of all complex functions on the unit interval [0, 1], topologized by the family of seminorms

$$p_x(f) = |f(x)| \qquad (0 \leq x \leq 1).$$

   This topology is called the *topology of pointwise convergence*. Justify this terminology.

   Show that there is a sequence $\{f_n\}$ in X such that (*a*) $\{f_n\}$ converges to 0 as $n \to \infty$, but (*b*) if $\{\gamma_n\}$ is any sequence of scalars such that $\gamma_n \to \infty$ then $\{\gamma_n f_n\}$

does not converge to 0. (Use the fact that the collection of all complex sequences converging to 0 has the same cardinality as [0, 1].)

This shows that metrizability cannot be omitted in (*b*) of Theorem 1.28.

8. (*a*) Suppose $\mathscr{P}$ is a separating family of seminorms on a vector space X. Let $\mathscr{Q}$ be the smallest family of seminorms on X that contains $\mathscr{P}$ and is closed under max. [This means: If $p_1 \in \mathscr{Q}$, $p_2 \in \mathscr{Q}$, and $p = \max(p_1, p_2)$, then $p \in \mathscr{Q}$.] If the construction of Theorem 1.37 is applied to $\mathscr{P}$ and to $\mathscr{Q}$, show that the two resulting topologies coincide. The main difference is that $\mathscr{Q}$ leads directly to a base, rather than to a subbase. [See Remark (*a*) of Section 1.38.]

(*b*) Suppose $\mathscr{Q}$ is as in part (*a*) and Λ is a linear functional on X. Show that Λ is continuous if and only if there exists a $p \in \mathscr{Q}$ such that $|\Lambda x| \leq Mp(x)$ for all $x \in X$ and some constant $M < \infty$.

9. Suppose

(*a*) X and Y are topological vector spaces,

(*b*) $\Lambda: X \to Y$ is linear,

(*c*) N is a closed subspace of X,

(*d*) $\pi: X \to X/N$ is the quotient map, and

(*e*) $\Lambda x = 0$ for every $x \in N$.

Prove that there is a unique $f: X/N \to Y$ which satisfies $\Lambda = f \circ \pi$, that is, $\Lambda x = f(\pi(x))$ for all $x \in X$. Prove that this f is linear and that Λ is continuous if and only if f is continuous. Also, Λ is open if and only if f is open.

10. Suppose X and Y are topological vector spaces, $\dim Y < \infty$, $\Lambda: X \to Y$ is linear, and $\Lambda(X) = Y$.

(*a*) Prove that Λ is an open mapping.

(*b*) Assume, in addition, that the null space of Λ is closed, and prove that Λ is then continuous.

11. If N is a subspace of a vector space X, the *codimension* of N in X is, by definition, the dimension of the quotient space X/N.

Suppose $0 < p < 1$ and prove that every subspace of finite codimension is dense in L^p. (See Section 1.47.)

12. Suppose $d_1(x, y) = |x - y|$, $d_2(x, y) = |\phi(x) - \phi(y)|$, where $\phi(x) = x/(1 + |x|)$. Prove that d_1 and d_2 are metrics on R which induce the same topology, although d_1 is complete and d_2 is not.

13. Let C be the vector space of all complex continuous functions on [0, 1]. Define

$$d(f, g) = \int_0^1 \frac{|f(x) - g(x)|}{1 + |f(x) - g(x)|} \, dx.$$

Let (C, σ) be C with the topology induced by this metric. Let (C, τ) be the topological vector space defined by the seminorms

$$p_x(f) = |f(x)| \qquad (0 \leq x \leq 1),$$

in accordance with Theorem 1.37.

(*a*) Prove that every τ-bounded set in C is also σ-bounded and that the identity map id: $(C, \tau) \to (C, \sigma)$ therefore carries bounded sets into bounded sets.

(*b*) Prove that id: $(C, \tau) \to (C, \sigma)$ is nevertheless not continuous, although it is sequentially continuous (by Lebesgue's dominated convergence theorem).

Hence (C, τ) is not metrizable. (See Appendix A6, or Theorem 1.32.) Show also directly that (C, τ) has no countable local base.

(c) Prove that every continuous linear functional on (C, τ) is of the form

$$f \to \sum_{i=1}^{n} c_i f(x_i)$$

for some choice of $x_1, \ldots, x_n$ in $[0, 1]$ and some $c_i \in \mathcal{C}$.

(d) Prove that (C, σ) contains no convex open sets other than $\varnothing$ and C.

(e) Prove that id: $(C, \sigma) \to (C, \tau)$ is not continuous.

14. Put $K = [0, 1]$ and define $\mathscr{D}_K$ as in Section 1.46. Show that the following three families of seminorms (where $n = 0, 1, 2, \ldots$) define the same topology on $\mathscr{D}_K$, if $D = d/dx$:

(a) $\|D^n f\|_\infty = \sup \{|D^n f(x)| : -\infty < x < \infty\}$.

(b) $\|D^n f\|_1 = \int_0^1 |D^n f(x)|\, dx$.

(c) $\|D^n f\|_2 = \left\{\int_0^1 |D^n f(x)|^2\, dx\right\}^{1/2}$.

15. Prove that the spaces $C(\Omega)$ (Section 1.44) do not have the Heine-Borel property.

16. Prove that the topology of $C(\Omega)$ does not depend on the particular choice of $\{K_n\}$, as long as this sequence satisfies the conditions specified in Section 1.44. Do the same for $C^\infty(\Omega)$ (Section 1.46).

17. In the setting of Section 1.46, prove that $f \to D^\alpha f$ is a continuous mapping of $C^\infty(\Omega)$ into $C^\infty(\Omega)$ and also of $\mathscr{D}_K$ into $\mathscr{D}_K$, for every multi-index α.

18. Prove the proposition concerning addition in the binary system which was used at the end of the proof of Theorem 1.24.

19. Suppose M is a dense subspace of a topological vector space X, Y is an F-space, and $\Lambda: M \to Y$ is continuous (relative to the topology that M inherits from X) and linear. Prove that Λ has a continuous linear extension $\tilde{\Lambda}: X \to Y$.

Suggestion: Let V_n be balanced neighborhoods of 0 in X such that $V_n + V_n \subset V_{n-1}$ and such that $d(0, \Lambda x) < 2^{-n}$ if $x \in M \cap V_n$. If $x \in X$ and $x_n \in (x + V_n) \cap M$, show that $\{\Lambda x_n\}$ is a Cauchy sequence in Y, and define $\tilde{\Lambda}x$ to be its limit. Show that $\tilde{\Lambda}$ is well defined, that $\tilde{\Lambda}x = \Lambda x$ if $x \in M$, and that $\tilde{\Lambda}$ is linear and continuous.

20. For each real number t and each integer n, define $e_n(t) = e^{int}$, and define

$$f_n = e_{-n} + ne_n \qquad (n = 1, 2, 3, \ldots).$$

Regard these functions as members of $L^2(-\pi, \pi)$. Let X_1 be the smallest closed subspace of L^2 that contains $e_0, e_1, e_2, \ldots$, and let X_2 be the smallest closed subspace of L^2 that contains $f_1, f_2, f_3, \ldots$. Show that $X_1 + X_2$ is dense in L^2 but not closed. For instance, the vector

$$x = \sum_{n=1}^{\infty} n^{-1} e_{-n}$$

is in L^2 but not in $X_1 + X_2$. (Compare with Theorem 1.42.)

21. Let V be a neighborhood of 0 in a topological vector space X. Prove that there is a real continuous function f on X such that $f(0) = 0$ and $f(x) = 1$ outside V. (Thus X is a *completely regular* topological space.) *Suggestion:* Let V_n be balanced neighborhoods of 0 such that $V_1 + V_1 \subset V$ and $V_{n+1} + V_{n+1} \subset V_n$. Construct f as in the proof of Theorem 1.24. Show that f is continuous at 0 and that

$$|f(x) - f(y)| \leq f(x - y).$$

22. If f is a complex function defined on the compact interval $I = [0, 1] \subset R$, define

$$\omega_\delta(f) = \sup \{|f(x) - f(y)| : |x - y| \leq \delta,\ x \in I,\ y \in I\}.$$

If $0 < \alpha \leq 1$, the corresponding *Lipschitz space* Lip α consists of all f for which

$$\|f\| = |f(0)| + \sup \{\delta^{-\alpha}\omega_\delta(f) : \delta > 0\}$$

is finite. Define

$$\text{lip } \alpha = \{f \in \text{Lip } \alpha : \lim_{\delta \to 0} \delta^{-\alpha}\omega_\delta(f) = 0\}.$$

Prove that Lip α is a Banach space and that lip α is a closed subspace of Lip α.

23. Let X be the vector space of all continuous functions on the open segment $(0, 1)$. For $f \in X$ and $r > 0$, let $V(f, r)$ consist of all $g \in X$ such that $|g(x) - f(x)| < r$ for all $x \in (0, 1)$. Let τ be the topology on X that these sets $V(f, r)$ generate. Show that addition is τ-continuous but scalar multiplication is not.

24. Show that the set W that occurs in the proof of Theorem 1.14 need not be convex, and that A need not be balanced unless U is convex.

CHAPTER 2

COMPLETENESS

The validity of many important theorems of analysis depends on the completeness of the systems with which they deal. This accounts for the inadequacy of the rational number system and of the Riemann integral (to mention just the two best-known examples) and for the success encountered by their replacements, the real numbers and the Lebesgue integral. Baire's theorem about complete metric spaces (often called the *category theorem*) is the basic tool in this area. In order to emphasize the role played by the concept of category, some theorems of this chapter (for instance, Theorems 2.7 and 2.11) are stated in a little more generality than is usually needed. When this is done, simpler versions (more easily remembered but sufficient for most applications) are also given.

Baire Category

2.1 Definition Let S be a topological space. A set $E \subset S$ is said to be *nowhere dense* if its closure $\bar{E}$ has an empty interior. The sets of the *first category in S* are those that are *countable* unions of nowhere dense sets. Any subset of S that is not of the first category is said to be of the *second category in S*.

This terminology (due to Baire) is admittedly rather bland and unsuggestive. *Meager* and *nonmeager* have been used instead in some texts. But "category arguments" are so entrenched in the mathematical literature and are so well known that it seems pointless to insist on a change.

Here are some obvious properties of category that will be freely used in the sequel:

(*a*) If $A \subset B$ and B is of the first category in S, so is A.
(*b*) Any countable union of sets of the first category is of the first category.
(*c*) Any closed set $E \subset S$ whose interior is empty is of the first category in S.
(*d*) If h is a homeomorphism of S onto S and if $E \subset S$, then E and $h(E)$ have the same category in S.

2.2 Baire's theorem *If S is either*

(*a*) *a complete metric space, or*
(*b*) *a locally compact Hausdorff space,*

then the intersection of every countable collection of dense open subsets of S is dense in S.

This is often called the *category theorem*, for the following reason. If $\{E_i\}$ is a countable collection of nowhere dense subsets of S, and if V_i is the complement of $\bar{E}_i$, then each V_i is dense, and the conclusion of Baire's theorem is that $\bigcap V_i \neq \emptyset$. Hence $S \neq \bigcup E_i$.

Therefore, complete metric spaces, as well as locally compact Hausdorff spaces, are of the second category in themselves.

PROOF. Suppose $V_1, V_2, V_3, \ldots$ are dense open subsets of S. Let B_0 be an arbitrary nonempty open set in S. If $n \geq 1$ and an open $B_{n-1} \neq \emptyset$ has been chosen, then (because V_n is dense) there exists an open $B_n \neq \emptyset$ with

$$\bar{B}_n \subset V_n \cap B_{n-1}.$$

In case (*a*), B_n may be taken to be a ball of radius $< 1/n$; in case (*b*) the choice can be made so that $\overline{B_n}$ is compact. Put

$$K = \bigcap_{n=1}^{\infty} \overline{B_n}.$$

In case (*a*), the centers of the nested balls B_n form a Cauchy sequence which converges to some point of K, and so $K \neq \emptyset$. In case (*b*), $K \neq \emptyset$ by compactness. Our construction shows that $K \subset B_0$ and $K \subset V_n$ for each n. Hence B_0 intersects $\bigcap V_n$. ////

The Banach-Steinhaus Theorem

2.3 Equicontinuity Suppose X and Y are topological vector spaces and Γ is a collection of linear mappings from X into Y. We say that Γ is *equicontinuous* if to every neighborhood W of 0 in Y there corresponds a neighborhood V of 0 in X such that $\Lambda(V) \subset W$ for all $\Lambda \in \Gamma$.

If Γ contains only one Λ, equicontinuity is, of course, the same as continuity (Theorem 1.17). We already saw (Theorem 1.32) that continuous linear mappings are bounded. Equicontinuous collections have this boundedness property in a uniform manner (Theorem 2.4). It is for this reason that the Banach-Steinhaus theorem (2.5) is often referred to as the *uniform boundedness principle.*

2.4 Theorem *Suppose X and Y are topological vector spaces, Γ is an equicontinuous collection of linear mappings from X into Y, and E is a bounded subset of X. Then Y has a bounded subset F such that $\Lambda(E) \subset F$ for every $\Lambda \in \Gamma$.*

PROOF. Let F be the union of the sets $\Lambda(E)$, for $\Lambda \in \Gamma$. Let W be a neighborhood of 0 in Y. Since Γ is equicontinuous, there is a neighborhood V of 0 in X such that $\Lambda(V) \subset W$ for all $\Lambda \in \Gamma$. Since E is bounded, $E \subset tV$ for all sufficiently large t. For these t,

$$\Lambda(E) \subset \Lambda(tV) = t\Lambda(V) \subset tW,$$

so that $F \subset tW$. Hence F is bounded. ////

2.5 Theorem (Banach-Steinhaus) *Suppose X and Y are topological vector spaces, Γ is a collection of continuous linear mappings from X into Y, and B is the set of all $x \in X$ whose orbits*

$$\Gamma(x) = \{\Lambda x: \Lambda \in \Gamma\}$$

are bounded in Y.

If B is of the second category in X, then $B = X$ and Γ is equicontinuous.

PROOF. Pick balanced neighborhoods W and U of 0 in Y such that $\bar{U} + \bar{U} \subset W$. Put

$$E = \bigcap_{\Lambda \in \Gamma} \Lambda^{-1}(\bar{U}).$$

If $x \in B$, then $\Gamma(x) \subset nU$ for some n, so that $x \in nE$. Consequently,

$$B \subset \bigcup_{n=1}^{\infty} nE.$$

At least one nE is of the second category in X, since this is true of B. Since $x \to nx$ is a homeomorphism of X onto X, E is itself of the second category in X. But E is closed because each Λ is continuous. Therefore E has an interior point x. Then $x - E$ contains a neighborhood V of 0 in X, and

$$\Lambda(V) \subset \Lambda x - \Lambda(E) \subset \bar{U} - \bar{U} \subset W$$

for every $\Lambda \in \Gamma$.

This proves that Γ is equicontinuous. By Theorem 2.4, Γ is uniformly bounded; in particular, each $\Gamma(x)$ is bounded in Y. Hence $B = X$. ////

In many applications, the hypothesis that B is of the second category is a consequence of Baire's theorem. For example, F-spaces are of the second category. This gives the following corollary of the Banach-Steinhaus theorem:

2.6 Theorem *If Γ is a collection of continuous linear mappings from an F-space X into a topological vector space Y, and if the sets*

$$\Gamma(x) = \{\Lambda x: \Lambda \in \Gamma\}$$

are bounded in Y, for every $x \in X$, then Γ is equicontinuous.

Briefly, pointwise boundedness implies uniform boundedness (Theorem 2.4).

As a special case of Theorem 2.6, let X and Y be Banach spaces, and suppose that

(1) $$\sup_{\Lambda \in \Gamma} \|\Lambda x\| < \infty \qquad \text{for every } x \in X.$$

The conclusion is that there exists $M < \infty$ such that

(2) $$\|\Lambda x\| \le M \qquad \text{if } \|x\| \le 1 \text{ and } \Lambda \in \Gamma.$$

Hence

(3) $$\|\Lambda x\| \le M\|x\| \qquad \text{if } x \in X \text{ and } \Lambda \in \Gamma.$$

The following theorem establishes the continuity of limits of sequences of continuous linear mappings:

2.7 Theorem *Suppose X and Y are topological vector spaces, and $\{\Lambda_n\}$ is a sequence of continuous linear mappings of X into Y.*

(*a*) *If C is the set of all $x \in X$ for which $\{\Lambda_n x\}$ is a Cauchy sequence in Y, and if C is of the second category in X, then $C = X$.*

(*b*) *If L is the set of all $x \in X$ at which*

$$\Lambda x = \lim_{n \to \infty} \Lambda_n x$$

exists, if L is of the second category in X, and if Y is an F-space, then $L = X$ and $\Lambda: X \to Y$ is continuous.

PROOF. (*a*) Since Cauchy sequences are bounded (Section 1.29), the Banach-Steinhaus theorem asserts that $\{\Lambda_n\}$ is equicontinuous.

One checks easily that C is a subspace of X. Hence C is dense. (Otherwise, $\bar{C}$ is a proper subspace of X; proper subspaces have empty interior; thus $\bar{C}$ would be of the first category.)

Fix $x \in X$; let W be a neighborhood of 0 in Y. Since $\{\Lambda_n\}$ is equicontinuous, there is a neighborhood V of 0 in X such that $\Lambda_n(V) \subset W$ for $n = 1, 2, 3, \ldots$. Since C is dense, there exists $x' \in C \cap (x + V)$. If n and m are so large that

$$\Lambda_n x' - \Lambda_m x' \in W,$$

the identity

$$(\Lambda_n - \Lambda_m)x = \Lambda_n(x - x') + (\Lambda_n - \Lambda_m)x' + \Lambda_m(x' - x)$$

shows that $\Lambda_n x - \Lambda_m x \in W + W + W$. Consequently, $\{\Lambda_n x\}$ is a Cauchy sequence in Y, and $x \in C$.

(*b*) The completeness of Y implies that $L = C$. Hence $L = X$, by (*a*). If V and W are as above, the inclusion $\Lambda_n(V) \subset W$, valid for all n, implies now that $\Lambda(V) \subset \bar{W}$. Thus Λ is continuous. ////

The hypotheses of (*b*) of Theorem 2.7 can be modified in various ways. Here is an easily remembered version:

2.8 Theorem *If $\{\Lambda_n\}$ is a sequence of continuous linear mappings from an F-space X into a topological vector space Y, and if*

$$\Lambda x = \lim_{n\to\infty} \Lambda_n x$$

exists for every $x \in X$, then Λ is continuous.

PROOF. Theorem 2.6 implies that $\{\Lambda_n\}$ is equicontinuous. Therefore if W is a neighborhood of 0 in Y, we have $\Lambda_n(V) \subset W$ for all n and for some neighborhood V of 0 in X. It follows that $\Lambda(V) \subset \overline{W}$; hence (being obviously linear) Λ is continuous. ////

In the following variant of the Banach-Steinhaus theorem the category argument is applied to a compact set, rather than to a complete metric one. Convexity also enters here in an essential way (Exercise 8).

2.9 Theorem *Suppose X and Y are topological vector spaces, K is a compact convex set in X, Γ is a collection of continuous linear mappings of X*

into Y, and the orbits

$$\Gamma(x) = \{\Lambda x : \Lambda \in \Gamma\}$$

are bounded subsets of Y, for every $x \in K$.

Then there is a bounded set $B \subset Y$ such that $\Lambda(K) \subset B$ for every $\Lambda \in \Gamma$.

PROOF. Let B be the union of all sets $\Gamma(x)$, for $x \in K$. Pick balanced neighborhoods W and U of 0 in Y such that $\bar{U} + \bar{U} \subset W$. Put

$$E = \bigcap_{\Lambda \in \Gamma} \Lambda^{-1}(\bar{U}). \tag{1}$$

If $x \in K$, then $\Gamma(x) \subset nU$ for some n, so that $x \in nE$. Consequently,

$$K = \bigcup_{n=1}^{\infty} (K \cap nE). \tag{2}$$

Since E is closed, Baire's theorem shows that $K \cap nE$ has nonempty interior (relative to K) for at least one n.

We fix such an n, we fix an interior point x_0 of $K \cap nE$, we fix a balanced neighborhood V of 0 in X such that

$$K \cap (x_0 + V) \subset nE, \tag{3}$$

and we fix a $p > 1$ such that

$$K \subset x_0 + pV. \tag{4}$$

Such a p exists since K is compact.

If now x is any point of K and

$$z = (1 - p^{-1})x_0 + p^{-1}x, \tag{5}$$

then $z \in K$, since K is convex. Also,

$$z - x_0 = p^{-1}(x - x_0) \in V, \tag{6}$$

by (4). Hence $z \in nE$, by (3). Since $\Lambda(nE) \subset n\bar{U}$ for every $\Lambda \in \Gamma$ and since $x = pz - (p-1)x_0$, we have

$$\Lambda x \in pn\bar{U} - (p-1)n\bar{U} \subset pn(\bar{U} + \bar{U}) \subset pnW.$$

Thus $B \subset pnW$, which proves that B is bounded. ////

The Open Mapping Theorem

2.10 Open mappings Suppose f maps S into T, where S and T are topological spaces. We say that f is *open at a point $p \in S$* if $f(V)$ contains a neighborhood of $f(p)$ whenever V is a neighborhood of p. We say that f is *open* if $f(U)$ is open in T whenever U is open in S.

It is clear that f is open if and only if f is open at every point of S. Because of the invariance of vector topologies, it follows that a linear mapping of one topological vector space into another is open if and only if it is open at the origin.

Let us also note that a one-to-one continuous mapping f of S onto T is a homeomorphism precisely when f is open.

2.11 The open mapping theorem *Suppose*

(a) *X is an F-space,*
(b) *Y is a topological vector space,*
(c) *$\Lambda: X \to Y$ is continuous and linear, and*
(d) *$\Lambda(X)$ is of the second category in Y.*

Then

(i) *$\Lambda(X) = Y$,*
(ii) *Λ is an open mapping, and*
(iii) *Y is an F-space.*

PROOF. Note that (ii) implies (i), since Y is the only open subspace of Y. To prove (ii), let V be a neighborhood of 0 in X. We have to show that $\Lambda(V)$ contains a neighborhood of 0 in Y.

Let d be an invariant metric on X that is compatible with the topology of X. Define

$$V_n = \{x: d(x, 0) < 2^{-n}r\} \qquad (n = 0, 1, 2, \ldots), \tag{1}$$

where $r > 0$ is so small that $V_0 \subset V$. We will prove that some neighborhood W of 0 in Y satisfies

$$W \subset \overline{\Lambda(V_1)} \subset \Lambda(V). \tag{2}$$

Since $V_1 \supset V_2 - V_2$, statement (b) of Theorem 1.13 implies

$$\overline{\Lambda(V_1)} \supset \overline{\Lambda(V_2) - \Lambda(V_2)} \supset \overline{\Lambda(V_2)} - \overline{\Lambda(V_2)}. \tag{3}$$

The first part of (2) will therefore be proved if we can show that $\overline{\Lambda(V_2)}$ has nonempty interior. But

$$\Lambda(X) = \bigcup_{k=1}^{\infty} k\Lambda(V_2), \tag{4}$$

because V_2 is a neighborhood of 0. At least one $k\Lambda(V_2)$ is therefore of the second category in Y. Since $y \to ky$ is a homeomorphism of Y onto Y, $\Lambda(V_2)$ is of the second category in Y. Its closure therefore has nonempty interior.

To prove the second inclusion in (2), fix $y_1 \in \overline{\Lambda(V_1)}$. Assume $n \geq 1$ and y_n has been chosen in $\overline{\Lambda(V_n)}$. What was just proved for V_1 holds equally well for V_{n+1}, so that $\overline{\Lambda(V_{n+1})}$ contains a neighborhood of 0. Hence

$$(5) \qquad (y_n - \overline{\Lambda(V_{n+1})}) \cap \Lambda(V_n) \neq \varnothing.$$

This says that there exists $x_n \in V_n$ such that

$$(6) \qquad \Lambda x_n \in y_n - \overline{\Lambda(V_{n+1})}.$$

Put $y_{n+1} = y_n - \Lambda x_n$. Then $y_{n+1} \in \overline{\Lambda(V_{n+1})}$, and the construction proceeds.

Since $d(x_n, 0) < 2^{-n}r$, for $n = 1, 2, 3, \ldots$, the sums $x_1 + \cdots + x_n$ form a Cauchy sequence which converges (by the completeness of X) to some $x \in X$, with $d(x, 0) < r$. Hence $x \in V$. Since

$$(7) \qquad \sum_{n=1}^{m} \Lambda x_n = \sum_{n=1}^{m} (y_n - y_{n+1}) = y_1 - y_{m+1},$$

and since $y_{m+1} \to 0$ as $m \to \infty$ (by the continuity of Λ), we conclude that $y_1 = \Lambda x \in \Lambda(V)$. This gives the second part of (2), and (*ii*) is proved.

Theorem 1.41 shows that X/N is an F-space, if N is the null space of Λ. Hence (*iii*) will follow as soon as we exhibit an isomorphism f of X/N onto Y which is also a homeomorphism. This can be done by defining

$$(8) \qquad f(x + N) = \Lambda x \qquad (x \in X).$$

It is trivial that this f is an isomorphism and that $\Lambda x = f(\pi(x))$, where π is the quotient map described in Section 1.40. If V is open in Y, then

$$(9) \qquad f^{-1}(V) = \pi(\Lambda^{-1}(V))$$

is open, since Λ is continuous and π is open. Hence f is continuous. If E is open in X/N, then

$$(10) \qquad f(E) = \Lambda(\pi^{-1}(E))$$

is open, since π is continuous and Λ is open. Consequently, f is a homeomorphism. ////

2.12 Corollaries

(*a*) *If Λ is a continuous linear mapping of an F-space X onto an F-space Y, then Λ is open.*

(*b*) *If Λ satisfies (a) and is one-to-one, then $\Lambda^{-1}: Y \to X$ is continuous.*

(c) *If X and Y are Banach spaces, and if $\Lambda: X \to Y$ is continuous, linear, one-to-one, and onto, then there exist positive real numbers a and b such that*

$$a\|x\| \le \|\Lambda x\| \le b\|x\|$$

for every $x \in X$.

(d) *If $\tau_1 \subset \tau_2$ are vector topologies on a vector space X and if both (X, τ_1) and (X, τ_2) are F-spaces, then $\tau_1 = \tau_2$.*

PROOF. Statement (*a*) follows from Theorem 2.11 and Baire's theorem, since Y is now of the second category in itself. Statement (*b*) is an immediate consequence of (*a*), and (*c*) follows from (*b*). The two inequalities in (*c*) simply express the continuity of Λ^{-1} and of Λ. Statement (*d*) is obtained by applying (*b*) to the identity mapping of (X, τ_2) onto (X, τ_1). ////

The Closed Graph Theorem

2.13 Graphs If X and Y are sets and f maps X into Y, the *graph* of f is the set of all points $(x, f(x))$ in the cartesian product $X \times Y$. If X and Y are topological spaces, if $X \times Y$ is given the usual product topology (the smallest topology that contains all sets $U \times V$ with U and V open in X and Y, respectively), and if $f: X \to Y$ is continuous, one would expect the graph of f to be closed in $X \times Y$ (Proposition 2.14). For linear mappings between F-spaces this trivial necessary condition is also sufficient to assure continuity. This important fact is proved in Theorem 2.15.

2.14 Proposition *If X is a topological space, Y is a Hausdorff space, and $f: X \to Y$ is continuous, then the graph G of f is closed.*

PROOF. Let Ω be the complement of G in $X \times Y$; fix $(x_0, y_0) \in \Omega$. Then $y_0 \neq f(x_0)$. Thus y_0 and $f(x_0)$ have disjoint neighborhoods V and W in Y. Since f is continuous, x_0 has a neighborhood U such that $f(U) \subset W$. The neighborhood $U \times V$ of (x_0, y_0) lies therefore in Ω. This proves that Ω is open. ////

Note: One cannot omit the hypothesis that Y is a Hausdorff space. To see this, consider an arbitrary topological space X, and let $f: X \to X$ be the identity. Its graph is the diagonal

$$D = \{(x, x): x \in X\} \subset X \times X.$$

The statement "D is closed in $X \times X$" is just a rewording of the Hausdorff separation axiom.

2.15 The closed graph theorem *Suppose*

(a) *X and Y are F-spaces,*
(b) *$\Lambda: X \to Y$ is linear,*
(c) *$G = \{(x, \Lambda x): x \in X\}$ is closed in $X \times Y$.*

Then Λ is continuous.

PROOF. $X \times Y$ is a vector space if addition and scalar multiplication are defined componentwise:

$$\alpha(x_1, y_1) + \beta(x_2, y_2) = (\alpha x_1 + \beta x_2, \alpha y_1 + \beta y_2).$$

There are complete invariant metrics d_X and d_Y on X and Y, respectively, which induce their topologies. If

$$d((x_1, y_1), (x_2, y_2)) = d_X(x_1, x_2) + d_Y(y_1, y_2),$$

then d is an invariant metric on $X \times Y$ which is compatible with its product topology and which makes $X \times Y$ into an F-space. (The easy but tedious verifications that are needed here are left as an exercise.)

Since Λ is linear, G is a subspace of $X \times Y$. Closed subsets of complete metric spaces are complete. Therefore *G is an F-space.*

Define $\pi_1: G \to X$ and $\pi_2: X \times Y \to Y$ by

$$\pi_1(x, \Lambda x) = x, \qquad \pi_2(x, y) = y.$$

Now π_1 is a continuous linear one-to-one mapping of the F-space G onto the F-space X. It follows from the open mapping theorem that

$$\pi_1^{-1}: X \to G$$

is continuous. But $\Lambda = \pi_2 \circ \pi_1^{-1}$ and π_2 is continuous. Hence Λ is continuous. ////

Remark. The crucial hypothesis (c), that G is closed, is often verified in applications by showing that Λ satisfies property (c′) below:

(c′) *If $\{x_n\}$ is a sequence in X such that the limits*

$$x = \lim_{n\to\infty} x_n \qquad \text{and} \qquad y = \lim_{n\to\infty} \Lambda x_n$$

exist, then $y = \Lambda x$.

Let us prove that (c′) implies (c). Pick a limit point (x, y) of G. Since $X \times Y$ is metrizable,

$$(x, y) = \lim_{n\to\infty} (x_n, \Lambda x_n)$$

for some sequence $\{x_n\}$. It follows from the definition of the product topology that $x_n \to x$ and $\Lambda x_n \to y$. Hence $y = \Lambda x$, by (c'), and so $(x, y) \in G$, and G is closed.

It is just as easy to prove that (c) implies (c').

Bilinear Mappings

2.16 Definitions Suppose X, Y, Z are vector spaces and B maps $X \times Y$ into Z. Associate to each $x \in X$ and to each $y \in Y$ the mappings

$$B_x: Y \to Z \quad \text{and} \quad B^y: X \to Z$$

by defining

$$B_x(y) = B(x, y) = B^y(x).$$

B is said to be *bilinear* if every B_x and every B^y are linear.

If X, Y, Z are topological vector spaces and if every B_x and every B^y is continuous, then B is said to be *separately continuous.* If B is continuous (relative to the product topology of $X \times Y$) then B is obviously separately continuous. In certain situations, the converse can be proved with the aid of the Banach-Steinhaus theorem.

2.17 Theorem *Suppose $B: X \times Y \to Z$ is bilinear and separately continuous, X is an F-space, and Y and Z are topological vector spaces. Then*

$$B(x_n, y_n) \to B(x_0, y_0) \text{ in } Z \tag{1}$$

whenever $x_n \to x_0$ in X and $y_n \to y_0$ in Y. If Y is metrizable, it follows that B is continuous.

PROOF. Let U and W be neighborhoods of 0 in Z such that $U + U \subset W$. Define

$$b_n(x) = B(x, y_n) \qquad (x \in X, n = 1, 2, 3, \ldots).$$

Since B is continuous as a function of y,

$$\lim_{n\to\infty} b_n(x) = B(x, y_0) \qquad (x \in X).$$

Thus $\{b_n(x)\}$ is a bounded subset of Z, for each $x \in X$. Since each b_n is a continuous linear mapping of the F-space X, the Banach-Steinhaus theorem 2.6 implies that $\{b_n\}$ is equicontinuous. Hence there is a neighborhood V of 0 in X such that

$$b_n(V) \subset U \qquad (n = 1, 2, 3, \ldots).$$

Note that

$$B(x_n, y_n) - B(x_0, y_0) = b_n(x_n - x_0) + B(x_0, y_n - y_0).$$

If n is sufficiently large, then (*i*) $x_n \in x_0 + V$, so that $b_n(x_n - x_0) \in U$, and (*ii*) $B(x_0, y_n - y_0) \in U$, since B is continuous in y and $B(x_0, 0) = 0$. Hence

$$B(x_n, y_n) - B(x_0, y_0) \in U + U \subset W$$

for all large n. This gives (1).

If Y is metrizable, so is $X \times Y$, and the continuity of B then follows from (1). (See Appendix A6.) ////

Exercises

1. If X is an infinite-dimensional topological vector space which is the union of countably many finite-dimensional subspaces, prove that X is of the first category in itself. Prove that therefore no infinite-dimensional F-space has a countable Hamel basis.

   (A set β is a *Hamel basis* for a vector space X if β is a maximal linearly independent subset of X. Alternatively, β is a Hamel basis if every $x \in X$ has a unique representation as a *finite* linear combination of elements of β.)
2. Sets of first and second category are "small" and "large" in a topological sense. These notions are different when "small" and "large" are understood in the sense of measure, even when the measure is intimately related to the topology. To see this, construct a subset of the unit interval which is of the first category but whose Lebesgue measure is 1.
3. Put $K = [-1, 1]$; define $\mathscr{D}_K$ as in Section 1.46 (with R in place of R^n). Suppose $\{f_n\}$ is a sequence of Lebesgue integrable functions such that

$$\Lambda\phi = \lim_{n\to\infty} \int_{-1}^{1} f_n(t)\phi(t)\, dt$$

   exists for every $\phi \in \mathscr{D}_K$. Show that Λ is a continuous linear functional on $\mathscr{D}_K$. Show that there is a positive integer p and a number $M < \infty$ such that

$$\left| \int_{-1}^{1} f_n(t)\phi(t)\, dt \right| \le M \|D^p\phi\|_\infty$$

   for all n. For example, if $f_n(t) = n^3 t$ on $[-1/n, 1/n]$ and 0 elsewhere, show that this can be done with $p = 1$. Construct an example where it can be done with $p = 2$ but not with $p = 1$.
4. Let L^1 and L^2 be the usual Lebesgue spaces on the unit interval. Prove that L^2 is of the first category in L^1, in three ways:
   (*a*) Show that $\{f : \int |f|^2 \le n\}$ is closed in L^1 but has empty interior.
   (*b*) Put $g_n = n$ on $[0, n^{-3}]$, and show that

$$\int f g_n \to 0$$

   for every $f \in L^2$ but not for every $f \in L^1$.

(c) Note that the inclusion map of L^2 into L^1 is continuous but not onto. Do the same for L^p and L^q if $p < q$.

5. Prove results analogous to those of Exercise 4 for the spaces ℓ^p, where ℓ^p is the Banach space of all complex functions x on $\{0, 1, 2, \ldots\}$ whose norm

$$\|x\|_p = \left\{\sum_{n=0}^{\infty} |x(n)|^p\right\}^{1/p}$$

is finite.

6. Define the Fourier coefficients $\hat{f}(n)$ of a function $f \in L^2(T)$ (T is the unit circle) by

$$\hat{f}(n) = \frac{1}{2\pi} \int_{-\pi}^{\pi} f(e^{i\theta}) e^{-in\theta}\, d\theta$$

for all $n \in Z$ (the integers). Put

$$\Lambda_n f = \sum_{k=-n}^{n} \hat{f}(k).$$

Prove that $\{f \in L^2(T): \lim_{n\to\infty} \Lambda_n f \text{ exists}\}$ is a dense subspace of $L^2(T)$ of the first category.

7. Let $C(T)$ be the set of all continuous complex functions on the unit circle T. Suppose $\{\gamma_n\}$ ($n \in Z$) is a complex sequence that associates to each $f \in C(T)$ a function $\Lambda f \in C(T)$ whose Fourier coefficients are

$$(\Lambda f)\hat{}(n) = \gamma_n \hat{f}(n) \qquad (n \in Z).$$

(The notation is as in Exercise 6.) Prove that $\{\gamma_n\}$ has this multiplier property if and only if there is a complex Borel measure μ on T such that

$$\gamma_n = \int e^{-in\theta}\, d\mu(\theta) \qquad (n \in Z).$$

Suggestion: With the supremum norm, $C(T)$ is a Banach space. Apply the closed graph theorem. Then consider the functional

$$f \to (\Lambda f)(1) = \sum_{-\infty}^{\infty} \gamma_n \hat{f}(n)$$

and apply the Riesz representation theorem ([23], Th. 6.19). (The above series may not converge; use it only for trigonometric polynomials.)

8. Define functionals Λ_m on ℓ^2 (see Exercise 5) by

$$\Lambda_m x = \sum_{n=1}^{m} n^2 x(n) \qquad (m = 1, 2, 3, \ldots).$$

Define $x_n \in \ell^2$ by $x_n(n) = 1/n$, $x_n(i) = 0$ if $i \neq n$. Let $K \subset \ell^2$ consist of $0, x_1, x_2, x_3, \ldots$. Prove that K is compact. Compute $\Lambda_m x_n$. Show that $\{\Lambda_m x\}$ is bounded for each $x \in K$ but $\{\Lambda_m x_m\}$ is not. Convexity can therefore not be omitted from the hypotheses of Theorem 2.9.

Choose $c_n > 0$ so that $\sum c_n = 1$, $\sum nc_n = \infty$. Take $x = \sum c_n x_n$. Show that x lies in the closed convex hull of K (by definition, this is the closure of the convex hull) and that $\{\Lambda_m x\}$ is not bounded.

Show that the convex hull of K is not closed.

9. Suppose X, Y, Z are Banach spaces and

$$B: X \times Y \to Z$$

is bilinear and continuous. Prove that there exists $M < \infty$ such that

$$\|B(x, y)\| \le M\|x\|\|y\| \qquad (x \in X, y \in Y).$$

Is completeness needed here?

10. Prove that a bilinear mapping is continuous if it is continuous at the origin $(0, 0)$.

11. Define $B(x_1, x_2; y) = (x_1 y, x_2 y)$. Show that B is a bilinear continuous mapping of $R^2 \times R$ onto R^2 which is *not* open at $(1, 1; 0)$. Find all points where this B is open.

12. Let X be the normed space of all real polynomials in one variable, with

$$\|f\| = \int_0^1 |f(t)|\, dt.$$

Put $B(f, g) = \int_0^1 f(t)g(t)\, dt$, and show that B is a bilinear functional on $X \times X$ which is separately continuous but is not continuous.

13. Suppose X is a topological vector space which is of the second category in itself. Let K be a closed, convex, absorbing subset of X. Prove that K contains a neighborhood of 0.

Suggestion: Show first that $H = K \cap (-K)$ is absorbing. By a category argument, H has interior. Then use

$$2H = H + H = H - H.$$

Show that the result is false without convexity of K, even if $X = R^2$. Show that the result is false if X is L^2 topologized by the L^1-norm (as in Exercise 4).

14. (*a*) Suppose X and Y are topological vector spaces, $\{\Lambda_n\}$ is an equicontinuous sequence of linear mappings of X into Y, and C is the set of all x at which $\{\Lambda_n(x)\}$ is a Cauchy sequence in Y. Prove that C is a closed subspace of X.

(*b*) Assume, in addition to the hypotheses of (*a*), that Y is an F-space and that $\{\Lambda_n(x)\}$ converges in some dense subset of X. Prove that then

$$\Lambda(x) = \lim_{n \to \infty} \Lambda_n(x)$$

exists for every $x \in X$ and that Λ is continuous.

15. Suppose X is an F-space and Y is a subspace of X whose complement is of the first category. Prove that $Y = X$. *Hint:* Y must intersect $x + Y$ for every $x \in X$.

16. Suppose that X and K are metric spaces, that K is compact, and that the graph of $f: X \to K$ is a closed subset of $X \times K$. Prove that f is continuous. (This is an analogue of Theorem 2.15 but is much easier.) Show that compactness of K cannot be omitted from the hypotheses, even when X is compact.

CHAPTER 3

CONVEXITY

This chapter deals primarily (though not exclusively) with the most important class of topological vector spaces, namely, the locally convex ones. The highlights, from the theoretical as well as the applied standpoints, are (*a*) the Hahn-Banach theorems (assuring a supply of continuous linear functionals that is adequate for a highly developed duality theory), (*b*) the Banach-Alaoglu compactness theorem in dual spaces, and (*c*) the Krein-Milman theorem about extreme points. Applications to various problems in analysis are postponed to Chapter 5.

The Hahn-Banach Theorems

The plural is used here because the term "Hahn-Banach theorem" is customarily applied to several closely related results. Among these are the *dominated extension theorems* 3.2 and 3.3 (in which no topology is involved), the *separation theorem* 3.4, and the *continuous extension theorem* 3.6. Another separation theorem (which implies 3.4) is stated as Exercise 3.

3.1 Definitions The *dual space* of a topological vector space X is the vector space X^* whose elements are the *continuous* linear functionals on X.

Note that addition and scalar multiplication are defined in X^* by

$$(\Lambda_1 + \Lambda_2)x = \Lambda_1 x + \Lambda_2 x, \qquad (\alpha\Lambda)x = \alpha \cdot \Lambda x.$$

It is clear that these operations do indeed make X^* into a vector space.

It will be necessary to use the obvious fact that every complex vector space is also a real vector space, and it will be convenient to use the following (temporary) terminology: An additive functional Λ on a complex vector space X is called *real-linear (complex-linear)* if $\Lambda(\alpha x) = \alpha\Lambda x$ for every $x \in X$ and for every real (complex) scalar α. Our standing rule that any statement about vector spaces in which no scalar field is mentioned applies to both cases is unaffected by this temporary terminology and is still in force.

If u is the real part of a complex-linear functional f on X, then u is real-linear and

$$f(x) = u(x) - iu(ix) \qquad (x \in X) \tag{1}$$

because $z = \text{Re } z - i \text{ Re } (iz)$ for every $z \in \mathbb{C}$.

Conversely, if $u: X \to R$ is real-linear on a complex vector space X and if f is defined by (1), a straightforward computation shows that f is complex-linear.

Suppose now that X is a complex topological vector space. The above facts imply that a complex-linear functional on X is in X^* if and only if its real part is continuous, and that every continuous real-linear $u: X \to R$ is the real part of a unique $f \in X^*$.

3.2 Theorem *Suppose*

(a) M is a subspace of a real vector space X,

(b) $p: X \to R$ satisfies

$$p(x + y) \leq p(x) + p(y) \qquad \textit{and} \qquad p(tx) = p(x)$$

if $x \in X$, $y \in X$, $t \geq 0$,

(c) $f: M \to R$ is linear and $f(x) \leq p(x)$ on M.

Then there exists a linear $\Lambda: X \to R$ such that

$$\Lambda x = f(x) \qquad (x \in M)$$

and

$$-p(-x) \leq \Lambda x \leq p(x) \qquad (x \in X).$$

PROOF. If $M \neq X$, choose $x_1 \in X$, $x_1 \notin M$, and define

$$M_1 = \{x + tx_1: x \in M, t \in R\}.$$

It is clear that M_1 is a vector space. Since

$$f(x) + f(y) = f(x + y) \leq p(x + y) \leq p(x - x_1) + p(x_1 + y),$$

we have

$$f(x) - p(x - x_1) \leq p(y + x_1) - f(y) \qquad (x, y \in M). \tag{1}$$

Let α be the least upper bound of the left side of (1), as x ranges over M. Then

$$f(x) - \alpha \leq p(x - x_1) \qquad (x \in M) \tag{2}$$

and

$$f(y) + \alpha \leq p(y + x_1) \qquad (y \in M). \tag{3}$$

Define f_1 on M_1 by

$$f_1(x + tx_1) = f(x) + t\alpha \qquad (x \in M, t \in R). \tag{4}$$

Then $f_1 = f$ on M, and f_1 is linear on M_1.

Take $t > 0$, replace x by $t^{-1}x$ in (2), replace y by $t^{-1}y$ in (3), and multiply the resulting inequalities by t. In combination with (4), this proves that $f_1 \leq p$ on M_1.

The second part of the proof can be done by whatever one's favorite method of transfinite induction is; one can use well-ordering, or Zorn's lemma, or Hausdorff's maximality theorem.

Let $\mathscr{P}$ be the collection of all ordered pairs (M', f'), where M' is a subspace of X that contains M and f' is a linear functional on M' that extends f and satisfies $f' \leq p$ on M'. Partially order $\mathscr{P}$ by declaring $(M', f') \leq (M'', f'')$ to mean that $M' \subset M''$ and $f'' = f'$ on M'. By Hausdorff's maximality theorem there exists a maximal totally ordered subcollection Ω of $\mathscr{P}$.

Let Φ be the collection of all M' such that $(M', f') \in \Omega$. Then Φ is totally ordered by set inclusion, and the union $\tilde{M}$ of all members of Φ is therefore a subspace of X. If $x \in \tilde{M}$ then $x \in M'$ for some $M' \in \Phi$; define $\Lambda x = f'(x)$, where f' is the function which occurs in the pair $(M', f') \in \Omega$.

It is now easy to check that Λ is well defined on $\tilde{M}$, that Λ is linear, and that $\Lambda \leq p$. If $\tilde{M}$ were a proper subspace of X, the first part of the proof would give a further extension of Λ, and this would contradict the maximality of Ω. Thus $\tilde{M} = X$.

Finally, the inequality $\Lambda \leq p$ implies that

$$-p(-x) \leq -\Lambda(-x) = \Lambda x$$

for all $x \in X$. This completes the proof. ////

3.3 Theorem *Suppose M is a subspace of a vector space X, p is a seminorm on X, and f is a linear functional on M such that*

$$|f(x)| \leq p(x) \qquad (x \in M).$$

Then f extends to a linear functional Λ on X that satisfies

$$|\Lambda x| \leq p(x) \qquad (x \in X).$$

PROOF. If the scalar field is R, this is contained in Theorem 3.2, since p now satisfies $p(-x) = p(x)$.

Assume that the scalar field is $\mathcal{C}$. Put $u = \text{Re} f$. By Theorem 3.2 there is a real-linear U on X such that $U = u$ on M and $U \leq p$ on X. Let Λ be the complex-linear functional on X whose real part is U. The discussion in Section 3.1 implies that $\Lambda = f$ on M.

Finally, to every $x \in X$ corresponds an $\alpha \in \mathcal{C}$, $|\alpha| = 1$, such that $\alpha\Lambda x = |\Lambda x|$. Hence

$$|\Lambda x| = \Lambda(\alpha x) = U(\alpha x) \leq p(\alpha x) = p(x). \qquad ////$$

Corollary. *If X is a normed space and $x_0 \in X$, there exists $\Lambda \in X^*$ such that*

$$\Lambda x_0 = \|x_0\| \qquad \textit{and} \qquad |\Lambda x| \leq \|x\| \qquad \textit{for all } x \in X.$$

PROOF. If $x_0 = 0$, take $\Lambda = 0$. If $x_0 \neq 0$, apply Theorem 3.3, with $p(x) = \|x\|$, M the one-dimensional space generated by x_0, and $f(\alpha x_0) = \alpha\|x_0\|$ on M. ////

3.4 Theorem *Suppose A and B are disjoint, nonempty, convex sets in a topological vector space X.*

(a) *If A is open there exist $\Lambda \in X^*$ and $\gamma \in R$ such that*

$$\text{Re } \Lambda x < \gamma \leq \text{Re } \Lambda y$$

for every $x \in A$ and for every $y \in B$.

(b) *If A is compact, B is closed, and X is locally convex, then there exist $\Lambda \in X^*$, $\gamma_1 \in R$, $\gamma_2 \in R$, such that*

$$\text{Re } \Lambda x < \gamma_1 < \gamma_2 < \text{Re } \Lambda y$$

for every $x \in A$ and for every $y \in B$.

Note that this is stated without specifying the scalar field; if it is R, then $\text{Re } \Lambda = \Lambda$, of course.

PROOF. It is enough to prove this for real scalars. For if the scalar field is $\mathcal{C}$ and the real case has been proved, then there is a continuous real-linear Λ_1 on X that gives the required separation; if Λ is the unique complex-linear functional on X whose real part is Λ_1, then $\Lambda \in X^*$. (See Section 3.1.) Assume real scalars.

(*a*) Fix $A_0 \in A$, $b_0 \in B$. Put $x_0 = b_0 - a_0$; put $C = A - B + x_0$. Then C is a convex neighborhood of 0 in X. Let p be the Minkowski functional of C. By Theorem 1.35, p satisfies hypothesis (*b*) of Theorem 3.2. Since $A \cap B = \varnothing$, $x_0 \notin C$, and so $p(x_0) \geq 1$.

Define $f(tx_0) = t$ on the subspace M of X generated by x_0. If $t \geq 0$ then

$$f(tx_0) = t \leq tp(x_0) = p(tx_0);$$

if $t < 0$ then $f(tx_0) < 0 \leq p(tx_0)$. Thus $f \leq p$ on M. By Theorem 3.2, f extends to a linear functional Λ on X that also satisfies $\Lambda \leq p$. In particular, $\Lambda \leq 1$ on C, hence $\Lambda \geq -1$ on $-C$, so that $|\Lambda| \leq 1$ on the neighborhood $C \cap (-C)$ of 0. By Theorem 1.18, $\Lambda \in X^*$.

If now $a \in A$ and $b \in B$, we have

$$\Lambda a - \Lambda b + 1 = \Lambda(a - b + x_0) \leq p(a - b + x_0) < 1$$

since $\Lambda x_0 = 1$, $a - b + x_0 \in C$, and C is open. Thus $\Lambda a < \Lambda b$.

It follows that $\Lambda(A)$ and $\Lambda(B)$ are disjoint convex subsets of R, with $\Lambda(A)$ to the left of $\Lambda(B)$. Also, $\Lambda(A)$ is an open set since A is open and since every nonconstant linear functional on X is an open mapping. Let γ be the right end point of $\Lambda(A)$ to get the conclusion of part (*a*).

(*b*) By Theorem 1.10 there is a convex neighborhood V of 0 in X such that $(A + V) \cap B = \varnothing$. Part (*a*), with $A + V$ in place of A, shows that there exists $\Lambda \in X^*$ such that $\Lambda(A + V)$ and $\Lambda(B)$ are disjoint convex subsets of R, with $\Lambda(A + V)$ open and to the left of $\Lambda(B)$. Since $\Lambda(A)$ is a compact subset of $\Lambda(A + V)$, we obtain the conclusion of (*b*). ////

Corollary. *If X is a locally convex space then X^* separates points on X.*

PROOF. If $x_1 \in X$, $x_2 \in X$, and $x_1 \neq x_2$, apply (*b*) of Theorem 3.4 with $A = \{x_1\}$, $B = \{x_2\}$. ////

3.5 Theorem *Suppose M is a subspace of a locally convex space X, and $x_0 \in X$. If x_0 is not in the closure of M, then there exists $\Lambda \in X^*$ such that $\Lambda x_0 = 1$ but $\Lambda x = 0$ for every $x \in M$.*

PROOF. By (*b*) of Theorem 3.4, with $A = \{x_0\}$ and $B = \bar{M}$, there exists $\Lambda \in X^*$ such that Λx_0 and $\Lambda(M)$ are disjoint. Thus $\Lambda(M)$ is a *proper* subspace of the scalar field. This forces $\Lambda(M) = \{0\}$ and $\Lambda x_0 \neq 0$. The desired functional is obtained by dividing Λ by Λx_0. ////

Remark. This theorem is the basis of a standard method of treating certain approximation problems: In order to prove that an $x_0 \in X$ lies in the closure of some subspace M of X it suffices (if X is locally convex) to show that $\Lambda x_0 = 0$ for every continuous linear functional Λ on X that vanishes on M.

3.6 Theorem *If f is a continuous linear functional on a subspace M of a locally convex space X, then there exists $\Lambda \in X^*$ such that $\Lambda = f$ on M.*

Remark. For normed spaces this is an immediate corollary of Theorem 3.3. The general case could also be obtained from 3.3, by relating the continuity of linear functionals to seminorms (see Exercise 8, Chapter 1). The proof given below shows that Theorem 3.6 depends only on the separation property of Theorem 3.5.

PROOF. Assume, without loss of generality, that f is not identically 0 on M. Put

$$M_0 = \{x \in M : f(x) = 0\}$$

and pick $x_0 \in M$ such that $f(x_0) = 1$. Since f is continuous, x_0 is not in the M-closure of M_0, and since M inherits its topology from X, it follows that x_0 is not in the X-closure of M_0.

Theorem 3.5 therefore assures the existence of a $\Lambda \in X^*$ such that $\Lambda x_0 = 1$ and $\Lambda = 0$ on M_0.

If $x \in M$, then $x - f(x)x_0 \in M_0$, since $f(x_0) = 1$. Hence

$$\Lambda x - f(x) = \Lambda x - f(x)\Lambda x_0 = \Lambda(x - f(x)x_0) = 0.$$

Thus $\Lambda = f$ on M. ////

We conclude this discussion with another useful corollary of the separation theorem.

3.7 Theorem *Suppose B is a convex, balanced, closed set in a locally convex space X, $x_0 \in X$, but $x_0 \notin B$. Then there exists $\Lambda \in X^*$ such that $|\Lambda x| \leq 1$ for all $x \in B$, but $\Lambda x_0 > 1$.*

PROOF. Since B is closed and convex, we can apply (*b*) of Theorem 3.4, with $A = \{x_0\}$, to obtain $\Lambda_1 \in X^*$ such that $\Lambda_1 x_0 = re^{i\theta}$ lies outside the closure K of $\Lambda_1(B)$. Since B is balanced, so is K. Hence there exists s, $0 < s < r$, so that $|z| \leq s$ for all $z \in K$. The functional $\Lambda = s^{-1}e^{-i\theta}\Lambda_1$ has the desired properties. ////

Weak Topologies

3.8 Topological preliminaries The purpose of this section is to explain and illustrate some of the phenomena that occur when a set is topologized in several ways.

Let τ_1 and τ_2 be two topologies on a set X, and assume $\tau_1 \subset \tau_2$; that is, every τ_1-open set is also τ_2-open. Then we say that τ_1 is *weaker* than τ_2, or that τ_2 is *stronger* than τ_1. [Note that (in accordance with the meaning of the inclusion symbol $\subset$) the terms "weaker" and "stronger" do not exclude equality.] In this situation, the identity mapping on X is *continuous* from (X, τ_2) to (X, τ_1) and is an *open mapping* from (X, τ_1) to (X, τ_2).

As a first illustration, let us prove that the topology of a compact Hausdorff space has a certain rigidity, in the sense that it cannot be weakened without losing the Hausdorff separation axiom and cannot be strengthened without losing compactness:

(*a*) *If $\tau_1 \subset \tau_2$ are topologies on a set X, if τ_1 is a Hausdorff topology, and if τ_2 is compact, then $\tau_1 = \tau_2$.*

To see this, let $F \subset X$ be τ_2-closed. Since X is τ_2-compact, so is F. Since $\tau_1 \subset \tau_2$, it follows that F is τ_1-compact. (Every τ_1-open cover of F is also a τ_2-open cover.) Since τ_1 is a Hausdorff topology, it follows that F is τ_1-closed.

As another illustration, consider the quotient topology τ_N of X/N, as defined in Section 1.40, and the quotient map $\pi: X \to X/N$. By its very definition, τ_N is the strongest topology on X/N that makes π continuous, and it is the weakest one that makes π an open mapping. Explicitly, if τ' and τ'' are topologies on X/N, and if π is continuous relative to τ' and open relative to τ'', then $\tau' \subset \tau_N \subset \tau''$.

Suppose next that X is a set and $\mathscr{F}$ is a nonempty family of mappings $f: X \to Y_f$, where each Y_f is a topological space. (In many important cases, Y_f is the same for all $f \in \mathscr{F}$.) Let τ be the collection of all unions of finite intersections of sets $f^{-1}(V)$, with $f \in \mathscr{F}$ and V open in Y_f. Then τ is a topology on X, and it is in fact the *weakest* topology on X that makes every $f \in \mathscr{F}$ continuous: If τ' is any other topology with that property, then $\tau \subset \tau'$. This τ is called *the weak topology on X induced by $\mathscr{F}$*, or, more succinctly, the *$\mathscr{F}$-topology of X*.

The best-known example of this situation is undoubtedly the usual way in which one topologizes the cartesian product X of a collection of topological spaces X_α. If $\pi_\alpha(x)$ denotes the αth coordinate of a point $x \in X$, then π_α maps X onto X_α, and the product topology τ of X is, by definition, its $\{\pi_\alpha\}$-topology, the weakest one that makes every π_α continuous. Assume now that every X_α is a *compact Hausdorff space.* Then τ is a compact topology on X (by Tychonoff's theorem), and proposition (*a*) implies that τ cannot be strengthened without spoiling Tychonoff's theorem.

In the last sentence a special case of the following proposition was tacitly used:

(*b*) *If $\mathscr{F}$ is a family of mappings $f: X \to Y_f$, where X is a set and each Y_f is a Hausdorff space, and if $\mathscr{F}$ separates points on X, then the $\mathscr{F}$-topology of X is a Hausdorff topology.*

For if $p \neq q$ are points of X, then $f(p) \neq f(q)$ for some $f \in \mathscr{F}$; the points $f(p)$ and $f(q)$ have disjoint neighborhoods in Y_f whose inverse images under f are open (by definition) and disjoint.

Here is an application of these ideas to a metrization theorem.

(*c*) *If X is a compact topological space and if some sequence $\{f_n\}$ of continuous real-valued functions separates points on X, then X is metrizable.*

Let τ be the given topology of X. Suppose, without loss of generality, that $|f_n| \leq 1$ for all n, and let τ_d be the topology induced on X by the metric

$$d(p, q) = \sum_{n=1}^{\infty} 2^{-n} |f_n(p) - f_n(q)|.$$

This is indeed a metric, since $\{f_n\}$ separates points. Since each f_n is τ-continuous and the series converges uniformly on $X \times X$, d is a τ-continuous function on $X \times X$. The balls

$$B_r(p) = \{q \in X: d(p, q) < r\}$$

are therefore τ-open. Thus $\tau_d \subset \tau$. Since τ_d is induced by a metric, τ_d is a Hausdorff topology, and now (*a*) implies that $\tau = \tau_d$.

The following lemma has applications in the study of vector topologies. In fact, the case $n = 1$ was needed (and proved) at the end of Theorem 3.6.

3.9 Lemma *Suppose $\Lambda_1, \ldots, \Lambda_n$ and Λ are linear functionals on a vector space X. Let*

$$N = \{x: \Lambda_1 x = \cdots = \Lambda_n x = 0\}.$$

The following three properties are then equivalent:

(*a*) *There are scalars $\alpha_1, \ldots, \alpha_n$ such that*

$$\Lambda = \alpha_1 \Lambda_1 + \cdots + \alpha_n \Lambda_n.$$

(*b*) *There exists* $\gamma < \infty$ *such that*

$$|\Lambda x| \le \gamma \max_{1 \le i \le n} |\Lambda_i x| \qquad (x \in X).$$

(*c*) $\Lambda x = 0$ *for every* $x \in N$.

PROOF. It is clear that (*a*) implies (*b*) and that (*b*) implies (*c*). Assume (*c*) holds. Let Φ be the scalar field. Define $\pi: X \to \Phi^n$ by

$$\pi(x) = (\Lambda_1 x, \ldots, \Lambda_n x).$$

If $\pi(x) = \pi(x')$, then (*c*) implies $\Lambda x = \Lambda x'$. Hence $f(\pi(x)) = \Lambda x$ defines a linear functional f on $\pi(X)$. Extend f to a linear functional F on Φ^n. This means that there exist $\alpha_i \in \Phi$ such that

$$F(u_1, \ldots, u_n) = \alpha_1 u_1 + \cdots + \alpha_n u_n.$$

Thus

$$\Lambda x = F(\pi(x)) = F(\Lambda_1 x, \ldots, \Lambda_n x) = \sum_{i=1}^{n} \alpha_i \Lambda_i x,$$

which is (*a*). ////

3.10 Theorem *Suppose X is a vector space and X′ is a separating vector space of linear functionals on X. Then the X′-topology τ′ makes X into a locally convex space whose dual space is X′.*

The assumptions on X' are, more explicitly, that X' is closed under addition and scalar multiplication and that $\Lambda x_1 \ne \Lambda x_2$ for some $\Lambda \in X'$ whenever x_1 and x_2 are distinct points of X.

PROOF. Since R and $\mathcal{C}$ are Hausdorff spaces, (*b*) of Section 3.8 shows that τ' is a Hausdorff topology. The linearity of the members of X' shows that τ' is translation-invariant. If $\Lambda_1, \ldots, \Lambda_n \in X'$, if $r_i > 0$, and if

$$V = \{x: |\Lambda_i x| < r_i \quad \text{for} \quad 1 \le i \le n\}, \tag{1}$$

then V is convex, balanced, and $V \in \tau'$. In fact, the collection of all V of the form (1) is a local base for τ'. Thus τ' is a locally convex topology on X.

If (1) holds, then $\frac{1}{2}V + \frac{1}{2}V = V$. This proves that addition is continuous. Suppose $x \in X$ and α is a scalar. Then $x \in sV$ for some $s > 0$. If $|\beta - \alpha| < r$ and $y - x \in rV$ then

$$\beta y - \alpha x = (\beta - \alpha)y + \alpha(y - x)$$

lies in V, provided that r is so small that

$$r(s + r) + |\alpha| r < 1.$$

Hence scalar multiplication is continuous.

We have now proved that τ' is a locally convex vector topology. Every $\Lambda \in X'$ is τ'-continuous. Conversely, suppose Λ is a τ'-continuous linear functional on X. Then $|\Lambda x| < 1$ for all x in some set V of the form (1). Condition (*b*) of Lemma 3.9 therefore holds; hence so does (*a*): $\Lambda = \sum \alpha_i \Lambda_i$. Since $\Lambda_i \in X'$ and X' is a vector space, $\Lambda \in X'$. This completes the proof. ////

Note: The first part of this proof could have been based on Theorem 1.37 and the separating family of seminorms $p_\Lambda(\Lambda \in X')$ given by $p_\Lambda(x) = |\Lambda x|$.

3.11 The weak topology of a topological vector space Suppose X is a topological vector space (with topology τ) whose dual X^* separates points on X. (We know that this happens in every locally convex X. It also happens in some others; see Exercise 5.) The X^*-topology of X is called *the weak topology of* X.

We shall let X_w denote X topologized by this weak topology τ_w. Theorem 3.10 implies that X_w is a locally convex space whose dual is also X^*.

Since every $\Lambda \in X^*$ is τ-continuous and since τ_w is the weakest topology on X with that property, we have $\tau_w \subset \tau$. In this context, the given topology τ will often be called the *original* topology of X.

Self-explanatory expressions such as original neighborhood, weak neighborhood, original closure, weak closure, originally bounded, weakly bounded, etc., will be used to make it clear with respect to which topology these terms are to be understood.[1]

For instance, let $\{x_n\}$ be a sequence in X. To say that $x_n \to 0$ originally means that every original neighborhood of 0 contains all x_n with sufficiently large n. To say that $x_n \to 0$ weakly means that every weak neighborhood of 0 contains all x_n with sufficiently large n. Since every weak neighborhood of

[1] When X is a Fréchet space (hence, in particular, when X is a Banach space) the original topology of X is usually called its *strong topology*. In that context, the terms "strong" and "strongly" will be used in place of "original" and "originally." For locally convex spaces in general, the term "strong topology" has been given a specific technical meaning. See [15], pp. 256–258; also [14], p. 169. It seems therefore advisable to use "original" in the present general discussion.

0 contains a neighborhood of the form

$$V = \{x: |\Lambda_i x| < r_i \quad \text{for} \quad 1 \le i \le n\}, \tag{1}$$

where $\Lambda_i \in X^*$ and $r_i > 0$, it is easy to see that $x_n \to 0$ *weakly if and only if* $\Lambda x_n \to 0$ *for every* $\Lambda \in X^*$.

Hence every originally convergent sequence converges weakly. (The converse is usually false; see Exercises 5 and 6.)

Similarly, a set $E \subset X$ is *weakly bounded* (that is, E is a bounded subset of X_w) if and only if every V as in (1) contains tE for some $t = t(V) > 0$. This happens if and only if there corresponds to each $\Lambda \in X^*$ a number $\gamma(\Lambda) < \infty$ such that $|\Lambda x| \le \gamma(\Lambda)$ for every $x \in E$. In other words, *a set* $E \subset X$ *is weakly bounded if and only if every* $\Lambda \in X^*$ *is a bounded function on* E.

Let V again be as in (1), and put

$$N = \{x: \Lambda_1 x = \cdots = \Lambda_n x = 0\}.$$

Since $x \to (\Lambda_1 x, \ldots, \Lambda_n x)$ maps X into $\mathcal{C}^n$ with null space N, we see that $\dim X \le n + \dim N$. Since $N \subset V$, this leads to the following conclusion.

If X is infinite-dimensional then every weak neighborhood of 0 contains an infinite-dimensional subspace; hence X_w is not locally bounded.

This implies in many cases that the weak topology is strictly weaker than the original one. Of course, the two may coincide: Theorem 3.10 implies that $(X_w)_w = X_w$.

We now come to a more interesting result.

3.12 Theorem *Suppose E is a convex subset of a locally convex space X. Then the weak closure $\bar{E}_w$ of E is equal to its original closure $\bar{E}$.*

PROOF. $\bar{E}_w$ is weakly closed, hence originally closed, so that $\bar{E} \subset \bar{E}_w$. To obtain the opposite inclusion, choose $x_0 \in X$, $x_0 \notin \bar{E}$. Part (*b*) of the separation theorem 3.4 shows that there exist $\Lambda \in X^*$ and $\gamma \in R$ such that, for every $x \in \bar{E}$,

$$\operatorname{Re} \Lambda x_0 < \gamma < \operatorname{Re} \Lambda x.$$

The set $\{x: \operatorname{Re} \Lambda x < \gamma\}$ is therefore a weak neighborhood of x_0 that does not intersect E. Thus x_0 is not in $\bar{E}_w$. This proves $\bar{E}_w \subset \bar{E}$. ////

Corollaries. *For convex subsets of a locally convex space,*

(*a*) *originally closed equals weakly closed, and*

(*b*) *originally dense equals weakly dense.*

The proofs are obvious. Here is another noteworthy consequence of Theorem 3.12.

3.13 Theorem *Suppose X is a metrizable locally convex space. If $\{x_n\}$ is a sequence in X that converges weakly to some $x \in X$, then there is a sequence $\{y_i\}$ in X such that*

(*a*) *each y_i is a convex combination of finitely many x_n, and*
(*b*) *$y_i \to x$ originally.*

Conclusion (*a*) says, more explicitly, that there exist numbers $\alpha_{in} \geq 0$, such that

$$\sum_{n=1}^{\infty} \alpha_{in} = 1, \qquad y_i = \sum_{n=1}^{\infty} \alpha_{in} x_n,$$

and, for each i, only finitely many α_{in} are $\neq 0$.

PROOF. Let H be the convex hull of the set of all x_n; let K be the weak closure of H. Then $x \in K$. By Theorem 3.12, x is also in the original closure of H. Since the original topology of X is assumed to be metrizable, it follows that there is a sequence $\{y_i\}$ in H that converges originally to x. ////

To get a feeling for what is involved here, consider the following example.

Let K be a compact Hausdorff space (the unit interval on the real line is a sufficiently interesting one), and assume that f and f_n ($n = 1, 2, 3, \ldots$) are continuous complex functions on K such that $f_n(x) \to f(x)$ for every $x \in K$, as $n \to \infty$, and such that $|f_n(x)| \leq 1$ for all n and all $x \in K$. Theorem 3.13 asserts that there are convex combinations of the f_n that converge *uniformly* to f.

To see this, let $C(K)$ be the Banach space of all complex continuous functions on K, normed by the supremum. Then strong convergence is the same as uniform convergence on K. If μ is any complex Borel measure on K, Lebesgue's dominated convergence theorem implies that $\int f_n \, d\mu \to \int f \, d\mu$. Hence $f_n \to f$ weakly, by the Riesz representation theorem which identifies the dual of $C(K)$ with the space of all regular complex Borel measures on K. Now Theorem 3.13 can be applied.

After this short detour we now return to our main line of development.

3.14 The weak*-topology of a dual space Let X again be a topological vector space whose dual is X^*. For the definitions that follow, it is irrelevant whether X^* separates points on X or not. The important observation to make is that *every* $x \in X$ *induces a linear functional* f_x *on* X^*, *defined by*

$$f_x \Lambda = \Lambda x,$$

and that $\{f_x : x \in X\}$ *separates points on* X^*.

The linearity of each f_x is obvious; if $f_x \Lambda = f_x \Lambda'$ for all $x \in X$, then $\Lambda x = \Lambda' x$ for all x, and so $\Lambda = \Lambda'$ by the very definition of what it means for two functions to be equal.

We are now in the situation described by Theorem 3.10, with X^* in place of X and with X in place of X'.

The X-topology of X^* is called *the weak*-topology of* X^* (pronunciation: weak star topology).

Theorem 3.10 implies that this is a locally convex vector topology on X^* and that *every linear functional on* X^* *that is weak*-continuous has the form* $\Lambda \to \Lambda x$ *for some* $x \in X$.

The weak*-topologies have a very important compactness property to which we now turn our attention. Various pathological features of the weak- and weak*-topologies are described in Exercises 9 and 10.

Compact Convex Sets

3.15 The Banach-Alaoglu theorem *If* V *is a neighborhood of* 0 *in a topological vector space* X *and if*

$$K = \{\Lambda \in X^* : |\Lambda x| \le 1 \quad \text{for every} \quad x \in V\}$$

then K *is weak*-compact.*

Note: K is sometimes called the *polar* of V. It is clear that K is convex and balanced, because this is true of the unit disc in $\mathbb{C}$ (and of the interval $[-1, 1]$ in R). There is some redundancy in the definition of K, since every linear functional on X that is bounded on V is continuous, hence is in X^*.

PROOF. Since neighborhoods of 0 are absorbing, there corresponds to each $x \in X$ a number $\gamma(x) < \infty$ such that $x \in \gamma(x)V$. Hence

$$(1) \qquad |\Lambda x| \le \gamma(x) \qquad (x \in X, \Lambda \in K).$$

Let D_x be the set of all scalars α such that $|\alpha| \le \gamma(x)$. Let τ be the product topology on P, the cartesian product of all D_x, one for each $x \in X$. Since each D_x is compact, so is P, by Tychonoff's theorem. The elements of P are the functions f on X (linear or not) that satisfy

$$(2) \qquad |f(x)| \le \gamma(x) \qquad (x \in X).$$

Thus $K \subset X^* \cap P$. It follows that K inherits two topologies: one from X^* (its weak*-topology, to which the conclusion of the theorem refers) and the other, τ, from P. We will see that

(a) these two topologies coincide on K, and

(b) K is a closed subset of P.

Since P is compact, (b) implies that K is τ-compact, and then (a) implies that K is weak*-compact.

Fix some $\Lambda_0 \in K$. Choose $x_i \in X$, for $1 \le i \le n$; choose $\delta > 0$. Put

$$W_1 = \{\Lambda \in X^*: |\Lambda x_i - \Lambda_0 x_i| < \delta \quad \text{for} \quad 1 \le i \le n\} \tag{3}$$

and

$$W_2 = \{f \in P: |f(x_i) - \Lambda_0 x_i| < \delta \quad \text{for} \quad 1 \le i \le n\}. \tag{4}$$

Let n, x_i, and δ range over all admissible values. The resulting sets W_1 then form a local base for the weak*-topology of X^* at Λ_0 and the sets W_2 form a local base for the product topology τ of P at Λ_0. Since $K \subset P \cap X^*$, we have

$$W_1 \cap K = W_2 \cap K.$$

This proves (a).

Next, suppose f_0 is in the τ-closure of K. Choose $x \in X$, $y \in X$, scalars α and β, and $\varepsilon > 0$. The set of all $f \in P$ such that $|f - f_0| < \varepsilon$ at x, at y, and at $\alpha x + \beta y$ is a τ-neighborhood of f_0. Therefore K contains such an f. Since this f is linear, we have

$$f_0(\alpha x + \beta y) - \alpha f_0(x) - \beta f_0(y)$$
$$= (f_0 - f)(\alpha x + \beta y) + \alpha(f - f_0)(x) + \beta(f - f_0)(y),$$

so that

$$|f_0(\alpha x + \beta y) - \alpha f_0(x) - \beta f_0(y)| < (1 + |\alpha| + |\beta|)\varepsilon.$$

Since ε was arbitrary, we see that f_0 is linear. Finally, if $x \in V$ and $\varepsilon > 0$, the same argument shows that there is an $f \in K$ such that $|f(x) - f_0(x)| < \varepsilon$. Since $|f(x)| \le 1$, by the definition of K, it follows that $|f_0(x)| \le 1$. We conclude that $f_0 \in K$. This proves (b) and hence the theorem. ////

When X is *separable* (i.e., when there is a countable dense set in X), then the conclusion of the Banach-Alaoglu theorem can be strengthened by combining it with the following fact:

3.16 Theorem *If X is a separable topological vector space, if $K \subset X^*$, and if K is weak*-compact, then K is metrizable, in the weak*-topology.*

Warning: It does not follow that X^* itself is metrizable in its weak*-topology. In fact, this is false whenever X is an infinite-dimensional Banach space. See Exercise 15.

PROOF. Let $\{x_n\}$ be a countable dense set in X. Put $f_n(\Lambda) = \Lambda x_n$, for $\Lambda \in X^*$. Each f_n is weak*-continuous, by the definition of the weak*-topology. If $f_n(\Lambda) = f_n(\Lambda')$ for all n, then $\Lambda x_n = \Lambda' x_n$ for all n, which implies that $\Lambda = \Lambda'$, since both are continuous on X and coincide on a dense set.

Thus $\{f_n\}$ is a countable family of continuous functions that separates points on X^*. The metrizability of K now follows from (*c*) of Section 3.8. ////

3.17 Theorem *If V is a neighborhood of 0 in a separable topological vector space X, and if $\{\Lambda_n\}$ is a sequence in X^* such that*

$$|\Lambda_n x| \le 1 \qquad (x \in V,\ n = 1, 2, 3, \ldots),$$

then there is a subsequence $\{\Lambda_{n_i}\}$ and there is a $\Lambda \in X^*$ such that

$$\Lambda x = \lim_{i \to \infty} \Lambda_{n_i} x \qquad (x \in X).$$

In other words, the polar of V is sequentially compact in the weak*-topology.

PROOF. Combine Theorems 3.15 and 3.16. ////

The next application of the Banach-Alaoglu theorem involves the Hahn-Banach theorem and a category argument.

3.18 Theorem *In a locally convex space X, every weakly bounded set is originally bounded, and vice versa.*

Part (*d*) of Exercise 5 shows that the local convexity of X cannot be omitted from the hypotheses.

PROOF. Since every weak neighborhood of 0 in X is an original neighborhood of 0, it is obvious from the definition of "bounded" that every originally bounded subset of X is weakly bounded. The converse is the nontrivial part of the theorem.

Suppose $E \subset X$ is weakly bounded and U is an original neighborhood of 0 in X.

Since X is locally convex, there is a convex, balanced, original neighborhood V of 0 in X such that $\bar{V} \subset U$. Let $K \subset X^*$ be the polar of V:

$$K = \{\Lambda \in X^*: |\Lambda x| \leq 1 \quad \text{for all} \quad x \in V\}. \tag{1}$$

We claim that

$$\bar{V} = \{x \in X: |\Lambda x| \leq 1 \quad \text{for all} \quad \Lambda \in K\}. \tag{2}$$

It is clear that V is a subset of the right side of (2) and hence so is $\bar{V}$, since the right side of (2) is closed. Suppose $x_0 \in X$ but $x_0 \notin \bar{V}$. Theorem 3.7 (with $\bar{V}$ in place of B) then shows that $\Lambda x_0 > 1$ for some $\Lambda \in K$. This proves (2).

Since E is weakly bounded, there corresponds to each $\Lambda \in X^*$ a number $\gamma(\Lambda) < \infty$ such that

$$|\Lambda x| \leq \gamma(\Lambda) \qquad (x \in E). \tag{3}$$

Since K is convex and weak*-compact (Theorem 3.15) and since the functions $\Lambda \to \Lambda x$ are weak*-continuous, we can apply Theorem 2.9 (with X^* in place of X and the scalar field in place of Y) to conclude from (3) that there is a constant $\gamma < \infty$ such that

$$|\Lambda x| \leq \gamma \qquad (x \in E, \Lambda \in K). \tag{4}$$

Now (2) and (4) show that $\gamma^{-1}x \in \bar{V} \subset U$ for all $x \in E$. Since V is balanced,

$$E \subset t\bar{V} \subset tU \qquad (t > \gamma). \tag{5}$$

Thus E is originally bounded. ////

Corollary. *If X is a normed space, if $E \subset X$, and if*

$$\sup_{x \in E} |\Lambda x| < \infty \qquad (\Lambda \in X^*) \tag{6}$$

then there exists $\gamma < \infty$ such that

$$\|x\| \leq \gamma \qquad (x \in E). \tag{7}$$

PROOF. Normed spaces are locally convex; (6) says that E is weakly bounded, and (7) says that E is originally bounded. ////

We now turn to the question: What can one say about the convex hull H of a compact set K? Even in a Hilbert space, H need not be closed, and there are situations in which $\bar{H}$ is not compact (Exercises 20, 22). In

Fréchet spaces the latter pathology does not occur (Theorem 3.20). The proof of this will depend on the fact that a subset of a complete metric space is compact if and only if it is closed and totally bounded (Appendix A4).

3.19 Definitions (*a*) If X is a vector space and $E \subset X$, the *convex hull* of E will be denoted by $co(E)$. Recall that $co(E)$ is the intersection of all convex subsets of X which contain E. Equivalently, $co(E)$ is the set of all finite convex combinations of members of E.

(*b*) If X is a topological vector space and $E \subset X$, the *closed convex hull* of E, written $\overline{co}(E)$, is the closure of $co(E)$.

(*c*) A subset E of a metric space X is said to be *totally bounded* if E lies in the union of finitely many open balls of radius ε, for every $\varepsilon > 0$.

The same concept can be defined in any topological vector space, metrizable or not:

(*d*) A set E in a topological vector space X is said to be *totally bounded* if to every neighborhood V of 0 in X corresponds a *finite* set F such that $E \subset F + V$.

If X happens to be a metrizable topological vector space, then these two notions of total boundedness coincide, provided we restrict ourselves to *invariant* metrics that are compatible with the topology of X. (The proof of this is as in Section 1.25.)

3.20 Theorem

(*a*) *If $A_1, \ldots, A_n$ are compact convex sets in a topological vector space X, then $co(A_1 \cup \cdots \cup A_n)$ is compact.*

(*b*) *If X is a locally convex topological vector space and $E \subset X$ is totally bounded, then $co(E)$ is totally bounded.*

(*c*) *If X is a Fréchet space and $K \subset X$ is compact, then $\overline{co}(K)$ is compact.*

(*d*) *If K is a compact set in R^n, then $co(K)$ is compact.*

PROOF. (*a*) Let S be the simplex in R^n consisting of all $s = (s_1, \ldots, s_n)$ with $s_i \geq 0$, $s_1 + \cdots + s_n = 1$. Put $A = A_1 \times \cdots \times A_n$. Define $f: S \times A \to X$ by

$$f(s, a) = s_1 a_1 + \cdots + s_n a_n \tag{1}$$

and put $K = f(S \times A)$.

It is clear that K is compact and that $K \subset co(A_1 \cup \cdots \cup A_n)$. We will see that this inclusion is actually an equality.

If (s, a) and (t, b) are in $S \times A$ and if $\alpha \geq 0$, $\beta \geq 0$, $\alpha + \beta = 1$, then

$$\alpha f(s, a) + \beta f(t, b) = f(u, c), \tag{2}$$

where $u = \alpha s + \beta t \in S$ and $c \in A$, because

$$c_i = \frac{\alpha s_i a_i + \beta t_i b_i}{\alpha s_i + \beta t_i} \in A_i \qquad (1 \leq i \leq n). \tag{3}$$

This shows that K is convex. Since $A_i \subset K$ for each i [take $s_i = 1$ in (1), $s_j = 0$ for $j \neq i$], the convexity of K implies that $co(A_1 \cup \cdots \cup A_n) \subset K$. This proves (*a*).

(*b*) Let U be a neighborhood of 0 in X. Choose a convex neighborhood V of 0 in X such that $V + V \subset U$. Then $E \subset F + V$ for some finite set $F \subset X$. Hence $E \subset co(F) + V$. The latter set is convex. It follows that

$$co(E) \subset co(F) + V. \tag{4}$$

But $co(F)$ is compact [a special case of (*a*)], and therefore $co(F) \subset F_1 + V$ for some finite set $F_1 \subset X$. Thus

$$co(E) \subset F_1 + V + V \subset F_1 + U. \tag{5}$$

Since U was arbitrary, $co(E)$ is totally bounded.

(*c*) Closures of totally bounded sets are totally bounded in every metric space, and hence are compact in every complete metric space (Appendix A4). So if K is compact in a Fréchet space, then K is obviously totally bounded; hence $co(K)$ is totally bounded, by (*b*), and therefore $\overline{co}(K)$ is compact.

(*d*) Let S be the simplex in R^{n+1} consisting of all $t = (t_1, \ldots, t_{n+1})$ with $t_i \geq 0$ and $\sum t_i = 1$. Let K be compact, $K \subset R^n$. By the proposition that follows, $x \in co(K)$ if and only if

$$x = t_1 x_1 + \cdots + t_{n+1} x_{n+1} \tag{6}$$

for some $t \in S$ and $x_i \in K$ $(1 \leq i \leq n + 1)$. In other words, $co(K)$ is the image of $S \times K^{n+1}$ under the continuous mapping

$$(t, x_1, \ldots, x_{n+1}) \to t_1 x_1 + \cdots + t_{n+1} x_{n+1}. \tag{7}$$

Hence $co(K)$ is compact. ////

Proposition. *If $E \subset R^n$ and $x \in co(E)$, then x lies in the convex hull of some subset of E which contains at most $n + 1$ points.*

PROOF. It is enough to show that if $k > n$ and $x = \sum_1^{k+1} t_i x_i$ is a convex combination of some $k + 1$ vectors $x_i \in R^n$, then x is actually a convex combination of some k of these vectors.

Assume, with no loss of generality, that $t_i > 0$ for $1 \le i \le k+1$. The null space of the linear map

$$(8) \qquad (a_1, \ldots, a_{k+1}) \to \left(\sum_1^{k+1} a_i x_i, \sum_1^{k+1} a_i \right),$$

which sends R^{k+1} into $R^n \times R$, has positive dimension, since $k > n$. Hence there exists $(a_1, \ldots, a_{k+1})$, with some $a_i \neq 0$, so that $\sum a_i x_i = 0$ and $\sum a_i = 0$. Since $t_i > 0$ for all i, there is a constant λ such that $|\lambda a_i| \le t_i$ for all i and $\lambda a_j = t_j$ for at least one j. Setting $c_i = t_i - \lambda a_i$, we conclude that $x = \sum c_i x_i$ and that at least one c_j is 0; note also that $\sum c_i = \sum t_i = 1$ and that $c_i \ge 0$ for all i. ////

The following analogue of part (*b*) of the separation theorem 3.4 will be used in the proof of the Krein-Milman theorem.

3.21 Theorem *Suppose X is a topological vector space on which X^* separates points. Suppose A and B are disjoint, nonempty, compact, convex sets in X. Then there exists $\Lambda \in X^*$ such that*

$$(1) \qquad \sup_{x \in A} \operatorname{Re} \Lambda x < \inf_{y \in B} \operatorname{Re} \Lambda y.$$

Note that part of the hypothesis is weaker than in (*b*) of Theorem 3.4 (since local convexity of X implies that X^* separates points on X); to make up for this, it is now assumed that *both* A and B are compact.

PROOF. Let X_w be X with its weak topology. The sets A and B are evidently compact in X_w. They are also closed in X_w (because X_w is a Hausdorff space). Since X_w is locally convex, (*b*) of Theorem 3.4 can be applied to X_w in place of X; it gives us a $\Lambda \in (X_w)^*$ that satisfies (1). But we saw in Section 3.11 (as a consequence of Theorem 3.10) that $(X_w)^* = X^*$. ////

3.22 Extreme points Let K be a subset of a vector space X. A nonempty set $S \subset K$ is called an *extreme set* of K if no point of S is an internal point of any line interval whose end points are in K, except when both end points are in S. Analytically, the condition can be expressed as follows: If $x \in K$, $y \in K$, $0 < t < 1$, and

$$(1-t)x + ty \in S,$$

then $x \in S$ and $y \in S$.

The *extreme points* of K are the extreme sets that consist of just one point.

The set of all extreme points of K will be denoted by $E(K)$.

The following two theorems show that under certain conditions $E(K)$ is quite a large set.

3.23 The Krein-Milman theorem *Suppose X is a topological vector space on which X^* separates points. If K is a nonempty compact convex set in X, then K is the closed convex hull of the set of its extreme points.*

In symbols, $K = \overline{co}(E(K))$.

PROOF. Let $\mathscr{P}$ be the collection of all compact extreme sets of K. Since $K \in \mathscr{P}$, $\mathscr{P} \neq \varnothing$. We shall use the following two properties of $\mathscr{P}$:

(*a*) *The intersection S of any nonempty subcollection of $\mathscr{P}$ is a member of $\mathscr{P}$, unless $S = \varnothing$.*

(*b*) *If $S \in \mathscr{P}$, $\Lambda \in X^*$, μ is the maximum of* Re Λ *on S, and*

$$S_\Lambda = \{x \in S: \text{Re } \Lambda x = \mu\},$$

then $S_\Lambda \in \mathscr{P}$.

The proof of (*a*) is immediate. To prove (*b*), suppose $tx + (1-t)y = z \in S_\Lambda$, $x \in K$, $y \in K$, $0 < t < 1$. Since $z \in S$ and $S \in \mathscr{P}$, we have $x \in S$ and $y \in S$. Hence Re $\Lambda x \leq \mu$, Re $\Lambda y \leq \mu$. Since Re $\Lambda z = \mu$ and Λ is linear, we conclude: Re $\Lambda x = \mu =$ Re Λy. Hence $x \in S_\Lambda$ and $y \in S_\Lambda$. This proves (*b*).

Choose some $S \in \mathscr{P}$. Let $\mathscr{P}'$ be the collection of all members of $\mathscr{P}$ that are subsets of S. Since $S \in \mathscr{P}'$, $\mathscr{P}'$ is not empty. Partially order $\mathscr{P}'$ by set inclusion, let Ω be a maximal totally ordered subcollection of $\mathscr{P}'$, and let M be the intersection of all members of Ω. Since Ω is a collection of compact sets with the finite intersection property, $M \neq \varnothing$. By (*a*), $M \in \mathscr{P}'$. The maximality of Ω implies that no proper subset of M belongs to $\mathscr{P}$. It now follows from (*b*) that every $\Lambda \in X^*$ is constant on M. Since X^* separates points on X, M has only one point. Therefore M is an extreme point of K.

We have now proved that

$$E(K) \cap S \neq \varnothing \tag{1}$$

for every $S \in \mathscr{P}$. In other words, *every compact extreme set of K contains on extreme point of K.*

Since K is compact and convex (the assumed convexity of K will now be used for the first time), we have

$$\overline{co}(E(K)) \subset K \tag{2}$$

and this shows that $\overline{co}(E(K))$ is compact.

Assume, to reach a contradiction, that some $x_0 \in K$ is not in $\overline{co}(E(K))$. Theorem 3.21 furnishes then a $\Lambda \in X^*$ such that

Re $\Lambda x <$ Re Λx_0 for every $x \in \overline{co}(E(K))$. If K_Λ is defined as in (*b*), then $K_\Lambda \in \mathscr{P}$. Our choice of Λ shows that K_Λ is disjoint from $\overline{co}(E(K))$, and this contradicts (1). ////

Remark. The convexity of K was used only to show that $\overline{co}(E(K))$ is compact. If X were assumed to be locally convex, the compactness of $\overline{co}(E(K))$ would not be needed, since one could use (*b*) of Theorem 3.4 in place of Theorem 3.21. The above argument proves then that $K \subset \overline{co}(E(K))$. The following version of the Krein-Milman theorem is thus obtained:

3.24 Theorem *If K is a compact subset of a locally convex space then $K \subset \overline{co}(E(K))$.*

Equivalently, $\overline{co}(K) = \overline{co}(E(K))$.

It may happen in this situation that $\overline{co}(K)$ has extreme points which are not in K. (See Exercise 33.) The next theorem shows that this pathology cannot occur if $\overline{co}(K)$ is compact. Therefore it occurs in no Fréchet space, by (*c*) of Theorem 3.20.

3.25 Milman's theorem *If K is a compact set in a locally convex space X, and if $\overline{co}(K)$ is also compact, then every extreme point of $\overline{co}(K)$ lies in K.*

PROOF. Assume that some extreme point p of $\overline{co}(K)$ is not in K. Then there is a convex balanced neighborhood V of 0 in X such that

$$(1)\qquad (p + \overline{V}) \cap K = \varnothing.$$

Choose $x_1, \ldots, x_n$ in K so that $K \subset \bigcup_1^n (x_i + V)$. Each set

$$(2)\qquad A_i = \overline{co}(K \cap (x_i + V)) \qquad (1 \le i \le n)$$

is convex and also compact, since $A_i \subset \overline{co}(K)$. Also, $K \subset A_1 \cup \cdots \cup A_n$. Part (*a*) of Theorem 3.20 shows therefore that

$$(3)\qquad \overline{co}(K) \subset \overline{co}(A_1 \cup \cdots \cup A_n) = co(A_1 \cup \cdots \cup A_n).$$

But the opposite inclusion holds also, because $A_i \subset \overline{co}(K)$ for each i. Thus

$$(4)\qquad \overline{co}(K) = co(A_1 \cup \cdots \cup A_n).$$

In particular, $p = t_1 y_1 + \cdots + t_N y_N$, where each y_j lies in some A_i, each t_j is positive, and $\sum t_j = 1$. The grouping

$$(5)\qquad p = t_1 y_1 + (1 - t_1) \frac{t_2 y_2 + \cdots + t_N y_N}{t_2 + \cdots + t_N}$$

exhibits p as a convex combination of two points of $\overline{co}(K)$, by (4). Since p is an extreme point of $\overline{co}(K)$, we conclude from (5) that $y_1 = p$. Thus, for some i,

(6) $$p \in A_i \subset x_i + \bar{V} \subset K + \bar{V},$$

which contradicts (1). [Note that $A_i \subset x_i + \bar{V}$, by (2), because V is convex.] ////

Vector-Valued Integration

Sometimes it is desirable to be able to integrate functions f that are defined on some measure space Q (with a real or complex measure μ) and whose values lie in some topological vector space X. The first problem is to associate with these data a vector in X that deserves to be called

$$\int_Q f\, d\mu,$$

i.e., which has at least some of the properties that integrals usually have. For instance, the equation

$$\Lambda\left(\int_Q f\, d\mu\right) = \int_Q (\Lambda f)\, d\mu$$

ought to hold for every $\Lambda \in X^*$, because it does hold for sums, and because integrals are (or ought to be) limits of sums in some sense or other. In fact, our definition will be based on this single requirement.

Many other approaches to vector-valued integration have been studied in great detail; in some of these, the integrals are defined more directly as limits of sums (see Exercise 23).

3.26 Definition Suppose μ is a measure on a measure space Q, X is a topological vector space on which X^* separates points, and f is a function from Q into X such that the scalar functions Λf are integrable with respect to μ, for every $\Lambda \in X^*$; note that Λf is defined by

(1) $$(\Lambda f)(q) = \Lambda(f(q)) \qquad (q \in Q).$$

If there exists a vector $y \in X$ such that

(2) $$\Lambda y = \int_Q (\Lambda f)\, d\mu$$

for every $\Lambda \in X^*$, then we define

(3) $$\int_Q f\, d\mu = y.$$

Remarks. It is clear that there is at most one such y, because X^* separates points on X. *Thus there is no uniqueness problem.*

Existence will be proved only in the rather special case (sufficient for many applications) in which Q is compact and f is continuous. In that case, $f(Q)$ is compact, and the only other requirement that will be imposed is that the closed convex hull of $f(Q)$ should be compact. By Theorem 3.20, *this additional requirement is automatically satisfied when X is a Fréchet space.*

Recall that a *Borel measure* on a compact (or locally compact) Hausdorff space Q is a measure defined on the σ-algebra of all Borel sets in Q; this is the smallest σ-algebra that contains all open subsets of Q. A *probability measure* is a positive measure of total mass 1.

3.27 Theorem *Suppose*

(*a*) *X is a topological vector space on which X^* separates points, and*

(*b*) *μ is a Borel probability measure on a compact Hausdorff space Q.*

If $f: Q \to X$ is continuous and if $\overline{co}(f(Q))$ is compact in X, then the integral

$$y = \int_Q f \, d\mu \tag{1}$$

exists, in the sense of Definition 3.26.

Moreover, $y \in \overline{co}(f(Q))$.

Remark. If ν is any positive Borel measure on Q, then some scalar multiple of ν is a probability measure. The theorem therefore holds (except for its last sentence) with ν in place of μ. It can then be extended to real-valued Borel measures (by the Jordan decomposition theorem) and (if the scalar field of X is $\mathcal{C}$) to complex ones.

Exercise 24 gives another generalization.

PROOF. Regard X as a real vector space. Put $H = co(f(Q))$. We have to prove that there exists $y \in H$ such that

$$\Lambda y = \int_Q (\Lambda f) \, d\mu \tag{2}$$

for *every* $\Lambda \in X^*$.

Let $L = \{\Lambda_1, \ldots, \Lambda_n\}$ be a finite subset of X^*. Let E_L be the set of all $y \in \bar{H}$ that satisfy (2) for every $\Lambda \in L$. Each E_L is closed (by the continuity of Λ) and is therefore *compact*, since $\bar{H}$ is compact. If no E_L

is empty, the collection of all E_L has the finite intersection property. The intersection of all E_L is therefore not empty, and any y in it satisfies (2) for every $\Lambda \in X^*$. It is therefore enough to prove $E_L \neq \varnothing$.

Regard $L = (\Lambda_1, \ldots, \Lambda_n)$ as a mapping from X into R^n, and put $K = L(f(Q))$. Define

(3)
$$m_i = \int_Q (\Lambda_i f)\, d\mu \qquad (1 \le i \le n).$$

We claim that the point $m = (m_1, \ldots, m_n)$ lies in the convex hull of K.

If $t = (t_1, \ldots, t_n) \in R^n$ is not in this hull, then [by Theorem 3.20 and (*b*) of Theorem 3.4 and the known form of the linear functionals on R^n] there are real numbers $c_1, \ldots, c_n$ such that

(4)
$$\sum_{i=1}^{n} c_i u_i < \sum_{i=1}^{n} c_i t_i$$

if $u = (u_1, \ldots, u_n) \in K$. Hence

(5)
$$\sum_{i=1}^{n} c_i \Lambda_i f(q) < \sum_{i=1}^{n} c_i t_i \qquad (q \in Q).$$

Since μ is a probability measure, integration of the left side of (5) gives $\sum c_i m_i < \sum c_i t_i$. Thus $t \neq m$.

This shows that m lies in the convex hull of K. Since $K = L(f(Q))$ and L is linear, it follows that $m = Ly$ for some y in the convex hull H of $f(Q)$. For this y we have

(6)
$$\Lambda_i y = m_i = \int_Q (\Lambda_i f)\, d\mu \qquad (1 \le i \le n).$$

Hence $y \in E_L$. This completes the proof. ////

3.28 Theorem *Suppose*

(*a*) *X is a topological vector space on which X^* separates points,*

(*b*) *Q is a compact subset of X, and*

(*c*) *the closed convex hull $\bar{H}$ of Q is compact.*

Then $y \in \bar{H}$ if and only if there is a regular Borel probability measure μ on Q such that

(1)
$$y = \int_Q x\, d\mu(x).$$

Remarks. The integral is to be understood as in Definition 3.26, with $f(x) = x$.

Recall that a positive Borel measure on Q is said to be *regular* if

(2) $$\mu(E) = \sup \{\mu(K): K \subset E\} = \inf \{\mu(G): E \subset G\}$$

for every Borel set $E \subset Q$, where K ranges over the compact subsets of E and G ranges over the open supersets of E.

The integral (1) represents every $y \in \bar{H}$ as a "weighted average" of Q, or as the "center of mass" of a certain unit mass distributed over Q.

We stress once more that (*c*) follows from (*b*) if X is a Fréchet space.

PROOF. Regard X again as a real vector space. Let $C(Q)$ be the Banach space of all real continuous functions on Q, with the supremum norm. The Riesz representation theorem identifies the dual space $C(Q)^*$ with the space of all real Borel measures on Q that are differences of regular positive ones. With this identification in mind, we define a mapping

(3) $$\phi: C(Q)^* \to X$$

by

(4) $$\phi(\mu) = \int_Q x \, d\mu(x).$$

Let P be the set of all regular Borel probability measures on Q. The theorem asserts that $\phi(P) = \bar{H}$.

For each $x \in Q$, the unit mass δ_x concentrated at x belongs to P. Since $\phi(\delta_x) = x$, we see that $Q \subset \phi(P)$. Since ϕ is linear and P is convex, it follows that $H \subset \phi(P)$, where H is the convex hull of Q. By Theorem 3.27, $\phi(P) \subset \bar{H}$. Therefore all that remains to be done is to show that $\phi(P)$ is closed in X.

This is a consequence of the following two facts:

(*i*) *P is weak*-compact in $C(Q)^*$.*
(*ii*) *The mapping ϕ defined by (4) is continuous if $C(Q)^*$ is given its weak*-topology and if X is given its weak topology.*

Once we have (*i*) and (*ii*), it follows that $\phi(P)$ is weakly compact, hence weakly closed, and since weakly closed sets are strongly closed, we have the desired conclusion.

To prove (*i*), note that

(5) $$P \subset \left\{\mu: \left|\int_Q h \, d\mu\right| \le 1 \text{ if } \|h\| < 1\right\}$$

and that this larger set is weak*-compact, by the Banach-Alaoglu theorem. It is therefore enough to show that P is weak*-closed.

If $h \in C(Q)$ and $h \geq 0$, put

(6) $$E_h = \left\{\mu: \int_Q h \, d\mu \geq 0\right\}.$$

Since $\mu \to \int h \, d\mu$ is continuous, by the definition of the weak*-topology, each E_h is weak*-closed. So is the set

(7) $$E = \left\{\mu: \int_Q 1 \, d\mu = 1\right\}.$$

Since P is the intersection of E and the sets E_h, P is weak*-closed.

To prove (*ii*) it is enough to prove that ϕ is continuous at the origin, since ϕ is linear. Every weak neighborhood of 0 in X contains a set of the form

(8) $$W = \{y \in X: |\Lambda_i y| < r_i \quad \text{for} \quad 1 \leq i \leq n\},$$

where $\Lambda_i \in X^*$ and $r_i > 0$. The restrictions of the Λ_i to Q lie in $C(Q)$. Hence

(9) $$V = \left\{\mu \in C(Q)^*: \left|\int_Q \Lambda_i \, d\mu\right| < r_i \quad \text{for} \quad 1 \leq i \leq n\right\}$$

is a weak*-neighborhood of 0 in $C(Q)^*$. But

(10) $$\int_Q \Lambda_i \, d\mu = \Lambda_i\left(\int_Q x \, d\mu(x)\right) = \Lambda_i \phi(\mu),$$

by Definition 3.26. It follows from (8), (9), and (10) that $\phi(V) \subset W$. Hence ϕ is continuous. ////

The following simple inequality sharpens the last assertion in the statement of Theorem 3.27.

3.29 Theorem *Suppose Q is a compact Hausdorff space, X is a Banach space, $f: Q \to X$ is continuous, and μ is a positive Borel measure on Q. Then*

$$\left\|\int_Q f \, d\mu\right\| \leq \int_Q \|f\| \, d\mu.$$

PROOF. Put $y = \int f \, d\mu$. By the corollary to Theorem 3.3, there is a $\Lambda \in X^*$ such that $\Lambda y = \|y\|$ and $|\Lambda x| \leq \|x\|$ for all $x \in X$. In particular,

$$|\Lambda f(s)| \leq \|f(s)\|$$

for all $s \in Q$. By Theorem 3.27, it follows that

$$\|y\| = \Lambda y = \int_Q (\Lambda f) \, d\mu \leq \int_Q \|f\| \, d\mu. \qquad ////$$

Holomorphic Functions

In the study of Banach algebras, as well as in some other contexts, it is useful to enlarge the concept of holomorphic function from complex-valued ones to vector-valued ones. (Of course, one can also generalize the domains, by going from $\mathcal{C}$ to $\mathcal{C}^n$ and even beyond. But this is another story.) There are at least two very natural definitions of "holomorphic" available in this general setting, a "weak" one and a "strong" one. They turn out to define the same class of functions if the values are assumed to lie in a Fréchet space.

3.30 Definition Let Ω be an open set in $\mathcal{C}$ and let X be a complex topological vector space.

(*a*) A function $f: \Omega \to X$ is said to be *weakly holomorphic in* Ω if Λf is holomorphic in the ordinary sense for every $\Lambda \in X^*$.

(*b*) A function $f: \Omega \to X$ is said to be *strongly holomorphic in* Ω if

$$\lim_{w \to z} \frac{f(w) - f(z)}{w - z}$$

exists (in the topology of X) for every $z \in \Omega$.

Note that the above quotient is the product of the scalar $(w - z)^{-1}$ and the vector $f(w) - f(z)$ in X.

The continuity of the functionals Λ that occur in (*a*) makes it obvious that every strongly holomorphic function is weakly holomorphic. The converse is true when X is a Fréchet space, but it is far from obvious. (Recall that weakly convergent sequences may very well fail to converge originally.) The Cauchy theorem will play an important role in this proof, as will Theorem 3.18.

The *index* of a point $z \in \mathcal{C}$ with respect to a closed path Γ that does not pass through z will be denoted by $\text{Ind}_\Gamma(z)$. We recall that

$$\text{Ind}_\Gamma(z) = \frac{1}{2\pi i} \int_\Gamma \frac{d\zeta}{\zeta - z}.$$

All paths considered here and later are assumed to be piecewise continuously differentiable, or at least rectifiable.

3.31 Theorem *Let Ω be open in $\mathcal{C}$, let X be a complex Fréchet space, and assume that*

$$f: \Omega \to X$$

is weakly holomorphic. The following conclusions hold:

(*a*) *f is strongly continuous in Ω.*

(*b*) *The Cauchy theorem and the Cauchy formula hold: If* Γ *is a closed path in* Ω *such that* $\operatorname{Ind}_\Gamma(w) = 0$ *for every* $w \notin \Omega$, *then*

$$\int_\Gamma f(\zeta)\, d\zeta = 0, \tag{1}$$

and

$$f(z) = \frac{1}{2\pi i} \int_\Gamma (\zeta - z)^{-1} f(\zeta)\, d\zeta \tag{2}$$

if $z \in \Omega$ *and* $\operatorname{Ind}_\Gamma(z) = 1$. *If* Γ_1 *and* Γ_2 *are closed paths in* Ω *such that*

$$\operatorname{Ind}_{\Gamma_1}(w) = \operatorname{Ind}_{\Gamma_2}(w)$$

for every $w \notin \Omega$, *then*

$$\int_{\Gamma_1} f(\zeta)\, d\zeta = \int_{\Gamma_2} f(\zeta)\, d\zeta. \tag{3}$$

(*c*) f *is strongly holomorphic in* Ω.

The integrals in (*b*) are to be understood in the sense of Theorem 3.27. Either one can regard $d\zeta$ as a complex measure on the range of Γ (a compact subset of $\mathcal{C}$), or one can parametrize Γ and integrate with respect to Lebesgue measure on a compact interval in R.

PROOF. (*a*) Assume $0 \in \Omega$. We shall prove that f is strongly continuous at 0. Define

$$\Delta_r = \{z \in \mathcal{C} : |z| \le r\}. \tag{4}$$

Then $\Delta_{2r} \subset \Omega$ for some $r > 0$. Let Γ be the positively oriented boundary of Δ_{2r}.

Fix $\Lambda \in X^*$. Since Λf is holomorphic,

$$\frac{(\Lambda f)(z) - (\Lambda f)(0)}{z} = \frac{1}{2\pi i} \int_\Gamma \frac{(\Lambda f)(\zeta)}{(\zeta - z)\zeta}\, d\zeta \tag{5}$$

if $0 < |z| < 2r$. Let $M(\Lambda)$ be the maximum of $|\Lambda f|$ on Δ_{2r}. If $0 < |z| \le r$, it follows that

$$|z^{-1}\Lambda[f(z) - f(0)]| \le r^{-1} M(\Lambda). \tag{6}$$

The set of all quotients

$$\left\{\frac{f(z) - f(0)}{z} : 0 < |z| \le r\right\} \tag{7}$$

is therefore weakly bounded in X. By Theorem 3.18, this set is also strongly bounded. Thus if V is any (strong) neighborhood of 0 in X, there exists $t < \infty$ such that

$$f(z) - f(0) \in ztV \qquad (0 < |z| \leq r). \tag{8}$$

Consequently, $f(z) \to f(0)$ strongly, as $z \to 0$. [It may be of some interest to observe that the proof of (*a*) used only the local convexity of X. Neither metrizability nor completeness has played a role so far.]

This was the crux of the matter. The rest is now almost automatic.

(*b*) By (*a*) and Theorem 3.27, the integrals in (1) to (3) exist. These three formulas are correct (by the theory of ordinary holomorphic functions) if f is replaced in them by Λf, where Λ is any member of X^*. The formulas are therefore correct as stated, by Definition 3.26.

(*c*) Assume, as in the proof of (*a*), that $\Delta_{2r} \subset \Omega$, and choose Γ as in (*a*). Define

$$y = \frac{1}{2\pi i} \int_\Gamma \zeta^{-2} f(\zeta)\, d\zeta. \tag{9}$$

The Cauchy formula (2) shows, after a small computation, that

$$\frac{f(z) - f(0)}{z} = y + zg(z) \tag{10}$$

if $0 < |z| < 2r$, where

$$g(z) = \frac{1}{2\pi} \int_{-\pi}^{\pi} [2re^{i\theta}(2re^{i\theta} - z)]^{-1} f(2re^{i\theta})\, d\theta. \tag{11}$$

Let V be a convex balanced neighborhood of 0 in X. Put $K = \{f(\zeta): |\zeta| = 2r\}$. Then K is compact, so that $K \subset tV$ for some $t < \infty$. If $s = tr^{-2}$ and $|z| \leq r$, it follows that the integrand (11) lies in sV for every θ. Thus $g(z) \in s\bar{V}$ if $|z| \leq r$. The left side of (10) therefore converges strongly to y, as $z \to 0$. ////

The following extension of Liouville's theorem concerning bounded entire functions does not even depend on Theorem 3.31. It can be used in the study of spectra in Banach algebras. (See Exercise 10, Chapter 10.)

3.32 Theorem *Suppose X is a complex topological vector space on which X^* separates points. Suppose $f: \mathbb{C} \to X$ is weakly holomorphic and $f(\mathbb{C})$ is a weakly bounded subset of X. Then f is constant.*

PROOF. For every $\Lambda \in X^*$, Λf is a bounded (complex-valued) entire function. If $z \in \mathcal{C}$, it follows from Liouville's theorem that

$$\Lambda f(z) = \Lambda f(0).$$

Since X^* separates points on X, this implies $f(z) = f(0)$, for every $z \in \mathcal{C}$. ////

Part (*d*) of Exercise 5 describes a weakly bounded set which is not originally bounded, in an F-space X on which X^* separates points. Compare with Theorem 3.18.

Exercises

1. Call a set $H \subset R^n$ a *hyperplane* if there exist real numbers $a_1, \ldots, a_n, c$ (with $a_i \neq 0$ for at least one i) such that H consists of all points $x = (x_1, \ldots, x_n)$ that satisfy $\sum a_i x_i = c$.

   Suppose E is a convex set in R^n, with nonempty interior, and y is a boundary point of E. Prove that there is a hyperplane H such that $y \in H$ and E lies entirely on one side of H. (State the conclusion more precisely.) *Suggestion:* Suppose 0 is an interior point of E, let M be the one-dimensional subspace that contains y, and apply Theorem 3.2.
2. Suppose $L^2 = L^2([-1, 1])$, with respect to Lebesgue measure. For each scalar α, let E_α be the set of all continuous functions f on $[-1, 1]$ such that $f(0) = \alpha$. Show that each E_α is convex and that each is dense in L^2. Thus E_α and E_β are disjoint convex sets (if $\alpha \neq \beta$) which cannot be separated by any continuous linear functional Λ on L^2. *Hint:* What is $\Lambda(E_\alpha)$?
3. Suppose X is a real vector space (without topology). Call a point $x_0 \in A \subset X$ an *internal* point of A if $A - x_0$ is an absorbing set.
   (*a*) Suppose A and B are disjoint convex sets in X, and A has an internal point. Prove that there is a nonconstant linear functional Λ on X such that $\Lambda(A) \cap \Lambda(B)$ contains at most one point. (The proof is similar to that of Theorem 3.4.)
   (*b*) Show (with $X = R^2$, for example) that it may not be possible to have $\Lambda(A)$ and $\Lambda(B)$ disjoint, under the hypotheses of (*a*).
4. Let ℓ^∞ be the space of all real bounded functions x on the positive integers. Let τ be the translation operator defined on ℓ^∞ by the equation

$$(\tau x)(n) = x(n+1) \qquad (n = 1, 2, 3, \ldots).$$

   Prove that there exists a linear functional Λ on ℓ^∞ (called a *Banach limit*) such that
   (*a*) $\Lambda \tau x = \Lambda x$, and
   (*b*) $\liminf_{n \to \infty} x(n) \leq \Lambda x \leq \limsup_{n \to \infty} x(n)$
   for every $x \in \ell^\infty$.

Suggestion: Define

$$\Lambda_n x = \frac{x(1) + \cdots + x(n)}{n}$$

$$M = \{x \in \ell^\infty : \lim_{n\to\infty} \Lambda_n x = \Lambda x \text{ exists}\}$$

$$p(x) = \limsup_{n\to\infty} \Lambda_n x$$

and apply Theorem 3.2.

5. For $0 < p < \infty$, let ℓ^p be the space of all functions x (real or complex, as the case may be) on the positive integers, such that

$$\sum_{n=1}^{\infty} |x(n)|^p < \infty.$$

For $1 \le p < \infty$, define $\|x\|_p = \{\sum |x(n)|^p\}^{1/p}$, and define $\|x\|_\infty = \sup_n |x(n)|$.

(*a*) Assume $1 \le p < \infty$. Prove that $\|x\|_p$ and $\|x\|_\infty$ make ℓ^p and ℓ^∞ into Banach spaces. If $p^{-1} + q^{-1} = 1$, prove that $(\ell^p)^* = \ell^q$, in the following sense: There is a one-to-one correspondence $\Lambda \leftrightarrow y$ between $(\ell^p)^*$ and ℓ^q, given by

$$\Lambda x = \sum x(n)y(n) \qquad (x \in \ell^p).$$

(*b*) Assume $1 < p < \infty$ and prove that ℓ^p contains sequences that converge weakly but not strongly.

(*c*) On the other hand, prove that every weakly convergent sequence in ℓ^1 converges strongly, in spite of the fact that the weak topology of ℓ^1 is different from its strong topology (which is induced by the norm).

(*d*) If $0 < p < 1$, prove that ℓ^p, metrized by

$$d(x, y) = \sum_{n=1}^{\infty} |x(n) - y(n)|^p,$$

is a locally bounded F-space which is not locally convex but that $(\ell^p)^*$ nevertheless separates points on ℓ^p. (Thus there are many convex open sets in ℓ^p but not enough to form a base for its topology.) Show that $(\ell^p)^* = \ell^\infty$, in the same sense as in (*a*). Show also that the set of all x with $\Sigma |x(n)| < 1$ is weakly bounded but not originally bounded.

(*e*) For $0 < p \le 1$, let τ_p be the weak*-topology induced on ℓ^∞ by ℓ^p; see (*a*) and (*d*). If $0 < p < r \le 1$, show that τ_p and τ_r are *different* topologies (is one weaker than the other?) but that they induce the same topology on each norm-bounded subset of ℓ^∞. *Hint:* The norm-closed unit ball of ℓ^∞ is weak*-compact.

6. Put $f_n(t) = e^{int}$ $(-\pi \le t \le \pi)$; let $L^p = L^p(-\pi, \pi)$, with respect to Lebesgue measure. If $1 \le p < \infty$, prove that $f_n \to 0$ weakly in L^p, but not strongly.

7. $L^\infty([0, 1])$ has its norm topology ($\|f\|_\infty$ is the essential supremum of $|f|$) and its weak*-topology as the dual of L^1. Show that C, the space of all continuous functions on $[0, 1]$, is dense in L^∞ in one of these topologies but not in the other. (Compare with the corollaries to Theorem 3.12.) Show the same with "closed" in place of "dense."

8. Let C be the Banach space of all complex continuous functions on $[0, 1]$, with the supremum norm. Let B be the closed unit ball of C. Show that there exist continuous linear functionals Λ on C for which $\Lambda(B)$ is an *open* subset of the complex plane; in particular, $|\Lambda|$ attains no maximum on B.

9. Let $E \subset L^2(-\pi, \pi)$ be the set of all functions

$$f_{m,n}(t) = e^{imt} + me^{int},$$

where m, n are integers and $0 \le m < n$. Let E_1 be the set of all $g \in L^2$ such that some sequence in E converges weakly to g. (E_1 is called the *weak sequential closure* of E.)

(*a*) Find all $g \in E_1$.

(*b*) Find all g in the weak closure $\bar{E}_w$ of E.

(*c*) Show that $0 \in \bar{E}_w$ but 0 is not in E_1, although 0 lies in the weak sequential closure of E_1.

This example shows that a weak sequential closure need not be weakly sequentially closed. The passage from a set to its weak sequential closure is therefore not a closure operation, in the sense in which that term is usually used in topology. (See also Exercise 28.)

10. Represent ℓ^1 as the space of all real functions x on $S = \{(m, n): m \ge 1, n \ge 1\}$, such that

$$\|x\|_1 = \sum |x(m, n)| < \infty.$$

Let c_0 be the space of all real functions y on S such that $y(m, n) \to 0$ as $m + n \to \infty$, with norm $\|y\|_\infty = \sup |y(m, n)|$.

Let M be the subspace of ℓ^1 consisting of all $x \in \ell^1$ that satisfy the equations

$$mx(m, 1) = \sum_{n=2}^{\infty} x(m, n) \qquad (m = 1, 2, 3, \ldots).$$

(*a*) Prove that $\ell^1 = (c_0)^*$. (See also Exercise 24, Chapter 4.)

(*b*) Prove that M is a norm-closed subspace of ℓ^1.

(*c*) Prove that M is weak*-dense in ℓ^1 [relative to the weak*-topology given by (*a*)].

(*d*) Let B be the norm-closed unit ball of ℓ^1. In spite of (*c*), prove that the weak*-closure of $M \cap B$ contains no ball. *Suggestion:* If $\delta > 0$ and $m > 2/\delta$, then

$$|x(m, 1)| \le \frac{\|x\|}{m} < \frac{\delta}{2}$$

if $x \in M \cap B$, although $x(m, 1) = \delta$ for some $x \in \delta B$. Thus δB is not in the weak*-closure of $M \cap B$. Extend this to balls with other centers.

(*e*) Put $x_0(m, 1) = m^{-2}$, $x_0(m, n) = 0$ when $n \ge 2$. Prove that no *sequence* in M is weak*-convergent to x_0, in spite of (*c*). *Hint:* Weak*-convergence of $\{x_j\}$ to x_0 implies that $x_j(m, n) \to x_0(m, n)$ for all m, n, as $j \to \infty$, and that $\{\|x_j\|_1\}$ is bounded.

11. Let X be an infinite-dimensional Fréchet space. Prove that X^*, with its weak*-topology, is of the first category in itself.

12. Show that the norm-closed unit ball of c_0 is not weakly compact; recall that $(c_0)^* = \ell^1$ (Exercise 10).

13. Put $f_N(t) = N^{-1} \sum_{n=1}^{N^2} e^{int}$. Prove that $f_N \to 0$ weakly in $L^2(-\pi, \pi)$.

By Theorem 3.13, some sequence of convex combinations of the f_N converges to 0 in the L^2-norm. Find such a sequence. Show that $g_N = N^{-1}(f_1 + \cdots + f_N)$ will not do.

14. (*a*) Suppose Ω is a locally compact Hausdorff space. For each compact $K \subset \Omega$ define a seminorm p_K on $C(\Omega)$, the space of all complex continuous functions on Ω, by

$$p_K(f) = \sup \{|f(x)| : x \in K\}.$$

Give $C(\Omega)$ the topology induced by this collection of seminorms. Prove that to every $\Lambda \in C(\Omega)^*$ correspond a compact $K \subset \Omega$ and a complex Borel measure μ on K such that

$$\Lambda f = \int_K f\, d\mu \qquad [f \in C(\Omega)].$$

(*b*) Suppose Ω is an open set in $\mathbb{C}$. Find a countable collection Γ of measures with compact support in Ω such that $H(\Omega)$ (the space of all holomorphic functions in Ω) consists of exactly those $f \in C(\Omega)$ which satisfy $\int f\, d\mu = 0$ for every $\mu \in \Gamma$.

15. Let X be a topological vector space on which X^* separates points. Prove that the weak*-topology of X^* is metrizable if and only if X has a finite or countable Hamel basis. (See Exercise 1, Chapter 2 for the definition.)

16. Prove that the closed unit ball of L^1 (relative to Lebesgue measure on the unit interval) has no extreme points but that every point on the "surface" of the unit ball in L^p $(1 < p < \infty)$ is an extreme point of the ball.

17. Determine the extreme points of the closed unit ball of C, the space of all continuous functions on the unit interval, with the supremum norm. (The answer depends on the choice of the scalar field.)

18. Let K be the smallest convex set in R^3 that contains the points (1, 0, 1), (1, 0, − 1), and $(\cos\theta, \sin\theta, 0)$, for $0 \le \theta \le 2\pi$. Show that K is compact but that the set of all extreme points of K is not compact. Does such an example exist in R^2?

19. Suppose K is a compact convex set in R^n. Prove that every $x \in K$ is a convex combination of at most $n + 1$ extreme points of K. *Suggestion:* Use induction on n. Draw a line from some extreme point of K through x to where it leaves K. Use Exercise 1.

20. Let $\{u_1, u_2, u_3, \ldots\}$ be a sequence of pairwise orthogonal unit vectors in a Hilbert space. Let K consist of the vectors 0 and $n^{-1}u_n$ $(n \ge 1)$. Show that (*a*) K is compact; (*b*) $co(K)$ is bounded; (*c*) $co(K)$ is not closed. Find all extreme points of $\overline{co}(K)$.

21. If $0 < p < 1$, every $f \in L^p$ (except $f = 0$) is the arithmetic mean of two functions whose distance from 0 is less than that of f. (See Section 1.47.) Use this to construct an explicit example of a countable compact set K in L^p (with 0 as its only limit point) which has no extreme point.

22. If $0 < p < 1$, show that ℓ^p contains a compact set K whose convex hull is unbounded. This happens in spite of the fact that $(\ell^p)^*$ separates points on ℓ^p;

see Exercise 5. *Suggestion:* Define $x_n \in \ell^p$ by

$$x_n(n) = n^{p-1}, \qquad x_n(m) = 0 \qquad \text{if } m \neq n.$$

Let K consist of $0, x_1, x_2, x_3, \ldots$. If

$$y_N = N^{-1}(x_1 + \cdots + x_N),$$

show that $\{y_N\}$ is unbounded in ℓ^p.

23. Suppose μ is a Borel probability measure on a compact Hausdorff space Q, X is a Fréchet space, and $f: Q \to X$ is continuous. A *partition* of Q is, by definition, a finite collection of disjoint Borel subsets of Q whose union is Q. Prove that to every neighborhood V of 0 in X there corresponds a partition $\{E_i\}$ such that the difference

$$z = \int_Q f\,d\mu - \sum_i \mu(E_i)f(s_i)$$

lies in V for every choice of $s_i \in E_i$. (This exhibits the integral as a strong limit of "Riemann sums.") *Suggestion:* Take V convex and balanced. If $\Lambda \in X^*$ and if $|\Lambda x| \leq 1$ for every $x \in V$, then $|\Lambda z| \leq 1$, provided that the sets E_i are chosen so that $f(s) - f(t) \in V$ whenever s and t lie in the same E_i.

24. In addition to the hypotheses of Theorem 3.27, assume that T is a continuous linear mapping of X into a topological vector space Y on which Y^* separates points, and prove that

$$T\int_Q f\,d\mu = \int_Q (Tf)\,d\mu.$$

Hint: $\Lambda T \in X^*$ for every $\Lambda \in Y^*$.

25. Let E be the set of all extreme points of a compact set K in a topological vector space X on which X^* separates points. Prove that to every $y \in K$ corresponds a regular Borel probability measure μ on $Q = \bar{E}$ such that

$$y = \int_Q x\,d\mu(x).$$

26. Suppose Ω is a region in $\mathbb{C}$, X is a Fréchet space, and $f: \Omega \to X$ is holomorphic.

(*a*) State and prove a theorem concerning the power series representation of f, that is, concerning the formula $f(z) = \sum (z-a)^n c_n$, where $c_n \in X$.

(*b*) Generalize Morera's theorem to X-valued holomorphic functions.

(*c*) For a sequence of complex holomorphic functions in Ω, uniform convergence on compact subsets of Ω implies that the limit is holomorphic. Does this generalize to X-valued holomorphic functions?

27. Suppose $\{\alpha_i\}$ is a bounded set of distinct complex numbers, $f(z) = \sum_0^\infty c_n z^n$ is an entire function with every $c_n \neq 0$, and

$$g_i(z) = f(\alpha_i z).$$

Prove that the vector space generated by the functions g_i is dense in the Fréchet space $H(\mathbb{C})$ defined in Section 1.45.

Suggestion: Assume μ is a measure with compact support such that $\int g_i \, d\mu = 0$ for all i. Put

$$\phi(w) = \int f(wz) \, d\mu(z) \qquad (w \in \mathcal{C}).$$

Prove that $\phi(w) = 0$ for all w. Deduce that $\int z^n \, d\mu(z) = 0$ for $n = 1, 2, 3, \ldots$. Use Exercise 14.

Describe the closed subspace of $H(\mathcal{C})$ generated by the functions g_i if some of the c_n are 0.

28. Suppose X is a Fréchet space (or, more generally, a metrizable locally convex space). Prove the following statements:
  (*a*) X^* is the union of countably many weak*-compact sets E_n.
  (*b*) If X is separable, each E_n is metrizable. The weak*-topology of X^* is therefore separable, and some countable subset of X^* separates points on X. (Compare with Exercise 15.)
  (*c*) If K is a weakly compact subset of X and if $x_0 \in K$ is a weak limit point of some countable set $E \subset K$, then there is a sequence $\{x_n\}$ in E which converges weakly to x_0. *Hint:* Let Y be the smallest closed subspace of X that contains E. Apply (*b*) to Y to conclude that the weak topology of $K \cap Y$ is metrizable.

  Remark: The point of (*c*) is the existence of convergent *subsequences* rather than *subnets*. Note that there exist compact Hausdorff spaces in which no sequence of distinct points converges. For an example, see Exercise 18, Chapter 11.

29. Let $C(K)$ be the Banach space of all continuous complex functions on the compact Hausdorff space K, with the supremum norm. For $p \in K$, define $\Lambda_p \in C(K)^*$ by $\Lambda_p f = f(p)$. Show that $p \to \Lambda_p$ is a homeomorphism of K into $C(K)^*$, equipped with its weak*-topology. Part (*c*) of Exercise 28 can therefore not be extended to weak*-compact sets.

30. Suppose that p is an extreme point of some convex set K, and that $p = t_1 x_1 + \cdots + t_n x_n$, where $\sum t_i = 1$, $t_i > 0$ and $x_i \in K$ for all i. Prove that $x_i = p$ for all i.

31. Suppose that $A_1, \ldots, A_n$ are convex sets in a vector space X. Prove that every $x \in co(A_1 \cup \cdots \cup A_n)$ can be represented in the form

$$x = t_1 a_1 + \cdots + t_n a_n,$$

with $a_i \in A_i$ and $t_i \geq 0$ for all i, $\sum t_i = 1$.

32. Let X be an infinite-dimensional Banach space and let $S = \{x \in X: \|x\| = 1\}$ be the unit sphere of X. We want to cover S with finitely many closed balls, none of which contains the origin of X. Can this be done in (*a*) every X, (*b*) some X, (*c*) no X?

33. Let $C(I)$ be the Banach space of all continuous complex functions on the closed unit interval I, with the supremum norm. Let $M = C(I)^*$, the space of all complex Borel measures on I. Give M the weak*-topology induced by $C(I)$.

For each $t \in I$, let $e_t \in M$ be the "evaluation functional" defined by $e_t f = f(t)$, and define $\Lambda \in M$ by $\Lambda f = \int_0^1 f(s) \, ds$.

(a) Show that $t \to e_t$ is a continuous map from I into M and that $K = \{e_t : t \in I\}$ is a compact set in M.
(b) Show that $\Lambda \in \overline{co}(K)$.
(c) Find all $\mu \in \overline{co}(K)$.
(d) Let X be the subspace of M consisting of all finite linear combinations

$$c_0 \Lambda + c_1 e_{t_1} + \cdots + c_n e_{t_n}$$

with complex coefficients c_j. Note that $co(K) \subset X$ and that $X \cap \overline{co}(K)$ is the closed convex hull of K *within* X. Prove that Λ is an extreme point of $X \cap \overline{co}(K)$, even though Λ is not in K.

CHAPTER 4

DUALITY IN BANACH SPACES

The Normed Dual of a Normed Space

Introduction If X and Y are topological vector spaces, $\mathscr{B}(X, Y)$ will denote the collection of all bounded linear mappings (or *operators*) of X into Y. For simplicity, $\mathscr{B}(X, X)$ will be abbreviated to $\mathscr{B}(X)$. Each $\mathscr{B}(X, Y)$ is itself a vector space, with respect to the usual definitions of addition and scalar multiplication of functions. (This depends only on the vector space structure of Y, not on that of X.) In general, there are many ways in which $\mathscr{B}(X, Y)$ can be made into a *topological* vector space.

In the present chapter, we shall deal only with normed spaces X and Y. In that case, $\mathscr{B}(X, Y)$ can itself be normed in a very natural way. When Y is specialized to be the scalar field, so that $\mathscr{B}(X, Y)$ is the dual space X^* of X, the above-mentioned norm on $\mathscr{B}(X, Y)$ defines a topology on X^* which turns out to be stronger than its weak*-topology. The relations between a Banach space X and its *normed dual* X^* form the main topic of this chapter.

4.1 Theorem *Suppose X and Y are normed spaces. Associate to each $\Lambda \in \mathscr{B}(X, Y)$ the number*

$$(1) \qquad \|\Lambda\| = \sup \{\|\Lambda x\| : x \in X, \|x\| \le 1\}.$$

This definition of $\|\Lambda\|$ makes $\mathscr{B}(X, Y)$ into a normed space. If Y is a Banach space, so is $\mathscr{B}(X, Y)$.

PROOF. Since subsets of normed spaces are bounded if and only if they lie in some multiple of the unit ball, $\|\Lambda\| < \infty$ for every $\Lambda \in \mathscr{B}(X, Y)$. If α is a scalar, then $(\alpha\Lambda)(x) = \alpha \cdot \Lambda x$, so that

$$\|\alpha\Lambda\| = |\alpha|\,\|\Lambda\|. \tag{2}$$

The triangle inequality in Y shows that

$$\begin{aligned}\|(\Lambda_1 + \Lambda_2)x\| &= \|\Lambda_1 x + \Lambda_2 x\| \le \|\Lambda_1 x\| + \|\Lambda_2 x\| \\ &\le (\|\Lambda_1\| + \|\Lambda_2\|)\|x\| \le \|\Lambda_1\| + \|\Lambda_2\|\end{aligned}$$

for every $x \in X$ with $\|x\| \le 1$. Hence

$$\|\Lambda_1 + \Lambda_2\| \le \|\Lambda_1\| + \|\Lambda_2\|. \tag{3}$$

If $\Lambda \neq 0$, then $\Lambda x \neq 0$ for some $x \in X$; hence $\|\Lambda\| > 0$. Thus $\mathscr{B}(X, Y)$ is a normed space.

Assume now that Y is complete and that $\{\Lambda_n\}$ is a Cauchy sequence in $\mathscr{B}(X, Y)$. Since

$$\|\Lambda_n x - \Lambda_m x\| \le \|\Lambda_n - \Lambda_m\|\,\|x\| \tag{4}$$

and since it is assumed that $\|\Lambda_n - \Lambda_m\| \to 0$ as n and m tend to ∞, $\{\Lambda_n x\}$ is a Cauchy sequence in Y for every $x \in X$. Hence

$$\Lambda x = \lim_{n\to\infty} \Lambda_n x \tag{5}$$

exists. It is clear that $\Lambda: X \to Y$ is linear. If $\varepsilon > 0$, the right side of (4) does not exceed $\varepsilon\|x\|$, provided that m and n are sufficiently large. It follows that

$$\|\Lambda x - \Lambda_m x\| \le \varepsilon\|x\| \tag{6}$$

for all large m. Hence $\|\Lambda x\| \le (\|\Lambda_m\| + \varepsilon)\|x\|$, so that $\Lambda \in \mathscr{B}(X, Y)$, and $\|\Lambda - \Lambda_m\| \le \varepsilon$. Thus $\Lambda_m \to \Lambda$ in the norm of $\mathscr{B}(X, Y)$. This establishes the completeness of $\mathscr{B}(X, Y)$. ////

4.2 Duality It will be convenient to designate elements of the dual space X^* of X by x^* and to write

$$\langle x, x^* \rangle \tag{1}$$

in place of $x^*(x)$. This notation is well adapted to the symmetry (or duality) that exists between the action of X^* on X on the one hand and the action of X on X^* on the other. The following theorem states some basic properties of this duality.

4.3 Theorem *Suppose B is the closed unit ball of a normed space X. Define*

$$\|x^*\| = \sup \{|\langle x, x^* \rangle| : x \in B\}$$

for every $x^* \in X^*$.

(*a*) *This norm makes X^* into a Banach space.*

(*b*) *Let B^* be the closed unit ball of X^*. For every $x \in X$,*

$$\|x\| = \sup \{|\langle x, x^* \rangle| : x^* \in B^*\}.$$

Consequently, $x^* \to \langle x, x^* \rangle$ is a bounded linear functional on X^*, of norm $\|x\|$.

(*c*) *B^* is weak*-compact.*

PROOF. Since $\mathscr{B}(X, Y) = X^*$, when Y is the scalar field, (*a*) is a corollary of Theorem 4.1.

Fix $x \in X$. The corollary to Theorem 3.3 shows that there exists $y^* \in B^*$ such that

$$\langle x, y^* \rangle = \|x\|. \tag{1}$$

On the other hand,

$$|\langle x, x^* \rangle| \le \|x\| \, \|x^*\| \le \|x\| \tag{2}$$

for every $x^* \in B^*$. Part (*b*) follows from (1) and (2).

Since the open unit ball U of X is dense in B, the definition of $\|x^*\|$ shows that $x^* \in B^*$ if and only if $|\langle x, x^* \rangle| \le 1$ for every $x \in U$. Part (*c*) now follows directly from Theorem 3.15. ////

Remark. The weak*-topology of X^* is, by definition, the weakest one that makes all functionals

$$x^* \to \langle x, x^* \rangle$$

continuous. Part (*b*) shows therefore that the norm topology of X^* is stronger than its weak*-topology; in fact, it is strictly stronger, unless $\dim X < \infty$, since the proposition stated at the end of Section 3.11 holds for the weak*-topology as well.

Unless the contrary is explicitly stated, X^* will from now on denote the normed dual of X (whenever X is normed), and all topological concepts relating to X^* will refer to its norm topology. This implies in no way that the weak*-topology will not play an important role.

We now give an alternative description of the operator norm defined in Theorem 4.1.

4.4 Theorem *If X and Y are normed spaces and if $\Lambda \in \mathscr{B}(X, Y)$, then*

$$\|\Lambda\| = \sup \{ |\langle \Lambda x, y^* \rangle| : \|x\| \le 1, \|y^*\| \le 1 \}.$$

PROOF. Apply (*b*) of Theorem 4.3 with Y in place of X. This gives

$$\|\Lambda x\| = \sup \{ |\langle \Lambda x, y^* \rangle| : \|y^*\| \le 1 \}$$

for every $x \in X$. To complete the proof, recall that

$$\|\Lambda\| = \sup \{ \|\Lambda x\| : \|x\| \le 1 \}. \qquad ////$$

4.5 The second dual of a Banach space The normed dual X^* of a Banach space X is itself a Banach space and hence has a normed dual of its own, denoted by X^{**}. Statement (*b*) of Theorem 4.3 shows that every $x \in X$ defines a unique $\phi x \in X^{**}$, by the equation

$$(1) \qquad \langle x, x^* \rangle = \langle x^*, \phi x \rangle \qquad (x^* \in X^*),$$

and that

$$(2) \qquad \|\phi x\| = \|x\| \qquad (x \in X).$$

It follows from (1) that $\phi: X \to X^{**}$ is linear; by (2), ϕ is an isometry. Since X is now assumed to be complete, $\phi(X)$ is closed in X^{**}.

Thus ϕ is an isometric isomorphism of X onto a closed subspace of X^{**}.

Frequently, X is identified with $\phi(X)$; then X is regarded as a subspace of X^{**}.

The members of $\phi(X)$ are exactly those linear functionals on X^* that are continuous relative to its weak*-topology. (See Section 3.14.) Since the norm topology of X^* is stronger, it may happen that $\phi(X)$ is a proper subspace of X^{**}. But there are many important spaces X (for example, all L^p-spaces with $1 < p < \infty$) for which $\phi(X) = X^{**}$; these are called *reflexive*. Some of their properties are given in Exercise 1.

It should be stressed that, in order for X to be reflexive, the existence of *some* isometric isomorphism ϕ of X onto X^{**} is not enough; it is crucial that the identity (1) be satisfied by ϕ.

4.6 Annihilators Suppose X is a Banach space, M is a subspace of X, and N is a subspace of X^*; neither M nor N is assumed to be closed. Their *annihilators* $M^\perp$ and ${}^\perp N$ are defined as follows:

$$M^\perp = \{x^* \in X^* : \langle x, x^* \rangle = 0 \text{ for all } x \in M\},$$

$${}^\perp N = \{x \in X : \langle x, x^* \rangle = 0 \text{ for all } x^* \in N\}.$$

Thus $M^\perp$ consists of all bounded linear functionals on X that vanish on M, and ${}^\perp N$ is the subset of X on which every member of N vanishes. It is clear that $M^\perp$ and ${}^\perp N$ are vector spaces. Since $M^\perp$ is the intersection of the null spaces of the functionals ϕx, where x ranges over M (see Section 4.5), $M^\perp$ is a weak*-closed subspace of X^*. The proof that ${}^\perp N$ is a norm-closed subspace of X is even more direct. The following theorem describes the duality between these two types of annihilators.

4.7 Theorem *Under the preceding hypotheses,*

(*a*) ${}^\perp(M^\perp)$ *is the norm-closure of M in X, and*
(*b*) $({}^\perp N)^\perp$ *is the weak*-closure of N in X^*.*

As regards (*a*), recall that the norm-closure of M equals its weak closure, by Theorem 3.12.

PROOF. If $x \in M$, then $\langle x, x^* \rangle = 0$ for every $x^* \in M^\perp$, so that $x \in {}^\perp(M^\perp)$. Since ${}^\perp(M^\perp)$ is norm-closed, it contains the norm-closure $\bar{M}$ of M. On the other hand, if $x \notin \bar{M}$ the Hahn-Banach theorem yields an $x^* \in M^\perp$ such that $\langle x, x^* \rangle \neq 0$. Thus $x \notin {}^\perp(M^\perp)$, and (*a*) is proved.

Similarly, if $x^* \in N$, then $\langle x, x^* \rangle = 0$ for every $x \in {}^\perp N$, so that $x^* \in ({}^\perp N)^\perp$. This weak*-closed subspace of X^* contains the weak*-closure $\tilde{N}$ of N. If $x^* \notin \tilde{N}$, the Hahn-Banach theorem (applied to the locally convex space X^* with its weak*-topology) implies the existence of an $x \in {}^\perp N$ such that $\langle x, x^* \rangle \neq 0$; thus $x^* \notin ({}^\perp N)^\perp$, which proves (*b*). ////

Observe, as a corollary, that every norm-closed subspace of X is the annihilator of its annihilator and that the same is true of every weak*-closed subspace of X^*.

4.8 Duals of subspaces and of quotient spaces If M is a closed subspace of a Banach space X, then X/M is also a Banach space, with respect to the *quotient norm.* This was defined in the proof of (*d*) of Theorem 1.41. The duals of M and of X/M can be described with the aid of the annihilator $M^\perp$ of M. Somewhat imprecisely, the result is that

$$M^* = X^*/M^\perp \qquad \text{and} \qquad (X/M)^* = M^\perp.$$

This is imprecise because the equalities should be replaced by isometric isomorphisms. The following theorem describes these explicitly.

4.9 Theorem *Let M be a closed subspace of a Banach space X.*

(a) The Hahn-Banach theorem extends each $m^* \in M^*$ to a functional $x^* \in X^*$. Define

$$\sigma m^* = x^* + M^\perp.$$

Then σ is an isometric isomorphism of M^* onto $X^*/M^\perp$.

(b) Let $\pi: X \to X/M$ be the quotient map. Put $Y = X/M$. For each $y^* \in Y^*$, define

$$\tau y^* = y^* \pi.$$

Then τ is an isometric isomorphism of Y^* onto $M^\perp$.

PROOF. (*a*) If x^* and x_1^* are extensions of m^*, then $x^* - x_1^*$ is in $M^\perp$; hence $x^* + M^\perp = x_1^* + M^\perp$. Thus σ is well defined. A trivial verification shows that σ is linear. Since the restriction of every $x^* \in X^*$ to M is a member of M^*, the range of σ is all of $X^*/M^\perp$.

Fix $m^* \in M^*$. If $x^* \in X^*$ extends m^*, it is obvious that $\|m^*\| \le \|x^*\|$. The greatest lower bound of the numbers $\|x^*\|$ so obtained is $\|x^* + M^\perp\|$, by the definition of the quotient norm. Hence

$$\|m^*\| \le \|\sigma m^*\| \le \|x^*\|.$$

But Theorem 3.3 furnishes an extension x^* of m^* with $\|x^*\| = \|m^*\|$. It follows that $\|\sigma m^*\| = \|m^*\|$. This completes (*a*).

(*b*) If $x \in X$ and $y^* \in Y^*$, then $\pi x \in Y$; hence $x \to y^*\pi x$ is a continuous linear functional on X which vanishes for $x \in M$. Thus $\tau y^* \in M^\perp$. The linearity of τ is obvious. Fix $x^* \in M^\perp$. Let N be the null space of x^*. Since $M \subset N$, there is a linear functional Λ on Y such that $\Lambda\pi = x^*$. The null space of Λ is $\pi(N)$, a closed subspace of Y, by the definition of the quotient topology in $Y = X/M$. By Theorem 1.18, Λ is continuous, that is, $\Lambda \in Y^*$. Hence $\tau\Lambda = \Lambda\pi = x^*$. The range of τ is therefore all of $M^\perp$.

It remains to be shown that τ is an isometry.

Let B be the open unit ball in X. Then πB is the open unit ball of $Y = \pi X$. Since $\tau y^* = y^*\pi$, we have

$$\begin{aligned}\|\tau y^*\| = \|y^*\pi\| &= \sup \{|\langle \pi x, y^*\rangle| : x \in B\}\\ &= \sup \{|\langle y, y^*\rangle| : y \in \pi B\} = \|y^*\|\end{aligned}$$

for every $y^* \in Y^*$. ////

Adjoints

We shall now associate with each $T \in \mathscr{B}(X, Y)$ its *adjoint*, an operator $T^* \in \mathscr{B}(Y^*, X^*)$, and will see how certain properties of T are reflected in the behavior of T^*. If X and Y are finite-dimensional, every $T \in \mathscr{B}(X, Y)$

can be represented by a matrix $[T]$; in that case, $[T^*]$ is the transpose of $[T]$, provided that the various vector space bases are properly chosen. No particular attention will be paid to the finite-dimensional case in what follows, but historically linear algebra did provide the background and much of the motivation that went into the construction of what is now known as *operator theory*.

Many of the nontrivial properties of adjoints depend on the completeness of X and Y (the open mapping theorem will play an important role). For this reason, it will be assumed throughout that X and Y are Banach spaces, except in Theorem 4.10, which furnishes the definition of T^*.

4.10 Theorem *Suppose X and Y are normed spaces. To each $T \in \mathscr{B}(X, Y)$ corresponds a unique $T^* \in \mathscr{B}(Y^*, X^*)$ that satisfies*

$$\langle Tx, y^* \rangle = \langle x, T^*y^* \rangle \tag{1}$$

for all $x \in X$ and all $y^* \in Y^*$. Moreover, T^* satisfies

$$\|T^*\| = \|T\|. \tag{2}$$

PROOF. If $y^* \in Y^*$ and $T \in \mathscr{B}(X, Y)$, define

$$T^*y^* = y^* \circ T. \tag{3}$$

Being the composition of two continuous linear mappings, $T^*y^* \in X^*$. Also,

$$\langle x, T^*y^* \rangle = (T^*y^*)(x) = y^*(Tx) = \langle Tx, y^* \rangle,$$

which is (1). The fact that (1) holds for every $x \in X$ obviously determines T^*y^* uniquely.

If $y_1^* \in Y^*$ and $y_2^* \in Y^*$, then

$$\begin{aligned}\langle x, T^*(y_1^* + y_2^*) \rangle &= \langle Tx, y_1^* + y_2^* \rangle \\ &= \langle Tx, y_1^* \rangle + \langle Tx, y_2^* \rangle \\ &= \langle x, T^*y_1^* \rangle + \langle x, T^*y_2^* \rangle \\ &= \langle x, T^*y_1^* + T^*y_2^* \rangle\end{aligned}$$

for every $x \in X$, so that

$$T^*(y_1^* + y_2^*) = T^*y_1^* + T^*y_2^*. \tag{4}$$

Similarly, $T^*(\alpha y^*) = \alpha T^*y^*$. Thus $T^*: Y^* \to X^*$ is linear. Finally, *(b)* of Theorem 4.3 leads to

$$\begin{aligned}\|T\| &= \sup\{|\langle Tx, y^* \rangle| : \|x\| \le 1, \|y^*\| \le 1\} \\ &= \sup\{|\langle x, T^*y^* \rangle| : \|x\| \le 1, \|y^*\| \le 1\} \\ &= \sup\{\|T^*y^*\| : \|y^*\| \le 1\} = \|T^*\|.\end{aligned}$$

////

4.11 Notation If T maps X into Y, the null space and the range of T will be denoted by $\mathscr{N}(T)$ and $\mathscr{R}(T)$, respectively:

$$\mathscr{N}(T) = \{x \in X\colon Tx = 0\},$$
$$\mathscr{R}(T) = \{y \in Y\colon Tx = y \text{ for some } x \in X\}.$$

The next theorem concerns annihilators; see Section 4.6 for the notation.

4.12 Theorem *Suppose X and Y are Banach spaces, and $T \in \mathscr{B}(X, Y)$. Then*

$$\mathscr{N}(T^*) = \mathscr{R}(T)^\perp \quad \text{and} \quad \mathscr{N}(T) = {}^\perp\mathscr{R}(T^*).$$

PROOF. In each of the following two columns, each statement is obviously equivalent to the one that immediately follows and/or precedes it.

$y^* \in \mathscr{N}(T^*)$.	$x \in \mathscr{N}(T)$.
$T^*y^* = 0$.	$Tx = 0$.
$\langle x, T^*y^*\rangle = 0$ for all x.	$\langle Tx, y^*\rangle = 0$ for all y^*.
$\langle Tx, y^*\rangle = 0$ for all x.	$\langle x, T^*y^*\rangle = 0$ for all y^*.
$y^* \in \mathscr{R}(T)^\perp$.	$x \in {}^\perp\mathscr{R}(T^*)$.

////

Corollaries

(*a*) *$\mathscr{N}(T^*)$ is weak*-closed in Y^*.*
(*b*) *$\mathscr{R}(T)$ is dense in Y if and only if T^* is one-to-one.*
(*c*) *T is one-to-one if and only if $\mathscr{R}(T^*)$ is weak*-dense in X^*.*

Recall that $M^\perp$ is weak*-closed in Y^* for every subspace M of Y. In particular, this is true of $\mathscr{R}(T)^\perp$. Thus (*a*) follows from the theorem.

As to (*b*), $\mathscr{R}(T)$ is dense in Y if and only if $\mathscr{R}(T)^\perp = \{0\}$; in that case, $\mathscr{N}(T^*) = \{0\}$.

Likewise, ${}^\perp\mathscr{R}(T^*) = \{0\}$ if and only if $\mathscr{R}(T^*)$ is annihilated by no $x \in X$ other than $x = 0$; this says that $\mathscr{R}(T^*)$ is weak*-dense in X^*.

Note that the Hahn-Banach theorem 3.5 was tacitly used in the proofs of (*b*) and (*c*).

There is a useful analogue of (*b*), namely, that $\mathscr{R}(T)$ is *all* of Y if and only if T^* is one-to-one *and* its inverse [mapping $\mathscr{R}(T^*)$ onto Y^*]

is bounded. The equivalence of (*a*) and (*d*) in the following theorem expresses this in slightly different terms. Theorem 4.15 is closely related. The union of the following three theorems is sometimes called the *closed range theorem.*

4.13 Theorem *Let U and V be the open unit balls in the Banach spaces X and Y, respectively. If $T \in \mathscr{B}(X, Y)$ and $\delta > 0$, then the implications*

$$(a) \to (b) \to (c) \to (d)$$

hold among the following statements:

(*a*) $\|T^*y^*\| \geq \delta\|y^*\|$ *for every* $y^* \in Y^*$.
(*b*) $\overline{T(U)} \supset \delta V$.
(c) $T(U) \supset \delta V$.
(*d*) $T(X) = Y$.

Moreover, if (d) holds, then (a) holds for some $\delta > 0$.

PROOF. Assume (*a*), and pick $y_0 \notin \overline{T(U)}$. Since $\overline{T(U)}$ is convex, closed, and balanced, Theorem 3.7 shows that there is a y^* such that $|\langle y, y^*\rangle| \leq 1$ for every $y \in \overline{T(U)}$, but $|\langle y_0, y^*\rangle| > 1$. If $x \in U$, it follows that

$$|\langle x, T^*y^*\rangle| = |\langle Tx, y^*\rangle| \leq 1.$$

Thus $\|T^*y^*\| \leq 1$, and now (*a*) gives

$$\delta < \delta|\langle y_0, y^*\rangle| \leq \delta\|y_0\| \|y^*\| \leq \|y_0\| \|T^*y^*\| \leq \|y_0\|.$$

It follows that $y \in \overline{T(U)}$ if $\|y\| \leq \delta$. Thus (*a*) $\to$ (*b*).

Next, assume (*b*). Take $\delta = 1$, without loss of generality. Then $\overline{T(U)} \supset \bar{V}$. To every $y \in Y$ and every $\varepsilon > 0$ corresponds therefore an $x \in X$ with $\|x\| \leq \|y\|$ and $\|y - Tx\| < \varepsilon$.

Pick $y_1 \in V$. Pick $\varepsilon_n > 0$ so that

$$\sum_{n=1}^{\infty} \varepsilon_n < 1 - \|y_1\|.$$

Assume $n \geq 1$ and y_n is picked. There exists x_n such that $\|x_n\| \leq \|y_n\|$ and $\|y_n - Tx_n\| < \varepsilon_n$. Put

$$y_{n+1} = y_n - Tx_n.$$

By induction, this process defines two sequences $\{x_n\}$ and $\{y_n\}$. Note that

$$\|x_{n+1}\| \leq \|y_{n+1}\| = \|y_n - Tx_n\| < \varepsilon_n.$$

Hence

$$\sum_{n=1}^{\infty} \|x_n\| \le \|x_1\| + \sum_{n=1}^{\infty} \varepsilon_n \le \|y_1\| + \sum_{n=1}^{\infty} \varepsilon_n < 1.$$

It follows that $x = \sum x_n$ is in U (see Exercise 23) and that

$$Tx = \lim_{N\to\infty} \sum_{n=1}^{N} Tx_n = \lim_{N\to\infty} \sum_{n=1}^{N} (y_n - y_{n+1}) = y_1$$

since $y_{N+1} \to 0$ as $N \to \infty$. Thus $y_1 = Tx \in T(U)$, which proves (*c*).

Note that the preceding argument is just a specialized version of part of the proof of the open mapping theorem 2.11.

That (*c*) implies (*d*) is obvious.

Assume (*d*). By the open mapping theorem, there is a $\delta > 0$ such that $T(U) \supset \delta V$. Hence

$$\begin{aligned}\|T^*y^*\| &= \sup\{|\langle x, T^*y^*\rangle| : x \in U\} \\ &= \sup\{|\langle Tx, y^*\rangle| : x \in U\} \\ &\ge \sup\{|\langle y, y^*\rangle| : y \in \delta V\} = \delta\|y^*\|\end{aligned}$$

for every $y^* \in Y^*$. This is (*a*). ////

4.14 Theorem *If X and Y are Banach spaces and if $T \in \mathscr{B}(X, Y)$, then each of the following three conditions implies the other two:*

(*a*) *$\mathscr{R}(T)$ is closed in Y.*
(*b*) *$\mathscr{R}(T^*)$ is weak*-closed in X^*.*
(*c*) *$\mathscr{R}(T^*)$ is norm-closed in X^*.*

Remark. Theorem 3.12 implies that (*a*) holds if and only if $\mathscr{R}(T)$ is weakly closed. However, norm-closed subspaces of X^* are not always weak*-closed (Exercise 7, Chapter 3).

PROOF. It is obvious that (*b*) implies (*c*). We will prove that (*a*) implies (*b*) and that (*c*) implies (*a*).

Suppose (*a*) holds. By Theorem 4.12 and (*b*) of Theorem 4.7, $\mathscr{N}(T)^{\perp}$ is the weak*-closure of $\mathscr{R}(T^*)$. To prove (*b*) it is therefore enough to show that $\mathscr{N}(T)^{\perp} \subset \mathscr{R}(T^*)$.

Pick $x^* \in \mathscr{N}(T)^{\perp}$. Define a linear functional Λ on $\mathscr{R}(T)$ by

$$\Lambda Tx = \langle x, x^*\rangle \qquad (x \in X).$$

Note that Λ is well defined, for if $Tx = Tx'$, then $x - x' \in \mathscr{N}(T)$; hence

$$\langle x - x', x^*\rangle = 0.$$

The open mapping theorem applies to

$$T: X \to \mathscr{R}(T)$$

since $\mathscr{R}(T)$ is assumed to be a closed subspace of the complete space Y and is therefore complete. It follows that there exists $K < \infty$ such that to each $y \in \mathscr{R}(T)$ corresponds an $x \in X$ with $Tx = y$, $\|x\| \leq K\|y\|$, and

$$|\Lambda y| = |\Lambda Tx| = |\langle x, x^*\rangle| \leq K\|y\| \, \|x^*\|.$$

Thus Λ is continuous. By the Hahn-Banach theorem, some $y^* \in Y^*$ extends Λ. Hence

$$\langle Tx, y^*\rangle = \Lambda Tx = \langle x, x^*\rangle \qquad (x \in X).$$

This implies $x^* = T^*y^*$. Since x^* was an arbitrary element of $\mathscr{N}(T)^\perp$, we have shown that $\mathscr{N}(T)^\perp \subset \mathscr{R}(T^*)$. Thus (*b*) follows from (*a*).

Suppose next that (*c*) holds. Let Z be the closure of $\mathscr{R}(T)$ in Y. Define $S \in \mathscr{B}(X, Z)$ by setting $Sx = Tx$. Since $\mathscr{R}(S)$ is dense in Z, Corollary (*b*) to Theorem 4.12 implies that

$$S^*: Z^* \to X^*$$

is one-to-one.

If $z^* \in Z^*$, the Hahn-Banach theorem furnishes an extension y^* of z^*; for every $x \in X$,

$$\langle x, T^*y^*\rangle = \langle Tx, y^*\rangle = \langle Sx, z^*\rangle = \langle x, S^*z^*\rangle.$$

Hence $S^*z^* = T^*y^*$. It follows that S^* and T^* have identical ranges. Since (*c*) is assumed to hold, $\mathscr{R}(S^*)$ is closed, hence complete.

Apply the open mapping theorem to

$$S^*: Z^* \to \mathscr{R}(S^*).$$

Since S^* is one-to-one, the conclusion is that there is a constant $c > 0$ which satisfies

$$c\|z^*\| \leq \|S^*z^*\|$$

for every $z^* \in Z^*$. Hence $S: X \to Z$ is an open mapping, by Theorem 4.13. In particular, $S(X) = Z$. But $\mathscr{R}(T) = \mathscr{R}(S)$, by the definition of S. Thus $\mathscr{R}(T) = Z$, a closed subspace of Y.

This completes the proof that (*c*) implies (*a*). ////

The following consequence is useful in applications.

4.15 Theorem *Suppose X and Y are Banach spaces, and $T \in \mathscr{B}(X, Y)$. Then*

(a) $\mathscr{R}(T) = Y$ *if and only if*
(b) T^* *is one-to-one and* $\mathscr{R}(T^*)$ *is norm-closed.*

PROOF. If (a) holds then T^* is one-to-one, by Theorem 4.12. The implication (d) → (a) of Theorem 4.13 shows that T^* is (a multiple of) a dilation; hence $\mathscr{R}(T^*)$ is closed, by Theorem 1.26.

If (b) holds, then $\mathscr{R}(T)$ is dense in Y, again by Theorem 4.12, and $\mathscr{R}(T)$ is closed by Theorem 4.14.

Compact Operators

4.16 Definition Suppose X and Y are Banach spaces and U is the open unit ball in X. A linear map $T: X \to Y$ is said to be *compact* if the closure of $T(U)$ is compact in Y. It is clear that T is then bounded. Thus $T \in \mathscr{B}(X, Y)$.

Since Y is a complete metric space, the subsets of Y whose closure is compact are precisely the totally bounded ones. Thus $T \in \mathscr{B}(X, Y)$ is compact if and only if $T(U)$ is totally bounded. Also, T is compact if and only if every bounded sequence $\{x_n\}$ in X contains a subsequence $\{x_{n_i}\}$ such that $\{Tx_{n_i}\}$ converges to a point of Y.

Many of the operators that arise in the study of integral equations are compact. This accounts for their importance from the standpoint of applications. They are in some respects as similar to linear operators on finite-dimensional spaces as one has any right to expect from operators on infinite-dimensional spaces. As we shall see, these similarities show up particularly strongly in their spectral properties.

4.17 Definitions (a) Suppose X is a Banach space. Then $\mathscr{B}(X)$ [which is an abbreviation for $\mathscr{B}(X, Y)$] is not merely a Banach space (see Theorem 4.1) but also an algebra: If $S \in \mathscr{B}(X)$ and $T \in \mathscr{B}(X)$, one defines $ST \in \mathscr{B}(X)$ by

$$(ST)(x) = S(T(x)) \qquad (x \in X).$$

The inequality

$$\|ST\| \le \|S\| \, \|T\|$$

is trivial to verify.

In particular, powers of $T \in \mathscr{B}(X)$ can be defined: $T^0 = I$, the identity mapping on X, given by $Ix = x$, and $T^n = TT^{n-1}$, for $n = 1, 2, 3, \ldots$.

(b) An operator $T \in \mathscr{B}(X)$ is said to be *invertible* if there exists $S \in \mathscr{B}(X)$ such that

$$ST = I = TS.$$

In this case, we write $S = T^{-1}$. By the open mapping theorem, this happens if and only if $\mathscr{N}(T) = \{0\}$ and $\mathscr{R}(T) = X$.

(c) The *spectrum* $\sigma(T)$ of an operator $T \in \mathscr{B}(X)$ is the set of all scalars λ such that $T - \lambda I$ is *not* invertible. Thus $\lambda \in \sigma(T)$ if and only if at least one of the following two statements is true:

(*i*) The range of $T - \lambda I$ is not all of X.

(*ii*) $T - \lambda I$ is not one-to-one.

If (*ii*) holds, λ is said to be an *eigenvalue* of T; the corresponding eigenspace is $\mathscr{N}(T - \lambda I)$; each $x \in \mathscr{N}(T - \lambda I)$ (except $x = 0$) is an *eigenvector* of T; it satisfies the equation

$$Tx = \lambda x.$$

Here are some very easy facts which will illustrate these concepts.

4.18 Theorem *Let X and Y be Banach spaces.*

(*a*) *If $T \in \mathscr{B}(X, Y)$ and* $\dim \mathscr{R}(T) < \infty$, *then T is compact.*

(*b*) *If $T \in \mathscr{B}(X, Y)$, T is compact, and $\mathscr{R}(T)$ is closed, then* $\dim \mathscr{R}(T) < \infty$.

(*c*) *The compact operators form a closed subspace of $\mathscr{B}(X, Y)$ in its norm-topology.*

(*d*) *If $T \in \mathscr{B}(X)$, T is compact, and $\lambda \neq 0$, then* $\dim \mathscr{N}(T - \lambda I) < \infty$.

(*e*) *If* $\dim X = \infty$, *$T \in \mathscr{B}(X)$, and T is compact, then $0 \in \sigma(T)$.*

(*f*) *If $S \in \mathscr{B}(X)$, $T \in \mathscr{B}(X)$, and T is compact, so are ST and TS.*

PROOF. Statement (*a*) is obvious. If $\mathscr{R}(T)$ is closed, then $\mathscr{R}(T)$ is complete (since Y is complete), so that T is an open mapping of X onto $\mathscr{R}(T)$; if T is compact, it follows that $\mathscr{R}(T)$ is locally compact; thus (*b*) is a consequence of Theorem 1.22.

Put $Y = \mathscr{N}(T - \lambda I)$ in (*d*). The restriction of T to Y is a compact operator whose range is Y. Thus (*d*) follows from (*b*), and so does (*e*), for if 0 is not in $\sigma(T)$, then $\mathscr{R}(T) = X$. The proof of (*f*) is trivial.

If S and T are compact operators from X into Y, so is $S + T$, because the sum of any two compact subsets of Y is compact. It follows that the compact operators form a subspace Σ of $\mathscr{B}(X, Y)$. To complete the proof of (*c*), we now show that Σ is closed. Let $T \in \mathscr{B}(X, Y)$ be the closure of Σ, choose $r > 0$, and let U be the open unit ball in X. There exists $S \in \Sigma$ with $\|S - T\| < r$. Since $S(U)$ is totally bounded, there are points $x_1, \ldots, x_n$ in U such that $S(U)$ is covered by the balls of radius r with centers at the points Sx_i. Since

$\|Sx - Tx\| < r$ for every $x \in U$, it follows that $T(U)$ is covered by the balls of radius $3r$ with centers at the points Tx_i. Thus $T(U)$ is totally bounded, which proves that $T \in \Sigma$. ////

The main objective of the rest of this chapter is to analyze the spectrum of a compact $T \in \mathscr{B}(X)$. Theorem 4.25 contains the principal results. Adjoints will play an important role in this investigation.

4.19 Theorem *Suppose X and Y are Banach spaces and $T \in \mathscr{B}(X, Y)$. Then T is compact if and only if T^* is compact.*

PROOF. Suppose T is compact. Let $\{y_n^*\}$ be a sequence in the unit ball of Y^*. Define

$$f_n(y) = \langle y, y_n^* \rangle \qquad (y \in Y).$$

Since $|f_n(y) - f_n(y')| \le \|y - y'\|$, $\{f_n\}$ is equicontinuous. Since $T(U)$ has compact closure in Y (as before, U is the unit ball of X), Ascoli's theorem implies that $\{f_n\}$ has a subsequence $\{f_{n_i}\}$ that converges uniformly on $T(U)$. Since

$$\begin{aligned} \|T^*y_{n_i}^* - T^*y_{n_j}^*\| &= \sup |\langle Tx, y_{n_i}^* - y_{n_j}^* \rangle| \\ &= \sup |f_{n_i}(Tx) - f_{n_j}(Tx)|, \end{aligned}$$

the supremum being taken over $x \in U$, the completeness of X^* implies that $\{T^*y_{n_i}^*\}$ converges. Hence T^* is compact.

The second half can be proved by the same method, but it may be more instructive to deduce it from the first half.

Let $\phi: X \to X^{**}$ and $\psi: Y \to Y^{**}$ be the isometric embeddings given by the formulas

$$\langle x, x^* \rangle = \langle x^*, \phi x \rangle \qquad \text{and} \qquad \langle y, y^* \rangle = \langle y^*, \psi y \rangle,$$

as in Section 4.5. Then

$$\langle y^*, \psi Tx \rangle = \langle Tx, y^* \rangle = \langle x, T^*y^* \rangle = \langle T^*y^*, \phi x \rangle = \langle y^*, T^{**}\phi x \rangle$$

for all $x \in X$ and $y^* \in Y^*$, so that

$$\psi T = T^{**}\phi.$$

If $x \in U$, then ϕx lies in the unit ball U^{**} of X^{**}. Thus

$$\psi T(U) \subset T^{**}(U^{**}).$$

Now assume that T^* is compact. The first half of the theorem shows that $T^{**}: X^{**} \to Y^{**}$ is compact. Hence $T^{**}(U^{**})$ is totally bounded, and so is its subset $\psi T(U)$. Since ψ is an isometry, $T(U)$ is also totally bounded. Hence T is compact. ////

4.20 Definition Suppose M is a closed subspace of a topological vector space X. If there exists a closed subspace N of X such that

$$X = M + N \qquad \text{and} \qquad M \cap N = \{0\},$$

then M is said to be *complemented in* X. In this case, X is said to be the *direct sum* of M and N, and the notation

$$X = M \oplus N$$

is sometimes used.

We shall see examples of uncomplemented subspaces in Chapter 5. At present we need only the following simple facts.

4.21 Lemma *Let M be a closed subspace of a topological vector space X.*

(*a*) *If X is locally convex and* $\dim M < \infty$, *then M is complemented in X.*
(*b*) *If* $\dim (X/M) < \infty$, *then M is complemented in X.*

The dimension of X/M is also called the *codimension* of M in X.

PROOF. (*a*) Let $\{e_1, \ldots, e_n\}$ be a basis for M. Every $x \in M$ has then a unique representation

$$x = \alpha_1(x)e_1 + \cdots + \alpha_n(x)e_n.$$

Each α_i is a continuous linear functional on M (Theorem 1.21 and Lemma 1.20) which extends to a member of X^*, by the Hahn-Banach theorem. Let N be the intersection of the null spaces of these extensions. Then $X = M \oplus N$.

(*b*) Let $\pi: X \to X/M$ be the quotient map, let $\{e_1, \ldots, e_n\}$ be a basis for X/M, pick $x_i \in X$ so that $\pi x_i = e_i$ $(1 \le i \le n)$, and let N be the vector space spanned by $\{x_1, \ldots, x_n\}$. Then $X = M \oplus N$. ////

4.22 Lemma *If M is a subspace of a normed space X, if M is not dense in X, and if $r > 1$, then there exists $x \in X$ such that*

$$\|x\| < r \qquad \textit{but} \qquad \|x - y\| \ge 1 \qquad \textit{for all } y \in M.$$

PROOF. There exists $x_1 \in X$ whose distance from M is 1, that is,

$$\inf \{\|x_1 - y\| : y \in M\} = 1.$$

Choose $y_1 \in M$ such that $\|x_1 - y_1\| < r$, and put $x = x_1 - y_1$. ////

4.23 Theorem *If X is a Banach space, $T \in \mathscr{B}(X)$, T is compact, and $\lambda \neq 0$, then $T - \lambda I$ has closed range.*

PROOF. By (*d*) of Theorem 4.18, $\dim \mathscr{N}(T - \lambda I) < \infty$. By (*a*) of Lemma 4.21, X is the direct sum of $\mathscr{N}(T - \lambda I)$ and a closed subspace M. Define an operator $S \in \mathscr{B}(M, X)$ by

$$Sx = Tx - \lambda x. \tag{1}$$

Then S is one-to-one on M. Also, $\mathscr{R}(S) = \mathscr{R}(T - \lambda I)$. To show that $\mathscr{R}(S)$ is closed, it suffices to show the existence of an $r > 0$ such that

$$r\|x\| \leq \|Sx\| \qquad \text{for all } x \in M, \tag{2}$$

by Theorem 1.26.

If (2) fails for every $r > 0$, there exists $\{x_n\}$ in M such that $\|x_n\| = 1$, $Sx_n \to 0$, and (after passage to a subsequence) $Tx_n \to x_0$ for some $x_0 \in X$. (This is where compactness of T is used.) It follows that $\lambda x_n \to x_0$. Thus $x_0 \in M$, and

$$Sx_0 = \lim (\lambda S x_n) = 0.$$

Since S is one-to-one, $x_0 = 0$. But $\|x_n\| = 1$ for all n, and $x_0 = \lim \lambda x_n$, and so $\|x_0\| = |\lambda| > 0$. This contradiction proves (2) for some $r > 0$. ////

4.24 Theorem *Suppose X is a Banach space, $T \in \mathscr{B}(X)$, T is compact, $r > 0$, and E is a set of eigenvalues λ of T such that $|\lambda| > r$. Then*

(*a*) *for each $\lambda \in E$, $\mathscr{R}(T - \lambda I) \neq X$, and*
(*b*) *E is a finite set.*

PROOF. We shall first show that if either (*a*) or (*b*) is false then there exist closed subspaces M_n of X and scalars $\lambda_n \in E$ such that

$$M_1 \subset M_2 \subset M_3 \subset \cdots, \qquad M_n \neq M_{n+1}, \tag{1}$$

$$T(M_n) \subset M_n \qquad \text{for } n \geq 1, \tag{2}$$

and

$$(T - \lambda_n I)(M_n) \subset M_{n-1} \qquad \text{for } n \geq 2. \tag{3}$$

The proof will be completed by showing that this contradicts the compactness of T.

Suppose (*a*) is false. Then $\mathscr{R}(T-\lambda_0 I)=X$ for some $\lambda_0 \in E$. Put $S=T-\lambda_0 I$, and define M_n to be the null space of S^n. (See Section 4.17.) Since λ_0 is an eigenvalue of T, there exists $x_1 \in M_1$, $x_1 \neq 0$. Since $\mathscr{R}(S)=X$, there is a sequence $\{x_n\}$ in X such that $Sx_{n+1}=x_n$, $n=1,2,3,\ldots$. Then

$$(4) \qquad S^n x_{n+1} = x_1 \neq 0 \quad \text{but} \quad S^{n+1}x_{n+1} = Sx_1 = 0.$$

Hence M_n is a proper closed subspace of M_{n+1}. It follows that (1) to (3) hold, with $\lambda_n=\lambda_0$. [Note that (2) holds because $ST=TS$.]

Suppose (*b*) is false. Then E contains a sequence $\{\lambda_n\}$ of distinct eigenvalues of T. Choose corresponding eigenvectors e_n, and let M_n be the (finite-dimensional, hence closed) subspace of X spanned by $\{e_1,\ldots,e_n\}$. Since the λ_n are distinct, $\{e_1,\ldots,e_n\}$ is a linearly independent set, so that M_{n-1} is a proper subspace of M_n. This gives (1). If $x \in M_n$, then

$$x = \alpha_1 e_1 + \cdots + \alpha_n e_n,$$

which shows that $Tx \in M_n$ and

$$(T-\lambda_n I)x = \alpha_1(\lambda_1-\lambda_n)e_1 + \cdots + \alpha_{n-1}(\lambda_{n-1}-\lambda_n)e_{n-1} \in M_{n-1}.$$

Thus (2) and (3) hold.

Once we have closed subspaces M_n satisfying (1) to (3), Lemma 4.22 gives us vectors $y_n \in M_n$, for $n=2,3,4,\ldots$, such that

$$(5) \qquad \|y_n\| \le 2 \quad \text{and} \quad \|y_n - x\| \ge 1 \quad \text{if} \quad x \in M_{n-1}.$$

If $2 \le m < n$, define

$$(6) \qquad z = Ty_m - (T-\lambda_n I)y_n.$$

By (2) and (3), $z \in M_{n-1}$. Hence (5) shows that

$$\|Ty_n - Ty_m\| = \|\lambda_n y_n - z\| = |\lambda_n|\,\|y_n - \lambda_n^{-1}z\| \ge |\lambda_n| > r.$$

The sequence $\{Ty_n\}$ has therefore no convergent subsequences, although $\{y_n\}$ is bounded. This is impossible if T is compact. ////

4.25 Theorem *Suppose X is a Banach space, $T \in \mathscr{B}(X)$, and T is compact.*

(*a*) *If $\lambda \neq 0$, then the four numbers*

$$\alpha = \dim \mathscr{N}(T-\lambda I)$$

$$\beta = \dim X/\mathscr{R}(T-\lambda I)$$

$$\alpha^* = \dim \mathscr{N}(T^*-\lambda I)$$

$$\beta^* = \dim X^*/\mathscr{R}(T^*-\lambda I)$$

are equal and finite.

(*b*) *If* $\lambda \neq 0$ *and* $\lambda \in \sigma(T)$ *then* λ *is an eigenvalue of* T *and of* T^*.

(*c*) $\sigma(T)$ *is compact, at most countable, and has at most one limit point, namely,* 0.

Note: The dimension of a vector space is here understood to be either a nonnegative integer or the symbol ∞. The letter I is used for the identity operators on both X and X^*; thus

$$(T - \lambda I)^* = T^* - \lambda I^* = T^* - \lambda I,$$

since the adjoint of the identity on X is the identity on X^*.

The spectrum $\sigma(T)$ of T was defined in Section 4.17. Theorem 4.24 contains a special case of (*a*): $\beta = 0$ implies $\alpha = 0$. This will be used in the proof of the inequality (4) below.

It should be noted that $\sigma(T)$ is compact even if T is not (Theorem 10.13). The compactness of T is needed for the other assertions in (*c*).

PROOF. Put $S = T - \lambda I$, to simplify the writing.

We begin with an elementary observation about quotient spaces. Suppose M_0 is a closed subspace of a locally convex space Y, and k is a positive integer such that $k \leq \dim Y/M_0$. Then there are vectors $y_1, \ldots, y_k$ in Y such that the vector space M_i generated by M_0 and $y_1, \ldots, y_i$ contains M_{i-1} as a proper subspace. By Theorem 1.42, each M_i is closed. By Theorem 3.5, there are continuous linear functionals $\Lambda_1, \ldots, \Lambda_k$ on Y such that $\Lambda_i y_i = 1$ but $\Lambda_i y = 0$ for all $y \in M_{i-1}$. These functionals are linearly independent. The following conclusion is therefore reached: If Σ *denotes the space of all continuous linear functionals on* Y *that annihilate* M_0, *then*

$$\dim Y/M_0 \leq \dim \Sigma. \tag{1}$$

Apply this with $Y = X$, $M_0 = \mathscr{R}(S)$. By Theorem 4.23, $\mathscr{R}(S)$ is closed. Also, $\Sigma = \mathscr{R}(S)^\perp = \mathscr{N}(S^*)$, by Theorem 4.12, so that (1) becomes

$$\beta \leq \alpha^*. \tag{2}$$

Next, take $Y = X^*$ with its weak*-topology; take $M_0 = \mathscr{R}(S^*)$. By Theorem 4.14, $\mathscr{R}(S^*)$ is weak*-closed. Since Σ now consists of all weak*-continuous linear functionals on X^* that annihilate $\mathscr{R}(S^*)$, Σ is isomorphic to ${}^\perp\mathscr{R}(S^*) = \mathscr{N}(S)$ (Theorem 4.12), and (1) becomes

$$\beta^* \leq \alpha. \tag{3}$$

Our next objective is to prove that

$$\alpha \leq \beta. \tag{4}$$

Once we have (4), the inequality

$$\alpha^* \leq \beta^* \tag{5}$$

is also true, since T^* is a compact operator (Theorem 4.19). Since $\alpha < \infty$ by (*d*) of Theorem 4.18, (*a*) is an obvious consequence of the inequalities (2) to (5).

Assume that (4) is false. Then $\alpha > \beta$. Since $\alpha < \infty$, Lemma 4.21 shows that X contains closed subspaces E and F such that $\dim F = \beta$ and

$$X = \mathscr{N}(S) \oplus E = \mathscr{R}(S) \oplus F. \tag{6}$$

Every $x \in X$ has a unique representation $x = x_1 + x_2$, with $x_1 \in \mathscr{N}(S)$, $x_2 \in E$. Define $\pi: X \to \mathscr{N}(S)$ by setting $\pi x = x_1$. It is easy to see (by the closed graph theorem, for instance) that π is continuous.

Since we assume that $\dim \mathscr{N}(S) > \dim F$, there is a linear mapping ϕ of $\mathscr{N}(S)$ onto F such that $\phi x_0 = 0$ for some $x_0 \neq 0$. Define

$$\Phi x = Tx + \phi\pi x \qquad (x \in X). \tag{7}$$

Then $\Phi \in \mathscr{B}(X)$. Since $\dim \mathscr{R}(\phi) < \infty$, $\phi\pi$ is a compact operator; hence so is Φ (Theorem 4.18).

Observe that

$$\Phi - \lambda I = S + \phi\pi. \tag{8}$$

If $x \in E$, then $\pi x = 0$, $(\Phi - \lambda I)x = Sx$; hence

$$(\Phi - \lambda I)(E) = \mathscr{R}(S). \tag{9}$$

If $x \in \mathscr{N}(S)$, then $\pi x = x$,

$$(\Phi - \lambda I)x = \varphi x, \tag{10}$$

and therefore

$$(\Phi - \lambda I)(\mathscr{N}(S)) = \varphi(\mathscr{N}(S)) = F. \tag{11}$$

It follows from (9) and (11) that

$$\mathscr{R}(\Phi - \lambda I) \supset \mathscr{R}(S) + F = X. \tag{12}$$

But if (10) is used with $x = x_0$, we see that λ is an eigenvalue of Φ, and since Φ is compact, Theorem 4.24 shows that the range of $\Phi - \lambda I$ cannot be all of X. This contradicts (12); hence (4) is true and (*a*) is proved.

Part (*b*) follows from (*a*), for if λ is not an eigenvalue of T, then $\alpha(T) = 0$, and (*a*) implies that $\beta(T) = 0$, that is, that $\mathscr{R}(T - \lambda I) = X$. Thus $T - \lambda I$ is invertible, so that $\lambda \notin \sigma(T)$.

It now follows from (*b*) of Theorem 4.24 that 0 is the only possible limit point of $\sigma(T)$, that $\sigma(T)$ is at most countable, and that

$\sigma(T) \cup \{0\}$ is compact. If $\dim X < \infty$, then $\sigma(T)$ is finite; if $\dim X = \infty$, then $0 \in \sigma(T)$, by (*e*) of Theorem 4.18. Thus $\sigma(T)$ is compact. This gives (*c*) and completes the proof of the theorem. ////

Exercises

Throughout this set of exercises, X and Y denote Banach spaces, unless the contrary is explicitly stated.

1. Let ϕ be the embedding of X into X^{**} described in Section 4.5. Let τ be the weak topology of X, and let σ be the weak*-topology of X^{**}—the one induced by X^*.
   (*a*) Prove that ϕ is a homeomorphism of (X, τ) onto a dense subspace of (X^{**}, σ).
   (*b*) If B is the closed unit ball of X, prove that $\phi(B)$ is σ-dense in the closed unit ball of X^{**}. (Use the Hahn-Banach separation theorem.)
   (*c*) Use (*a*), (*b*), and the Banach-Alaoglu theorem to prove that X is reflexive if and only if B is weakly compact.
   (*d*) Deduce from (*c*) that every norm-closed subspace of a reflexive space X is reflexive.
   (*e*) If X is reflexive and Y is a closed subspace of X, prove that X/Y is reflexive.
   (*f*) Prove that X is reflexive if and only if X^* is reflexive.
   Suggestion: One half follows from (*c*); for the other half, apply (*d*) to the subspace $\phi(X)$ of X^{**}.
2. Which of the spaces c_0, ℓ^1, ℓ^p, ℓ^∞ are reflexive? Prove that every finite-dimensional normed space is reflexive. Prove that C, the supremum-normed space of all complex continuous functions, on the unit interval, is not reflexive.
3. Prove that a subset E of $\mathscr{B}(X, Y)$ is equicontinuous if and only if there exists $M < \infty$ such that $\|\Lambda\| \leq M$ for every $\Lambda \in E$.
4. Recall that $X^* = \mathscr{B}(X, \mathcal{C})$, if $\mathcal{C}$ is the scalar field. Hence $\Lambda^* \in \mathscr{B}(\mathcal{C}, X^*)$ for every $\Lambda \in X^*$. Identify the range of Λ^*.
5. Prove that $T \in \mathscr{B}(X, Y)$ is an isometry of X onto Y if and only if T^* is an isometry of Y^* onto X^*.
6. Let σ and τ be the weak*-topologies of X^* and Y^*, respectively, and prove that S is a continuous linear mapping of (Y^*, τ) into (X^*, σ) if and only if $S = T^*$ for some $T \in \mathscr{B}(X, Y)$.
7. Let L^1 be the usual space of integrable functions on the closed unit interval J, relative to Lebesgue measure. Suppose $T \in \mathscr{B}(L^1, Y)$, so that $T^* \in \mathscr{B}(Y^*, L^\infty)$. Suppose $\mathscr{R}(T^*)$ contains every continuous function on J. What can you deduce about T?
8. Prove that $(ST)^* = T^*S^*$. Supply the hypotheses under which this makes sense.
9. Suppose $S \in \mathscr{B}(X)$, $T \in \mathscr{B}(X)$.
   (*a*) Show, by an example, that $ST = I$ does not imply $TS = I$.
   (*b*) However, assume T is compact, show that
   $$S(I - T) = I \text{ if and only if } (I - T)S = I,$$
   and show that either of these equalities implies that $I - (I - T)^{-1}$ is compact.

10. Assume $T \in \mathscr{B}(X)$ is compact, and assume either that $\dim X = \infty$ or that the scalar field is $\mathcal{C}$. Prove that $\sigma(T)$ is not empty. However, $\sigma(T)$ may be empty if $\dim X < \infty$ and the scalar field is R.

11. Suppose $\dim X < \infty$ and show that the equality $\beta^* = \beta$ of Theorem 4.25 reduces to the statement that the row rank of a square matrix is equal to its column rank.

12. Suppose $T \in \mathscr{B}(X, Y)$ and $\mathscr{R}(T)$ is closed in Y. Prove that

$$\dim \mathscr{N}(T) = \dim X^*/\mathscr{R}(T^*),$$
$$\dim \mathscr{N}(T^*) = \dim Y/\mathscr{R}(T).$$

This generalizes the assertions $\alpha = \beta^*$ and $\alpha^* = \beta$ of Theorem 4.25.

13. (*a*) Suppose $T \in \mathscr{B}(X, Y)$, $T_n \in \mathscr{B}(X, Y)$ for $n = 1, 2, 3, \ldots$, each T_n has finite-dimensional range, and $\lim \|T - T_n\| = 0$. Prove that T is compact.

(*b*) Assume Y is a Hilbert space, and prove the converse of (*a*): Every compact $T \in \mathscr{B}(X, Y)$ can be approximated in the operator norm by operators with finite-dimensional ranges. *Hint:* In a Hilbert space there are linear projections of norm 1 onto any closed subspace. (See Theorems 5.16, 12.4.)

14. Define a shift operator S and a multiplication operator M on ℓ^2 by

$$(Sx)(n) = \begin{cases} 0 & \text{if } n = 0, \\ x(n-1) & \text{if } n \geq 1, \end{cases}$$

$$(Mx)(n) = (n+1)^{-1}x(n) \qquad \text{if } n \geq 0.$$

Put $T = MS$. Show that T is a compact operator which has no eigenvalue and whose spectrum consists of exactly one point. Compute $\|T^n\|$, for $n = 1, 2, 3, \ldots$, and compute $\lim_{n\to\infty} \|T^n\|^{1/n}$.

15. Suppose μ is a finite (or σ-finite) positive measure on a measure space Ω, $\mu \times \mu$ is the corresponding product measure on $\Omega \times \Omega$, and $K \in L^2(\mu \times \mu)$. Define

$$(Tf)(s) = \int_\Omega K(s, t)f(t)\, d\mu(t) \qquad [f \in L^2(\mu)].$$

(*a*) Prove that $T \in \mathscr{B}(L^2(\mu))$ and that

$$\|T\|^2 \leq \iint_{\Omega\Omega} |K(s, t)|^2\, d\mu(s)\, d\mu(t).$$

(*b*) Suppose a_i, b_i are members of $L^2(\mu)$, for $1 \leq i \leq n$, put $K_1(s, t) = \sum a_i(s)b_i(t)$, and define T_1 in terms of K_1 as T was defined in terms of K. Prove that $\dim \mathscr{R}(T_1) \leq n$.

(*c*) Deduce that T is a compact operator on $L^2(\mu)$. *Hint:* Use Exercise 13.

(*d*) Suppose $\lambda \in \mathcal{C}$, $\lambda \neq 0$. Prove: Either the equation

$$Tf - \lambda f = g$$

has a unique solution $f \in L^2(\mu)$ for every $g \in L^2(\mu)$ or there are infinitely many solutions for some g and none for others. (This is known as the *Fredholm alternative.*)

(*e*) Describe the adjoint of T.

16. Define

$$K(s, t) = \begin{cases} (1-s)t & \text{if } 0 \le t \le s \\ (1-t)s & \text{if } s \le t \le 1 \end{cases}$$

and define $T \in \mathscr{B}(L^2(0, 1))$ by

$$(Tf)(s) = \int_0^1 K(s, t) f(t)\, dt \qquad (0 \le s \le 1).$$

(*a*) Show that the eigenvalues of T are $(n\pi)^{-2}$, $n = 1, 2, 3, \ldots$, that the corresponding eigenfunctions are $\sin n\pi x$, and that each eigenspace is one-dimensional. *Hint:* If $\lambda \neq 0$, the equation $Tf = \lambda f$ implies that f is infinitely differentiable, that $\lambda f'' + f = 0$, and that $f(0) = f(1) = 0$. The case $\lambda = 0$ can be treated separately.

(*b*) Show that the above eigenfunctions form an orthogonal basis for $L^2(0, 1)$.

(*c*) Suppose $g(t) = \sum c_n \sin n\pi t$. Discuss the equation $Tf - \lambda f = g$.

(*d*) Show that T is also a compact operator on C, the space of all continuous functions on $[0, 1]$. *Hint:* If $\{f_i\}$ is uniformly bounded, then $\{Tf_i\}$ is equicontinuous.

17. If $L^2 = L^2(0, \infty)$ relative to Lebesgue measure, and if

$$(Tf)(s) = \frac{1}{s}\int_0^s f(t)\, dt \qquad (0 < s < \infty),$$

prove that $T \in \mathscr{B}(L^2)$ and that T is not compact. (The fact that $\|T\| \le 2$ is a special case of Hardy's inequality. See p. 72 of [23].)

18. Prove the following statements:

(*a*) If $\{x_n\}$ is a weakly convergent sequence in X, then $\{\|x_n\|\}$ is bounded.

(*b*) If $T \in \mathscr{B}(X, Y)$ and $x_n \to x$ weakly, then $Tx_n \to Tx$ weakly.

(*c*) If $T \in \mathscr{B}(X, Y)$, if $x_n \to x$ weakly, and if T is compact, then $\|Tx_n - Tx\| \to 0$.

(*d*) Conversely, if X is reflexive, if $T \in \mathscr{B}(X, Y)$, and if $\|Tx_n - Tx\| \to 0$ whenever $x_n \to x$ weakly, then T is compact. *Hint:* Use (*c*) of Exercise 1, and part (*c*) of Exercise 28 in Chapter 3.

(*e*) If X is reflexive and $T \in \mathscr{B}(X, \ell^1)$, then T is compact. Hence $\mathscr{R}(T) \neq \ell^1$. *Hint:* Use (*c*) of Exercise 5 of Chapter 3.

(*f*) If Y is reflexive and $T \in \mathscr{B}(c_0, Y)$, then T is compact.

19. Suppose Y is a closed subspace of X, and $x_0^* \in X^*$. Put

$$\mu = \sup \{|\langle x, x_0^*\rangle| : x \in Y, \|x\| \le 1\},$$

$$\delta = \inf \{\|x^* - x_0^*\| : x^* \in Y^\perp\}.$$

In other words, μ is the norm of the restriction of x_0^* to Y, and δ is the distance from x_0^* to the annihilator of Y. Prove that $\mu = \delta$. Prove also that $\delta = \|x^* - x_0^*\|$ for at least one $x^* \in Y^\perp$.

20. Extend Sections 4.6 to 4.9 to locally convex spaces. (The word "isometric" must of course be deleted from the statement of Theorem 4.9.)

21. Let B and B^* be the closed unit balls in X and X^*, respectively. The following is a converse of the Banach-Alaoglu theorem: *If E is a convex set in X^* such that*

$E \cap (rB^*)$ is weak-compact for every $r > 0$, then E is weak*-closed.* (Corollary: A subspace of X^* is weak*-closed if and only if its intersection with B^* is weak*-compact.)

Complete the following outline of the proof.

(*i*) E is norm-closed.

(*ii*) Associated to each $F \subset X$ its polar

$$P(F) = \{x^*: |\langle x, x^*\rangle| \le 1 \text{ for all } x \in F\}.$$

The intersection of all sets $P(F)$, as F ranges over the collection of all finite subsets of $r^{-1}B$, is exactly rB^*.

(*iii*) The theorem is a consequence of the following proposition: *If, in addition to the stated hypotheses, $E \cap B^* = \varnothing$, then there exists $x \in X$ such that* $\text{Re}\,\langle x, x^*\rangle \ge 1$ *for every $x^* \in E$.*

(*iv*) Proof of the proposition: Put $F_0 = \{0\}$. Assume finite sets $F_0, \ldots, F_{k-1}$ have been chosen so that $iF_i \subset B$ and so that

$$P(F_0) \cap \cdots \cap P(F_{k-1}) \cap E \cap kB^* = \varnothing. \tag{1}$$

Note that (1) is true for $k = 1$. Put

$$Q = P(F_0) \cap \cdots \cap P(F_{k-1}) \cap E \cap (k+1)B^*.$$

If $P(F) \cap Q \ne \varnothing$ for every finite set $F \subset k^{-1}B$, the weak*-compactness of Q, together with (*ii*), implies that $(kB^*) \cap Q \ne \varnothing$, which contradicts (1). Hence there is a finite set $F_k \subset k^{-1}B$ such that (1) holds with $k + 1$ in place of k. The construction can thus proceed. It yields

$$E \cap \bigcap_{k=1}^{\infty} P(F_k) = \varnothing. \tag{2}$$

Arrange the members of $\bigcup F_k$ in a sequence $\{x_n\}$. Then $\|x_n\| \to 0$. Define $T: X^* \to c_0$ by

$$Tx^* = \{\langle x_n, x^*\rangle\}.$$

Then $T(E)$ is a convex subset of c_0. By (2),

$$\|Tx^*\| = \sup_n |\langle x_n, x^*\rangle| \ge 1$$

for every $x^* \in E$. Hence there is a scalar sequence $\{\alpha_n\}$, with $\sum |\alpha_n| < \infty$, such that

$$\text{Re} \sum_{n=1}^{\infty} \alpha_n \langle x_n, x^*\rangle \le 1$$

for every $x^* \in E$. To complete the proof, put $x = \sum \alpha_n x_n$.

22. Suppose $T \in \mathscr{B}(X)$, T is compact, $\lambda \ne 0$, and $S = T - \lambda I$.

(*a*) If $\mathscr{N}(S^n) = \mathscr{N}(S^{n+1})$ for some nonnegative integer n, prove that $\mathscr{N}(S^n) = \mathscr{N}(S^{n+k})$ for $k = 1, 2, 3, \ldots$.

(*b*) Prove that (*a*) must happen for some n. (*Hint:* Consider the proof of Theorem 4.24.)

(c) Let n be the smallest nonnegative integer for which (a) holds. Prove that $\dim \mathscr{N}(S^n)$ is finite, that

$$X = \mathscr{N}(S^n) \oplus \mathscr{R}(S^n),$$

and that the restriction of S to $\mathscr{R}(S^n)$ is a one-to-one mapping of $\mathscr{R}(S^n)$ onto $\mathscr{R}(S^n)$.

23. Suppose $\{x_n\}$ is a sequence in a Banach space X, and

$$\sum_{n=1}^{\infty} \|x_n\| = M < \infty.$$

Prove that the series $\sum x_n$ converges to some $x \in X$. Explicitly, prove that

$$\lim_{n \to \infty} \|x - (x_1 + \cdots + x_n)\| = 0.$$

Prove also that $\|x\| \le M$. (These facts were used in the proof of Theorem 4.13.)

24. Let c be the space of all complex sequences

$$x = \{x_1, x_2, x_3, \ldots\}$$

for which $x_\infty = \lim x_n$ exists (in $\mathcal{C}$). Put $\|x\| = \sup |x_n|$. Let c_0 be the subspace of c that consists of all x with $x_\infty = 0$.

(a) Describe explicitly two isometric isomorphisms u and v, such that u maps c^* onto ℓ^1 and v maps c_0^* onto ℓ^1.

(b) Define $S: c_0 \to c$ by $Sf = f$. Describe the operator vS^*u^{-1} that maps ℓ^1 to ℓ^1.

(c) Define $T: c \to c_0$ by setting

$$y_1 = x_\infty, \quad y_{n+1} = x_n - x_\infty \qquad \text{if } n \ge 1.$$

Prove that T is one-to-one and that $Tc = c_0$. Find $\|T\|$ and $\|T^{-1}\|$. Describe the operator uT^*v^{-1} that maps ℓ^1 to ℓ^1.

25. If $T \in \mathscr{B}(X, Y)$ and $\mathscr{R}(T^*) = \mathscr{N}(T)^\perp$, prove that $\mathscr{R}(T)$ is closed.

26. Assume $T \in \mathscr{B}(X, Y)$ and $T(X) = Y$. Show that there exists $\delta > 0$ such that $S(X) = Y$ for all $S \in \mathscr{B}(X, Y)$ with $\|S - T\| < \delta$.

27. Suppose $T \in \mathscr{B}(X)$. Prove that $\lambda \in \sigma(T)$ if and only if there is a sequence $\{x_n\}$ in X, $\|x_n\| = 1$, for which

$$\lim_{n \to \infty} \|Tx_n - \lambda x_n\| = 0.$$

[Thus every $\lambda \in \sigma(T)$ which is not an eigenvalue of T is an "approximate" eigenvalue.]

CHAPTER 5

SOME APPLICATIONS

This chapter contains some applications of the preceding abstract material to more concrete problems in analysis. Most of these applications depend only on a small part of the contents of Chapters 1 through 4. Here is a partial list of the theorems, ordered more or less according to prerequisites.

Theorems	Prerequisites
5.23	Vector topologies
5.27	Minkowski functionals (and Brouwer's fixed point theorem)
5.1, 5.2	Closed graph theorem
5.4	Hahn-Banach theorem
5.5, 5.7, 5.10, 5.11	Banach-Alaoglu and Krein-Milman theorems
5.18	Banach-Steinhaus theorem and vector-valued integrals
5.9, 5.21	Closed range theorem

A Continuity Theorem

One of the very early theorems in functional analysis (Hellinger and Toeplitz, 1910) states that *if T is a linear operator on a Hilbert space H which is symmetric in the sense that*

$$(Tx, y) = (x, Ty)$$

for all $x \in H$ and $y \in H$, then T is continuous. Here (x, y) denotes the usual Hilbert space inner product. (See Section 12.1.)

If $\{x_n\}$ is a sequence in H such that $\|x_n\| \to 0$, the symmetry of T implies that $Tx_n \to 0$ weakly. (This depends on knowing that all continuous linear functionals on H are given by inner products.) The Hellinger-Toeplitz theorem is therefore a consequence of the following one.

5.1 Theorem *Suppose X and Y are F-spaces, Y^* separates points on Y, $T: X \to Y$ is linear, and $\Lambda Tx_n \to 0$ for every $\Lambda \in Y^*$ whenever $x_n \to 0$. Then T is continuous.*

PROOF. Suppose $x_n \to x$ and $Tx_n \to y$. If $\Lambda \in Y^*$, then

$$\Lambda T(x_n - x) \to 0$$

so that

$$\Lambda y = \lim \Lambda Tx_n = \Lambda Tx.$$

Consequently, $y = Tx$, and the closed graph theorem can be applied. ////

In the context of Banach spaces, Theorem 5.1 can be stated as follows: *If $T: X \to Y$ is linear, if $\|x_n\| \to 0$ implies that $Tx_n \to 0$ weakly, then $\|x_n\| \to 0$ actually implies that $\|Tx_n\| \to 0$.*

To see that completeness is important here, let X be the vector space of all complex infinitely differentiable functions on $(-\infty, \infty)$ which vanish outside the unit interval, put

$$(f, g) = \int_0^1 f\bar{g}, \qquad \|f\| = (f, f)^{1/2},$$

and define $T: X \to X$ by $(Tf)(x) = if'(x)$. Then $(Tf, g) = (f, Tg)$, but T is not continuous.

Closed Subspaces of L^p-Spaces

The proof of the following theorem of Grothendieck also involves the closed graph theorem.

5.2 Theorem *Suppose $0 < p < \infty$, and*

(a) μ is a probability measure on a measure space Ω.
(b) S is a closed subspace of $L^p(\mu)$.
(c) $S \subset L^\infty(\mu)$.

Then S is finite-dimensional.

PROOF. Let j be the identity map that takes S into L^∞, where S is given the L^p-topology, so that S is complete. If $\{f_n\}$ is a sequence in S such that $f_n \to f$ in S and $f_n \to g$ in L^∞, it is obvious that $f = g$ a.e. Hence j satisfies the hypotheses of the closed graph theorem, and we conclude that there is a constant $K < \infty$ such that

$$\|f\|_\infty \le K\|f\|_p \tag{1}$$

for all $f \in S$. As usual, $\|f\|_p$ means $(\int |f|^p \, d\mu)^{1/p}$, and $\|f\|_\infty$ is the essential supremum of $|f|$. If $p \le 2$ then $\|f\|_p \le \|f\|_2$. If $2 < p < \infty$, integration of the inequality

$$|f|^p \le \|f\|_\infty^{p-2} |f|^2$$

leads to $\|f\|_\infty \le K^{p/2}\|f\|_2$. In either case, we have a constant $M < \infty$ such that

$$\|f\|_\infty \le M\|f\|_2 \qquad (f \in S). \tag{2}$$

In the rest of the proof we shall deal with individual functions, not with equivalence classes modulo null sets.

Let $\{\phi_1, \ldots, \phi_n\}$ be an orthonormal set in S, regarded as a subspace of L^2. Let Q be a countable dense subset of the euclidean unit ball B of $\mathbb{C}^n$. If $c = (c_1, \ldots, c_n) \in B$, define $f_c = \sum c_i \phi_i$. Then $\|f_c\|_2 \le 1$, and so $\|f_c\|_\infty \le M$. Since Q is countable, there is a set $\Omega' \subset \Omega$, with $\mu(\Omega') = 1$, such that $|f_c(x)| \le M$ for every $c \in Q$ and for *every* $x \in \Omega'$. If x is fixed, $c \to |f_c(x)|$ is a continuous function on B. Hence $|f_c(x)| \le M$ whenever $c \in B$ and $x \in \Omega'$. It follows that $\sum |\phi_i(x)|^2 \le M^2$ for every $x \in \Omega'$. Integration of this inequality gives $n \le M^2$. We conclude that $\dim S \le M^2$. This proves the theorem. ////

It is crucial in this theorem that L^∞ occurs in the hypothesis (*c*). To illustrate this we will now construct an infinite-dimensional closed subspace of L^1 which lies in L^4. For our probability measure we take Lebesgue measure on the circle, divided by 2π.

5.3 Theorem *Let E be an infinite set of integers such that no integer has more than one representation as a sum of two members of E. Let P_E be the vector space of all finite sums f of the form*

$$f(e^{i\theta}) = \sum_{n=-\infty}^{\infty} c(n)e^{in\theta} \tag{1}$$

in which $c(n) = 0$ whenever n is not in E. Let S_E be the L^1-closure of P_E. Then S_E is a closed subspace of L^4.

An example of such a set is furnished by 2^k, $k = 1, 2, 3, \ldots$. Much slower growth can also be achieved.

PROOF. If f is as in (1), then

$$f^2(e^{i\theta}) = \sum_n c(n)^2 e^{2in\theta} + \sum_{n \neq m} c(n)c(m)e^{i(n+m)\theta}.$$

Our combinatorial hypothesis about E implies that

$$\int |f|^4 = \int |f^2|^2 = \sum_n |c(n)|^4 + 4 \sum_{m<n} |c(m)|^2 |c(n)|^2$$

so that

$$\text{(2)} \qquad \int |f|^4 \leq 2\left(\sum |c(n)|^2\right)^2 = 2\left(\int |f|^2\right)^2.$$

Hölder's inequality, with 3 and $\frac{3}{2}$ as conjugate exponents, gives

$$\text{(3)} \qquad \int |f|^2 \leq \left(\int |f|^4\right)^{1/3} \left(\int |f|\right)^{2/3}.$$

It follows from (2) and (3) that

$$\text{(4)} \qquad \|f\|_4 \leq 2^{1/4} \|f\|_2 \qquad \text{and} \qquad \|f\|_2 \leq 2^{1/2} \|f\|_1$$

for every $f \in P_E$. Every L^1-Cauchy sequence in P_E is therefore also a Cauchy sequence in L^4. Hence $S_E \subset L^4$. The obvious inequality $\|f\|_1 \leq \|f\|_4$ then shows that S_E is closed in L^4. ////

An interesting result can be obtained by applying a duality argument to the second inequality (4). Recall that the Fourier coefficients $\hat{g}(n)$ of every $g \in L^\infty$ satisfy $\sum |\hat{g}(n)|^2 < \infty$. The next theorem shows that nothing more can be said about the restriction of $\hat{g}$ to E.

5.4 Theorem *If E is as in Theorem 5.3 and if*

$$\sum_{-\infty}^{\infty} |a(n)|^2 = A^2 < \infty$$

then there exists $g \in L^\infty$ such that $\hat{g}(n) = a(n)$ for every $n \in E$.

PROOF. If $f \in P_E$, the preceding proof shows that

$$\left|\sum \hat{f}(n)a(n)\right| \leq A\left\{\sum |\hat{f}(n)|^2\right\}^{1/2} = A\|f\|_2 \leq 2^{1/2} A \|f\|_1.$$

Hence $f \to \sum \hat{f}(n)a(n)$ is a linear functional on P_E which is continuous relative to the L^1-norm. By the Hahn-Banach theorem, this functional has a continuous linear extension to L^1. Hence there exists $g \in L^\infty$

(with $\|g\|_\infty \le 2^{1/2}A$) such that

$$\sum_{-\infty}^{\infty} \hat{f}(n)a(n) = \frac{1}{2\pi}\int_{-\pi}^{\pi} f(e^{-i\theta})g(e^{i\theta})\,d\theta \qquad (f \in P_E).$$

With $f(e^{i\theta}) = e^{in\theta}$ $(n \in E)$, this shows that $\hat{g}(n) = a(n)$. ////

The Range of a Vector-Valued Measure

We now give a rather striking application of the theorems of Krein-Milman and Banach-Alaoglu.

Let $\mathfrak{M}$ be a σ-algebra. A real-valued measure λ on $\mathfrak{M}$ is said to be *nonatomic* if every set $E \in \mathfrak{M}$ with $|\lambda|(E) > 0$ contains a set $A \in \mathfrak{M}$ with $0 < |\lambda|(A) < |\lambda|(E)$. Here $|\lambda|$ denotes the total variation measure of λ; the terminology is as in [23].

5.5 Theorem *Suppose $\mu_1, \ldots, \mu_n$ are real-valued nonatomic measures on a σ-algebra $\mathfrak{M}$. Define*

$$\mu(E) = (\mu_1(E), \ldots, \mu_n(E)) \qquad (E \in \mathfrak{M}).$$

Then μ is a function with domain $\mathfrak{M}$ whose range is a compact convex subset of R^n.

PROOF. Associate to each bounded measurable real function g the vector

$$\Lambda g = \left(\int g\,d\mu_1, \ldots, \int g\,d\mu_n\right)$$

in R^n. Put $\sigma = |\mu_1| + \cdots + |\mu_n|$. If $g_1 = g_2$ a.e. $[\sigma]$, then $\Lambda g_1 = \Lambda g_2$. Hence Λ may be regarded as a linear mapping of $L^\infty(\sigma)$ into R^n.

Each μ_i is absolutely continuous with respect to σ. The Radon-Nikodym theorem [23] shows therefore that there are functions $h_i \in L^1(\sigma)$ such that $d\mu_i = h_i\,d\sigma$ $(1 \le i \le n)$. Hence Λ is a weak*-continuous linear mapping of $L^\infty(\sigma)$ into R^n; recall that $L^\infty(\sigma) = L^1(\sigma)^*$. Put

$$K = \{g \in L^\infty(\sigma) : 0 \le g \le 1\}.$$

It is obvious that K is convex. Since $g \in K$ if and only if

$$0 \le \int fg\,d\sigma \le \int f\,d\sigma$$

for every nonnegative $f \in L^1(\sigma)$, K is weak*-closed. And since K lies in the closed unit ball of $L^\infty(\sigma)$, the Banach-Alaoglu theorem shows that K is weak*-compact. Hence $\Lambda(K)$ is a compact convex set in R^n.

We shall prove that $\mu(\mathfrak{M}) = \Lambda(K)$.

If χ_E is the characteristic function of a set $E \in \mathfrak{M}$, then $\chi_E \in K$ and $\mu(E) = \Lambda g$. Thus $\mu(\mathfrak{M}) \subset \Lambda(K)$. To obtain the opposite inclusion, pick a point $p \in \Lambda(K)$ and define

$$K_p = \{g \in K: \Lambda g = p\}.$$

We have to show that K_p contains some χ_E, for then $p = \mu(E)$.

Note that K_p is convex; since Λ is continuous, K_p is weak*-compact. By the Krein-Milman theorem, K_p has an extreme point.

Suppose $g_0 \in K_p$ and g_0 is not a characteristic function in $L^\infty(\sigma)$. Then there is a set $E \in \mathfrak{M}$ and an $r > 0$ such that $\sigma(E) > 0$ and $r \leq g_0 \leq 1 - r$ on E. Put $Y = \chi_E \cdot L^\infty(\sigma)$. Since $\sigma(E) > 0$ and σ is non-atomic, dim $Y > n$. Hence there exists $g \in Y$, not the zero element of $L^\infty(\sigma)$, such that $\Lambda g = 0$, and such that $-r < g < r$. It follows that $g_0 + g$ and $g_0 - g$ are in K_p. Thus g_0 is not an extreme point of K_p.

Every extreme point of K_p is therefore a characteristic function. This completes the proof. ////

A Generalized Stone-Weierstrass Theorem

The theorems of Krein-Milman, Hahn-Banach, and Banach-Alaoglu will now be applied to an approximation problem.

5.6 Definitions Let $C(S)$ be the familiar sup-normed Banach space of all continuous complex functions on the compact Hausdorff space S. A subspace A of $C(S)$ is an *algebra* if $fg \in A$ whenever $f \in A$ and $g \in A$. A set $E \subset S$ is said to be *A-antisymmetric* if every $f \in A$ which is real on E is constant on E; in other words, the algebra A_E which consists of the restrictions $f|_E$ of the functions $f \in A$ to E contains no nonconstant real functions.

For example, if S is a compact set in $\mathcal{C}$ and if A consists of all $f \in C(S)$ that are holomorphic in the interior of S, then every component of the interior of S is *A-antisymmetric.*

Suppose $A \subset C(S)$, $p \in S$, $q \in S$, and write $p \sim q$ provided that there is an A-antisymmetric set E which contains both p and q. It is easily verified that this defines an equivalence relation in S and that each equivalence class is a closed set. These equivalence classes are the *maximal* A-antisymmetric sets.

5.7 Bishop's theorem *Let A be a closed subalgebra of $C(S)$. Suppose $g \in C(S)$ and $g|_E \in A_E$ for every maximal A-antisymmetric set E. Then $g \in A$.*

Stated differently, the hypothesis on g is that to every maximal A-antisymmetric set E corresponds a function $f \in A$ which coincides with g on

E; the conclusion is that one f exists which does this for every E, namely, $f = g$.

A special case of Bishop's theorem is the Stone-Weierstrass theorem: *Suppose that*

(a) *A is a closed subalgebra of $C(S)$,*
(b) *A is self-adjoint (i.e., $\bar{f} \in A$ for all $f \in A$),*
(c) *A separates points on S, and*
(d) *at every $p \in S, f(p) \neq 0$ for some $f \in A$. Then $A = C(S)$.*

For in this case the real-valued members $f + \bar{f}$ of A separate points on S. Therefore no A-antisymmetric set contains more than one point. It follows that every $g \in C(S)$ satisfies the hypothesis of Bishop's theorem.

PROOF. The annihilator $A^{\perp}$ of A consists of all regular complex Borel measures μ on S such that $\int f \, d\mu = 0$ for every $f \in A$. Define

$$K = \{\mu \in A^{\perp} : \|\mu\| \leq 1\},$$

where $\|\mu\| = |\mu|(S)$. Then K is convex, balanced, and weak*-compact, by (c) of Theorem 4.3. If $K = \{0\}$, then $A^{\perp} = \{0\}$; hence $A = C(S)$, and there is nothing to prove.

Assume $K \neq \{0\}$, and let μ be an extreme point of K. Clearly, $\|\mu\| = 1$. Let E be the *support* of μ; this means that E is compact, that $|\mu|(E) = \|\mu\|$, and that E is the smallest set with these two properties.

We claim: E is antisymmetric.

Consider an $f \in A$ with $f|_E$ real; without loss of generality, $-1 < f < 1$ on E. Define measures σ and τ by

$$d\sigma = \tfrac{1}{2}(1 + f) \, d\mu, \qquad d\tau = \tfrac{1}{2}(1 - f) \, d\mu.$$

Since A is an algebra, $\sigma \in A^{\perp}$ and $\tau \in A^{\perp}$. Since $1 + f$ and $1 - f$ are positive on E, $\|\sigma\| > 0$, $\|\tau\| > 0$, and

$$\|\sigma\| + \|\tau\| = \frac{1}{2} \int_E (1 + f) \, d|\mu| + \frac{1}{2} \int_E (1 - f) \, d|\mu| = |\mu|(E) = 1.$$

This shows that μ is a convex combination of the measures $\sigma_1 = \sigma/\|\sigma\|$ and $\tau_1 = \tau/\|\tau\|$. Both of these are in K. Since μ is extreme in K, $\mu = \sigma_1$. In other words,

$$\tfrac{1}{2}(1 + f) \, d\mu = \|\sigma\| \, d\mu.$$

Therefore $f = 2\|\sigma\| - 1$ on E, i.e., $f|_E$ is constant.

This proves our claim.

If g satisfies the hypothesis of the theorem, it follows that $\int g \, d\mu = 0$ for every μ that is extreme in K, hence for every μ in the convex hull of these extreme points. Since $\mu \to \int g \, d\mu$ is a weak*-

continuous function on K, the Krein-Milman theorem implies that $\int g\, d\mu = 0$ for every $\mu \in K$, hence for every $\mu \in A^{\perp}$.

Every continuous linear functional on $C(S)$ that annihilates A thus also annihilates g. Hence $g \in A$, by the Hahn-Banach separation theorem. ////

Note: If (*d*) is dropped from the hypotheses of the Stone-Weierstrass theorem, then (*c*) implies that there is at most one $p_0 \in S$, where $f(p_0) = 0$ for every $f \in A$. If this is the case, then the proof shows that $A = \{f \in C(S): f(p_0) = 0\}$.

Here is an example that illustrates Bishop's theorem:

5.8 Theorem *Suppose*

(*a*) *K is a compact subset of $R^n \times \mathbb{C}$ and*

(*b*) *if $t = (t_1, \ldots, t_n) \in R^n$, the set*

$$K_t = \{z \in \mathbb{C}: (t, z) \in K\}$$

does not separate $\mathbb{C}$. If $g \in C(K)$, define g_t on K_t by $g_t(z) = g(t, z)$.

Assume that $g \in C(K)$, that each g_t is holomorphic in the interior of K_t and that $\varepsilon > 0$. Then there is a polynomial P in the variables $t_1, \ldots, t_n, z$ such that

$$|P(t, z) - g(t, z)| < \varepsilon$$

for every $(t, z) \in K$.

PROOF. Let A be the closure in $C(K)$ of the set of all polynomials $P(t, z)$. Since the real polynomials on R^n separate points, every A-antisymmetric set lies in some K_t. By Theorem 5.7 it is therefore enough to show that to every $t \in R^n$ corresponds an $f \in A$ such that $f_t = g_t$.

Fix $t \in R^n$. By Mergelyan's theorem [23] there are polynomials $P_i(z)$ such that

$$g_t(z) = \sum_{i=1}^{\infty} P_i(z) \qquad (z \in K_t)$$

and $|P_i| < 2^{-i}$ if $i > 1$. There is a polynomial Q on R^n that peaks at t, in the sense that $Q(t) = 1$ but $|Q(s)| < 1$ if $s \neq t$ and $K_s \neq \varnothing$. Consider a fixed $i > 1$. The functions ϕ_m defined on K by

$$\phi_m(s, z) = |Q^m(s)P_i(z)|$$

form a monotonically decreasing sequence of continuous functions whose limit is $<2^{-i}$ at every point of K. Since K is compact, it follows that there is a positive integer m_i such that $\phi_{m_i}(s, z) < 2^{-i}$ at every point of K. The series

$$f(s, z) = \sum_{i=1}^{\infty} Q^{m_i}(s)P_i(z)$$

converges uniformly on K. Hence $f \in A$, and obviously $f_t = g_t$. ////

Two Interpolation Theorems

The proof of the first of these theorems involves the adjoint of an operator. The second furnishes another application of the Krein-Milman theorem.

The first one (due to Bishop) again concerns $C(S)$. Our notation is as in Theorem 5.7.

5.9 Theorem *Suppose Y is a closed subspace of $C(S)$, K is a compact subset of S, and $|\mu|(K) = 0$ for every $\mu \in Y^\perp$. If $g \in C(K)$ and $|g| < 1$, it follows that there exists $f \in Y$ such that $f|_K = g$ and $|f| < 1$ on S.*

Thus every continuous function on K extends to a member of Y. In other words, the restriction map $f \to f|_K$ maps Y onto $C(K)$.

This theorem generalizes the following special case.

Let A be the *disc algebra*, i.e., the set of all continuous functions on the closure of the unit disc U in $\mathcal{C}$ which are holomorphic in U. Take $S = T$, the unit circle. Let Y consist of the restrictions to T of the members of A. By the maximum modulus theorem, Y is a closed subspace of $C(T)$. If $K \subset T$ is compact and has Lebesgue measure 0, the theorem of F. and M. Riesz [23] states precisely that K satisfies the hypothesis of Theorem 5.9. Consequently, *to every $g \in C(K)$ corresponds an $f \in A$ such that $f = g$ on K.*

PROOF. Let $\rho: Y \to C(K)$ be the restriction map defined by $\rho f = f|_K$. We have to prove that ρ maps the open unit ball of Y *onto* the open unit ball of $C(K)$.

Consider the adjoint $\rho^*: M(K) \to Y^*$, where $M(K) = C(K)^*$ is the Banach space of all regular complex Borel measures on K, with the total variation norm $\|\mu\| = |\mu|(K)$. For each $\mu \in M(K)$, $\rho^*\mu$ is a bounded linear functional on Y; by the Hahn-Banach theorem, $\rho^*\mu$ extends to a linear functional on $C(S)$, of the same norm. In other words, there exists $\sigma \in M(S)$, with $\|\sigma\| = \|\rho^*\mu\|$, such that

$$\int_S f\, d\sigma = \langle f, \rho^*\mu \rangle = \langle \rho f, \mu \rangle = \int_K f\, d\mu$$

for every $f \in Y$. Regard μ as a member of $M(S)$, with support in K. Then $\sigma - \mu \in Y^\perp$, and our hypothesis about K implies that $\sigma(E) = \mu(E)$ for every Borel set $E \subset K$. Hence $\|\mu\| \leq \|\sigma\|$. We conclude that $\|\mu\| \leq \|\rho^*\mu\|$. By (*b*) of Theorem 4.13, this inequality proves the theorem. ////

Note: Since $\|\rho^*\| = \|\rho\| \leq 1$, we also have $\|\sigma\| \leq \|\mu\|$ in the preceding proof. It follows that $\sigma = \mu$. Hence $\rho^*\mu$ has a *unique* norm-preserving extension to $C(S)$.

Our second interpolation theorem concerns finite *Blaschke products*, i.e., functions B of the form

$$B(z) = c \prod_{k=1}^{N} \frac{z - \alpha_k}{1 - \bar{\alpha}_k z},$$

where $|c| = 1$ and $|\alpha_k| < 1$ for $1 \leq k \leq N$. It is easy to see that the finite Blaschke products are precisely those members of the disc algebra whose absolute value is 1 at every point of the unit circle.

The *data* of the *Pick-Nevanlinna interpolation problem* are two finite sets of complex numbers, $\{z_0, \ldots, z_n\}$ and $\{w_0, \ldots, w_n\}$, all of absolute value less than 1, with $z_i \neq z_j$ if $i \neq j$. The *problem* is to find a holomorphic function f in the open unit disc U, such that $|f(z)| < 1$ for all $z \in U$, and such that

$$f(z_i) = w_i \qquad (0 \leq i \leq n).$$

The data may very well admit no solution. For example, if $\{z_0, z_1\} = \{0, \frac{1}{2}\}$ and $\{w_0, w_1\} = \{0, \frac{2}{3}\}$, the Schwarz lemma shows this. But if the problem has solutions, then among them there must be some very nice ones. The next theorem shows this.

5.10 Theorem *Let $\{z_0, \ldots, z_n\}$, $\{w_0, \ldots, w_n\}$ be Pick-Nevanlinna data. Let E be the set of all holomorphic functions f in U such that $|f| < 1$ and $f(z_i) = w_i$ for $0 \leq i \leq n$. If E is not empty, then E contains a finite Blaschke product.*

PROOF. Without loss of generality, assume $z_0 = w_0 = 0$. We will show that there is a holomorphic function F in U which satisfies

(1) $$\operatorname{Re} F(z) > 0 \qquad \text{for } z \in U,\ F(0) = 1,$$

(2) $$F(z_i) = \beta_i = \frac{1 + w_i}{1 - w_i} \qquad \text{for } 1 \leq i \leq n,$$

and which has the form

$$F(z) = \sum_{k=1}^{N} c_k \frac{a_k + z}{a_k - z}, \tag{3}$$

where $c_k > 0$, $\sum c_k = 1$, and $|a_k| = 1$. Once such an F is found, put $B = (F - 1)/(F + 1)$. This is a finite Blaschke product that satisfies $B(z_i) = w_i$ for $0 \le i \le n$.

Let K be the set of all holomorphic functions F in U that satisfy (1).

Associate to each $\mu \in M(T) = C(T)^*$ the function

$$F_\mu(z) = \int_{-\pi}^{\pi} \frac{e^{i\theta} + z}{e^{i\theta} - z}\, d\mu(e^{i\theta}) \qquad (z \in U). \tag{4}$$

If P is the set of all Borel *probability* measures on T, then $\mu \leftrightarrow F_\mu$ is a one-to-one correspondence between P and K (Theorems 11.9 and 11.30 of [23]). Define $\Lambda: M(T) \to \mathcal{C}^n$ by

$$\Lambda\mu = (F_\mu(z_1), \ldots, F_\mu(z_n)). \tag{5}$$

Since E is assumed to be nonempty, there exists $\mu_0 \in P$ such that

$$\Lambda\mu_0 = \beta = (\beta_1, \ldots, \beta_n). \tag{6}$$

Since P is convex and weak*-compact, and since Λ is linear and weak*-continuous, $\Lambda(P)$ is a convex compact set in $\mathcal{C} = R^{2n}$. Since $\beta \in \Lambda(P)$, β is a convex combination of $N \le 2n + 1$ extreme points of $\Lambda(P)$ (Exercise 19, Chapter 3). If γ is an extreme point of $\Lambda(P)$, then $\Lambda^{-1}(\gamma)$ is an extreme set of K, and every extreme point of $\Lambda^{-1}(\gamma)$ (their existence follows from the Krein-Milman theorem) is an extreme point of P. It follows that there are extreme points $\mu_1, \ldots, \mu_N$ of P and positive numbers c_k with $\sum c_k = 1$, such that

$$\Lambda(c_1\mu_1 + \cdots + c_N\mu_N) = \beta. \tag{7}$$

Being an extreme point of P, each μ_k that occurs in (7) has a single point $a_k \in T$ for its support; hence

$$F_{\mu_k}(z) = \frac{a_k + z}{a_k - z}. \tag{8}$$

If F is now defined by (3), it follows from (7) and (8) that F satisfies (1) and (2). ////

Kakutani's Fixed Point Theorem

Fixed point theorems play an important role in many parts of analysis and topology. The one we shall now prove will be used to establish the existence of a Haar measure on every compact group. Rather than state it for linear maps, we shall state it in terms of affine maps. These are essentially linear maps followed by a translation (Exercise 17), but in the present context they need not be defined globally. The following definition makes this precise:

If K is a convex set, Y is a vector space, and $T: K \to Y$ satisfies

$$T((1-\lambda)x + \lambda y) = (1-\lambda)Tx + \lambda Ty$$

whenever $x \in K$, $y \in K$, $0 < \lambda < 1$, then T is said to be *affine*.

5.11 Theorem *Suppose that*

(*a*) *K is a nonempty compact convex set in a locally convex space X, and*

(*b*) *G is an equicontinuous group of affine maps taking K into K.*

Then G has a common fixed point in K.

More explicitly, the conclusion is that there exists $p \in K$ such that $Tp = p$ for very $T \in G$.

Part (*b*) of the hypothesis may need some explanation. To say that G is a group means that every $T \in G$ is a one-to-one map of K into K whose inverse T^{-1} also belongs to G (so T maps K *onto* K!) and that $T_1 T_2 \in G$ whenever $T_i \in G$ for $i = 1, 2$. Here $(T_1 T_2)x = T_1(T_2 x)$, of course; note that the composition of two affine maps is affine.

To say that G is equicontinuous (compare with Section 2.3) means now that to every neighborhood W of 0 in X corresponds a neighborhood V of 0 in X such that $Tx - Ty \in W$ whenever $x \in K$, $y \in K$, $x - y \in V$, and $T \in G$.

Hypothesis (*b*) is satisfied, for instance, when G is a group of linear isometries on a normed space X.

PROOF. Let Ω be the collection of all nonempty compact convex sets $H \subset K$ such that $T(H) \subset H$ for every $T \in G$. Partially order Ω by set inclusion. Note that $\Omega \neq \emptyset$, since $K \in \Omega$. By Hausdorff's maximality theorem, Ω contains a maximal totally ordered subcollection Ω_0. The intersection Q of all members of Ω_0 is a minimal member of Ω. The theorem will be proved by showing that Q contains only one point.

Assume, to the contrary, that there exist $x \in Q$, $y \in Q$, $x \neq y$. Then there is a neighborhood W of 0 in X such that $x - y \notin W$. Let V be associated to W as in the preceding definition of equicontinuity. If $Tx - Ty$ were in V, for some $T \in G$, then

$$x - y = T^{-1}(Tx) - T^{-1}(Ty)$$

would be in W, a contradiction. We conclude:

For no $T \in G$ is $Tx - Ty$ in V.

Put $z = \frac{1}{2}(x + y)$. Then $z \in Q$. Define $G(z) = \{Tz: T \in G\}$. This "G-orbit of z" is G-invariant (i.e., every $T \in G$ maps it into itself), hence so is its closure $K_0 = \overline{G(z)}$, and therefore $\overline{co}(K_0)$ is a nonempty G-invariant compact convex subset of Q. The minimality of Q implies that $\overline{co}(K_0) = Q$.

Let p be an extreme point of Q. (It exists, by the Krein-Milman theorem.) Since Q is compact and $Q = \overline{co}(K_0)$, Theorem 3.25 shows *that p lies in the closure K_0 of $G(z)$.*

Define a set

$$E = \{(Tz, Tx, Ty): T \in G\} \subset Q \times Q \times Q.$$

Since $p \in \overline{G(z)}$ and $Q \times Q$ is compact, the lemma that is stated below shows that there exists a point $(x^*, y^*) \in Q \times Q$ so that (p, x^*, y^*) lies in the closure of E. Since $2Tz = Tx + Ty$ for every $T \in G$, it follows that $2p = x^* + y^*$, and this implies that $x^* = y^*$ because p is an extreme point of Q.

But $Tx - Ty \notin V$, for every $T \in G$; hence $x^* - y^* \notin V$; hence $x^* \neq y^*$, and we have our contradiction. ////

Lemma *Suppose that A and B are topological spaces, B is compact, π is the natural projection of $A \times B$ onto A, and $E \subset A \times B$.*

If $p \in A$ lies in the closure of $\pi(E)$, then (p, q) lies in the closure of E for some $q \in B$.

PROOF. If the conclusion fails, then every $q \in B$ has a neighborhood $W_q \subset B$ so that $(V_q \times W_q) \cap E = \emptyset$ for some neighborhood V_q of p in A. The compactness of B implies that $B \subset W_{q_1} \cup \cdots \cup W_{q_n}$ for some finite set $\{q_1, \ldots, q_n\}$. Then $V_{q_1} \cap \cdots \cap V_{q_n}$ is a neighborhood of p which does not intersect $\pi(E)$, contrary to the assumption that p lies in the closure of $\pi(E)$. ////

Haar Measure on Compact Groups

5.12 Definitions A *topological group* is a group G in which a topology is defined that makes the group operations continuous. The most concise way to express this requirement is to postulate the continuity of the mapping $\phi: G \times G \to G$ defined by

$$\phi(x, y) = xy^{-1}.$$

For each $a \in G$, the mappings $x \to ax$ and $x \to xa$ are homeomorphisms of G onto G; so is $x \to x^{-1}$. The topology of G is therefore completely determined by any local base at the identity element e.

If we require (as we shall from now on) that *every point of G is a closed set*, then the analogues of Theorems 1.10 to 1.12 hold (with exactly the same proofs, except for changes in notation); in particular, the Hausdorff separation axiom holds.

If f is any function with domain G, its *left translates* $L_s f$ and its *right translates* $R_s f$ are defined, for every $s \in G$, by

$$(L_s f)(x) = f(sx), \qquad (R_s f)(x) = f(xs) \qquad (x \in G).$$

A complex function f on G is said to be *uniformly continuous* if to every $\varepsilon > 0$ corresponds a neighborhood V of e in G such that

$$|f(t) - f(s)| < \varepsilon$$

whenever $s \in G$, $t \in G$, and $s^{-1}t \in V$.

A topological group G whose topology is compact is called a *compact group*; in this case, $C(G)$ is, as usual, the Banach space of all complex continuous functions on G, with the supremum norm.

5.13 Theorem *Let G be a compact group, suppose $f \in C(G)$, and define $H_L(f)$ to be the convex hull of the set of all left translates of f. Then*

(a) *$s \to L_s f$ is a continuous map from G into $C(G)$, and*

(b) *the closure of $H_L(f)$ is compact in $C(G)$.*

PROOF. Fix $\varepsilon > 0$. Since f is continuous, there corresponds to each $a \in G$ a neighborhood W_a of e such that $|f(x) - f(a)| < \varepsilon$ if $xa^{-1} \in W_a$. The continuity of the group operations gives neighborhoods V_a of e which satisfy $V_a^{-1}V_a \subset W_a$. Since G is compact, there is a finite set $A \subset G$ such that

(1) $$G = \bigcup_{a \in A} V_a \cdot a.$$

Put

(2) $$V = \bigcap_{a \in A} V_a.$$

Choose $x, y \in G$ so that $yx^{-1} \in V$, and choose $a \in A$ so that $ya^{-1} \in V_a$. Then $|f(y) - f(a)| < \varepsilon$, and since $xa^{-1} = (xy^{-1})(ya^{-1}) \in V^{-1}V_a \subset W_a$, we also have $|f(x) - f(a)| < \varepsilon$.

Thus $|f(x) - f(y)| < 2\varepsilon$ whenever $yx^{-1} \in V$.

For any $s \in G$, $(ys)(xs)^{-1} = yx^{-1}$. Hence $yx^{-1} \in V$ implies that $|f(xs) - f(ys)| < 2\varepsilon$. This is just another way of saying that

(3) $$\|L_x f - L_y f\| < 2\varepsilon$$

whenever y lies in the neighborhood Vx of x. This proves (a).

As a consequence of (*a*), $\{L_x f\colon x \in G\}$ is compact in the Banach space $C(G)$. Hence (*b*) follows from part (*c*) of Theorem 3.20. ////

5.14 Theorem *On every compact group G exists a unique regular Borel probability measure m which is left-invariant, in the sense that*

$$\int_G f\,dm = \int_G (L_s f)\,dm \qquad [s \in G, f \in C(G)]. \tag{1}$$

This m is also right-invariant:

$$\int_G f\,dm = \int_G (R_s f)\,dm \qquad [s \in G, f \in C(G)] \tag{2}$$

and it satisfies the relation

$$\int_G f(x)\,dm(x) = \int_G f(x^{-1})\,dm(x) \qquad [f \in C(G)]. \tag{3}$$

This m is called the *Haar measure* of G.

PROOF. The operators L_s satisfy $L_s L_t = L_{ts}$, because

$$(L_s L_t f)(x) = (L_t f)(sx) = f(tsx) = (L_{ts} f)(x).$$

Since each L_s is an isometry of $C(G)$ onto itself, $\{L_s\colon s \in G\}$ is an equicontinuous group of linear operators on $C(G)$. If $f \in C(G)$, let K_f be the closure of $H_L(f)$. By Theorem 5.13, K_f is compact. It is obvious that $L_s(K_f) = K_f$ for every $s \in G$. The fixed point theorem 5.11 now implies that K_f contains a function ϕ such that $L_s \phi = \phi$ for every $s \in G$. In particular, $\phi(s) = \phi(e)$, so that ϕ is constant. By the definition of K_f, this constant can be uniformly approximated by functions in $H_L(f)$.

So far we have proved that to each $f \in C(G)$ corresponds at least one constant c which can be uniformly approximated on G by convex combinations of left translates of f. Likewise, there is a constant c' which bears the same relation to the right translates of f. We claim that $c' = c$.

To prove this, pick $\varepsilon > 0$. There exist finite sets $\{a_i\}$ and $\{b_j\}$ in G, and there exist numbers $\alpha_i > 0$, $\beta_j > 0$, with $\sum \alpha = 1 = \sum \beta_j$, such that

$$\left| c - \sum_i \alpha_i f(a_i x) \right| < \varepsilon \qquad (x \in G) \tag{4}$$

and

$$\left| c' - \sum_j \beta_j f(xb_j) \right| < \varepsilon \qquad (x \in G). \tag{5}$$

Put $x = b_j$ in (4); multiply (4) by β_j, and add with respect to j. The result is

$$\left| c - \sum_{i,j} \alpha_i \beta_j f(a_i b_j) \right| < \varepsilon. \tag{6}$$

Put $x = a_i$ in (5), multiply (5) by α_i, and add with respect to i, to obtain

$$\left| c' - \sum_{i,j} \alpha_i \beta_j f(a_i b_j) \right| < \varepsilon. \tag{7}$$

Now (6) and (7) imply that $c = c'$.

It follows that to each $f \in C(G)$ corresponds a *unique* number, which we shall write Mf, which can be uniformly approximated by convex combinations of left translates of f; the same Mf is also the unique number that can be uniformly approximated by convex combinations of right translates of f. The following properties of M are obvious:

$$Mf \geq 0 \qquad \text{if } f \geq 0. \tag{8}$$

$$M1 = 1. \tag{9}$$

$$M(\alpha f) = \alpha Mf \qquad \text{if } \alpha \text{ is a scalar.} \tag{10}$$

$$M(L_s f) = Mf = M(R_s f) \qquad \text{for every } s \in G. \tag{11}$$

We now prove that

$$M(f + g) = Mf + Mg. \tag{12}$$

Pick $\varepsilon > 0$. Then

$$\left| Mf - \sum_i \alpha_i f(a_i x) \right| < \varepsilon \qquad (x \in G) \tag{13}$$

for some finite set $\{a_i\} \subset G$ and for some numbers $\alpha_i > 0$ with $\sum \alpha_i = 1$. Define

$$h(x) = \sum_i \alpha_i g(a_i x). \tag{14}$$

Then $h \in K_g$, hence $K_h \subset K_g$, and since each of these sets contains a *unique* constant function, we have $Mh = Mg$. Hence there is a finite set $\{b_j\} \subset G$, and there are numbers $\beta_j > 0$ with $\sum \beta_j = 1$, such that

$$\left| Mg - \sum_j \beta_j h(b_j x) \right| < \varepsilon \qquad (x \in G); \tag{15}$$

by (14), this gives

$$(16) \qquad \left| Mg - \sum_{i,j} \alpha_i \beta_j g(a_i b_j x) \right| < \varepsilon \qquad (x \in G).$$

Replace x by $b_j x$ in (13), multiply (13) by β_j, and add with respect to j, to obtain

$$(17) \qquad \left| Mf - \sum_{i,j} \alpha_i \beta_j f(a_i b_j x) \right| < \varepsilon \qquad (x \in G).$$

Thus

$$(18) \qquad \left| Mf + Mg - \sum_{i,j} \alpha_i \beta_j (f + g)(a_i b_j x) \right| < 2\varepsilon \qquad (x \in G).$$

Since $\sum \alpha_i \beta_j = 1$, (18) implies (12).

The Riesz representation theorem, combined with (8), (9), (10), and (12), yields a regular Borel probability measure m that satisfies

$$(19) \qquad Mf = \int_G f \, dm \qquad (f \in C(G));$$

properties (1) and (2) follow now from (11).

To prove uniqueness, let μ be a regular Borel probability measure on G which is left-invariant. Since m is right-invariant, we have, for every $f \in C(G)$,

$$\begin{aligned} \int_G f \, d\mu &= \int_G dm(y) \int_G f(yx) \, d\mu(x) \\ &= \int_G d\mu(x) \int_G f(yx) \, dm(y) = \int_G f \, dm. \end{aligned}$$

Hence $\mu = m$.

The proof of (3) is similar. Put $g(x) = f(x^{-1})$. Then

$$\int_G dm(y) \int_G g(xy^{-1}) \, dm(x) = \int_G dm(x) \int_G f(yx^{-1}) \, dm(y).$$

The two inner integrals are independent of y and x, respectively. Hence $\int g \, dm = \int f \, dm$. ////

Uncomplemented Subspaces

Complemented subspaces of a topological vector space were defined in Section 4.20; Lemma 4.21 furnished some examples. It is also very easy to see that every closed subspace of a Hilbert space is complemented (Theorem 12.4). We will now show that some very familiar closed subspaces of certain other Banach spaces are, in fact, not complemented. These examples will be derived from a rather general theorem about compact groups of operators that have an invariant subspace; its proof uses vector-valued integration with respect to Haar measure.

We begin by looking at some relations that exist between complemented subspaces on the one hand and projections on the other.

5.15 Projections Let X be a vector space. A linear mapping $P: X \to X$ is called a *projection* in X if

$$P^2 = P,$$

i.e., if $P(Px) = Px$ for every $x \in X$.

Suppose P is a projection in X, with null space $\mathscr{N}(P)$ and range $\mathscr{R}(P)$. The following facts are almost obvious:

(*a*) $\mathscr{R}(P) = \mathscr{N}(I - P) = \{x \in X: Px = x\}$.
(*b*) $\mathscr{N}(P) = \mathscr{R}(I - P)$.
(*c*) $\mathscr{R}(P) \cap \mathscr{N}(P) = \{0\}$ and $X = \mathscr{R}(P) + \mathscr{N}(P)$.
(*d*) If A and B are subspaces of X such that $A \cap B = \{0\}$ and $X = A + B$, then there is a unique projection P in X with $A = \mathscr{R}(P)$ and $B = \mathscr{N}(P)$.

Since $(I - P)P = 0$, $\mathscr{R}(P) \subset \mathscr{N}(I - P)$. If $x \in \mathscr{N}(I - P)$, then $x - Px = 0$, and so $x = Px \in \mathscr{R}(P)$. This gives (*a*); (*b*) follows by applying (*a*) to $I - P$. If $x \in \mathscr{R}(P) \cap \mathscr{N}(P)$, then $x = Px = 0$; if $x \in X$, then $x = Px + (x - Px)$, and $x - Px \in \mathscr{N}(P)$. This proves (*c*). If A and B satisfy (*d*), every $x \in X$ has a unique decomposition $x = x' + x''$, with $x' \in A$, $x'' \in B$. Define $Px = x'$. Trivial verifications then prove (*d*).

5.16 Theorem

(*a*) *If P is a continuous projection in a topological vector space X, then*

$$X = \mathscr{R}(P) \oplus \mathscr{N}(P).$$

(*b*) *Conversely, if X is an F-space and if $X = A \oplus B$, then the projection P with range A and null space B is continuous.*

Recall that we use the notation $X = A \oplus B$ only when A and B are *closed* subspaces of X such that $A \cap B = \{0\}$ and $A + B = X$.

PROOF. Statement (*a*) is contained in (*c*) of Section 5.15, except for the assertion that $\mathscr{R}(P)$ is closed. To see the latter, note that $\mathscr{R}(P) = \mathscr{N}(I - P)$ and that $I - P$ is continuous.

Next, suppose P is the projection with range A and null space B, as in (*b*). To prove that P is continuous we verify that P satisfies the hypotheses of the closed graph theorem: Suppose $x_n \to x$ and $Px_n \to y$. Since $Px_n \in A$ and A is closed, we have $y \in A$, hence $y = Py$. Since $x_n - Px_n \in B$ and B is closed, we have $x - y \in B$, hence $Py = Px$. It follows that $y = Px$. Hence P is continuous. ////

Corollary. *A closed subspace of an F-space X is complemented in X if and only if it is the range of some continuous projection in X.*

5.17 Groups of linear operators Suppose that a topological vector space X and a topological group G are related in the following manner: To every $s \in G$ corresponds a continuous linear operator $T_s\colon X \to X$ such that

$$T_e = I, \qquad T_{st} = T_s T_t \qquad (s \in G, t \in G);$$

also, the mapping $(s, x) \to T_s x$ of $G \times X$ into X is continuous.

Under these conditions, G is said to act as a group of continuous linear operators on X.

5.18 Theorem *Suppose*

(*a*) *X is a Fréchet space,*

(*b*) *Y is a complemented subspace of X,*

(*c*) *G is a compact group which acts as a group of continuous linear operators on X, and*

(*d*) *$T_s(Y) \subset Y$ for every $s \in G$.*

Then there is a continuous projection Q of X onto Y which commutes with every T_s.

PROOF. For simplicity, write sx in place of $T_s x$. By (*b*) and Theorem 5.16, there is a continuous projection P of X onto Y. The desired projection Q is to satisfy $s^{-1}Qs = Q$ for all $s \in G$. The idea of the proof is to obtain Q by averaging the operators $s^{-1}Ps$ with respect to the Haar measure m of G: define

$$Qx = \int_G s^{-1}Psx\, dm(s) \qquad (x \in X). \tag{1}$$

To show that this integral exists, in accordance with Definition 3.26, put

$$f_x(s) = s^{-1}Psx \qquad (s \in G). \tag{2}$$

By Theorem 3.27, it suffices to show that $f_x: G \to X$ is continuous. Fix $s_0 \in G$; let U be a neighborhood of $f_x(s_0)$ in X. Put $y = Ps_0 x$, so that

(3) $$s_0^{-1} y = f_x(s_0).$$

Since $(s, z) \to sz$ is assumed to be continuous, s_0 has a neighborhood V_1 and y has a neighborhood W such that

(4) $$s^{-1}(W) \subset U \qquad \text{if } s \in V_1.$$

Also, s_0 has a neighborhood V_2 such that

(5) $$Psx \in W \qquad \text{if } s \in V_2.$$

The continuity of P was used here. If $s \in V_1 \cap V_2$, it follows from (2), (4), and (5) that $f_x(s) \in U$. Thus f_x is continuous.

Since G is compact, each f_x has compact range in X. The Banach-Steinhaus theorem 2.6 implies therefore that $\{s^{-1}Ps: s \in G\}$ is an equicontinuous collection of linear operators on X. To every convex neighborhood U_1 of 0 in X corresponds therefore a neighborhood U_2 of 0 such that $s^{-1}Ps(U_2) \subset U_1$. It now follows from (1) and the convexity of U_1 that $Q(U_2) \subset \bar{U}_1$. (See Theorem 3.27.) Hence Q is continuous. The linearity of Q is obvious.

If $x \in X$, then $Psx \in Y$, hence $s^{-1}Psx \in Y$ by (*d*), for every $s \in G$. Since Y is closed, $Qx \in Y$.

If $x \in Y$, then $sx \in Y$, $Psx = sx$, and so $s^{-1}Psx = x$, for every $s \in G$. Hence $Qx = x$.

These two statements prove that Q is a projection of X onto Y. To complete the proof, we have to show that

(6) $$Qs_0 = s_0 Q \qquad \text{for every } s_0 \in G.$$

Note that $s^{-1}Pss_0 = s_0(ss_0)^{-1}P(ss_0)$. It now follows from (1) and (2) that

$$\begin{aligned} Qs_0 x &= \int_G s^{-1}Pss_0 x \, dm(s) \\ &= \int_G s_0 f_x(ss_0) \, dm(s) \\ &= \int_G s_0 f_x(s) \, dm(s) \\ &= s_0 \int_G f_x(s) \, dm(s) = s_0 Qx. \end{aligned}$$

The third equality is due to the translation-invariance of m; for the fourth (moving s_0 across the integral sign), see Exercise 24 of Chapter 3. ////

5.19 Examples In our first example, we take $X = L^1$, $Y = H^1$. Here L^1 is the space of all integrable functions on the unit circle, and H^1 consists of those $f \in L^1$ that satisfy $\hat{f}(n) = 0$ for all $n < 0$. Recall that $\hat{f}(n)$ denotes the nth Fourier coefficient of f:

$$\hat{f}(n) = \frac{1}{2\pi}\int_{-\pi}^{\pi} f(\theta)e^{-in\theta}\,d\theta \qquad (n = 0, \pm 1, \pm 2, \ldots). \tag{1}$$

Note that we write $f(\theta)$ in place of $f(e^{i\theta})$, for simplicity.

For G we take the unit circle, i.e., the multiplicative group of all complex numbers of absolute value 1, and we associate to each $e^{is} \in G$ the translation operators τ_s defined by

$$(\tau_s f)(\theta) = f(s + \theta). \tag{2}$$

It is a simple matter to verify that G then acts on L^1 as described in Section 5.17 and that

$$(\tau_s f)\hat{}\,(n) = e^{ins}\hat{f}(n). \tag{3}$$

Hence $\tau_s(H^1) = H^1$ for every real s. (See Exercise 12.)

If H^1 were complemented in L^1, Theorem 5.18 would imply that there is a continuous projection Q of L^1 onto H^1 such that

$$\tau_s Q = Q\tau_s \qquad \text{for all } s. \tag{4}$$

Let us see what such a Q would have to be.

Put $e_n(\theta) = e^{in\theta}$. Then $\tau_s e_n = e^{ins}e_n$, and

$$\tau_s Qe_n = Q\tau_s e_n = e^{ins}Qe_n, \tag{5}$$

since Q is linear. The first equality in

$$e^{iks}(Qe_n)\hat{}\,(k) = (\tau_s Qe_n)\hat{}\,(k) = e^{ins}(Qe_n)\hat{}\,(k) \tag{6}$$

follows from (3), the second from (5). Hence $(Qe_n)\hat{}\,(k) = 0$ when $k \neq n$. Since L^1-functions are determined by their Fourier coefficients, it follows that there are constants c_n such that

$$Qe_n = c_n e_n \qquad (n = 0, \pm 1, \pm 2, \ldots). \tag{7}$$

So far we have just used (4). Since $Qe_n \in H^1$ for all n, $c_n = 0$ when $n < 0$. Since $Qf = f$ for every $f \in H^1$, $c_n = 1$ when $n \geq 0$. Thus Q (if it exists at all) is the "natural" projection of L^1 onto H^1, the one that replaces $\hat{f}(n)$ by 0 when $n < 0$. In terms of Fourier series,

$$Q\left(\sum_{-\infty}^{\infty} a_n e^{in\theta}\right) = \sum_{0}^{\infty} a_n e^{in\theta}. \tag{8}$$

To get our contradiction, consider the functions

$$f_r(\theta) = \sum_{-\infty}^{\infty} r^{|n|}e^{in\theta} \qquad (0 < r < 1). \tag{9}$$

These are the well-known Poisson kernels. Explicit summation of the series (9) shows that $f_r \geq 0$. Hence

(10) $$\|f_r\|_1 = \frac{1}{2\pi}\int_{-\pi}^{\pi} |f_r(\theta)|\,d\theta = \frac{1}{2\pi}\int_{-\pi}^{\pi} f_r(\theta)\,d\theta = 1$$

for all r. But

(11) $$(Qf_r)(\theta) = \sum_0^\infty r^n e^{in\theta} = \frac{1}{1 - re^{i\theta}}.$$

Fatou's lemma implies that $\|Qf_r\|_1 \to \infty$ as $r \to 1$, since $\int |1 - e^{i\theta}|^{-1}\,d\theta = \infty$. By (10), this contradicts the continuity of Q.

Hence H^1 is not complemented in L^1.

The same analysis can be applied to A and C, where C is the space of all continuous functions on the unit circle, and A consists of those $f \in C$ that have $\hat{f}(n) = 0$ for all $n < 0$. If A were complemented in C, the operator Q described by (8) would be a continuous projection from C onto A. Application of Q to real-valued $f \in C$ shows that there is a constant $M < \infty$ that satisfies

(12) $$\sup_\theta |f(\theta)| \leq M \cdot \sup_\theta |\mathrm{Re}\, f(\theta)|$$

for every $f \in A$. To see that no such M can exist, consider conformal mappings of the closed unit disc onto tall thin ellipses.

Hence A is not complemented in C.

However, the projection (8) is continuous as an operator in L^p, if $1 < p < \infty$. Hence H^p is then a complemented subspace of L^p. This is a theorem of M. Riesz (Th. 17.26 of [23]).

We conclude with an analogue of (*b*) of Theorem 5.16; it will be used in the proof of Theorem 11.31.

5.20 Theorem *Suppose X is a Banach space, A and B are closed subspaces of X, and $X = A + B$. Then there exists a constant $\gamma < \infty$ such that every $x \in X$ has a representation $x = a + b$, where $a \in A$, $b \in B$, and $\|a\| + \|b\| \leq \gamma\|x\|$.*

This differs from (*b*) of Theorem 5.16 inasmuch as it is not assumed that $A \cap B = \{0\}$.

PROOF. Let Y be the vector space of all ordered pairs (a, b), with $a \in A$, $b \in B$, and componentwise addition and scalar multiplication, normed by

$$\|(a, b)\| = \|a\| + \|b\|.$$

Since A and B are complete, Y is a Banach space. The mapping $\Lambda\colon Y \to X$ defined by

$$\Lambda(a, b) = a + b$$

is continuous, since $\|a + b\| \le \|(a, b)\|$, and maps Y *onto* X. By the open mapping theorem, there exists $\gamma < \infty$ such that each $x \in X$ is $\Lambda(a, b)$ for some (a, b) with $\|(a, b)\| \le \gamma\|x\|$. ////

Sums of Poisson Kernels

Let U and T be the open unit disc and the unit circle in $\mathbb{C}$. Let $L^1 = L^1(T)$ be as in Theorem 5.19, with norm

$$\|f\|_1 = \frac{1}{2\pi}\int_{-\pi}^{\pi} |f(e^{i\theta})|\, d\theta.$$

Associate to each $z \in U$ the Poisson kernel $P_z \in L^1(T)$:

$$P_z(e^{i\theta}) = \frac{1 - |z|^2}{|e^{i\theta} - z|^2}.$$

It is easy to check that $\|P_z\|_1 = 1$ for every $z \in U$.

Call a set $E \subset U$ *nontangentially dense on T* if to every $e^{i\theta} \in T$ and to every $\varepsilon > 0$ there is a point $z \in E$ such that

$$|z - e^{i\theta}| < \min(\varepsilon, 2(1 - |z|)).$$

There are such sets which have no limit point in U. To construct one, let $0 < r_1 < r_2 < \cdots$, $\lim r_n = 1$, and place m_n equally spaced points on the circle $r_n T$, taking $m_n > 2/(1 - r_n)$.

It is a rather surprising fact that every $f \in L^1(T)$ can be represented as the sum of a convergent series of multiples of Poisson kernels. This was proved by F. F. Bonsall as an application of the closed range theorem. Here is his more precise statement:

5.21 Theorem *If $\{z_1, z_2, z_3, \ldots\} \subset U$ is nontangentially dense on T, then to every $f \in L^1(T)$ and every $\varepsilon > 0$ correspond scalars c_n such that $\sum |c_n| \le \|f\|_1 + \varepsilon$ and*

$$f = \sum_1^\infty c_n P_{z_n}.$$

This turns out to be a special case of the following abstract result:

5.22 Theorem *Let $\{x_n\}$ be a sequence in a Banach space X, with $\|x_n\| \le 1$ for all n, and suppose that there is a $\delta > 0$ such that*

$$\sup_n |\langle x_n, x^*\rangle| \ge \delta\|x^*\|$$

for every $x^* \in X^*$. *If* $\varepsilon > 0$, *every* $x \in X$ *can then be represented in the form*

$$x = \sum_{n=1}^{\infty} c_n x_n,$$

with $\delta \sum_1^\infty |c_n| \le \|x\| + \varepsilon$.

PROOF. Define $T: \ell^1 \to X$ by $Tc = \sum_1^\infty c_n x_n$, for $c = \{c_1, c_2, c_3, \ldots\} \in l^1$.

Then, for every $x^* \in X^*$,

$$\langle c, T^*x^* \rangle = \langle Tc, x^* \rangle = \sum_1^\infty c_n \langle x_n, x^* \rangle$$

so that

$$\left| \sum_1^\infty c_n \langle x_n, x^* \rangle \right| \le \|T^*x^*\|$$

if $\|c\|_1 \le 1$. The supremum of the left side, over all such c, is $\sup_n |\langle x_n, x^* \rangle|$, which is $\ge \delta \|x^*\|$ by assumption. Theorem 4.13 asserts therefore that T maps the set of all c with $\sum |c_n| < 1/\delta$ onto a set that contains the open unit ball of X.

This proves Theorem 5.22. Let us apply it with $X = L^1(T)$, $x_n = P_{z_n}$, where $\{z_n\}$ is nontangentially dense on T. Every $g \in L^\infty(T) = L^1(T)^*$ has a harmonic extension

$$G(z) = \frac{1}{2\pi} \int_{-\pi}^{\pi} P_z(e^{i\theta}) g(e^{i\theta}) \, d\theta = \langle P_z, g \rangle.$$

Since $\{z_n\}$ is nontangentially dense on T, Fatou's theorem concerning nontangential limits of bounded harmonic functions implies that

$$\sup_n |\langle P_{z_n}, g \rangle| = \sup_n |G(z_n)| = \|g\|_\infty.$$

Therefore Theorem 5.21 is a consequence of Theorem 5.22, with $\delta = 1$. ////

Two More Fixed Point Theorems

It is a well-known consequence of the axiom of choice that there is no measure on the real line R which is finite on compact sets, not identically zero, translation-invariant, and defined on the σ-algebra of *all* subsets of R. The usual proof that nonmeasurable sets exist shows this. However, if *countable* additivity—a property that, by definition, measures have—is weakened to *finite* additivity, i.e., to the requirement that

$$\mu(E_1 \cup \cdots \cup E_n) = \mu(E_1) + \cdots + \mu(E_n)$$

for all finite unions of pairwise disjoint sets E_i, then there do exist such "finitely additive" measures μ which have all the other above-mentioned properties. Moreover, one can have $0 \le \mu(E) \le 1$ for every $E \subset R$.

Theorem 5.25 will prove this, with any abelian group G in place of R, as an application of an "invariant" version of the Hahn-Banach theorem. The latter will be derived from the surprisingly elementary fixed point theorem 5.23, due to Markov and Kakutani:

5.23 Theorem *If K is a nonempty compact convex set in a topological vector space X and $\mathscr{F}$ is a commuting family of continuous affine maps taking K into K, then there exists a point $p \in K$ such that $Tp = p$ for every $T \in \mathscr{F}$.*

PROOF. For $T \in \mathscr{F}$, put $T^1 = T$, $T^{n+1} = T \circ T^n$, for $n = 1, 2, 3, \ldots$. The fact that the averages

$$T_n = \frac{1}{n}(I + T + T^2 + \cdots + T^{n-1}) \tag{1}$$

are also affine maps of K into K leads to the conclusion that any two of them (with possibly different T's and different n's) commute with each other.

Let $\mathscr{F}^*$ be the semigroup generated by the maps (1). Thus $\mathscr{F}^*$ is the collection of all compositions of finitely many averages (1). If f, $g \in \mathscr{F}^*$, and $h = f \circ g = g \circ f$, then $h \in \mathscr{F}^*$. Since $f(g(K)) \subset f(K)$ and $g(f(K)) \subset g(K)$, we see that

$$f(K) \cap g(K) \supset h(K). \tag{2}$$

Induction shows therefore that the collection $\{f(K): f \in \mathscr{F}^*\}$ has the finite intersection property. Since each $f(K)$ is compact, there is a point $p \in K$ which lies in $f(K)$ for every $f \in \mathscr{F}^*$.

Now fix $T \in \mathscr{F}$ and let V be a neighborhood of 0 in X. For every $n \ge 1$, $p \in T_n(K)$, since $T_n \in \mathscr{F}^*$. This means that there exist points $x_n \in K$ such that

$$p = \frac{1}{n}(x_n + Tx_n + \cdots + T^{n-1}x_n). \tag{3}$$

But then

$$p - Tp = \frac{1}{n}(x_n - T^n x_n) \in \frac{1}{n}(K - K), \tag{4}$$

and $K - K \subset nV$ for all sufficiently large n, because $K - K$ is compact and therefore bounded. Thus $p - Tp \in V$, for every neighborhood V of 0. This forces $p - Tp$ to be 0. ////

5.24 An invariant Hahn-Banach theorem *Suppose that Y is a subspace of a normed linear space X, $f \in Y^*$, $\Gamma \subset \mathscr{B}(X)$, and that*

(*a*) $T(Y) \subset Y$ *and* $ST = TS$ *for all* $S, T \in \Gamma$,
(*b*) $f \circ T = f$, *for every* $T \in \Gamma$.

Then there exists $F \in X^*$ *such that* $F = f$ *on* Y, $\|F\| = \|f\|$, *and* $F \circ T = F$ *for every* $T \in \Gamma$.

Briefly, the given Γ-invariant f has a Γ-invariant Hahn-Banach extension F.

PROOF. Assume that $\|f\| = 1$, without loss of generality. Define

$$K = \{\Lambda \in X^*: \|\Lambda\| \leq 1, \Lambda = f \text{ on } Y\}. \tag{1}$$

It is clear that K is convex. The Hahn-Banach theorem implies that K is not empty. Since K is weak*-closed, the Banach-Alaoglu theorem shows that K is a weak*-compact subset of X^*. For every $T \in \Gamma$, the map

$$\Lambda \to \Lambda \circ T \tag{2}$$

is an affine map of K into K which is weak*-continuous (as we will see in a moment). Theorem 5.23 shows therefore that some $F \in K$ satisfies $F \circ T = F$ for every $T \in \Gamma$.

To finish, we show *that* (2) *is a weak*-continuous map of* X^* *into* X^*, *for every* $T \in \mathscr{B}(X)$. Fix $\Lambda_1 \in X^*$, let

$$V = \{L \in X^*: |Lx_i - (\Lambda_1 T)x_i| < \varepsilon, 1 \leq i \leq n\} \tag{3}$$

be a typical weak*-neighborhood of $\Lambda_1 T$, determined by $x_1, \ldots, x_n \in X$ and $\varepsilon > 0$. Then

$$W = \{\Lambda \in X^*: |\Lambda(Tx_i) - \Lambda_1(Tx_i)| < \varepsilon, 1 \leq i \leq n\} \tag{4}$$

is a weak*-neighborhood of Λ_1, and if $\Lambda \in W$, it is clear that $\Lambda T \in V$.

////

5.25 Theorem *If G is an abelian group (with + as group operation) and $\mathscr{M}$ is the collection of all subsets of G (the "power set" of G), then there is a function $\mu: \mathscr{M} \to [0, 1]$ such that*

(*a*) $\mu(E_1 \cup E_2) = \mu(E_1) + \mu(E_2)$ *if* $E_1 \cap E_2 = \varnothing$,
(*b*) $\mu(E + a) = \mu(E)$ *for all* $E \in \mathscr{M}$, $a \in G$, *and*
(*c*) $\mu(G) = 1$.

PROOF. This is trivial if G is finite. So assume G is infinite, and let $\ell^\infty(G)$ be the Banach space of all bounded complex functions on G, with the supremum norm.

Let Y be the space of all $f \in \ell^\infty(G)$ which have a limit, call it Λf, at ∞. This means that, if $f \in Y$ and $\varepsilon > 0$, then there is a finite set $E \subset G$ so that $|\Lambda f - f(x)| < \varepsilon$ for all x outside E. Note that $\Lambda \in Y^*$ and that $\|\Lambda\| = 1$.

Let Γ be the set of all translation operators τ_a, for $a \in G$, defined by

$$(\tau_a f)(x) = f(x - a). \tag{1}$$

Since G is abelian, any two members of Γ commute; each τ_a is a linear isometry of $\ell^\infty(G)$; and it is clear that $\tau_a(Y) \subset Y$ and that $\Lambda\tau_a = \Lambda$ on Y.

The hypotheses of Theorem 5.24 are thus satisfied, with $X = \ell^\infty(G)$. We conclude that there exists an extension L of Λ, a linear functional of norm 1 on $\ell^\infty(G)$, which satisfies

$$Lf = \Lambda f \qquad \text{for every } f \in Y \tag{2}$$

and

$$L\tau_a f = Lf \qquad \text{for every } f \in l^\infty(G). \tag{3}$$

If we now define $\mu(E) = L\chi_E$ (where χ_E is the characteristic function of $E \subset G$), then (a) holds because $\chi_{E_1} + \chi_{E_2} = \chi_{E_1 \cup E_2}$ if $E_1 \cap E_2 = \varnothing$, and L is linear, and (b) holds because

$$\chi_{E+a}(x) = \chi_E(x - a) = \tau_a \chi_E(x). \tag{4}$$

It remains to be shown that $0 \le \mu(E) \le 1$ for every $E \subset G$. This is done by the following lemma, since Λ (hence also L) preserves constants: If $f(x) = c$ for all $x \in G$, then $f \in Y$ and $\Lambda f = c$. ////

5.26 Lemma *Suppose that X is a normed linear space of bounded functions, with the supremum norm, and that L is a linear functional on X, such that*

$$\|L\| = L(1) = 1.$$

Then $0 \le Lf \le 1$ if $f \in X$ and $0 \le f \le 1$.

PROOF. Put $Lf = \alpha + i\beta$. For every real t,

$$L(f - \tfrac{1}{2} + it) = \alpha - \tfrac{1}{2} + i(\beta + t).$$

Since $\|f - \frac{1}{2}\| \le \frac{1}{2}$, it follows that

$$(\alpha - \tfrac{1}{2})^2 + (\beta + t)^2 \le \|f - \tfrac{1}{2} + it\|^2 \le \tfrac{1}{4} + t^2,$$

so that $\alpha^2 - \alpha + \beta^2 + 2\beta t \leq 0$ for every real t. This forces $\beta = 0$, hence $\alpha^2 \leq \alpha$, hence $0 \leq \alpha \leq 1$. ////

5.27 Example Commutativity cannot be dropped from the hypotheses of the preceding three theorems. To see this, let G be the free group on two generators, a and b. Except for the identity element, G is the union of four disjoint sets, say I, II, III, IV, consisting of those reduced words that start with a, a^{-1}, b, b^{-1}, respectively. If μ is a finitely additive measure on the power set of G, with $0 \leq \mu \leq 1$ and $\mu(aE) = \mu(E) = \mu(bE)$ for all $E \subset G$, then we see that $\mu(\text{I} \cup \text{III} \cup \text{IV}) = \mu(\text{I})$ and $\mu(\text{I} \cup \text{II} \cup \text{III}) = \mu(\text{III})$. The first of these shows that $\mu(\text{III}) = \mu(\text{IV}) = 0$, the second that $\mu(\text{I}) = \mu(\text{II}) = 0$. Since singletons must have measure 0, $\mu \equiv 0$. Thus Theorem 5.25 fails for this group.

We conclude this chapter with the Schauder-Tychonoff fixed point theorem. This is an infinite-dimensional version of Brouwer's theorem concerning the fixed point property of closed balls in R^n. It is nonlinear, and its proof is therefore not really an application of any of the preceding material, except that it will involve a Minkowski functional.

5.28 Theorem *If K is a nonempty compact convex set in a locally convex space X, and $f: K \to K$ is continuous, then $f(p) = p$ for some $p \in K$.*

PROOF. Assume f fixes no point of K. Its graph

$$G = \{(x, f(x)) \in X \times X : x \in K\} \tag{1}$$

is then disjoint from the diagonal Δ of $X \times X$ and is compact. Hence there is a convex balanced neighborhood V of 0 in X such that $G + (V \times V)$ misses Δ. In particular,

$$f(x) \notin x + V \qquad (x \in K). \tag{2}$$

Let μ be the Minkowski functional of V. Theorem 1.36 shows that μ is continuous on X and that $\mu(x) < 1$ if and only if $x \in V$. Define

$$\alpha(x) = \max\{0, 1 - \mu(x)\} \qquad (x \in X). \tag{3}$$

Choose $x_1, \ldots, x_n \in K$ so that the sets $x_i + V$ $(1 \leq i \leq n)$ cover K, put $\alpha_i(x) = \alpha(x - x_i)$, and define

$$\beta_i(x) = \frac{\alpha_i(x)}{\alpha_1(x) + \cdots + \alpha_n(x)} \qquad (x \in K, 1 \leq i \leq n), \tag{4}$$

noting that the denominator in (4) is positive for every $x \in K$.

Let $H = co\{x_1, \ldots, x_n\}$. Then g, defined by

$$g(x) = \sum_1^n \beta_i(x)x_i \qquad (x \in K), \tag{5}$$

is a continuous map from K into the compact finite-dimensional simplex $H \subset K$. The same is true of $g \circ f$. Brouwer's fixed point theorem asserts therefore that there is an $x^* \in H$ such that

$$g(f(x^*)) = x^*. \tag{6}$$

Since $\beta_i(x) = 0$ outside $x_i + V$, we see that

$$x - g(x) = \sum_1^n \beta_i(x)(x - x_i) \qquad (x \in K) \tag{7}$$

is a convex combination of vectors $x - x_i \in V$. Thus $x - g(x) \in V$, for every $x \in K$. In particular, this is true for $x = f(x^*)$. We conclude that

$$f(x^*) \in g(f(x^*)) + V = x^* + V, \tag{8}$$

contrary to (2). ////

Exercises

1. Define measures μ_1, μ_2 on the unit circle by
$$d\mu_1 = \cos\theta\, d\theta, \qquad d\mu_2 = \sin\theta\, d\theta$$
and find the range of the measure $\mu = (\mu_1, \mu_2)$.
2. Construct two functions f and g on $[0, 1]$ with the following property: If
$$d\mu_1 = f(x)\,dx, \qquad d\mu_2 = g(x)\,dx, \qquad \mu = (\mu_1, \mu_2),$$
then the range of μ is the square with vertices at $(1, 0)$, $(0, 1)$, $(-1, 0)$, $(0, -1)$.
3. Suppose that the hypotheses of Theorem 5.9 are satisfied, that $\phi \in C(S)$, $\phi > 0$, $g \in C(K)$, and $|g| < \phi|_K$. Prove that there exists $f \in Y$ such that $f|_K = g$ and $|f| < \phi$ on S. *Hint:* Apply Theorem 5.9 to the space of all functions f/ϕ, with $f \in Y$.
4. Supply the details of the proof that every extreme point of P has its support at a single point. (This refers to the end of proof of Theorem 5.10.)
5. Prove the analogues of Theorems 1.10 to 1.12 that are alluded to in Section 5.12. (Do not assume that G is commutative.)
6. Suppose G is a topological group and H is the largest connected subset of G that contains the identity element e. Prove that H is a normal subgroup of G, that is, a subgroup that satisfies $x^{-1}Hx = H$ for every $x \in G$. *Hint:* If A and B are connected subsets of G, so are AB and A^{-1}.
7. Prove that every open subgroup of a topological group is closed. (The converse is obviously false.)
8. Suppose m is the Haar measure of a compact group G, and V is a nonempty open set in G. Prove that $m(V) > 0$.

9. Put $e_n(\theta) = e^{in\theta}$. Let L^2 refer to the Haar measure of the unit circle. Let A be the smallest closed subspace of L^2 that contains e_n for $n = 0, 1, 2, \ldots$, let B be the smallest closed subspace of L^2 that contains $e_{-n} + ne_n$ for $n = 1, 2, 3, \ldots$. Prove the following:

(*a*) $A \cap B = \{0\}$.

(*b*) If $X = A + B$ then X is dense in L^2, but $X \neq L^2$.

(*c*) Although $X = A \oplus B$, the projection in X with range A and null space B is not continuous. (The topology of X is, of course, the one that X inherits from L^2. Compare with Theorem 5.16.)

10. Suppose X is a Banach space, $P \in \mathscr{B}(X^*)$, $Q \in \mathscr{B}(X)$, and P and Q are projections.

(*a*) Show that the adjoint P^* of P is a projection in X^*.

(*b*) Show that $\|P - Q\| \geq 1$ if $PQ = QP$ and $P \neq Q$.

11. Suppose P and Q are projections in a vector space X.

(*a*) Prove that $P + Q$ is a projection if and only if $PQ = QP = 0$. In that case,

$$\mathscr{N}(P + Q) = \mathscr{N}(P) \cap \mathscr{N}(Q),$$
$$\mathscr{R}(P + Q) = \mathscr{R}(P) + \mathscr{R}(Q),$$
$$\mathscr{R}(P) \cap \mathscr{R}(Q) = \{0\}.$$

(*b*) If $PQ = QP$, prove that PQ is a projection and that

$$\mathscr{N}(PQ) = \mathscr{N}(P) + \mathscr{N}(Q),$$
$$\mathscr{R}(PQ) = \mathscr{R}(P) \cap \mathscr{R}(Q).$$

(*c*) What do the matrices

$$\begin{pmatrix} 1 & 0 \\ 0 & 0 \end{pmatrix} \quad \text{and} \quad \begin{pmatrix} 1 & -1 \\ 0 & 0 \end{pmatrix}$$

show about part (*b*)?

12. Prove that the translation operators τ_s used in Example 5.19 satisfy the continuity property described in Section 5.17. Explicitly, prove that

$$\|\tau_r g - \tau_s f\|_1 \to 0$$

if $r \to s$ and $g \to f$ in L^1.

13. Use the following example to show that the compactness of G cannot be omitted from the hypotheses of Theorem 5.18. Take $X = L^1$ on the real line R, relative to Lebesgue measure; $f \in Y$ if and only if $\int_R f = 0$; $G = R$ with the usual topology; G acts on L^1 by translation: $(\tau_s f)(x) = f(s + x)$. The joint continuity property is satisfied (see Exercise 12), $\tau_s Y = Y$ for every s, and Y is complemented in X. Yet there is no projection of X onto Y (continuous or not) that commutes with every τ_s.

14. Suppose S and T are continuous linear operators in a topological vector space, and

$$T = TST.$$

Prove that T has closed range. (See Theorem 5.16 for the case $S = I$.)

15. Suppose A is a closed *subspace* of $C(S)$, where S is a compact Hausdorff space; suppose μ is an extreme point of the unit ball of $A^\perp$; and suppose $f \in C(S)$ is a *real* function such that

$$\int_S gf\, d\mu = 0$$

for every $g \in A$. Prove that f is then constant on the support of μ. (Compare with Theorem 5.7.) Show, by an example, that the conclusion is false if the word "real" is omitted from the hypotheses.

16. Suppose X is a vector space, $E \subset X$, $T\colon co(E) \to X$ is affine, and $T(E) \subset E$. Prove that $T(co(E)) \subset co(E)$. (This was tacitly used in the proof of Theorem 5.11.)

17. If X and Y are vector spaces and $T\colon X \to Y$ is affine, prove that $T - T(0)$ is linear.

18. Suppose K is a compact set in a Fréchet space X and $f\colon X \to K$ is continuous. Prove that f fixes some point of K.

Do the same if Ω is a convex open set in X, $\Omega \supset K$, and $f\colon \Omega \to K$ is continuous.

19. Prove the existence of a continuous function f on $I = [0, 1]$ which satisfies the equation

$$f(x) = \int_0^1 \sin\,(x + f^2(t))\, dt$$

for all $x \in I$. *Hint:* Denoting the right side by $(Tf)(x)$, show that the set $\{Tf\colon f \in C(I)\}$ is uniformly bounded and equicontinuous and that its closure is therefore compact in $C(I)$. Apply Schauder's fixed point theorem (via Exercise 18).

PART II

DISTRIBUTIONS AND FOURIER TRANSFORMS

CHAPTER 6

TEST FUNCTIONS AND DISTRIBUTIONS

Introduction

6.1 The theory of distributions frees differential calculus from certain difficulties that arise because nondifferentiable functions exist. This is done by extending it to a class of objects (called *distributions* or *generalized functions*) which is much larger than the class of differentiable functions to which calculus applies in its original form.

Here are some features that any such extension ought to have in order to be useful; our setting is some open subset of R^n:

(*a*) Every continuous function should be a distribution.
(*b*) Every distribution should have partial derivatives which are again distributions. For differentiable functions, the new motion of derivative should coincide with the olc one. (Every distribution should therefore be infinitely differentiable.)
(*c*) The usual formal rules of calculus should hold.

(*d*) There should be a supply of convergence theorems that is adequate for handling the usual limit processes.

To motivate the definitions to come, let us temporarily restrict our attention to the case $n = 1$. The integrals that follow are taken with respect to Lebesgue measure, and they extend over the whole line R, unless the contrary is indicated.

A complex function f is said to be *locally integrable* if f is measurable and $\int_K |f| < \infty$ for every compact $K \subset R$. The idea is to reinterpret f as being something that assigns the number $\int f\phi$ to every suitably chosen "test function" ϕ, rather than as being something that assigns the number $f(x)$ to each $x \in R$. (This point of view is particularly appropriate for functions that arise in physics, since measured quantities are almost always averages. In fact, distributions were used by physicists long before their mathematical theory was constructed.) Of course, a well-chosen class of test functions must be specified.

We let $\mathscr{D} = \mathscr{D}(R)$ be the vector space of all $\phi \in C^\infty(R)$ whose support is compact. Then $\int f\phi$ exists for every locally integrable f and for every $\phi \in \mathscr{D}$. Moreover, $\mathscr{D}$ is sufficiently large to assure that f is determined (a.e.) by the integrals $\int f\phi$. (To see this, note that the uniform closure of $\mathscr{D}$ contains every continuous function with compact support.) If f happens to be continuously differentiable, then

$$\int f'\phi = -\int f\phi' \qquad (\phi \in \mathscr{D}). \tag{1}$$

If $f \in C^\infty(R)$, then

$$\int f^{(k)}\phi = (-1)^k \int f\phi^{(k)} \qquad (\phi \in \mathscr{D}, k = 1, 2, 3, \ldots). \tag{2}$$

The compactness of the support of ϕ was used in these integrations by parts.

Observe that the integrals on the right sides of (1) *and* (2) *make sense whether f is differentiable or not and that they define linear functionals on* $\mathscr{D}$.

We can therefore assign a "kth derivative" to every f that is locally integrable: $f^{(k)}$ is the linear functional on $\mathscr{D}$ that sends ϕ to $(-1)^k \int f\phi^{(k)}$. Note that f itself corresponds to the functional $\phi \to \int f\phi$.

The *distributions* will be those linear functionals on $\mathscr{D}$ that are continuous with respect to a certain topology. (See Definition 6.7.) The preceding discussion suggests that we associate to each distribution Λ its "derivative" Λ' by the formula

$$\Lambda'(\phi) = -\Lambda(\phi') \qquad (\phi \in \mathscr{D}). \tag{3}$$

It turns out that this definition (when extended to n variables) has all the desirable properties that were listed earlier. One of the most important features of the resulting theory is that it makes it possible to apply Fourier transform techniques to many problems in partial differential equations where this cannot be done by more classical methods.

Test Function Spaces

6.2 The space $\mathscr{D}(\Omega)$ Consider a nonempty open set $\Omega \subset R^n$. For each compact $K \subset \Omega$, the Fréchet space $\mathscr{D}_K$ was described in Section 1.46. The union of the spaces $\mathscr{D}_K$, as K ranges over all compact subsets of Ω, is the *test function space* $\mathscr{D}(\Omega)$. It is clear that $\mathscr{D}(\Omega)$ is a vector space, with respect to the usual definitions of addition and scalar multiplication of complex functions. Explicitly, $\phi \in \mathscr{D}(\Omega)$ if and only if $\phi \in C^\infty(\Omega)$ and the support of ϕ is a compact subset of Ω.

Let us introduce the norms

(1) $$\|\phi\|_N = \max \{|D^\alpha \phi(x)|: x \in \Omega, |\alpha| \le N\},$$

for $\phi \in \mathscr{D}(\Omega)$ and $N = 0, 1, 2, \ldots$; see Section 1.46 for the notations D^α and $|\alpha|$.

The restrictions of these norms to any fixed $\mathscr{D}_K \subset \mathscr{D}(\Omega)$ induce the same topology on $\mathscr{D}_K$ as do the seminorms p_N of Section 1.46. To see this, note that to each K corresponds an integer N_0 such that $K \subset K_N$ for all $N \ge N_0$. For these N, $\|\phi\|_N = p_N(\phi)$ if $\phi \in \mathscr{D}_K$. Since

(2) $$\|\phi\|_N \le \|\phi\|_{N+1} \qquad \text{and} \qquad p_N(\phi) \le p_{N+1}(\phi),$$

the topologies induced by either sequence of seminorms are unchanged if we let N start at N_0 rather than at 1. These two topologies of $\mathscr{D}_K$ coincide therefore; a local base is formed by the sets

(3) $$V_N = \left\{\phi \in \mathscr{D}_K: \|\phi\|_N < \frac{1}{N}\right\} \qquad (N = 1, 2, 3, \ldots).$$

The same norms (1) can be used to define a locally convex metrizable topology on $\mathscr{D}(\Omega)$; see Theorem 1.37 and (*b*) of Section 1.38. However, this topology has the disadvantage of not being complete. For example, take $n = 1$, $\Omega = R$, pick $\phi \in \mathscr{D}(R)$ with support in $[0, 1]$, $\phi > 0$ in $(0, 1)$, and define

$$\psi_m(x) = \phi(x-1) + \frac{1}{2}\phi(x-2) + \cdots + \frac{1}{m}\phi(x-m).$$

Then $\{\psi_m\}$ is a Cauchy sequence in the suggested topology of $\mathscr{D}(R)$, but $\lim \psi_m$ does not have compact support, hence is not in $\mathscr{D}(R)$.

We shall now define another locally convex topology τ on $\mathscr{D}(\Omega)$ in which Cauchy sequences do converge. The fact that this τ is not metrizable is only a minor inconvenience, as we shall see.

6.3 Definitions Let Ω be a nonempty open set in R^n.

(*a*) For every compact $K \subset \Omega$, τ_K denotes the Fréchet space topology of $\mathscr{D}_K$, as described in Sections 1.46 and 6.2.

(*b*) β is the collection of all convex balanced sets $W \subset \mathscr{D}(\Omega)$ such that $\mathscr{D}_K \cap W \in \tau_K$ for every compact $K \subset \Omega$.

(*c*) τ is the collection of all unions of sets of the form $\phi + W$, with $\phi \in \mathscr{D}(\Omega)$ and $W \in \beta$.

Throughout this chapter, K will aways denote a compact subset of Ω.

The following two theorems establish the basic properties of the topology τ, which is quite different from the one discussed in Section 6.2. For example, if $\{x_m\}$ is a sequence in Ω, without limit point in Ω, and if $\{c_m\}$ is a sequence of positive numbers, then the set

$$\{\varphi \in \mathscr{D}(\Omega) : |\varphi(x_m)| < c_m \text{ for } m = 1, 2, 3, \ldots\}$$

belongs to β, i.e., is a τ-neighborhood of 0 in $\mathscr{D}(\Omega)$. It is this fact (see Theorem 6.5) which forces τ-bounded sets (and hence τ-Cauchy sequences) to be concentrated on a common compact set $K \subset \Omega$, and therefore τ-Cauchy sequences converge.

6.4 Theorem

(*a*) *τ is a topology in $\mathscr{D}(\Omega)$, and β is a local base for τ.*

(*b*) *τ makes $\mathscr{D}(\Omega)$ into a locally convex topological vector space.*

PROOF. Suppose $V_1 \in \tau$, $V_2 \in \tau$, $\phi \in V_1 \cap V_2$. To prove (*a*), it is clearly enough to show that

$$\phi + W \subset V_1 \cap V_2 \tag{1}$$

for some $W \in \beta$.

The definition of τ shows that there exist $\phi_i \in \mathscr{D}(\Omega)$ and $W_i \in \beta$ such that

$$\phi \in \phi_i + W_i \subset V_i \qquad (i = 1, 2). \tag{2}$$

Choose K so that $\mathscr{D}_K$ contains ϕ_1, ϕ_2, and ϕ. Since $\mathscr{D}_K \cap W_i$ is open in $\mathscr{D}_K$, we have

$$\phi - \phi_i \in (1 - \delta_i)W_i \tag{3}$$

for some $\delta_i > 0$. The convexity of W_i implies therefore that

$$\phi - \phi_i + \delta_i W_i \subset (1 - \delta_i)W_i + \delta_i W_i = W_i, \tag{4}$$

so that

(5) $$\phi + \delta_i W_i \subset \phi_i + W_i \subset V_i \qquad (i = 1, 2).$$

Hence (1) holds with $W = (\delta_1 W_1) \cap (\delta_2 W_2)$, and (*a*) is proved.

Suppose next that ϕ_1 and ϕ_2 are distinct elements of $\mathscr{D}(\Omega)$, and put

(6) $$W = \{\phi \in \mathscr{D}(\Omega): \|\phi\|_0 < \|\phi_1 - \phi_2\|_0\},$$

where $\|\phi\|_0$ is as in (1) in Section 6.2. Then $W \in \beta$ and ϕ_1 is not in $\phi_2 + W$. It follows that the singleton $\{\phi_1\}$ is a closed set, relative to τ.

Addition is τ-continuous, since the convexity of every $W \in \beta$ implies that

(7) $$(\psi_1 + \tfrac{1}{2}W) + (\psi_2 + \tfrac{1}{2}W) = (\psi_1 + \psi_2) + W$$

for any $\psi_1 \in \mathscr{D}(\Omega)$, $\psi_2 \in \mathscr{D}(\Omega)$.

To deal with scalar multiplication, pick a scalar α_0 and a $\phi_0 \in \mathscr{D}(\Omega)$. Then

(8) $$\alpha\phi - \alpha_0\phi_0 = \alpha(\phi - \phi_0) + (\alpha - \alpha_0)\phi_0.$$

If $W \in \beta$, there exists $\delta > 0$ such that $\delta\phi_0 \in \tfrac{1}{2}W$. Choose c so that $2c(|\alpha_0| + \delta) = 1$. Since W is convex and balanced, it follows that

(9) $$\alpha\phi - \alpha_0\phi_0 \in W$$

whenever $|\alpha - \alpha_0| < \delta$ and $\phi - \phi_0 \in cW$.

This completes the proof. ////

Note: From now on, the symbol $\mathscr{D}(\Omega)$ will denote the topological vector space $(\mathscr{D}(\Omega), \tau)$ that has just been described. All topological concepts related to $\mathscr{D}(\Omega)$ will refer to this topology τ.

6.5 Theorem

(*a*) *A convex balanced subset V of $\mathscr{D}(\Omega)$ is open if and only if $V \in \beta$.*

(*b*) *The topology τ_K of any $\mathscr{D}_K \subset \mathscr{D}(\Omega)$ coincides with the subspace topology that $\mathscr{D}_K$ inherits from $\mathscr{D}(\Omega)$.*

(*c*) *If E is a bounded subset of $\mathscr{D}(\Omega)$, then $E \subset \mathscr{D}_K$ for some $K \subset \Omega$, and there are numbers $M_N < \infty$ such that every $\phi \in E$ satisfies the inequalities*

$$\|\phi\|_N \le M_N \qquad (N = 0, 1, 2, \ldots).$$

(*d*) *$\mathscr{D}(\Omega)$ has the Heine-Borel property.*

(e) *If* $\{\phi_i\}$ *is a Cauchy sequence in* $\mathscr{D}(\Omega)$, *then* $\{\phi_i\} \subset \mathscr{D}_K$ *for some compact* $K \subset \Omega$, *and*

$$\lim_{i, j \to \infty} \|\phi_i - \phi_j\|_N = 0 \qquad (N = 0, 1, 2, \ldots).$$

(f) *If* $\phi_i \to 0$ *in the topology of* $\mathscr{D}(\Omega)$, *then there is a compact* $K \subset \Omega$ *which contains the support of every* ϕ_i, *and* $D^\alpha \phi_i \to 0$ *uniformly, as* $i \to \infty$, *for every multi-index* α.

(g) *In* $\mathscr{D}(\Omega)$, *every Cauchy sequence converges.*

Remark. In view of (*b*), the necessary conditions expressed by (*c*), (*e*), and (*f*) are also sufficient. For example, if $E \subset \mathscr{D}_K$ and $\|\phi\|_N \le M_N < \infty$ for every $\phi \in E$, then E is a bounded subset of $\mathscr{D}_K$ (Section 1.46), and now (*b*) implies that E is also bounded in $\mathscr{D}(\Omega)$.

PROOF. Suppose first that $V \in \tau$. Pick $\phi \in \mathscr{D}_K \cap V$. By Theorem 6.4, $\phi + W \subset V$ for some $W \in \beta$. Hence

$$\phi + (\mathscr{D}_K \cap W) \subset \mathscr{D}_K \cap V.$$

Since $\mathscr{D}_K \cap W$ is open in $\mathscr{D}_K$, we have proved that

$$\mathscr{D}_K \cap V \in \tau_K \quad \text{if } V \in \tau \text{ and } K \subset \Omega. \tag{1}$$

Statement (*a*) is an immediate consequence of (1), since it is obvious that $\beta \subset \tau$.

One half of (*b*) is proved by (1). For the other half, suppose $E \in \tau_K$. We have to show that $E = \mathscr{D}_K \cap V$ for some $V \in \tau$. The definition of τ_K implies that to every $\phi \in E$ correspond N and $\delta > 0$ such that

$$\{\psi \in \mathscr{D}_K : \|\psi - \phi\|_N < \delta\} \subset E. \tag{2}$$

Put $W_\phi = \{\psi \in \mathscr{D}(\Omega) : \|\psi\|_N < \delta\}$. Then $W_\phi \in \beta$, and

$$\mathscr{D}_K \cap (\phi + W_\phi) = \phi + (\mathscr{D}_K \cap W_\phi) \subset E. \tag{3}$$

If V is the union of these sets $\phi + W_\phi$, one for each $\phi \in E$, then V has the desired property.

For (*c*), consider a set $E \subset \mathscr{D}(\Omega)$ which lies in no $\mathscr{D}_K$. Then there are functions $\phi_m \in E$ and there are distinct points $x_m \in \Omega$, without limit point in Ω, such that $\phi_m(x_m) \ne 0$ $(m = 1, 2, 3, \ldots)$. Let W be the set of all $\phi \in \mathscr{D}(\Omega)$ that satisfy

$$|\phi(x_m)| < m^{-1} |\phi_m(x_m)| \qquad (m = 1, 2, 3, \ldots). \tag{4}$$

Since each K contains only finitely many x_m, it is easy to see that $\mathscr{D}_K \cap W \in \tau_K$. Thus $W \in \beta$. Since $\phi_m \notin mW$, no multiple of W contains E. This shows that E is not bounded.

It follows that every bounded subset E of $\mathscr{D}(\Omega)$ lies in some $\mathscr{D}_K$. By (*b*), E is then a bounded subset of $\mathscr{D}_K$. Consequently (see Section 1.46)

$$\sup \{\|\phi\|_N \colon \phi \in E\} < \infty \qquad (N = 0, 1, 2, \ldots). \tag{5}$$

This completes the proof of (*c*).

Statement (*d*) follows from (*c*), since $\mathscr{D}_K$ has the Heine-Borel property.

Since Cauchy sequences are bounded (Section 1.29), (*c*) implies that every Cauchy sequence $\{\phi_i\}$ in $\mathscr{D}(\Omega)$ lies in some $\mathscr{D}_K$. By (*b*), $\{\phi_i\}$ is then also a Cauchy sequence relative to τ_K. This proves (*e*).

Statement (*f*) is just a restatement of (*e*).

Finally, (*g*) follows from (*b*), (*e*), and the completeness of $\mathscr{D}_K$. (Recall that $\mathscr{D}_K$ is a Fréchet space.) ////

6.6 Theorem *Suppose Λ is a linear mapping of $\mathscr{D}(\Omega)$ into a locally convex space Y. Then each of the following four properties implies the others:*

(*a*) *Λ is continuous.*
(*b*) *Λ is bounded.*
(*c*) *If $\phi_i \to 0$ in $\mathscr{D}(\Omega)$ then $\Lambda\phi_i \to 0$ in Y.*
(*d*) *The restrictions of Λ to every $\mathscr{D}_K \subset \mathscr{D}(\Omega)$ are continuous.*

PROOF. The implication (*a*) → (*b*) is contained in Theorem 1.32.

Assume Λ is bounded and $\phi_i \to 0$ in $\mathscr{D}(\Omega)$. By Theorem 6.5, $\phi_i \to 0$ in some $\mathscr{D}_K$, and the restriction of Λ to this $\mathscr{D}_K$ is bounded. Theorem 1.32, applied to $\Lambda\colon \mathscr{D}_K \to Y$, shows that $\Lambda\phi_i \to 0$ in Y. Thus (*b*) implies (*c*).

Assume (*c*) holds, $\{\phi_i\} \subset \mathscr{D}_K$, and $\phi_i \to 0$ in $\mathscr{D}_K$. By (*b*) of Theorem 6.5, $\phi_i \to 0$ in $\mathscr{D}(\Omega)$. Hence (*c*) implies that $\Lambda\phi_i \to 0$ in Y, as $i \to \infty$. Since $\mathscr{D}_K$ is metrizable, (*d*) follows.

To prove that (*d*) implies (*a*), let U be a convex balanced neighborhood of 0 in Y, and put $V = \Lambda^{-1}(U)$. Then V is convex and balanced. By (*a*) of Theorem 6.5, V is open in $\mathscr{D}(\Omega)$ if and only if $\mathscr{D}_K \cap V$ is open in $\mathscr{D}_K$, for every $\mathscr{D}_K \subset \mathscr{D}(\Omega)$. This proves the equivalence of (*a*) and (*d*). ////

Corollary. *Every differential operator D^α is a continuous mapping of $\mathscr{D}(\Omega)$ into $\mathscr{D}(\Omega)$.*

PROOF. Since $\|D^\alpha\phi\|_N \le \|\phi\|_{N+|\alpha|}$ for $N = 0, 1, 2, \ldots$, D^α is continuous on each $\mathscr{D}_K$. ////

6.7 Definition A linear functional on $\mathscr{D}(\Omega)$ which is continuous (with respect to the topology τ described in Definition 6.3) is called a *distribution in* Ω.

The space of all distributions in Ω is denoted by $\mathscr{D}'(\Omega)$.

Note that Theorem 6.6 applies to linear functionals on $\mathscr{D}(\Omega)$. It leads to the following useful characterization of distributions.

6.8 Theorem *If Λ is a linear functional on $\mathscr{D}(\Omega)$, the following two conditions are equivalent:*

(*a*) $\Lambda \in \mathscr{D}'(\Omega)$.

(*b*) *To every compact $K \subset \Omega$ corresponds a nonnegative integer N and a constant $C < \infty$ such that the inequality*

$$|\Lambda\phi| \le C\|\phi\|_N$$

holds for every $\phi \in \mathscr{D}_K$.

PROOF. This is precisely the equivalence of (*a*) and (*d*) in Theorem 6.6, combined with the description of the topology of $\mathscr{D}_K$ by means of the seminorms $\|\phi\|_N$ given in Section 6.2. ////

Note: If Λ is such that one N will do for all K (but not necessarily with the same C), then the smallest such N is called the *order* of Λ. If no N will do for all K, then Λ is said to have *infinite order.*

6.9 Remark Each $x \in \Omega$ determines a linear functional δ_x on $\mathscr{D}(\Omega)$, by the formula

$$\delta_x(\phi) = \phi(x).$$

Theorem 6.8 shows that δ_x is a distribution, of order 0.

If $x = 0$, the origin of R^n, the functional $\delta = \delta_0$ is frequently called the *Dirac measure* on R^n.

Since $\mathscr{D}_K$, for $K \subset \Omega$, is the intersection of the null spaces of these δ_x, as x ranges over the complement of K, it follows that each $\mathscr{D}_K$ is a *closed* subspace of $\mathscr{D}(\Omega)$. [This follows also from Theorem 1.27 and part (*b*) of Theorem 6.5, since each $\mathscr{D}_K$ is complete.] It is obvious that each $\mathscr{D}_K$ has empty interior, relative to $\mathscr{D}(\Omega)$. Since there is a countable collection of sets $K_i \subset \Omega$ such that $\mathscr{D}(\Omega) = \bigcup \mathscr{D}_{K_i}$, $\mathscr{D}(\Omega)$ is of the first category in itself. Since Cauchy sequences converge in $\mathscr{D}(\Omega)$ (Theorem 6.5), Baire's theorem implies that $\mathscr{D}(\Omega)$ is not metrizable.

Calculus with Distributions

6.10 Notations As before, Ω will denote a nonempty open set in R^n. If $\alpha = (\alpha_1, \ldots, \alpha_n)$ and $\beta = (\beta_1, \ldots, \beta_n)$ are multi-indices (see Section 1.46) then

(1) $$|\alpha| = \alpha_1 + \cdots + \alpha_n,$$

(2) $$D^\alpha = D_1^{\alpha_1} \cdots D_n^{\alpha_n}, \qquad \text{where } D_j = \frac{\partial}{\partial x_j},$$

(3) $$\beta \le \alpha \text{ means } \beta_i \le \alpha_i \text{ for } 1 \le i \le n,$$

(4) $$\alpha \pm \beta = (\alpha_1 \pm \beta_1, \ldots, \alpha_n \pm \beta_n).$$

If $x \in R^n$ and $y \in R^n$, then

(5) $$x \cdot y = x_1 y_1 + \cdots + x_n y_n,$$

(6) $$|x| = (x \cdot x)^{1/2} = (x_1^2 + \cdots + x_n^2)^{1/2}.$$

The fact that the absolute value sign has different meanings in (1) and in (6) should cause no confusion.

If $x \in R^n$ and α is a multi-index, the *monomial* x^α is defined by

(7) $$x^\alpha = x_1^{\alpha_1} \cdots x_n^{\alpha_n}.$$

6.11 Functions and measures as distributions Suppose f is a locally integrable complex function in Ω. This means that f is Lebesgue measurable and $\int_K |f(x)|\, dx < \infty$ for every compact $K \subset \Omega$; dx denotes Lebesgue measure. Define

(1) $$\Lambda_f(\phi) = \int_\Omega \phi(x) f(x)\, dx \qquad [\phi \in \mathscr{D}(\Omega)].$$

Since

(2) $$|\Lambda_f(\phi)| \le \left(\int_K |f| \right) \cdot \|\phi\|_0 \qquad (\phi \in \mathscr{D}_K),$$

Theorem 6.8 shows that $\Lambda_f \in \mathscr{D}'(\Omega)$.

It is customary to identify the distribution Λ_f with the function f and to say that such distributions "are" functions.

Similarly, if μ is a complex Borel measure on Ω, or if μ is a positive measure on Ω with $\mu(K) < \infty$ for every compact $K \subset \Omega$, the equation

(3) $$\Lambda_\mu(\phi) = \int_\Omega \phi\, d\mu \qquad [\phi \in \mathscr{D}(\Omega)]$$

defines a distribution Λ_μ in Ω, which is usually identified with μ.

6.12 Differentiation of distributions If α is a multi-index and $\Lambda \in \mathscr{D}'(\Omega)$, the formula

$$(1) \qquad (D^\alpha\Lambda)(\phi) = (-1)^{|\alpha|}\Lambda(D^\alpha\phi) \qquad [\phi \in \mathscr{D}(\Omega)]$$

(motivated in Section 6.1) deffnes a linear functional $D^\alpha\Lambda$ on $\mathscr{D}(\Omega)$. If

$$(2) \qquad |\Lambda\phi| \le C\|\phi\|_N$$

for all $\phi \in \mathscr{D}_K$, then

$$(3) \qquad |(D^\alpha\Lambda)(\phi)| \le C\|D^\alpha\phi\|_N \le C\|\phi\|_{N+|\alpha|}.$$

Theorem 6.8 shows therefore that $D^\alpha\Lambda \in \mathscr{D}'(\Omega)$.

Note that the formula

$$(4) \qquad D^\alpha D^\beta\Lambda = D^{\alpha+\beta}\Lambda = D^\beta D^\alpha\Lambda$$

holds for every distribution Λ and for all multi-indices α and β, simply because the operators D^α and D^β commute on $C^\infty(\Omega)$:

$$\begin{aligned}(D^\alpha D^\beta\Lambda)(\phi) &= (-1)^{|\alpha|}(D^\beta\Lambda)(D^\alpha\phi)\\ &= (-1)^{|\alpha|+|\beta|}\Lambda(D^\beta D^\alpha\phi)\\ &= (-1)^{|\alpha+\beta|}\Lambda(D^{\alpha+\beta}\phi)\\ &= (D^{\alpha+\beta}\Lambda)(\phi).\end{aligned}$$

6.13 Distribution derivatives of functions The αth distribution derivative of a locally integrable function f in Ω is, by definition, the distribution $D^\alpha\Lambda_f$.

If $D^\alpha f$ also exists in the classical sense and is locally integrable, then $D^\alpha f$ is also a distribution in the sense of Section 6.11. The obvious consistency problem is whether the equation

$$(1) \qquad D^\alpha\Lambda_f = \Lambda_{D^\alpha f}$$

always holds under these conditions.

More explicitly, the question is whether

$$(2) \qquad (-1)^{|\alpha|}\int_\Omega f(x)(D^\alpha\phi)(x)\,dx = \int_\Omega (D^\alpha f)(x)\phi(x)\,dx$$

for every $\phi \in \mathscr{D}(\Omega)$.

If f has *continuous* partial derivatives of all orders up to N, integrations by part give (2) without difficulty, if $|\alpha| \le N$.

In general, (1) may be false. The following example illustrates this, in the case $n = 1$.

6.14 Example Suppose Ω is a segment in R, and f is a left-continuous function of bounded variation in Ω. If $D = d/dx$, it is well known that

$(Df)(x)$ exists a.e. and that $Df \in L^1$. We claim that

(1) $$D\Lambda_f = \Lambda_\mu$$

where μ is the measure defined in Ω by

(2) $$\mu([a, b)) = f(b) - f(a).$$

Thus $D\Lambda_f = \Lambda_{Df}$ if and only if f is absolutely continuous.

To prove (1), we have to show that

$$(\Lambda_\mu)(\phi) = (D\Lambda_f)(\phi) = -\Lambda_f(D\phi)$$

for every $\phi \in \mathscr{D}(\Omega)$, that is, that

(3) $$\int_\Omega \phi \, d\mu = -\int_\Omega \phi'(x) f(x) \, dx.$$

But (3) is a simple consequence of Fubini's theorem, since each side of (3) is equal to the integral of $\phi'(x)$ over the set

(4) $$\{(x, y): x \in \Omega, y \in \Omega, x < y\}$$

with respect to the product measure of dx and $d\mu$. The fact that ϕ has compact support in Ω is used in this computation.

6.15 Multiplication by functions Suppose $\Lambda \in \mathscr{D}'(\Omega)$ and $f \in C^\infty(\Omega)$. The right side of the equation

(1) $$(f\Lambda)(\phi) = \Lambda(f\phi) \qquad [\phi \in \mathscr{D}(\Omega)]$$

makes sense because $f\phi \in \mathscr{D}(\Omega)$ when $\phi \in \mathscr{D}(\Omega)$. Thus (1) defines a linear functional $f\Lambda$ on $\mathscr{D}(\Omega)$. We shall see that $f\Lambda$ is, in fact, a distribution in Ω.

Observe that the notation must be handled with care: If $f \in \mathscr{D}(\Omega)$, then Λf is a *number*, whereas $f\Lambda$ is a *distribution*.

The proof that $f\Lambda \in \mathscr{D}'(\Omega)$ depends on the Leibniz formula

(2) $$D^\alpha(fg) = \sum_{\beta \le \alpha} c_{\alpha\beta}(D^{\alpha-\beta}f)(D^\beta g),$$

valid for all f and g in $C^\infty(\Omega)$ and all multi-indices α, which is obtained by iteration of the familiar formula

(3) $$(uv)' = u'v + uv'.$$

The numbers $c_{\alpha\beta}$ are positive integers whose exact value is easily computed but is irrelevant to our present needs.

To each compact $K \subset \Omega$ correspond C and N such that $|\Lambda\phi| \le C\|\phi\|_N$ for all $\phi \in \mathscr{D}_K$. By (2), there is a constant C', depending on f, K, and N, such that $\|f\phi\|_N \le C'\|\phi\|_N$ for $\phi \in \mathscr{D}_K$. Hence

(4) $$|(f\Lambda)(\phi)| \le CC'\|\phi\|_N \qquad (\phi \in \mathscr{D}_K).$$

By Theorem 6.8, $f\Lambda \in \mathscr{D}'(\Omega)$.

Now we want to show that the Leibniz formula (2) holds with Λ in place of g, so that

(5) $$D^\alpha(f\Lambda) = \sum_{\beta \le \alpha} c_{\alpha\beta}(D^{\alpha-\beta}f)(D^\beta \Lambda).$$

The proof is a purely formal calculation. Associate to each $u \in R^n$ the function h_u defined by

$$h_u(x) = \exp(u \cdot x).$$

Then $D^\alpha h_u = u^\alpha h_u$. If (2) is applied to h_u and h_v in place of f and g, the identity

(6) $$(u+v)^\alpha = \sum_{\beta \le \alpha} c_{\alpha\beta}\, u^{\alpha-\beta} v^\beta \qquad (u \in R^n,\ v \in R^n)$$

is obtained. In particular,

$$\begin{aligned} u^\alpha &= [v + (-v + u)]^\alpha \\ &= \sum_{\beta \le \alpha} c_{\alpha\beta}\, v^{\alpha-\beta} \sum_{\gamma \le \beta} c_{\beta\gamma}(-1)^{|\beta-\gamma|} v^{\beta-\gamma} u^\gamma \\ &= \sum_{\gamma \le \alpha} (-1)^{|\gamma|} v^{\alpha-\gamma} u^\gamma \sum_{\gamma \le \beta \le \alpha} (-1)^{|\beta|} c_{\alpha\beta}\, c_{\beta\gamma}. \end{aligned}$$

Hence

(7) $$\sum_{\gamma \le \beta \le \alpha} (-1)^{|\beta|} c_{\alpha\beta}\, c_{\beta\gamma} = \begin{cases} (-1)^{|\alpha|} & \text{if } \gamma = \alpha, \\ 0 & \text{otherwise.} \end{cases}$$

Apply (2) to $D^\beta(\phi D^{\alpha-\beta}f)$, and use (7), to obtain the identity

(8) $$\sum_{\beta \le \alpha} (-1)^{|\beta|} c_{\alpha\beta}\, D^\beta(\phi D^{\alpha-\beta} f) = (-1)^{|\alpha|} f D^\alpha \phi.$$

The point of all this is that (8) gives (5). For if $\phi \in \mathscr{D}(\Omega)$, then

$$\begin{aligned} D^\alpha(f\Lambda)(\phi) &= (-1)^{|\alpha|}(f\Lambda)(D^\alpha\phi) = (-1)^{|\alpha|}\Lambda(fD^\alpha\phi) \\ &= \sum_{\beta \le \alpha} (-1)^{|\beta|} c_{\alpha\beta}\, \Lambda(D^\beta(\phi D^{\alpha-\beta}f)) \\ &= \sum_{\beta \le \alpha} c_{\alpha\beta}(D^\beta \Lambda)(\phi D^{\alpha-\beta} f) \\ &= \sum_{\beta \le \alpha} c_{\alpha\beta}[(D^{\alpha-\beta}f)(D^\beta\Lambda)](\phi). \end{aligned}$$

6.16 Sequences of distributions Since $\mathscr{D}'(\Omega)$ is the space of all continuous linear functions on $\mathscr{D}(\Omega)$, the general considerations made in Section 3.14 provide a topology for $\mathscr{D}'(\Omega)$—its weak*-topology induced by $\mathscr{D}(\Omega)$—which makes $\mathscr{D}'(\Omega)$ into a locally convex space. If $\{\Lambda_i\}$ is a sequence of distributions in Ω, the statement

(1) $$\Lambda_i \to \Lambda \text{ in } \mathscr{D}'(\Omega)$$

refers to this weak*-topology and means, explicitly, that

$$\text{(2)} \qquad \lim_{i\to\infty} \Lambda_i \phi = \Lambda\phi \qquad [\phi \in \mathscr{D}(\Omega)].$$

In particular, if $\{f_i\}$ is a sequence of locally integrable functions in Ω, the statements "$f_i \to \Lambda$ in $\mathscr{D}'(\Omega)$" or "$\{f_i\}$ converges to Λ in the distribution sense" mean that

$$\text{(3)} \qquad \lim_{i\to\infty} \int_\Omega \phi(x) f_i(x)\, dx = \Lambda\phi$$

for every $\phi \in \mathscr{D}(\Omega)$.

The simplicity of the next theorem, concerning termwise differentiation of a sequence, is rather striking.

6.17 Theorem *Suppose $\Lambda_i \in \mathscr{D}'(\Omega)$ for $i = 1, 2, 3, \ldots$, and*

$$\text{(1)} \qquad \Lambda\phi = \lim_{i\to\infty} \Lambda_i \phi$$

exists (as a complex number) for every $\phi \in \mathscr{D}(\Omega)$. Then $\Lambda \in \mathscr{D}'(\Omega)$, and

$$\text{(2)} \qquad D^\alpha \Lambda_i \to D^\alpha \Lambda \text{ in } \mathscr{D}'(\Omega),$$

for every multi-index α.

PROOF. Let K be an arbitrary compact subset of Ω. Since (1) holds for every $\phi \in \mathscr{D}_K$, and since $\mathscr{D}_K$ is a Fréchet space, the Banach-Steinhaus theorem 2.8 implies that the restriction of Λ to $\mathscr{D}_K$ is continuous. It follows from Theorem 6.6 that Λ is continuous on $\mathscr{D}(\Omega)$; in other words, $\Lambda \in \mathscr{D}'(\Omega)$. Consequently (1) implies that

$$\begin{aligned}(D^\alpha\Lambda)(\phi) &= (-1)^{|\alpha|}\Lambda(D^\alpha\phi)\\ &= (-1)^{|\alpha|} \lim_{i\to\infty} \Lambda_i(D^\alpha\phi) = \lim_{i\to\infty} (D^\alpha\Lambda_i)(\phi).\end{aligned}$$

////

6.18 Theorem *If $\Lambda_i \to \Lambda$ in $\mathscr{D}'(\Omega)$ and $g_i \to g$ in $C^\infty(\Omega)$, then $g_i\Lambda_i \to g\Lambda$ in $\mathscr{D}'(\Omega)$.*

Note: The statement "$g_i \to g$ in $C^\infty(\Omega)$" refers to the Fréchet space topology of $C^\infty(\Omega)$ described in Section 1.46.

PROOF. Fix $\phi \in \mathscr{D}(\Omega)$. Define a bilinear functional B on $C^\infty(\Omega) \times \mathscr{D}'(\Omega)$ by

$$B(g, \Lambda) = (g\Lambda)(\phi) = \Lambda(g\phi).$$

Then B is separately continuous, and Theorem 2.17 implies that

$$B(g_i, \Lambda_i) \to B(g, \Lambda) \qquad \text{as } i \to \infty.$$

Hence

$$(g_i \Lambda_i)(\phi) \to (g\Lambda)(\phi). \qquad ////$$

Localization

6.19 Local equality Suppose $\Lambda_i \in \mathscr{D}'(\Omega)$ $(i = 1, 2)$ and ω is an open subset of Ω. The statement

(1) $$\Lambda_1 = \Lambda_2 \text{ in } \omega$$

means, by definition, that $\Lambda_1 \phi = \Lambda_2 \phi$ for every $\phi \in \mathscr{D}(\omega)$.

For example, if f is a locally integrable function and μ is a measure, then $\Lambda_f = 0$ in ω if and only if $f(x) = 0$ for almost every $x \in \omega$, and $\Lambda_\mu = 0$ in ω if and only if $\mu(E) = 0$ for every Borel set $E \subset \omega$.

This definition makes it possible to discuss distributions locally. On the other hand, it is also possible to describe a distribution globally if its local behavior is known. This is stated precisely in Theorem 6.21. The proof uses partitions of unity, which we now construct.

6.20 Theorem *If Γ is a collection of open sets in R^n whose union is Ω, then there exists a sequence $\{\psi_i\} \subset \mathscr{D}(\Omega)$, with $\psi_i \geq 0$, such that*

(a) *each ψ_i has its support in some member of Γ,*

(b) $\sum_{i=1}^{\infty} \psi_i(x) = 1$ *for every $x \in \Omega$,*

(c) *to every compact $K \subset \Omega$ correspond an integer m and an open set $W \supset K$ such that*

(1) $$\psi_1(x) + \cdots + \psi_m(x) = 1$$

for all $x \in W$.

Such a collection $\{\psi_i\}$ is called a *locally finite partition of unity in* Ω, subordinate to the open cover Γ of Ω. Note that it follows from (b) and (c) that every point of Ω has a neighborhood which intersects the supports of only finitely many ψ_i. This is the reason for calling $\{\psi_i\}$ locally finite.

PROOF. Let S be a countable dense subset of Ω. Let $\{B_1, B_2, B_3, \ldots\}$ be a sequence that contains every closed ball B_i whose center p_i lies in S, whose radius r_i is rational, and which lies in some member of Γ. Let

V_i be the open ball with center p_i and radius $r_i/2$. It is easy to see that $\Omega = \bigcup V_i$.

The construction described in Section 1.46 shows that there are functions $\phi_i \in \mathscr{D}(\Omega)$ such that $0 \le \phi \le 1$, $\phi_i = 1$ in V_i, $\phi_i = 0$ off B_i. Define $\psi_1 = \phi_1$, and, inductively,

(2) $$\psi_{i+1} = (1 - \phi_1) \cdots (1 - \phi_i)\phi_{i+1} \qquad (i \ge 1).$$

Obviously, $\psi_i = 0$ outside B_i. This gives (*a*). The relation

(3) $$\psi_1 + \cdots + \psi_i = 1 - (1 - \phi_1) \cdots (1 - \phi_i)$$

is trivial when $i = 1$. If (3) holds for some i, addition of (2) and (3) yields (3) with $i + 1$ in place of i. Hence (3) holds for every i. Since $\phi_i = 1$ in V_i, it follows that

(4) $$\psi_1(x) + \cdots + \psi_m(x) = 1 \qquad \text{if } x \in V_1 \cup \cdots \cup V_m.$$

This gives (*b*). Moreover, if K is compact, then $K \subset V_1 \cup \cdots \cup V_m$ for some m, and (*c*) follows. ////

6.21 Theorem *Suppose Γ is an open cover of an open set $\Omega \subset R^n$, and suppose that to each $\omega \in \Gamma$ corresponds a distribution $\Lambda_\omega \in \mathscr{D}'(\omega)$ such that*

(1) $$\Lambda_{\omega'} = \Lambda_{\omega''} \text{ in } \omega' \cap \omega''$$

whenever $\omega' \cap \omega'' \neq \varnothing$.

Then there exists a unique $\Lambda \in \mathscr{D}'(\Omega)$ such that

(2) $$\Lambda = \Lambda_\omega \text{ in } \omega$$

for every $\omega \in \Gamma$.

PROOF. Let $\{\psi_i\}$ be a locally finite partition of unity, subordinate to Γ, as in Theorem 6.20, and associate to each i a set $\omega_i \in \Gamma$ such that ω_i contains the support of ψ_i.

If $\phi \in \mathscr{D}(\Omega)$, then $\phi = \sum \psi_i \phi$. Only finitely many terms in this sum are different from 0, since ϕ has compact support. Define

(3) $$\Lambda\phi = \sum_{i=1}^{\infty} \Lambda_{\omega_i}(\psi_i \phi).$$

It is clear that Λ is a linear functional on $\mathscr{D}(\Omega)$.

To show that Λ is continuous, suppose $\phi_j \to 0$ in $\mathscr{D}(\Omega)$. There is a compact $K \subset \Omega$ which contains the support of every ϕ_j. If m is chosen as in part (*c*) of Theorem 6.20, then

(4) $$\Lambda\phi_j = \sum_{i=1}^{m} \Lambda_{\omega_i}(\psi_i \phi_j) \qquad (j = 1, 2, 3, \ldots).$$

Since $\psi_i \phi_j \to 0$ in $\mathscr{D}(\omega_i)$, as $j \to \infty$, it follows from (4) that $\Lambda\phi_j \to 0$. By Theorem 6.6, $\Lambda \in \mathscr{D}'(\Omega)$.

To prove (2), pick $\phi \in \mathscr{D}(\omega)$. Then

(5) $$\psi_i \phi \in \mathscr{D}(\omega_i \cap \omega) \qquad (i = 1, 2, 3, \ldots)$$

so that (1) implies $\Lambda_{\omega_i}(\psi_i \phi) = \Lambda_\omega(\psi_i \phi)$. Hence

(6) $$\Lambda\phi = \sum \Lambda_\omega(\psi_i \phi) = \Lambda_\omega(\sum \psi_i \phi) = \Lambda_\omega \phi,$$

which proves (2).

This gives the existence of Λ. The uniqueness is trivial since (2) (with ω_i in place of ω) implies that Λ must satisfy (3). ////

Supports of Distributions

6.22 Definition Suppose $\Lambda \in \mathscr{D}'(\Omega)$. If ω is an open subset of Ω and if $\Lambda\phi = 0$ for every $\phi \in \mathscr{D}(\omega)$, we say that Λ *vanishes in* ω. Let W be the union of all open $\omega \subset \Omega$ in which Λ vanishes. The complement of W (relative to Ω) is the *support of* Λ.

6.23 Theorem *If W is as above, then Λ vanishes in W.*

PROOF. W is the union of open sets ω in which Λ vanishes. Let Γ be the collection of these ω's, and let $\{\psi_i\}$ be a locally finite partition of unity in W, subordinate to Γ, as in Theorem 6.20. If $\phi \in \mathscr{D}(W)$, then $\phi = \sum \psi_i \phi$. Only finitely many terms of this sum are different from 0. Hence

$$\Lambda\phi = \sum \Lambda(\psi_i \phi) = 0$$

since each ψ_i has its support in some $\omega \in \Gamma$. ////

The most significant part of the next theorem is (*d*). Exercise 20 complements it.

6.24 Theorem *Suppose $\Lambda \in \mathscr{D}'(\Omega)$ and S_Λ is the support of Λ.*

(*a*) *If the support of some $\phi \in \mathscr{D}(\Omega)$ does not intersect S_Λ, then $\Lambda\phi = 0$.*
(*b*) *If S_Λ is empty, then $\Lambda = 0$.*
(*c*) *If $\psi \in C^\infty(\Omega)$ and $\psi = 1$ in some open set V containing S_Λ, then $\psi\Lambda = \Lambda$.*
(*d*) *If S_Λ is a compact subset of Ω, then Λ has finite order; in fact, there is a constant $C < \infty$ and a nonnegative integer N such that*

$$|\Lambda\phi| \le C\|\phi\|_N$$

for every $\phi \in \mathscr{D}(\Omega)$. Furthermore, Λ extends in a unique way to a continuous linear functional on $C^\infty(\Omega)$.

PROOF. Parts (*a*) and (*b*) are obvious. If ψ is as in (*c*) and if $\phi \in \mathscr{D}(\Omega)$, then the support of $\phi - \psi\phi$ does not intersect S_Λ. Thus $\Lambda\phi = \Lambda(\psi\phi) = (\psi\Lambda)(\phi)$, by (*a*).

If S_Λ is compact, it follows from Theorem 6.20 that there exists $\psi \in \mathscr{D}(\Omega)$ that satisfies (*c*). Fix such a ψ; call its support K. By Theorem 6.8, there exist c_1 and N such that $|\Lambda\phi| \leq c_1 \|\phi\|_N$ for all $\phi \in \mathscr{D}_K$. The Leibniz formula shows that there is a constant c_2 such that $\|\psi\phi\|_N \leq c_2 \|\phi\|_N$ for every $\phi \in \mathscr{D}(\Omega)$. Hence

$$|\Lambda\phi| = |\Lambda(\psi\phi)| \leq c_1 \|\psi\phi\|_N \leq c_1 c_2 \|\phi\|_N$$

for every $\phi \in \mathscr{D}(\Omega)$.

Since $\Lambda\phi = \Lambda(\psi\phi)$ for all $\phi \in \mathscr{D}(\Omega)$, the formula

(1) $$\Lambda f = \Lambda(\psi f) \qquad [f \in C^\infty(\Omega)]$$

defines an extension of Λ. This extension is continuous, for if $f_i \to 0$ in $C^\infty(\Omega)$, then each derivative of f_i tends to 0, uniformly on compact subsets of Ω; the Leibniz formula shows therefore that $\psi f_i \to 0$ in $\mathscr{D}(\Omega)$; since $\Lambda \in \mathscr{D}'(\Omega)$, it follows that $\Lambda f_i \to 0$.

If $f \in C^\infty(\Omega)$ and if K_0 is any compact subset of Ω, there exists $\phi \in \mathscr{D}(\Omega)$ such that $\phi = f$ on K_0. It follows that $\mathscr{D}(\Omega)$ is dense in $C^\infty(\Omega)$. Each $\Lambda \in \mathscr{D}'(\Omega)$ has therefore at most one continuous extension to $C^\infty(\Omega)$. ////

Note: In (*a*) it is assumed that ϕ vanishes in some open set containing S_Λ, not merely that ϕ vanishes on S_Λ.

In view of (*b*), the next simplest case is the one in which S_Λ consists of a single point. These distributions will now be completely described.

6.25 Theorem *Suppose $\Lambda \in \mathscr{D}'(\Omega)$, $p \in \Omega$, $\{p\}$ is the support of Λ, and Λ has order N. Then there are constants c_α such that*

(1) $$\Lambda = \sum_{|\alpha| \leq N} c_\alpha D^\alpha \delta_p,$$

where δ_p is the evaluation functional defined by

(2) $$\delta_p(\phi) = \phi(p).$$

Conversely, every distribution of the form (1) *has $\{p\}$ for its support (unless $c_\alpha = 0$ for all α).*

PROOF. It is clear that the support of $D^\alpha \delta_p$ is $\{p\}$, for every multi-index α. This proves the converse.

To prove the nontrivial half of the theorem, assume that $p = 0$ (the origin of R^n), and consider a $\phi \in \mathscr{D}(\Omega)$ that satisfies

$$(D^\alpha \phi)(0) = 0 \qquad \text{for all } \alpha \text{ with } |\alpha| \le N. \tag{3}$$

Our first objective is to prove that (3) implies $\Lambda\phi = 0$.

If $\eta > 0$, there is a compact ball $K \subset \Omega$, with center at 0, such that

$$|D^\alpha \phi| \le \eta \text{ in } K, \qquad \text{if } |\alpha| = N. \tag{4}$$

We claim that

$$|D^\alpha \phi(x)| \le \eta n^{N-|\alpha|} |x|^{N-|\alpha|} \qquad (x \in K, |\alpha| \le N). \tag{5}$$

When $|\alpha| = N$, this is (4). Suppose $1 \le i \le N$, assume (5) is proved for all α with $|\alpha| = i$, and suppose $|\beta| = i - 1$. The gradient of $D^\beta \phi$ is the vector

$$\text{grad } D^\beta \phi = (D_1 D^\beta \phi, \ldots, D_n D^\beta \phi). \tag{6}$$

Our induction hypothesis implies that

$$|(\text{grad } D^\beta \phi)(x)| \le n \cdot \eta n^{N-i} |x|^{N-i} \qquad (x \in K), \tag{7}$$

and since $(D^\beta \phi)(0) = 0$ the mean value theorem now shows that (5) holds with β in place of α. Thus (5) is proved.

Choose an auxiliary function $\psi \in \mathscr{D}(R^n)$, which is 1 in some neighborhood of 0 and whose support is in the unit ball B of R^n. Define

$$\psi_r(x) = \psi\left(\frac{x}{r}\right) \qquad (r > 0, x \in R^n). \tag{8}$$

If r is small enough, the support of ψ_r lies in $rB \subset K$. By Leibniz' formula

$$D^\alpha(\psi_r \phi)(x) = \sum_{\beta \le \alpha} c_{\alpha\beta} (D^{\alpha-\beta}\psi)\left(\frac{x}{r}\right)(D^\beta \phi)(x) r^{|\beta|-|\alpha|}. \tag{9}$$

It now follows from (5) that

$$\|\psi_r \phi\|_N \le \eta C \|\psi\|_N \tag{10}$$

as soon as r is small enough; here C depends on n and N.

Since Λ has order N, there is a constant C_1 such that $|\Lambda\psi| \le C_1 \|\psi\|_N$ for all $\psi \in \mathscr{D}_K$. Since $\psi_r = 1$ in a neighborhood of the support of Λ, it now follows from (10) and (*c*) of Theorem 6.24 that

$$|\Lambda\phi| = |\Lambda(\psi_r \phi)| \le C_1 \|\psi_r \phi\|_N \le \eta C C_1 \|\psi\|_N.$$

Since η was arbitrary, we have proved that $\Lambda\phi = 0$ whenever (3) holds.

In other words, Λ vanishes on the intersection of the null spaces of the functionals $D^\alpha\delta_0$ ($|\alpha| \le N$), since

(11) $$(D^\alpha\delta_0)\phi = (-1)^{|\alpha|}\delta_0(D^\alpha\phi) = (-1)^{|\alpha|}(D^\alpha\phi)(0).$$

The representation (1) follows now from Lemma 3.9. ////

Distributions as Derivatives

It was pointed out in the introduction to this chapter that one of the aims of the theory of distributions is to enlarge the concept of function in such a way that partial differentiations can be carried out unrestrictedly. The distributions do satisfy this requirement. Conversely—as we shall now see—every distribution is (at least locally) $D^\alpha f$ for some continuous function f and some multi-index α. If every continuous function is to have partial derivatives of all orders, no proper subclass of the distributions can therefore be adequate. In this sense, the distribution extension of the function concept is as economical as it possibly can be.

6.26 Theorem *Suppose $\Lambda \in \mathscr{D}'(\Omega)$, and K is a compact subset of Ω. Then there is a continuous function f in Ω and there is a multi-index α such that*

(1) $$\Lambda\phi = (-1)^{|\alpha|}\int_\Omega f(x)(D^\alpha\phi)(x)\,dx$$

for every $\phi \in \mathscr{D}_K$.

PROOF. Assume, without loss of generality, that $K \subset Q$, where Q is the unit cube in R^n, consisting of all $x = (x_1, \ldots, x_n)$ with $0 \le x_i \le 1$ for $i = 1, \ldots, n$. The mean value theorem shows that

(2) $$|\psi| \le \max_{x \in Q} |(D_i\psi)(x)| \qquad (\psi \in \mathscr{D}_Q)$$

for $i = 1, \ldots, n$. Put $T = D_1 D_2 \cdots D_n$. For $y \in Q$, let $Q(y)$ denote the subset of Q in which $x_i \le y_i$ ($1 \le i \le n$). Then

(3) $$\psi(y) = \int_{Q(y)} (T\psi)(x)\,dx \qquad (\psi \in \mathscr{D}_Q).$$

If N is a nonnegative integer and if (2) is applied to successive derivatives of ψ, (3) leads to the inequality

(4) $$\|\psi\|_N \le \max_{x \in Q} |(T^N\psi)(x)| \le \int_Q |(T^{N+1}\psi)(x)|\,dx,$$

for every $\psi \in \mathscr{D}_Q$.

Since $\Lambda \in \mathscr{D}'(\Omega)$, there exist N and C such that

$$(5) \qquad |\Lambda\phi| \le C\|\phi\|_N \qquad (\phi \in \mathscr{D}_K).$$

Hence (4) shows that

$$(6) \qquad |\Lambda\phi| \le \int_K |(T^{N+1}\phi)(x)|\,dx \qquad (\phi \in \mathscr{D}_K).$$

By (3), T is one-to-one on $\mathscr{D}_Q$, hence on $\mathscr{D}_K$. Consequently, $T^{N+1}\colon \mathscr{D}_K \to \mathscr{D}_K$ is one-to-one. A functional Λ_1 can therefore be defined on the range Y of T^{N+1} by setting

$$(7) \qquad \Lambda_1 T^{N+1}\phi = \Lambda\phi \qquad (\phi \in \mathscr{D}_K),$$

and (6) shows that

$$(8) \qquad |\Lambda_1\psi| \le C \int_K |\psi(x)|\,dx \qquad (\psi \in Y).$$

The Hahn-Banach theorem therefore extends Λ_1 to a bounded linear functional on $L^1(K)$. In other words, there is a bounded Borel function g on K such that

$$(9) \qquad \Lambda\phi = \Lambda_1 T^{N+1}\phi = \int_K g(x)(T^{N+1}\phi)(x)\,dx \qquad (\phi \in \mathscr{D}_K).$$

Define $g(x) = 0$ outside K and put

$$(10) \qquad f(y) = \int_{-\infty}^{y_1} \cdots \int_{-\infty}^{y_n} g(x)\,dx_n \cdots dx_1 \qquad (y \in R^n).$$

Then f is continuous, and n integrations by parts show that (9) gives

$$(11) \qquad \Lambda\phi = (-1)^n \int_\Omega f(x)(T^{N+2}\phi)(x)\,dx \qquad (\phi \in \mathscr{D}_K).$$

This is (1), with $\alpha = (N+2, \ldots, N+2)$, except for a possible change in sign. ////

When Λ has compact support, the local result just proved can be turned into a global one:

6.27 Theorem *Suppose K is compact, V and Ω are open in R^n, and $K \subset V \subset \Omega$. Suppose also that $\Lambda \in \mathscr{D}'(\Omega)$, that K is the support of Λ, and that Λ has order N. Then there exist finitely many continuous functions f_β in Ω (one for each multi-index β with $\beta_i \le N+2$ for $i = 1, \ldots, n$) with supports in V, such that*

$$(1) \qquad \Lambda = \sum_\beta D^\beta f_\beta.$$

These derivatives are, of course, to be understood in the distribution sense: (1) means that

$$\Lambda\phi = \sum_{\beta} (-1)^{|\beta|} \int_{\Omega} f_{\beta}(x)(D^{\beta}\phi)(x)\,dx \qquad [\phi \in \mathscr{D}(\Omega)]. \tag{2}$$

PROOF. Choose an open set W with compact closure $\bar{W}$, such that $K \subset W$ and $\bar{W} \subset V$. Apply Theorem 6.26 with $\bar{W}$ in place of K. Put $\alpha = (N+2, \ldots, N+2)$. The proof of Theorem 6.26 shows that there is a continuous function f in Ω such that

$$\Lambda\phi = (-1)^{|\alpha|} \int_{\Omega} f(x)(D^{\alpha}\phi)(x)\,dx \qquad [\phi \in \mathscr{D}(W)]. \tag{3}$$

We may multiply f by a continuous function which is 1 on $\bar{W}$ and whose support lies in V, without disturbing (3).

Fix $\psi \in \mathscr{D}(\Omega)$, with support in W, such that $\psi = 1$ on some open set containing K. Then (3) implies, for every $\phi \in \mathscr{D}(\Omega)$, that

$$\begin{aligned}\Lambda\phi = \Lambda(\psi\phi) &= (-1)^{|\alpha|} \int_{\Omega} f \cdot D^{\alpha}(\psi\phi) \\ &= (-1)^{|\alpha|} \int_{\Omega} f \sum_{\beta \le \alpha} c_{\alpha\beta} D^{\alpha-\beta}\psi D^{\beta}\phi.\end{aligned}$$

This is (2), with

$$f_{\beta} = (-1)^{|\alpha-\beta|} c_{\alpha\beta} f \cdot D^{\alpha-\beta}\psi \qquad (\beta \le \alpha). \qquad ////$$

Our next theorem describes the global structure of distributions.

6.28 Theorem *Suppose $\Lambda \in \mathscr{D}'(\Omega)$. There exist continuous functions g_{α} in Ω, one for each multi-index α, such that*

(a) each compact $K \subset \Omega$ intersects the supports of only finitely many g_{α}, and
(b) $\Lambda = \sum_{\alpha} D^{\alpha} g_{\alpha}$.

If Λ has finite order, then the functions g_{α} can be chosen so that only finitely many are different from 0.

PROOF. There are compact cubes Q_i and open sets V_i $(i = 1, 2, 3, \ldots)$ such that $Q_i \subset V_i \subset \Omega$, Ω is the union of the Q_i, and no compact subset of Ω intersects infinitely many V_i. There exist $\phi_i \in \mathscr{D}(V_i)$ such that $\phi_i = 1$ on Q_i. Use this sequence $\{\phi_i\}$ to construct a partition of unity $\{\psi_i\}$, as in Theorem 6.20; each ψ_i has its support in V_i.

Theorem 6.27 applies to each $\psi_i \Lambda$. It shows that there are finitely many continuous functions $f_{i,\alpha}$ with supports in V_i, such that

(1) $$\psi_i \Lambda = \sum_\alpha D^\alpha f_{i,\alpha}.$$

Define

(2) $$g_\alpha = \sum_{i=1}^{\infty} f_{i,\alpha}.$$

These sums are locally finite, in the sense that each compact $K \subset \Omega$ intersects the supports of only finitely many $f_{i,\alpha}$. It follows that each g_α is continuous in Ω and that (*a*) holds.

Since $\phi = \sum \psi_i \phi$, for every $\phi \in \mathscr{D}(\Omega)$, we have $\Lambda = \sum \psi_i \Lambda$, and therefore (1) and (2) give (*b*).

The final assertion follows from Theorem 6.27. ////

Convolutions

Starting from convolutions of two functions, we shall now define the convolution of a distribution and a test function and then (under certain conditions) the convolution of two distributions. These are important in the applications of Fourier transforms to differential equations. A characteristic property of convolutions is that they commute with translations and with differentiations (Theorems 6.30, 6.33, 6.37). Also, differentiations may be regarded as convolutions with derivatives of the Dirac measure (Theorem 6.37).

It will be convenient to make a small change in notation and to use the letters $u, v, \ldots$ for distributions as well as for functions.

6.29 Definitions In the rest of this chapter, we shall write $\mathscr{D}$ and $\mathscr{D}'$ in place of $\mathscr{D}(R^n)$ and $\mathscr{D}'(R^n)$. If u is a function in R^n, and $x \in R^n$, $\tau_x u$ and $\check{u}$ are the functions defined by

(1) $$(\tau_x u)(y) = u(y - x), \qquad \check{u}(y) = u(-y) \qquad (y \in R^n).$$

Note that

(2) $$(\tau_x \check{u})(y) = \check{u}(y - x) = u(x - y).$$

If u and v are complex functions in R^n, their convolution $u * v$ is defined by

(3) $$(u * v)(x) = \int_{R^n} u(y)v(x - y)\, dy,$$

provided that the integral exists for all (or at least for almost all) $x \in R^n$, in the Lebesgue sense. Because of (2),

$$(u * v)(x) = \int_{R^n} u(y)(\tau_x \check{v})(y)\, dy. \tag{4}$$

This makes it natural to define

$$(u * \phi)(x) = u(\tau_x \check{\phi}) \qquad (u \in \mathscr{D}',\ \phi \in \mathscr{D},\ x \in R^n), \tag{5}$$

for if u is a locally integrable function, (5) agrees with (4). Note that $u * \phi$ is a function.

The relation $\int (\tau_x u) \cdot v = \int u \cdot (\tau_{-x} v)$, valid for functions u and v, makes it natural to define the translate $\tau_x u$ of $u \in \mathscr{D}'$ by

$$(\tau_x u)(\phi) = u(\tau_{-x} \phi) \qquad (\phi \in \mathscr{D},\ x \in R^n). \tag{6}$$

Then, for each $x \in R^n$, $\tau_x u \in \mathscr{D}'$; we leave the verification of the appropriate continuity requirement as an exercise.

6.30 Theorem *Suppose $u \in \mathscr{D}'$, $\phi \in \mathscr{D}$, $\psi \in \mathscr{D}$. Then*

(a) $\tau_x(u * \phi) = (\tau_x u) * \phi = u * (\tau_x \phi)$ for all $x \in R^n$;
(b) $u * \phi \in C^\infty$ and

$$D^\alpha(u * \phi) = (D^\alpha u) * \phi = u * (D^\alpha \phi)$$

for every multi-index α;
(c) $u * (\phi * \psi) = (u * \phi) * \psi$.

PROOF. For any $y \in R^n$,

$$(\tau_x(u * \phi))(y) = (u * \phi)(y - x) = u(\tau_{y-x} \check{\phi}),$$
$$((\tau_x u) * \phi)(y) = (\tau_x u)(t_y \check{\phi}) = u(\tau_{y-x} \check{\phi}),$$
$$(u * (\tau_x \phi))(y) = u(\tau_y(\tau_x \phi)^\vee) = u(\tau_{y-x} \check{\phi}),$$

which gives (*a*); the relations

$$\tau_y \tau_{-x} = \tau_{y-x} \qquad \text{and} \qquad (\tau_x \phi)^\vee = \tau_{-x} \check{\phi}$$

were used. In the sequel, purely formal calculations such as the preceding ones will sometimes be omitted.

If u is applied to both sides of the identity

$$\tau_x((D^\alpha \phi)^\vee) = (-1)^{|\alpha|} D^\alpha(\tau_x \check{\phi}) \tag{1}$$

one obtains part of (*b*), namely,

$$(u * (D^\alpha \phi))(x) = ((D^\alpha u) * \phi)(x).$$

To prove the rest of (*b*), let e be a unit vector in R^n, and put

(2) $$\eta_r = r^{-1}(\tau_0 - \tau_{re}) \qquad (r > 0).$$

Then (*a*) gives

(3) $$\eta_r(u * \phi) = u * (\eta_r \phi).$$

As $r \to 0$, $\eta_r \phi \to D_e \phi$ in $\mathscr{D}$, where D_e denotes the directional derivative in the direction e. Hence

$$\tau_x((\eta_r \phi)^\vee) \to \tau_x(D_e \phi)^\vee \text{ in } \mathscr{D},$$

for each $x \in R^n$, so that

(4) $$\lim_{r \to 0} (u * (\eta_r \phi))(x) = (u * (D_e \phi))(x).$$

By (3) and (4) we have

(5) $$D_e(u * \phi) = u * (D_e \phi),$$

and iteration of (5) gives (*b*).

To prove (*c*), we begin with the identity

(6) $$(\phi * \psi)^\vee(t) = \int_{R^n} \check{\psi}(s)(\tau_s \check{\phi})(t)\, ds.$$

Let K_1 and K_2 be the supports of $\check{\phi}$ and $\check{\psi}$. Put $K = K_1 + K_2$. Then

$$s \to \check{\psi}(s)\tau_s \check{\phi}$$

is a continuous mapping of R^n into $\mathscr{D}_K$, which is 0 outside K_2. Therefore (6) may be written as a $\mathscr{D}_K$-valued integral, namely,

(7) $$(\phi * \psi)^\vee = \int_{K_2} \check{\psi}(s)\tau_s \check{\phi}\, ds,$$

and now Theorem 3.27 shows that

$$\begin{aligned}(u * (\phi * \psi))(0) &= u((\phi * \psi)^\vee) \\ &= \int_{K_2} \check{\psi}(s) u(\tau_s \check{\phi})\, ds = \int_{R^n} \psi(-s)(u * \phi)(s)\, ds,\end{aligned}$$

or

(8) $$(u * (\phi * \psi))(0) = ((u * \phi) * \psi(0).$$

To obtain (8) with x in place of 0, apply (8) to $\tau_{-x}\psi$ in place of ψ, and appeal to (*a*). This proves (*c*). ////

6.31 Definition The term *approximate identity on R^n* will denote a sequence of functions h_j of the form

$$h_j(x) = j^n h(jx) \qquad (j = 1, 2, 3, \ldots),$$

where $h \in \mathscr{D}(R^n)$, $h \geq 0$, and $\int_{R^n} h(x)\, dx = 1$.

6.32 Theorem *Suppose $\{h_j\}$ is an approximate identity on R^n, $\phi \in \mathscr{D}$, and $u \in \mathscr{D}'$. Then*

(a) $\lim_{j\to\infty} \phi * h_j = \phi$ *in* $\mathscr{D}$,

(b) $\lim_{j\to\infty} u * h_j = u$ *in* $\mathscr{D}'$.

Note that (*b*) implies that every distribution is a limit, in the topology of $\mathscr{D}'$, of a sequence of infinitely differentiable functions.

PROOF. It is a trivial exercise to check that $f * h_j \to f$ uniformly on compact sets, if f is any continuous function on R^n. Applying this to $D^\alpha\phi$ in place of f, we see that $D^\alpha(\phi * h_j) \to D^\alpha\phi$ uniformly. Also, the supports of all $\phi * h_j$ lie in some compact set, since the supports of the h_j shrink to $\{0\}$. This gives (*a*).

Next, (*a*) and statement (*c*) of Theorem 6.30 give (*b*), because

$$\begin{aligned} u(\check{\phi}) &= (u * \phi)(0) = \lim\, (u * (h_j * \phi))(0) \\ &= \lim\, ((u * h_j) * \phi)(0) = \lim\, (u * h_j)(\check{\phi}). \end{aligned}$$ ////

6.33 Theorem

(a) *If $u \in \mathscr{D}'$ and*

(1) $$L\phi = u * \phi \qquad (\phi \in \mathscr{D}),$$

then L is a continuous linear mapping of $\mathscr{D}$ into C^∞ which satisfies

(2) $$\tau_x L = L\tau_x \qquad (x \in R^n).$$

(b) *Conversely, if L is a continuous linear mapping of $\mathscr{D}$ into $C(R^n)$, and if L satisfies* (2), *then there is a unique $u \in \mathscr{D}'$ such that* (1) *holds*

Note that (*b*) implies that the range of L actually lies in C^∞.

PROOF. (*a*) Since $\tau_x(u * \phi) = u * (\tau_x\phi)$, (1) implies (2). To prove that L is continuous, we have to show that the restriction of L to each $\mathscr{D}_K$ is a continuous mapping into C^∞. Since these are Fréchet spaces, the closed graph theorem can be applied. Suppose that $\phi_i \to \phi$ in $\mathscr{D}_K$ and that $u * \phi_i \to f$ in C^∞; we have to prove that $f = u * \phi$.

Fix $x \in R^n$. Then $\tau_x \check{\phi}_i \to \tau_x \check{\phi}$ in $\mathscr{D}$, so that

$$f(x) = \lim (u * \phi_i)(x) = \lim u(\tau_x \check{\phi}_i) = u(\tau_x \check{\phi}) = (u * \phi)(x).$$

(*b*) Define $u(\phi) = (L\check{\phi})(0)$. Since $\phi \to \check{\phi}$ is a continuous operator on $\mathscr{D}$, and since evaluation at 0 is a continuous linear functional on C, u is continuous on $\mathscr{D}$. Thus $u \in \mathscr{D}'$. Since L satisfies (2),

$$\begin{aligned}(L\phi)(x) &= (\tau_{-x} L\phi)(0) = (L\tau_{-x}\phi)(0)\\ &= u((\tau_{-x}\phi)^\vee) = u(\tau_x \check{\phi}) = (u * \phi)(x).\end{aligned}$$

The uniqueness of u is obvious, for if $u \in \mathscr{D}'$ and $u * \phi = 0$ for every $\phi \in \mathscr{D}$, then

$$u(\check{\phi}) = (u * \phi)(0) = 0$$

for every $\phi \in \mathscr{D}$; hence $u = 0$. ////

6.34 Definition Suppose now that $u \in \mathscr{D}'$ and that u has compact support. By Theorem 6.24, u extends then in a unique fashion to a continuous linear functional on C^∞. One can therefore define the convolution of u and any $\phi \in C^\infty$ by the same formula as before, namely,

$$(u * \phi)(x) = u(\tau_x \check{\phi}) \qquad (x \in R^n).$$

6.35 Theorem *Suppose $u \in \mathscr{D}'$ has compact support, and $\phi \in C^\infty$. Then*

(*a*) $\tau_x(u * \phi) = (\tau_x u) * \phi = u * (\tau_x \phi)$ *if* $x \in R^n$,

(*b*) $u * \phi \in C^\infty$ *and*

$$D^\alpha(u * \phi) = (D^\alpha u) * \phi = u * (D^\alpha \phi).$$

If, in addition, $\psi \in \mathscr{D}$, *then*

(*c*) $u * \psi \in \mathscr{D}$, *and*

(*d*) $u * (\phi * \psi) = (u * \phi) * \psi = (u * \psi) * \phi$.

PROOF. The proofs of (*a*) and (*b*) are so similar to those given in Theorem 6.30 that they need not be repeated. To prove (*c*), let K and H be the supports of u and ψ, respectively. The support of $\tau_x \check{\psi}$ is $x - H$. Therefore

$$(u * \psi)(x) = u(\tau_x \check{\psi}) = 0$$

unless K intersects $x - H$, that is, unless $x \in K + H$. The support of $u * \psi$ thus lies in the compact set $K + H$.

To prove (*d*), let W be a bounded open set that contains K, and choose $\phi_0 \in \mathscr{D}$ so that $\check{\phi}_0 = \check{\phi}$ in $W + H$. Then $(\phi * \psi)^\vee = (\phi_0 * \psi)^\vee$ in W, so that

$$(u * (\phi * \psi))(0) = (u * (\phi_0 * \psi))(0). \tag{1}$$

If $-s \in H$, then $\tau_s \check{\phi} = \tau_s \check{\phi}_0$ in W; hence $u * \phi = u * \phi_0$ in $-H$. This gives

(2) $$((u * \phi) * \psi)(0) = ((u * \phi_0) * \psi)(0).$$

Since the support of $u * \psi$ lies in $K + H$,

(3) $$((u * \psi) * \phi)(0) = ((u * \psi) * \phi_0(0).$$

The right sides of (1) to (3) are equal, by Theorem 6.30; hence so are their left sides. This proves that the three convolutions in (*d*) are equal at the origin. The general case follows by translation, as at the end of the proof of Theorem 6.30. ////

6.36 Definition If $u \in \mathscr{D}'$, $v \in \mathscr{D}'$, *and at least one of these two distributions has compact support, define*

(1) $$L\phi = u * (v * \phi) \qquad (\phi \in \mathscr{D}).$$

Note that this is well defined. For if v has compact support, then $v * \phi \in \mathscr{D}$, and $L\phi \in C^\infty$; if u has compact support, then again $L\phi \in C^\infty$, since $v * \phi \in C^\infty$. Also, $\tau_x L = L\tau_x$, for all $x \in R^n$. These assertions follow from Theorems 6.30 and 6.35.

The functional $\phi \to (L\check{\phi})(0)$ is in fact a distribution. To see this, suppose $\phi_i \to 0$ in $\mathscr{D}$. By (*a*) of Theorem 6.33, $v * \check{\phi}_i \to 0$ in C^∞; if, in addition, v has compact support then $v * \check{\phi}_i \to 0$ in $\mathscr{D}$. It follows, in either case, that $(L\check{\phi}_i)(0) \to 0$.

The proof of (*b*) of Theorem 6.33 now shows that this distribution, which we shall denote by $u * v$, is related to L by the formula

(2) $$L\phi = (u * v) * \phi \qquad (\phi \in \mathscr{D}).$$

In other words, $u * v \in \mathscr{D}'$ is characterized by

(3) $$(u * v) * \phi = u * (v * \phi) \qquad (\phi \in \mathscr{D}).$$

6.37 Theorem *Suppose* $u \in \mathscr{D}'$, $v \in \mathscr{D}'$, $w \in \mathscr{D}'$.

(*a*) *If at least one of* u, v *has compact support, then* $u * v = v * u$.

(*b*) *If* S_u *and* S_v *are the supports of* u *and* v, *and if at least one of these is compact, then*
$$S_{u*v} \subset S_u + S_v.$$

(*c*) *If at least two of the supports* S_u, S_v, S_w *are compact, then* $(u * v) * w = u * (v * w)$.

(*d*) *If* δ *is the Dirac measure and* α *is a multi-index, then*
$$D^\alpha u = (D^\alpha \delta) * u.$$
In particular, $u = \delta * u$.

(*e*) *If at least one of the sets* S_u, S_v *is compact, then*

$$D^\alpha(u * v) = (D^\alpha u) * v = u * (D^\alpha v)$$

for every multi-index α.

Note: The associative law (*c*) depends strongly on the stated hypotheses; see Exercise 24.

PROOF. (*a*) Pick $\phi \in \mathscr{D}$, $\psi \in \mathscr{D}$. Since convolution of functions is commutative, (*c*) of Theorem 6.30 implies that

$$\begin{aligned}(u * v) * (\phi * \psi) &= u * (v * (\phi * \psi)) \\ &= u * ((v * \phi) * \psi) = u * (\psi * (v * \phi)).\end{aligned}$$

If S_v is compact, apply (*c*) of Theorem 6.30 once more; if S_u is compact, apply (*d*) of Theorem 6.35; in either case

$$(1) \qquad (u * v) * (\phi * \psi) = (u * \psi) * (v * \phi).$$

Since $\phi * \psi = \psi * \phi$, the same computation gives

$$(2) \qquad (v * u) * (\phi * \psi) = (v * \phi) * (u * \psi).$$

The two right members of (1) and (2) are convolutions of functions (one in $\mathscr{D}$, one in C^∞); hence they are equal. Thus

$$(3) \qquad ((u * v) * \phi) * \psi = ((v * u) * \phi) * \psi.$$

Two applications of the uniqueness argument used at the end of the proof of Theorem 6.33 now give $u * v = v * u$.

(*b*) If $\phi \in \mathscr{D}$, a simple computation gives

$$(4) \qquad (u * v)(\phi) = u((v * \check{\phi})^\vee).$$

By (*a*) we may assume, without loss of generality, that S_v is compact. The proof of (*c*) of Theorem 6.35 shows that the support of $v * \check{\phi}$ lies in $S_v - S_\phi$. By (4), $(u * v)(\phi) = 0$ unless S_u intersects $S_\phi - S_v$, that is, unless S_ϕ intersects $S_u + S_v$.

(*c*) We conclude from (*b*) that both

$$(u * v) * w \qquad \text{and} \qquad u * (v * w)$$

are defined if at most one of the sets S_u, S_v, S_w fails to be compact. If $\phi \in \mathscr{D}$, it follows directly from Definition 6.36 that

$$(5) \qquad (u * (v * w)) * \phi = u * ((v * w) * \phi) = u * (v * (w * \phi)).$$

If S_w is compact, then

$$(6) \qquad ((u * v) * w) * \phi = (u * v) * (w * \phi) = u * (v * (w * \phi))$$

because $w * \phi \in \mathscr{D}$, by (c) of Theorem 6.35. Comparison of (5) and (6) gives (c) whenever S_w is compact.

If S_w is not compact, then S_u is compact, and the preceding case, combined with the commutative law (a), gives

$$u * (v * w) = u * (w * v) = (w * v) * u$$
$$= w * (v * u) = w * (u * v) = (u * v) * w.$$

(d) If $\phi \in \mathscr{D}$, then $\delta * \phi = \phi$, because

$$(\delta * \phi)(x) = \delta(\tau_x \check{\phi}) = (\tau_x \check{\phi})(0) = \check{\phi}(-x) = \phi(x).$$

Hence (c) above and (b) of Theorem 6.30 give

$$(D^\alpha u) * \phi = u * D^\alpha \phi = u * D^\alpha(\delta * \phi) = u * (D^\alpha \delta) * \phi.$$

Finally, (e) follows from (d), (c), and (a):

$$D^\alpha(u * v) = (D^\alpha \delta) * (u * v) = ((D^\alpha \delta) * u) * v = (D^\alpha u) * v$$

and

$$((D^\alpha \delta) * u) * v = (u * D^\alpha \delta) * v = u * ((D^\alpha \delta) * v) = u * D^\alpha v. \quad ////$$

Exercises

1. Suppose f is a complex continuous function in R^n, with compact support. Prove that $\psi P_j \to f$ uniformly on R^n, for some $\psi \in \mathscr{D}$ and for some sequence $\{P_j\}$ of polynomials.
2. Show that the metrizable topology for $\mathscr{D}(\Omega)$ that was rejected in Section 6.2 is not complete for any Ω.
3. If E is an arbitrary closed subset of R^n, show that there is an $f \in C^\infty(R^n)$ such that $f(x) = 0$ for every $x \in E$ and $f(x) > 0$ for every other $x \in R^n$.
4. Suppose $\Lambda \in \mathscr{D}'(\Omega)$ and $\Lambda\phi \geq 0$ whenever $\phi \in \mathscr{D}(\Omega)$ and $\phi \geq 0$. Prove that Λ is then a positive measure in Ω (which is finite on compact sets).
5. Prove that the numbers $c_{\alpha\beta}$ in the Leibniz formula are
$$c_{\alpha\beta} = \prod_{i=1}^{n} \frac{\alpha_i\,!}{\beta_i\,!(\alpha_i - \beta_i)!}$$
6. (a) Suppose $c_m = \exp\{-(m!)!\}$, $m = 0, 1, 2, \ldots$. Does the series
$$\sum_{m=0}^{\infty} c_m (D^m \phi)(0)$$
converge for every $\phi \in C^\infty(R)$?
(b) Let Ω be open in R^n, suppose $\Lambda_i \in \mathscr{D}'(\Omega)$, and suppose that all Λ_i have their supports in some fixed compact $K \subset \Omega$. Prove that the sequence $\{\Lambda_i\}$ cannot converge in $\mathscr{D}'(\Omega)$ unless the orders of the Λ_j are bounded. *Hint:* Use the Banach-Steinhaus theorem.
(c) Can the assumption about the supports be dropped in (b)?

7. Let $\Omega = (0, \infty)$. Define

$$\Lambda\phi = \sum_{m=1}^{\infty} (D^m\phi)\left(\frac{1}{m}\right) \qquad [\phi \in \mathscr{D}(\Omega)].$$

Prove that Λ is a distribution of infinite order in Ω. Prove that Λ cannot be extended to a distribution in R; that is, there exists no $\Lambda_0 \in \mathscr{D}'(R)$ such that $\Lambda_0 = \Lambda$ in $(0, \infty)$.

8. Characterize all distributions whose supports are finite sets.

9. (*a*) Prove that a set $E \subset \mathscr{D}(\Omega)$ is bounded if and only if

$$\sup \{|\Lambda\phi|: \phi \in E\} < \infty$$

for every $\Lambda \in \mathscr{D}'(\Omega)$.

(*b*) Suppose $\{\phi_j\}$ is a sequence in $\mathscr{D}(\Omega)$ such that $\{\Lambda\phi_j\}$ is a bounded sequence of numbers, for every $\Lambda \in \mathscr{D}'(\Omega)$. Prove that some subsequence of $\{\phi_j\}$ converges, in the topology of $\mathscr{D}(\Omega)$.

(*c*) Suppose $\{\Lambda_j\}$ is a sequence in $\mathscr{D}'(\Omega)$ such that $\{\Lambda_j\phi\}$ is bounded, for every $\phi \in \mathscr{D}(\Omega)$. Prove that some subsequence of $\{\Lambda_j\}$ converges in $\mathscr{D}'(\Omega)$ and that the convergence is uniform on every bounded subset of $\mathscr{D}(\Omega)$. *Hint:* By the Banach-Steinhaus theorem, the restrictions of the Λ_j to $\mathscr{D}_K$ are equicontinuous. Apply Ascoli's theorem.

10. Suppose $\{f_i\}$ is a sequence of locally integrable functions in Ω (an open set in R^n) and

$$\lim_{i\to\infty} \int_K |f_i(x)|\, dx = 0$$

for every compact $K \subset \Omega$. Prove that then $D^\alpha f_i \to 0$ in $\mathscr{D}'(\Omega)$, as $i \to \infty$, for every multi-index α.

11. Suppose Ω is open in R^2, and $\{f_i\}$ is a sequence of harmonic functions in Ω that converges in the distribution sense to some $\Lambda \in \mathscr{D}'(\Omega)$; explicitly, the assumption is that

$$\Lambda\phi = \lim_{i\to\infty} \int_\Omega f_i(x)\phi(x)\, dx \qquad [\phi \in \mathscr{D}(\Omega)].$$

Prove then that $\{f_i\}$ converges uniformly on every compact subset of Ω and that Λ is a harmonic function. *Hint:* If f is harmonic, $f(x)$ is the average of f over small circles centered at x.

12. Recall that δ (the Dirac measure) is the distribution defined by $\delta(\phi) = \phi(0)$, for $\phi \in \mathscr{D}(R)$. For which $f \in C^\infty(R)$ is it true that $f\delta' = 0$? Answer the same question for $f\delta''$. Conclude that a function $f \in C^\infty(R)$ may vanish on the support of a distribution $\Lambda \in \mathscr{D}'(R)$ although $f\Lambda \neq 0$.

13. If $\phi \in \mathscr{D}(\Omega)$ and $\Lambda \in \mathscr{D}'(\Omega)$, does either of the statements

$$\phi\Lambda = 0, \quad \Lambda\phi = 0$$

imply the other?

14. Suppose K is the closed unit ball in R^n, $\Lambda \in \mathscr{D}'(R^n)$ has its support in K, and $f \in C^\infty(R^n)$ vanishes on K. Prove that $f\Lambda = 0$. Find other sets K for which this is true. (Compare with Exercise 12.)

15. Suppose $K \subset V \subset \Omega$, K is compact, V and Ω are open in R^n, $\Lambda \in \mathscr{D}'(\Omega)$ has its support in K, and $\{\phi_i\} \subset \mathscr{D}(\Omega)$ satisfies

$$(a) \qquad \lim_{i\to\infty}\left[\sup_{x\in V} |(D^\alpha \phi_i)(x)|\right] = 0$$

for every multi-index α. Prove that then

$$\lim_{i\to\infty} \Lambda(\phi_i) = 0.$$

16. The preceding statement becomes false if V is replaced by K in the hypothesis (*a*). Show this by means of the following example, in which $\Omega = R$. Choose $c_1 > c_2 > \cdots > 0$, such that $\sum c_j < \infty$; define

$$\Lambda\phi = \sum_{j=1}^{\infty} (\phi(c_j) - \phi(0)) \qquad (\phi \in \mathscr{D}(R));$$

and consider functions $\phi_i \in \mathscr{D}(R)$ such that $\phi_i(x) = 0$ if $x \le c_{i+1}$, $\phi_i(x) = 1/i$ if $c_i \le x \le c_1$. Show also that this Λ is a distribution of order 1.

However, for certain K, V *can* be replaced by K in the hypothesis (*a*) of Exercise 15. Show that this is so when K is the closed unit ball of R^n. Find other sets K for which this is true.

17. If $\Lambda \in \mathscr{D}'(R)$ has order N, show that $\Lambda = D^{N+2}f$, for some continuous function f. If $\Lambda = \delta$, what are the possibilities for f?

18. Express $\delta \in \mathscr{D}'(R^2)$ in the form given by Theorem 6.27, as explicitly as you can.

19. Suppose $\Lambda \in \mathscr{D}'(\Omega)$, $\phi \in \mathscr{D}(\Omega)$, and $(D^\alpha\phi)(x) = 0$ for every x in the support of Λ and for every multi-index α. Prove that $\Lambda\phi = 0$. *Suggestion:* Do it first for distributions with compact support, by the method used in Theorem 6.25.

20. Prove that every continuous linear functional on $C^\infty(\Omega)$ is of the form $f \to \Lambda f$, where Λ is a distribution with compact support in Ω; this is a converse to (*d*) of Theorem 6.24.

21. Let $C^\infty(T)$ be the space of all infinitely differentiable complex functions on the unit circle T in $\mathbb{C}$. One may regard $C^\infty(T)$ as the subspace of $C^\infty(R)$ consisting of those functions that have period 2π. Suppose

$$f(z) = \sum_{n=0}^{\infty} a_n z^n$$

converges in the open unit disc U in $\mathbb{C}$. Prove that each of the following three properties of f implies the other two:

(*a*) There exist $p < \infty$ and $\gamma < \infty$ such that

$$|a_n| \le \gamma \cdot n^p \qquad (n = 1, 2, 3, \ldots).$$

(*b*) There exist $p < \infty$ and $\gamma < \infty$ such that

$$|f(z)| \le \gamma \cdot (1 - |z|)^{-p} \qquad (z \in U).$$

(*c*) $\lim_{r\to 1} \int_{-\pi}^{\pi} f(re^{i\theta})\phi(e^{i\theta})\, d\theta$ exists (as a complex number) for every $\phi \in C^\infty(T)$.

22. For $u \in \mathscr{D}'(R)$, show that

$$\frac{u - \tau_x u}{x} \to Du \quad \text{in } \mathscr{D}'(R),$$

as $x \to 0$. (The derivative of u may thus still be regarded as a limit of quotients.)

23. Suppose $\{f_i\}$ is a sequence of locally integrable functions in R^n, such that

$$\lim_{i \to \infty} (f_i * \phi)(x)$$

exists, for each $\phi \in \mathscr{D}(R^n)$ and each $x \in R^n$. Prove that then $\{D^\alpha(f_i * \phi)\}$ converges uniformly on compact sets, for each multi-index α.

24. Let H be the *Heaviside function* on R, defined by

$$H(x) = \begin{cases} 1 & \text{if } x > 0, \\ 0 & \text{if } x \le 0, \end{cases}$$

and let δ be the Dirac measure.

(*a*) Show that $(H * \phi)(x) = \int_{-\infty}^{x} \phi(s)\, ds$, if $\phi \in \mathscr{D}(R)$.

(*b*) Show that $\delta' * H = \delta$.

(*c*) Show that $1 * \delta' = 0$. (Here 1 denotes the locally integrable function whose value is 1 at every point and which is thought of as a distribution.)

(*d*) It follows that the associative law fails:

$$1 * (\delta' * H) = 1 * \delta = 1,$$

but

$$(1 * \delta') * H = 0 * H = 0.$$

25. Here is another characterization of convolutions analogous to Theorem 6.33. Suppose L is a continuous linear mapping of $\mathscr{D}$ into C^∞ which commutes with every D^α, that is,

(*a*) $$LD^\alpha\phi = D^\alpha L\phi \qquad (\phi \in \mathscr{D}).$$

Then there is a $u \in \mathscr{D}'$ such that

$$L\phi = u * \phi.$$

Suggestion: Fix $\phi \in \mathscr{D}$, put

$$h(x) = (\tau_{-x} L \tau_x \phi)(0) = (L\tau_x\phi)(x) \qquad (x \in R^n),$$

let D_e be the directional derivative used in the proof of Theorem 6.30, and show that

$$(D_e h)(x) = (D_e L\tau_x\phi)(x) - (L\tau_x D_e\phi)(x),$$

which is 0 if (*a*) holds. Thus $h(x) = h(0)$, which implies that $\tau_x L = L\tau_x$.

Can the assumption that the range of L is in C^∞ be weakened?

26. If $f \in L^1((-\infty, -\delta) \cup (\delta, \infty))$ for every $\delta > 0$, define its *principal value integral* to be

$$PV \int_{-\infty}^{\infty} f(x)\, dx = \lim_{\delta \to 0} \left(\int_{-\infty}^{-\delta} + \int_{\delta}^{\infty} \right) f(x)\, dx,$$

if the limit exists. For $\phi \in \mathscr{D}(R)$, put

$$\Lambda\phi = \int_{-\infty}^{\infty} \phi(x) \log |x| \, dx.$$

Show that

$$\Lambda'\phi = PV \int_{-\infty}^{\infty} \phi(x) \frac{dx}{x},$$

$$\Lambda''\phi = -PV \int_{-\infty}^{\infty} \frac{\phi(x) - \phi(0)}{x^2} \, dx.$$

27. Find all distributions $u \in \mathscr{D}'(R^n)$ that satisfy at least one of the following two conditions:
(*a*) $\tau_x u = u$ for every $x \in R^n$,
(*b*) $D^\alpha u = 0$ for every α with $|\alpha| = 1$.

CHAPTER 7

FOURIER TRANSFORMS

Basic Properties

7.1 Notations (*a*) The *normalized Lebesgue measure* on R^n is the measure m_n defined by

$$dm_n(x) = (2\pi)^{-n/2}\, dx.$$

The factor $(2\pi)^{-n/2}$ simplifies the appearance of the inversion theorem 7.7 and the Plancherel theorem 7.9. The usual Lebesgue spaces L^p, or $L^p(R^n)$, will be normed by means of m_n:

$$\|f\|_p = \left\{\int_{R^n} |f|^p\, dm_n\right\}^{1/p} \qquad (1 \le p < \infty).$$

It is also convenient to redefine the convolution of two functions on R^n by

$$(f * g)(x) = \int_{R^n} f(x-y)g(y)\, dm_n(y)$$

whenever the integral exists.

(*b*) For each $t \in R^n$, the *character* e_t is the function defined by

$$e_t(x) = e^{it\cdot x} = \exp\{i(t_1x_1 + \cdots + t_nx_n)\} \qquad (x \in R^n).$$

Each e_t satisfies the functional equation

$$e_t(x+y) = e_t(x)e_t(y).$$

Thus e_t is a homomorphism of the additive group R^n into the multiplicative group of the complex numbers of absolute value 1.

(*c*) The *Fourier transform* of a function $f \in L^1(R^n)$ is the function $\hat{f}$ defined by

$$\hat{f}(t) = \int_{R^n} fe_{-t}\, dm_n \qquad (t \in R^n).$$

The term "Fourier transform" is often also used for the *mapping* that takes f to $\hat{f}$. Note that

$$\hat{f}(t) = (f * e_t)(0).$$

(*d*) If α is a multi-index, then

$$D_\alpha = (i)^{-|\alpha|}D^\alpha = \left(\frac{1}{i}\frac{\partial}{\partial x_1}\right)^{\alpha_1} \cdots \left(\frac{1}{i}\frac{\partial}{\partial x_n}\right)^{\alpha_n}$$

The use of D_α in place of D^α simplifies some of the formalism. Note that

$$D_\alpha e_t = t^\alpha e_t$$

where, as before, $t^\alpha = t_1^{\alpha_1} \cdots t_n^{\alpha_n}$. If P is a polynomial of n variables, with complex coefficients, say

$$P(\xi) = \sum c_\alpha \xi^\alpha = \sum c_\alpha \xi_1^{\alpha_1} \cdots \xi_n^{\alpha_n},$$

the differential operators $P(D)$ and $P(-D)$ are defined by

$$P(D) = \sum c_\alpha D_\alpha, \qquad P(-D) = \sum (-1)^{|\alpha|} c_\alpha D_\alpha.$$

It follows that

$$P(D)e_t = P(t)e_t \qquad (t \in R^n).$$

(*e*) The translation operators τ_x are defined, as before, by

$$(\tau_x f)(y) = f(y-x) \qquad (x, y \in R^n).$$

7.2 Theorem *Suppose $f, g \in L^1(R^n)$, $x \in R^n$. Then*

(*a*) $(\tau_x f)^\wedge = e_{-x}\hat{f}$;
(*b*) $(e_x f)^\wedge = \tau_x \hat{f}$;
(*c*) $(f * g)^\wedge = \hat{f}\hat{g}$.
(*d*) *If $\lambda > 0$ and $h(x) = f(x/\lambda)$, then $\hat{h}(t) = \lambda^n \hat{f}(\lambda t)$.*

PROOF. It follows from the definitions that

$$(\tau_x f)^\wedge(t) = \int (\tau_x f) \cdot e_{-t} = \int f \cdot \tau_{-x} e_{-t} = \int f \cdot e_{-t}(x) e_{-t} = e_{-x}(t)\hat{f}(t)$$

and

$$(e_x f)^\wedge(t) = \int e_x f e_{-t} = \int f e_{-(t-x)} = (\tau_x \hat{f})(t).$$

An application of Fubini's theorem gives (*c*); (*d*) is obtained by a linear change of variables in the definition of $\hat{f}$. ////

7.3 Rapidly decreasing functions This name is sometimes given to those $f \in C^\infty(R^n)$ for which

(1) $$\sup_{|\alpha| \le N} \sup_{x \in R^n} (1 + |x|^2)^N |(D_\alpha f)(x)| < \infty$$

for $N = 0, 1, 2, \ldots$. (Recall that $|x|^2 = \sum x_i^2$.) In other words, the requirement is that $P \cdot D_\alpha f$ is a bounded function on R^n, for every polynomial P and for every multi-index α. Since this is true with $(1 + |x|^2)^N P(x)$ in place of $P(x)$, it follows that every $P \cdot D_\alpha f$ lies in $L^1(R^n)$.

These functions form a vector space, denoted by $\mathscr{S}_n$, in which the countable collection of norms (1) defines a locally convex topology, as described in Theorem 1.37.

It is clear that $\mathscr{D}(R^n) \subset \mathscr{S}_n$.

7.4 Theorem

(*a*) *$\mathscr{S}_n$ is a Fréchet space.*

(*b*) *If P is a polynomial, $g \in \mathscr{S}_n$, and α is a multi-index, then each of the three mappings*

$$f \to Pf, \qquad f \to gf, \qquad f \to D_\alpha f$$

is a continuous linear mapping of $\mathscr{S}_n$ into $\mathscr{S}_n$.

(*c*) *If $f \in \mathscr{S}_n$ and P is a polynomial, then*

$$(P(D)f)^\wedge = P\hat{f} \qquad \text{and} \qquad (Pf)^\wedge = P(-D)\hat{f}.$$

(*d*) *The Fourier transform is a continuous linear mapping of $\mathscr{S}_n$ into $\mathscr{S}_n$.*

[Part (*d*) will be strengthened in Theorem 7.7.]

PROOF. (*a*) Suppose $\{f_i\}$ is a Cauchy sequence in $\mathscr{S}_n$. For every pair of multi-indices α and β the functions $x^\beta D^\alpha f_i(x)$ converge then (uniformly on R^n) to a bounded function $g_{\alpha\beta}$, as $i \to \infty$. It follows that

$$g_{\alpha\beta}(x) = x^\beta D^\alpha g_{00}(x)$$

and hence that $f_i \to g_{00}$ in $\mathscr{S}_n$. Thus $\mathscr{S}_n$ is complete.

(*b*) If $f \in \mathscr{S}_n$, it is obvious that $D_\alpha f \in \mathscr{S}_n$, and the Leibniz formula implies that Pf and gf are also in $\mathscr{S}_n$. The continuity of the three mappings is now an easy consequence of the closed graph theorem.

(*c*) If $f \in \mathscr{S}_n$, so is $P(D)f$, by (*b*), and

$$(P(D)f) * e_t = f * P(D)e_t = f * P(t)e_t = P(t)[f * e_t].$$

Evaluation of these functions at the origin of R_n gives the first part of (*c*), namely,

$$(P(D)f)\hat{}(t) = P(t)\hat{f}(t).$$

If $t = (t_1, \ldots, t_n)$ and $t' = (t_1 + \varepsilon, t_2, \ldots, t_n)$, $\varepsilon \neq 0$, then

$$\frac{\hat{f}(t') - \hat{f}(t)}{i\varepsilon} = \int_{R^n} x_1 f(x) \frac{e^{-ix_1\varepsilon} - 1}{ix_1\varepsilon} e^{-ix \cdot t}\, dm_n(x).$$

The dominated convergence theorem can be applied, since $x_1 f \in L^1$, and yields

$$-\frac{1}{i}\frac{\partial}{\partial t_1}\hat{f}(t) = \int_{R^n} x_1 f(x) e^{-ix \cdot t}\, dm_n(x).$$

This is the case $P(x) = x_1$ of the second part of (*c*); the general case follows by iteration.

(*d*) Suppose $f \in \mathscr{S}_n$ and $g(x) = (-1)^{|\alpha|}x^\alpha f(x)$. Then $g \in \mathscr{S}_n$; now (*c*) implies that $\hat{g} = D_\alpha \hat{f}$ and $P \cdot D_\alpha \hat{f} = P \cdot \hat{g} = (P(D)g)\hat{}$, which is a bounded function, since $P(D)g \in L^1(R^n)$. This proves that $\hat{f} \in \mathscr{S}_n$. If $f_i \to f$ in $\mathscr{S}_n$, then $f_i \to f$ in $L^1(R^n)$. Therefore $\hat{f}_i(t) \to \hat{f}(t)$ for all $t \in R^n$. That $f \to \hat{f}$ is a *continuous* mapping of $\mathscr{S}_n$ into $\mathscr{S}_n$ follows now from the closed graph theorem. ////

7.5 Theorem *If $f \in L^1(R^n)$, then $\hat{f} \in C_0(R^n)$, and $\|\hat{f}\|_\infty \leq \|f\|_1$.*

Here $C_0(R^n)$ is the supremum-normed Banach space of all complex continuous functions on R^n that vanish at infinity.

PROOF. Since $|e_t(x)| = 1$, it is clear that

(1) $$|\hat{f}(t)| \leq \|f\|_1 \qquad (f \in L^1, t \in R^n).$$

Since $\mathscr{D}(R^n) \subset \mathscr{S}_n$, $\mathscr{S}_n$ is dense in $L^1(R^n)$. To each $f \in L^1(R^n)$ correspond functions $f_i \in \mathscr{S}_n$ such that $\|f - f_i\|_1 \to 0$. Since $\hat{f}_i \in \mathscr{S}_n \subset$

$C_0(R^n)$ and since (1) implies that $\hat{f}_i \to \hat{f}$ uniformly on R^n, the proof is complete. ////

The following lemma will be used in the proof of the inversion theorem. It depends on the particular normalization that was chosen for m_n.

7.6 Lemma *If ϕ_n is defined on R^n by*

(1) $$\phi_n(x) = \exp\{-\tfrac{1}{2}|x|^2\}$$

then $\phi_n \in \mathscr{S}_n$, $\hat{\phi}_n = \phi_n$, and

(2) $$\phi_n(0) = \int_{R^n} \hat{\phi}_n \, dm_n.$$

PROOF. It is clear that $\phi_n \in \mathscr{S}_n$. Since ϕ_1 satisfies the differential equation

(3) $$y' + xy = 0,$$

a short computation, or an appeal to (*c*) of Theorem 7.4, shows that $\hat{\phi}_1$ also satisfies (3). Hence $\hat{\phi}_1/\phi_1$ is a constant. Since $\phi_1(0) = 1$ and

$$\hat{\phi}_1(0) = \int_R \phi_1 \, dm_1 = (2\pi)^{-1/2} \int_{-\infty}^{\infty} \exp\{-\tfrac{1}{2}x^2\} \, dx = 1,$$

we conclude that $\hat{\phi}_1 = \phi_1$. Next,

(4) $$\phi_n(x) = \phi_1(x_1) \cdots \phi_1(x_n) \qquad (x \in R^n)$$

so that

(5) $$\hat{\phi}_n(t) = \hat{\phi}_1(t_1) \cdots \hat{\phi}_1(t_n) \qquad (t \in R^n).$$

It follows that $\hat{\phi}_n = \phi_n$ for all n. Since $\hat{\phi}_n(0) = \int \phi_n \, dm_n$, by definition, and since $\hat{\phi}_n = \phi_n$, we obtain (2). ////

7.7 The inversion theorem

(*a*) *If $g \in \mathscr{S}_n$, then*

(1) $$g(x) = \int_{R^n} \hat{g} e_x \, dm_n \qquad (x \in R^n).$$

(*b*) ***The Fourier transform is a continuous, linear, one-to-one mapping of $\mathscr{S}_n$ onto $\mathscr{S}_n$, of period 4, whose inverse is also continuous.***

(c) If $f \in L^1(R^n)$, $\hat{f} \in L^1(R^n)$, *and*

(2) $$f_0(x) = \int_{R^n} \hat{f} e_x \, dm_n \qquad (x \in R^n),$$

then $f(x) = f_0(x)$ *for almost every* $x \in R^n$.

PROOF. If f and g are in $L^1(R^n)$, Fubini's theorem can be applied to the double integral

$$\int_{R^n} \int_{R^n} f(x) g(y) e^{-ix \cdot y} \, dm_n(x) \, dm_n(y)$$

to yield the identity

(3) $$\int_{R^n} \hat{f} g \, dm_n = \int_{R^n} f \hat{g} \, dm_n.$$

To prove part (*a*), take $g \in \mathscr{S}_n$, $\phi \in \mathscr{S}_n$, $f(x) = \phi(x/\lambda)$, where $\lambda > 0$. By (*d*) of Theorem 7.2, (3) becomes

$$\int_{R^n} g(t) \lambda^n \hat{\phi}(\lambda t) \, dm_n(t) = \int_{R^n} \phi\left(\frac{y}{\lambda}\right) \hat{g}(y) \, dm_n(y),$$

or

(4) $$\int_{R^n} g\left(\frac{t}{\lambda}\right) \hat{\phi}(t) \, dm_n(t) = \int_{R^n} \phi\left(\frac{y}{\lambda}\right) \hat{g}(y) \, dm_n(y).$$

As $\lambda \to \infty$, $g(t/\lambda) \to g(0)$ and $\phi(y/\lambda) \to \phi(0)$, boundedly, so that the dominated convergence theorem can be applied to the two integrals in (4). The result is

(5) $$g(0) \int_{R^n} \hat{\phi} \, dm_n = \phi(0) \int_{R^n} \hat{g} \, dm_n \qquad (g, \phi \in \mathscr{S}_n).$$

If we specialize ϕ to be the function ϕ_n of Lemma 7.6, (5) gives the case $x = 0$ of the inversion formula (1). The general case follows from this, since (*a*) of Theorem 7.2 yields

$$g(x) = (\tau_{-x} g)(0) = \int_{R^n} (\tau_{-x} g)^\wedge \, dm_n = \int_{R^n} \hat{g} e_x \, dm_n.$$

This completes part (*a*).

To prove part (*b*), we introduce the temporary notation $\Phi g = \hat{g}$. The inversion formula (1) shows that Φ is one-to-one on $\mathscr{S}_n$, since $\hat{g} = 0$ obviously implies $g = 0$. It also shows that

(6) $$\Phi^2 g = \check{g}$$

where, we recall, $\check{g}(x) = g(-x)$, and hence that $\Phi^4 g = g$. It follows that Φ maps $\mathscr{S}_n$ *onto* $\mathscr{S}_n$. The continuity of Φ has already been proved in Theorem 7.4. To prove the continuity of Φ^{-1}, one can now either refer to the open mapping theorem or to the fact that $\Phi^{-1} = \Phi^3$.

To prove (*c*), we return to the identity (3), with $g \in \mathscr{S}_n$. Insert the inversion formula (1) into (3) and use Fubini's theorem, to obtain

$$\int_{R^n} f_0 \hat{g}\, dm_n = \int_{R^n} f \hat{g}\, dm_n \qquad (g \in \mathscr{S}_n). \tag{7}$$

By (*b*), the functions $\hat{g}$ cover all of $\mathscr{S}_n$. Since $\mathscr{D}(R^n) \subset \mathscr{S}_n$, (7) implies that

$$\int_{R^n} (f_0 - f)\phi\, dm_n = 0 \tag{8}$$

for every $\phi \in \mathscr{D}(R^n)$, hence (by a uniform approximation described in Exercise 1 of Chapter 6) for every continuous ϕ with compact support. It follows that $f_0 - f = 0$ a.e. ////

7.8 Theorem *If $f \in \mathscr{S}_n$ and $g \in \mathscr{S}_n$, then*

(*a*) $f * g \in \mathscr{S}_n$, *and*
(*b*) $(fg)^\wedge = \hat{f} * \hat{g}$.

PROOF. By (*c*) of Theorem 7.2, $(f * g)^\wedge = \hat{f}\hat{g}$, or

$$\Phi(f * g) = \Phi f \cdot \Phi g, \tag{1}$$

in the notation used in the proof of (*b*) of Theorem 7.7. With $\hat{f}$ and $\hat{g}$ in place of f and g, (1) becomes

$$\Phi(\hat{f} * \hat{g}) = \Phi^2 f \cdot \Phi^2 g = \check{f}\check{g} = (fg)^\vee = \Phi^2(fg). \tag{2}$$

Now apply Φ^{-1} to both sides of (2) to obtain (*b*). Note that $fg \in \mathscr{S}_n$; hence (*b*) implies that $\hat{f} * \hat{g} \in \mathscr{S}_n$, and this gives (*a*), since the Fourier transform maps $\mathscr{S}_n$ *onto* $\mathscr{S}_n$. ////

7.9 The Plancherel theorem *There is a linear isometry Ψ of $L^2(R^n)$ onto $L^2(R^n)$ which is uniquely determined by the requirement that*

$$\Psi f = \hat{f} \qquad \text{for every } f \in \mathscr{S}_n.$$

Observe that the equality $\Psi f = \hat{f}$ extends from $\mathscr{S}_n$ to $L^1 \cap L^2$, since $\mathscr{S}_n$ is dense in L^2 as well as in L^1. This gives consistency: The domain of Ψ is L^2, $\hat{f}$ was defined in Section 7.1 for all $f \in L^1$, and $\Psi f = \hat{f}$ whenever both definitions are applicable. Thus Ψ extends the Fourier transform from $L^1 \cap L^2$ to

L^2. This extension Ψ is still called the Fourier transform (sometimes the *Fourier-Plancherel* transform), and the notation $\hat{f}$ will continue to be used in place of Ψf, for any $f \in L^2(R^n)$.

PROOF. If f and g are in $\mathscr{S}_n$, the inversion theorem yields

$$\int_{R^n} f\bar{g}\, dm_n = \int_{R^n} \bar{g}(x)\, dm_n(x) \int_{R^n} \hat{f}(t)e^{ix\cdot t}\, dm_n(t)$$

$$= \int_{R^n} \hat{f}(t)\, dm_n(t) \int_{R^n} \bar{g}(x)e^{ix\cdot t}\, dm_n(x).$$

The last inner integral is the complex conjugate of $\hat{g}(t)$. We thus get the *Parseval formula*

$$\int_{R^n} f\bar{g}\, dm_n = \int_{R^n} \hat{f}\bar{\hat{g}}\, dm_n \qquad (f, g \in \mathscr{S}_n). \tag{1}$$

If $g = f$, (1) specializes to

$$\|f\|_2 = \|\hat{f}\|_2 \qquad (f \in \mathscr{S}_n). \tag{2}$$

Note that $\mathscr{S}_n$ is dense in $L^2(R^n)$, for the same reason that $\mathscr{S}_n$ is dense in $L^1(R^n)$. Thus (2) shows that $f \to \hat{f}$ is an isometry (relative to the L^2-metric) of the dense subspace $\mathscr{S}_n$ of $L^2(R^n)$ *onto* $\mathscr{S}_n$. (The mapping is onto by the inversion theorem.) It follows, by elementary metric space arguments, that $f \to \hat{f}$ has a unique continuous extension $\Psi: L^2(R^n) \to L^2(R^n)$ and that this Ψ is a linear isometry onto $L^2(R^n)$. Some details of this are given in Exercise 13. ////

It should be noted that the Parseval formula (1) remains true for arbitrary f and g in $L^2(R^n)$.

That the Fourier transform is an L^2-isometry is one of the most important features of the whole subject.

Tempered Distributions

Before we define these, we establish the following relation between $\mathscr{S}_n$ and $\mathscr{D}(R^n)$.

7.10 Theorem

(*a*) *$\mathscr{D}(R^n)$ is dense in $\mathscr{S}_n$.*

(*b*) *The identity mapping of $\mathscr{D}(R^n)$ into $\mathscr{S}_n$ is continuous.*

These statements refer, of course, to the usual topologies of $\mathscr{D}(R^n)$ and $\mathscr{S}_n$, as defined in Sections 6.3 and 7.3.

PROOF. (*a*) Choose $f \in \mathscr{S}_n$, $\psi \in \mathscr{D}(R^n)$ so that $\psi = 1$ on the unit ball of R^n, and put

(1) $$f_r(x) = f(x)\psi(rx) \qquad (x \in R^n, r > 0).$$

Then $f_r \in \mathscr{D}(R^n)$. If P is a polynomial and α is a multi-index, then

$$P(x)D^\alpha(f - f_r)(x) = P(x) \sum_{\beta \le \alpha} c_{\alpha\beta}(D^{\alpha-\beta}f)(x)r^{|\beta|}D^\beta[1 - \psi](rx).$$

Our choice of ψ shows that $D^\beta[1 - \psi](rx) = 0$ for every multi-index β when $|x| \le 1/r$. Since $f \in \mathscr{S}_n$, we have $P \cdot D^{\alpha-\beta}f \in C_0(R^n)$ for all $\beta \le \alpha$. It follows that the above sum tends to 0, uniformly on R^n, when $r \to 0$. Thus $f_r \to f$ in $\mathscr{S}_n$, and (*a*) is proved.

(*b*) If K is a compact set in R^n, the topology induced on $\mathscr{D}_K$ by $\mathscr{S}_n$ is clearly the same as its usual one (as defined in Section 1.46), since each $(1 + |x|^2)^N$ is bounded on K. The identity mapping of $\mathscr{D}_K$ into $\mathscr{S}_n$ is therefore continuous (actually, a homeomorphism), and now (*b*) follows from Theorem 6.6. ////

7.11 Definition If $i: \mathscr{D}(R^n) \to \mathscr{S}_n$ is the identity mapping, if L is a continuous linear functional on $\mathscr{S}_n$, and if

(1) $$u_L = L \circ i$$

then the continuity of i (Theorem 7.10) shows that $u_L \in \mathscr{D}'(R^n)$; the denseness of $\mathscr{D}(R^n)$ in $\mathscr{S}_n$ shows that two distinct L's cannot give rise to the same u. Thus (1) describes a vector space isomorphism between the dual space $\mathscr{S}'_n$ of $\mathscr{S}_n$, on the one hand, and a certain space of distribution on the other. The distributions that arise in this way are called *tempered*:

The tempered distributions are precisely those $u \in \mathscr{D}'(R^n)$ that have continuous extensions to $\mathscr{S}_n$.

In view of the preceding remarks, it is customary and natural to identify u_L with L. *The tempered distributions on R^n are then precisely the members of $\mathscr{S}'_n$.*

The following examples will explain the use of the word "tempered" in this connection; it indicates a growth restriction at infinity. (See also Exercise 3.)

7.12 Examples (*a*) *Every distribution with compact support is tempered.* Suppose K is the compact support of some $u \in \mathscr{D}'(R^n)$, fix $\psi \in \mathscr{D}(R^n)$ so that $\psi = 1$ in some open set containing K, and define

(1) $$\tilde{u}(f) = u(\psi f) \qquad (f \in \mathscr{S}_n).$$

If $f_i \to 0$ in $\mathscr{S}_n$, then all $D^\alpha f_i \to 0$ uniformly on R^n, hence all $D^\alpha(\psi f_i) \to 0$ uniformly on R^n, so that $\psi f_i \to 0$ in $\mathscr{D}(R^n)$. It follows that $\tilde{u}$ is continuous on $\mathscr{S}_n$. Since $\tilde{u}(\phi) = u(\phi)$ for $\phi \in \mathscr{D}(R^n)$, $\tilde{u}$ is an extension of u.

(*b*) *Suppose μ is a positive Borel measure on R^n such that*

$$\int_{R^n} (1 + |x|^2)^{-k}\, d\mu(x) < \infty \tag{2}$$

for some positive integer k. Then μ is a tempered distribution. The assertion is, more explicitly, that the formula

$$\Lambda f = \int_{R^n} f\, d\mu \tag{3}$$

defines a continuous linear functional on $\mathscr{S}_n$.

To see this, suppose $f_i \to 0$ in $\mathscr{S}_n$. Then

$$\varepsilon_i = \sup_{x \in R^n} (1 + |x|^2)^k |f_i(x)| \to 0. \tag{4}$$

Since $|\Lambda f_i|$ is at most ε_i times the integral in (2), $\Lambda f_i \to 0$. This proves the continuity of Λ.

(*c*) *Suppose $1 \le p < \infty$, $N > 0$, and g is a measurable function on R^n such that*

$$\int_{R^n} |(1 + |x|^2)^{-N} g(x)|^p\, dm_n(x) = C < \infty. \tag{5}$$

Then g is a tempered distribution.

As in (*b*), define

$$\Lambda f = \int_{R^n} fg\, dm_n. \tag{6}$$

Assume first that $p > 1$; let q be the conjugate exponent. Then Hölder's inequality gives

$$\begin{aligned} |\Lambda f| &\le C^{1/p} \left\{ \int_{R^n} |(1 + |x|^2)^N f(x)|^q\, dm_n(x) \right\}^{1/q} \\ &\le C^{1/p} B^{1/q} \sup_{x \in R^n} |(1 + |x|^2)^M f(x)|, \end{aligned} \tag{7}$$

where M is taken so large that

$$\int_{R^n} (1 + |x|^2)^{(N-M)q}\, dm_n(x) = B < \infty.$$

The inequality (7) proves that Λ is continuous on $\mathscr{S}_n$. The case $p = 1$ is even easier.

(*d*) It follows from (*c*) that every $g \in L^p(R^n)$ $(1 \le p \le \infty)$ is a tempered distribution. So is every polynomial and, more generally, every measurable function whose absolute value is majorized by some polynomial.

7.13 Theorem *If α is a multi-index, P is a polynomial, $g \in \mathscr{S}_n$, and u is a tempered distribution, then the distributions $D^\alpha u$, Pu, and gu are also tempered.*

PROOF. This follows directly from (*b*) of Theorem 7.4 and the definitions

$$(D^\alpha u)(f) = (-1)^{|\alpha|}u(D^\alpha f),$$
$$(Pu)(f) = u(Pf),$$
$$(gu)(f) = u(gf). \qquad ////$$

7.14 Definition For $u \in \mathscr{S}'_n$, define

(1) $$\hat{u}(\phi) = u(\hat{\phi}) \qquad (\phi \in \mathscr{S}_n).$$

Since $\phi \to \hat{\phi}$ is a continuous linear mapping of $\mathscr{S}_n$ into $\mathscr{S}_n$ [(*d*) of Theorem 7.4], and since u is continuous on $\mathscr{S}_n$, it follows that $\hat{u} \in \mathscr{S}'_n$.

We have thus associated with each tempered distribution u its *Fourier transform* $\hat{u}$, which is again a tempered distribution. Our next theorem will show that the formal properties of Fourier transforms of rapidly decreasing functions are preserved in the larger setting of tempered distributions.

But first there arises a consistency question that ought to be settled. If $f \in L^1(R^n)$, then f may also be regarded as a tempered distribution, say u_f, so that two definitions of the Fourier transform are available, namely, (*c*) of Section 7.1 and Definition 7.14. The question is whether they agree, i.e., whether the distribution $(u_f)^\wedge$ corresponds to the function $\hat{f}$. The answer is affirmative, because

$$(u_f)^\wedge(\phi) = u_f(\hat{\phi}) = \int f\hat{\phi} = \int \hat{f}\phi = (u_{\hat{f}})(\phi)$$

for every $\phi \in \mathscr{S}_n$. The third of these equalities is the identity (3) of Section 7.7; the others are definitions.

Since $L^2(R^n) \subset \mathscr{S}'_n$, the same question arises for the Fourier-Plancherel transform. The answer is again affirmative, by the same proof, since the identity $\int f\hat{\phi} = \int \hat{f}\phi$ persists for $f \in L^2(R^n)$ and $\phi \in \mathscr{S}_n$.

7.15 Theorem

(*a*) *The Fourier transform is a continuous, linear, one-to-one mapping of $\mathscr{S}'_n$ onto $\mathscr{S}'_n$, of period 4, whose inverse is also continuous.*

(*b*) *If $u \in \mathscr{S}'_n$ and P is a polynomial, then*

$$(P(D)u)^\wedge = P\hat{u} \quad \textit{and} \quad (Pu)^\wedge = P(-D)\hat{u}.$$

Note that these are the analogues of (*b*) of Theorem 7.7 and (*c*) of Theorem 7.4. The topology to which (*a*) refers is the weak*-topology that $\mathscr{S}_n$ induces on $\mathscr{S}'_n$. Note also that the differential operators $P(D)$ and $P(-D)$ are defined in terms of D_α, not D^α; see (*d*) of Section 7.1.

PROOF. Let W be a neighborhood of 0 in $\mathscr{S}'_n$. Then there exist functions $\phi_1, \ldots, \phi_k \in \mathscr{S}_n$ such that

$$\{u \in \mathscr{S}'_n : |u(\phi_i)| < 1 \quad \text{for} \quad 1 \le i \le k\} \subset W. \tag{1}$$

Define

$$V = \{u \in \mathscr{S}'_n : |u(\hat{\phi}_i)| < 1 \quad \text{for} \quad 1 \le i \le k\}. \tag{2}$$

Then V is a neighborhood of 0 in $\mathscr{S}'_n$, and since

$$\hat{u}(\phi) = u(\hat{\phi}) \qquad (\phi \in \mathscr{S}_n, u \in \mathscr{S}'_n), \tag{3}$$

we see that $\hat{u} \in W$ whenever $u \in V$. This proves the continuity of Φ, where we write $\Phi u = \hat{u}$. Since Φ has period 4 on $\mathscr{S}_n$, (3) shows that Φ has period 4 on $\mathscr{S}'_n$, that is, that $\Phi^4 u = u$ for every $u \in \mathscr{S}'_n$. Hence Φ is one-to-one and onto, and since $\Phi^{-1} = \Phi^3$, Φ^{-1} is continuous.

Statement (*b*) follows from (*c*) of Theorem 7.4 and from Theorem 7.13, by the computations

$$\begin{aligned}(P(D)u)\hat{}(\phi) &= (P(D)u)(\hat{\phi}) = u(P(-D)\hat{\phi}) \\ &= u((P\phi)\hat{}) = \hat{u}(P\phi) = (P\hat{u})(\phi)\end{aligned}$$

and

$$\begin{aligned}(P(-D)\hat{u})(\phi) &= \hat{u}(P(D)\phi) = u((P(D)\phi)\hat{}) \\ &= u(P\hat{\phi}) = (Pu)(\hat{\phi}) = (Pu)\hat{}(\phi),\end{aligned}$$

where ϕ is an arbitrary function in $\mathscr{S}_n$. ////

7.16 Examples We saw in (*d*) of Section 7.12 that polynomials are tempered distributions. Their Fourier transforms are easily computed. We begin with the polynomial 1; regarded as a distribution on R^n, 1 acts on test functions ϕ by the formula

$$1(\phi) = \int_{R^n} 1\phi \, dm_n = \int_{R^n} \phi \, dm_n. \tag{1}$$

Hence

$$\hat{1}(\phi) = 1(\hat{\phi}) = \int_{R^n} \hat{\phi} \, dm_n = \phi(0) = \delta(\phi), \tag{2}$$

where δ is the Dirac measure on R^n. Likewise,

$$(3)\qquad \hat{\delta}(\phi) = \delta(\hat{\phi}) = \hat{\phi}(0) = \int_{R^n} \phi \, dm_n = 1(\phi).$$

Thus (2) and (3) give the results

$$(4)\qquad \hat{1} = \delta \quad \text{and} \quad \hat{\delta} = 1.$$

If P is now an arbitrary polynomial on R^n, and if we apply (*b*) of Theorem 7.15 with $u = \delta$ and with $u = 1$, the results in (4) show that

$$(5)\qquad (P(D)\delta)\hat{} = P \quad \text{and} \quad \hat{P} = P(-D)\delta.$$

The two formulas in (4) [as well as those in (5)] can also be derived from each other by the inversion theorem, which may be stated for tempered distributions in the following way:

If $u \in \mathscr{S}'_n$, then $(\hat{u})\hat{} = \check{u}$, where $\check{u}$ is defined by

$$(6)\qquad \check{u}(\phi) = u(\check{\phi}) \qquad (\phi \in \mathscr{S}_n).$$

The proof is trivial, since $(\hat{\phi})\hat{} = \check{\phi}$, by (*a*) of Theorem 7.7:

$$(\hat{u})\hat{}(\phi) = \hat{u}(\hat{\phi}) = u((\hat{\phi})\hat{}) = u(\check{\phi}) = \check{u}(\phi).$$

Note that $\check{\delta} = \delta$.

If we combine (5) with Theorem 6.25, we find that *a distribution is the Fourier transform of a polynomial if and only if its support is the origin* (or the empty set).

The following lemma will be used in the proof of Theorem 7.19. Its analogue, with $\mathscr{D}(R^n)$ in place of $\mathscr{S}_n$, is much easier and was used without comment in the proof of Theorem 6.30.

7.17 Lemma *If $w = (1, 0, \ldots, 0) \in R^n$, if $\phi \in \mathscr{S}_n$, and if*

$$(1)\qquad \phi_\varepsilon(x) = \frac{\phi(x + \varepsilon w) - \phi(x)}{\varepsilon} \qquad (x \in R^n, \varepsilon > 0),$$

then $\phi_\varepsilon \to \partial\phi/\partial x_1$ in the topology of $\mathscr{S}_n$, as $\varepsilon \to 0$.

PROOF. The conclusion can be obtained by showing that the Fourier transform of $\phi_\varepsilon - \partial\phi/\partial x_1$ tends to 0 in $\mathscr{S}_n$, that is, by showing that

$$(2)\qquad \psi_\varepsilon \hat{\phi} \to 0 \text{ in } \mathscr{S}_n, \qquad \text{as } \varepsilon \to 0,$$

where

$$(3)\qquad \psi_\varepsilon(y) = \frac{\exp(i\varepsilon y_1) - 1}{\varepsilon} - iy_1 \qquad (y \in R^n, \varepsilon > 0).$$

If P is a polynomial and α is a multi-index, then

$$(4) \qquad P \cdot D^\alpha(\psi_\varepsilon \hat{\phi}) = \sum_{\beta \le \alpha} c_{\alpha\beta} P \cdot (D^{\alpha-\beta}\hat{\phi}) \cdot (D^\beta \psi_\varepsilon).$$

A simple computation shows that

$$(5) \qquad |D^\beta \psi_\varepsilon(y)| \le \begin{cases} \varepsilon y_1^2 & \text{if } |\beta| = 0, \\ \varepsilon |y_1| & \text{if } |\beta| = 1, \\ \varepsilon^{|\beta|-1} & \text{if } |\beta| > 1. \end{cases}$$

The left side of (4) tends therefore to 0, uniformly on R^n, as $\varepsilon \to 0$. The definition of the topology of $\mathscr{S}_n$ (Section 7.3) shows now that (2) holds. ////

7.18 Definition If $u \in \mathscr{S}'_n$ and $\phi \in \mathscr{S}_n$, then

$$(u * \phi)(x) = u(\tau_x \check{\phi}) \qquad (x \in R^n).$$

Note that this is well defined, since $\tau_x \check{\phi} \in \mathscr{S}_n$ for every $x \in R^n$.

7.19 Theorem *Suppose $\phi \in \mathscr{S}_n$ and u is a tempered distribution. Then*

(a) $u * \phi \in C^\infty(R^n)$, *and*

$$D^\alpha(u * \phi) = (D^\alpha u) * \phi = u * (D^\alpha \phi)$$

for every multi-index α,

(b) $u * \phi$ *has polynomial growth, hence is a tempered distribution,*

(c) $(u * \phi)^\wedge = \hat{\phi}\hat{u}$,

(d) $(u * \phi) * \psi = u * (\phi * \psi)$, *for every* $\psi \in \mathscr{S}_n$,

(e) $\hat{u} * \hat{\phi} = (\phi u)^\wedge$.

PROOF. The second equality in (a) is proved exactly as in Theorem 6.30, since convolution obviously still commutes with translations. This also shows that

$$(1) \qquad \left(\frac{\tau_{-\varepsilon w} - \tau_0}{\varepsilon}\right)(u * \phi) = u * \left(\frac{\tau_{-\varepsilon w} - \tau_0}{\varepsilon}\right)\phi.$$

Lemma 7.17 now gives $D^\alpha(u * \phi) = u * (D^\alpha \phi)$ if $\alpha = (1, 0, \ldots, 0)$. Iteration of this special case gives (a).

Let $p_N(f)$ denote the norm (1) of Section 7.3, for $f \in \mathscr{S}_n$. The inequality

$$(2) \qquad 1 + |x + y|^2 \le 2(1 + |x|^2)(1 + |y|^2) \qquad (x, y \in R^n)$$

shows that

(3) $$p_N(\tau_x f) \le 2^N(1 + |x|^2)^N p_N(f) \qquad (x \in R^n, f \in \mathscr{S}_n).$$

Since u is a continuous linear functional on $\mathscr{S}_n$ and since the norms p_N determine the topology of $\mathscr{S}_n$, there is an N and a $C < \infty$ such that

(4) $$|u(f)| \le Cp_N(f) \qquad (f \in \mathscr{S}_n);$$

see Chapter 1, Exercise 8. By (3) and (4),

(5) $$|(u * \phi)(x)| = |u(\tau_x \check{\phi})| \le 2^N C p_N(\phi)(1 + |x|^2)^N,$$

which proves (*b*).

Thus $u * \phi$ has a Fourier transform, in $\mathscr{S}'_n$. If $\psi \in \mathscr{D}(R^n)$, with support K, then

$$\begin{aligned}(u * \phi)^\wedge(\hat{\psi}) = (u * \phi)(\check{\psi}) &= \int_{R^n} (u * \phi)(x)\psi(-x)\, dm_n(x) \\ &= \int_{-K} u[\psi(-x)\tau_x \check{\phi}]\, dm_n(x) = u \int_{-K} \psi(-x)\tau_x \check{\phi}\, dm_n(x) \\ &= u((\phi * \psi)^\vee) = \hat{u}((\phi * \psi)^\wedge) = \hat{u}(\hat{\phi}\hat{\psi})\end{aligned}$$

so that

(6) $$(u * \phi)^\wedge(\hat{\psi}) = (\hat{\phi}\hat{u})(\hat{\psi}).$$

In the preceding calculation, Theorem 3.27 was applied to an $\mathscr{S}_n$-valued integral, when u was moved across the integral sign. So far, (6) has been proved for $\psi \in \mathscr{D}(R^n)$. Since $\mathscr{D}(R^n)$ is dense in $\mathscr{S}_n$, the Fourier transforms of members of $\mathscr{D}(R^n)$ are also dense in $\mathscr{S}_n$, by (*b*) of Theorem 7.7. Hence (6) holds for every $\hat{\psi} \in \mathscr{S}_n$. The distributions $(u * \phi)^\wedge$ and $\hat{\phi}\hat{u}$ are therefore equal. This proves (*c*).

In the computation that precedes (6), the two end terms are now seen to be equal for any $\psi \in \mathscr{S}_n$. Hence

(7) $$(u * \phi)(\check{\psi}) = u((\phi * \psi)^\vee),$$

which is the same as

(8) $$((u * \phi) * \psi)(0) = (u * (\phi * \psi))(0).$$

If we replace ψ by $\tau_x \psi$ in (8), we obtain (*d*).

Finally, $(\hat{u} * \hat{\phi})^\wedge = \check{\phi}\check{u} = (\phi u)^\vee$, by (*c*) above and (6) of Section 7.16; this gives (*e*), since $(\phi u)^\vee = ((\phi u)^\wedge)^\wedge$. ////

Paley-Wiener Theorems

One of the classical theorems of Paley and Wiener characterizes the entire functions of exponential type (of one complex variable), whose restriction to the real axis is in L^2, as being exactly the Fourier transforms of L^2-functions with compact support; see, for instance, Theorem 19.3 of [23]. We shall give two analogues of this (in several variables), one for C^∞-functions with compact support and one for distributions with compact support.

7.20 Definitions If Ω is an open set in $\mathcal{C}^n$, and if f is a continuous complex function in Ω, then f is said to be *holomorphic in* Ω if it is holomorphic in each variable separately. This means that if $(a_1, \ldots, a_n) \in \Omega$ and if

$$g_i(\lambda) = f(a_1, \ldots, a_{i-1}, a_i + \lambda, a_{i+1}, \ldots, a_n),$$

each of the functions $g_1, \ldots, g_n$ is to be holomorphic in some neighborhood of 0 in $\mathcal{C}$. A function that is holomorphic in all of $\mathcal{C}^n$ is said to be *entire*.

Points of $\mathcal{C}^n$ will be denoted by $z = (z_1, \ldots, z_n)$, where $z_k \in \mathcal{C}$. If $z_k = x_k + iy_k$, $x = (x_1, \ldots, x_n)$, $y = (y_1, \ldots, y_n)$, then we write $z = x + iy$. The vectors

$$x = \text{Re } z \qquad \text{and} \qquad y = \text{Im } z$$

are the *real* and *imaginary* parts of z, respectively; R^n will be thought of as the set of all $z \in \mathcal{C}^n$ with Im $z = 0$. The notations

$$|z| = (|z_1|^2 + \cdots + |z_n|^2)^{1/2}$$

$$|\text{Im } z| = (y_1^2 + \cdots + y_n^2)^{1/2}$$

$$z^\alpha = z_1^{\alpha_1} \cdots z_n^{\alpha_n}$$

$$z \cdot t = z_1 t_1 + \cdots + z_n t_n$$

$$e_z(t) = \exp(iz \cdot t)$$

will be used for any multi-index α and any $t \in R^n$.

7.21 Lemma *If f is an entire function in $\mathcal{C}^n$ that vanishes on R^n, then* $f = 0$.

PROOF. We consider the case $n = 1$ as known. Let P_k be the following property of f: If $z \in \mathcal{C}^n$ has at least k real coordinates, then $f(z) = 0$. P_n is given; P_0 is to be proved. Assume $1 \le i \le n$ and P_i is true. Take $a_1, \ldots, a_i$ real. The function g_i considered in Section 7.20 is then 0 on the real axis, hence is 0 for all $\lambda \in \mathcal{C}$. It follows that P_{i-1} is true. ////

In the following two theorems,

$$rB = \{x \in R^n : |x| \le r\}.$$

7.22 Theorem

(a) If $\phi \in \mathscr{D}(R^n)$ has its support in rB, and if

$$(1) \qquad f(z) = \int_{R^n} \phi(t) e^{-iz \cdot t}\, dm_n(t) \qquad (z \in \mathcal{C}^n),$$

then f is entire, and there are constants $\gamma_N < \infty$ such that

$$(2) \qquad |f(z)| \le \gamma_N (1 + |z|)^{-N} e^{r|\text{Im } z|} \qquad (z \in \mathcal{C}^n,\ N = 0, 1, 2, \ldots).$$

(b) Conversely, if an entire function f satisfies the conditions (2), then there exists $\phi \in \mathscr{D}(R^n)$, with support in rB, such that (1) holds.

PROOF. *(a)* If $t \in rB$ then

$$|e^{-iz \cdot t}| = e^{y \cdot t} \le e^{|y||t|} \le e^{r|\text{Im } z|}.$$

The integrand in (1) is therefore in $L^1(R^n)$, for every $z \in \mathcal{C}^n$, and f is well defined on $\mathcal{C}^n$. The continuity of f is trivial, and an application of Morera's theorem, to each variable separately, shows that f is entire. Integrations by part give

$$z^\alpha f(z) = \int_{R^n} (D_\alpha \phi)(t) e^{-iz \cdot t}\, dm_n(t).$$

Hence

$$(3) \qquad |z^\alpha| |f(z)| \le \|D_\alpha \phi\|_1 e^{r|\text{Im } z|},$$

and (2) follows from the inequalities (3).

(b) Suppose f is an entire function that satisfies (2), and define

$$(4) \qquad \phi(t) = \int_{R^n} f(x) e^{it \cdot x}\, dm_n(x) \qquad (t \in R^n).$$

Note first that $(1 + |x|)^N f(x)$ is in $L^1(R^n)$ for every N, by (2). Hence $\phi \in C^\infty(R^n)$, by the argument that proved (c) of Theorem 7.4.

Next, we claim that the integral

$$(5) \qquad \int_{-\infty}^{\infty} f(\xi + i\eta, z_2, \ldots, z_n) \exp\{i[t_1(\xi + i\eta) + t_2 z_2 + \cdots + t_n z_n]\}\, d\xi$$

is independent of η, for arbitrary real $t_1, \ldots, t_n$ and complex $z_2, \ldots, z_n$. To see this, let Γ be a rectangular path in the $(\xi + i\eta)$-plane, with one edge on the real axis, one on the line $\eta = \eta_1$, whose vertical edges move off to infinity. By Cauchy's theorem, the integral of the integrand (5) over Γ is 0. By (2), the contribution of the vertical edges to this integral tend to 0. It follows that (5) is the same for $\eta = 0$ as for $\eta = \eta_1$. This establishes our claim.

The same can be done for the other coordinates. Hence we conclude from (4) that

$$\phi(t) = \int_{R^n} f(x + iy)e^{it \cdot (x+iy)}\, dm_n(x) \tag{6}$$

for *every* $y \in R^n$.

Given $t \in R^n$, $t \neq 0$, choose $y = \lambda t/|t|$, where $\lambda > 0$. Then $t \cdot y = \lambda |t|, |y| = \lambda$,

$$|f(x + iy)e^{it \cdot (x+iy)}| \le \gamma_N(1 + |x|)^{-N}e^{(r-|t|)\lambda},$$

and therefore

$$|\phi(t)| \le \gamma_N e^{(r-|t|)\lambda} \int_{R^n} (1 + |x|)^{-N}\, dm_n(x), \tag{7}$$

where N is chosen so large that the last integral is finite. Now let $\lambda \to \infty$. If $|t| > r$, (7) shows that $\phi(t) = 0$. Thus ϕ has its support in rB.

Now (1) follows, for real z, from (4) and the inversion theorem. Since both sides of (1) are entire functions, they coincide on $\mathbb{C}^n$, by Lemma 7.21. This completes the proof. ////

The following remarks will motivate the next theorem.

Let u be a distribution in R^n, with *compact* support. Then $\hat{u}$ is defined, as a tempered distribution, by $\hat{u}(\phi) = u(\hat{\phi})$. However, the definition $\hat{f}(x) = \int fe_{-x}\, dm_n$, made for $f \in L^1(R^n)$, suggests that $\hat{u}$ ought to be a function, namely,

$$\hat{u}(x) = u(e_{-x}) \qquad (x \in R^n),$$

because $e_{-x} \in C^\infty(R^n)$ and $u(\phi)$ makes sense for every $\phi \in C^\infty(R^n)$, as shown by (*d*) of Theorem 6.24. Moreover, $e_{-z} \in C^\infty(R^n)$ for every $z \in \mathbb{C}^n$, and $u(e_{-z})$ therefore looks like an entire function, whose restriction to R^n is $\hat{u}$.

That all this is correct is part of the content of the next theorem, which also characterizes the resulting entire functions by certain growth conditions.

7.23 Theorem

(*a*) *If $u \in \mathscr{D}'(R^n)$ has its support in rB, if u has order N, and if*

$$f(z) = u(e_{-z}) \qquad (z \in \mathbb{C}^n), \tag{1}$$

then f is entire, the restriction of f to R^n is the Fourier transform of u, and there is a constant $\gamma < \infty$ such that

$$|f(z)| \le \gamma(1 + |z|)^N e^{r|\text{Im}\, z|} \qquad (z \in \mathbb{C}^n). \tag{2}$$

(b) *Conversely, if f is an entire function in $\mathbb{C}^n$ which satisfies* (2) *for some N and some γ, then there exists $u \in \mathscr{D}'(R^n)$, with support in rB, such that* (1) *holds.*

Note: The notation $\hat{u}$ will sometimes be used to denote the extension to $\mathbb{C}^n$ given by (1). Thus

$$\hat{u}(z) = u(e_{-z})$$

for $z \in \mathbb{C}^n$. This extension is sometimes called the Fourier-Laplace transform of u.

PROOF. (a) Suppose $u \in \mathscr{D}'(R^n)$ has its support in rB. Pick $\psi \in \mathscr{D}(R^n)$ so that $\psi = 1$ on $(r+1)B$. Then $u = \psi u$, and (e) of Theorem 7.19 shows that

$$\hat{u} = (\psi u)^\wedge = \hat{u} * \hat{\psi}. \tag{3}$$

Thus $\hat{u} \in C^\infty(R^n)$. Pick $\phi \in \mathscr{S}_n$ so that $\hat{\phi} = \psi$. Then

$$\begin{aligned}(\hat{u} * \hat{\psi})(x) &= (\hat{u} * \check{\phi})(x) = \hat{u}(\tau_x \phi) = u((\tau_x \phi)^\wedge) \\ &= u(e_{-x}\hat{\phi}) = u(\psi e_{-x}) = u(e_{-x}),\end{aligned}$$

so that (3) gives

$$\hat{u}(x) = u(e_{-x}) \qquad (x \in R^n). \tag{4}$$

Our next aim is to show that the function f defined by (1) is entire. Choose $a \in \mathbb{C}^n$, $b \in \mathbb{C}^n$, and put

$$g(\lambda) = f(a + \lambda b) = u(e_{-a-\lambda b}) \qquad (\lambda \in \mathbb{C}). \tag{5}$$

The continuity of f poses no problem: If $w \to z$ in $\mathbb{C}^n$, then $e_{-w} \to e_{-z}$ in $C^\infty(R^n)$, and u is continuous on $C^\infty(R^n)$. To prove that f is entire it is therefore enough to show that each of the functions g defined by (5) is entire.

Let Γ be a rectangular path in $\mathbb{C}$. Since $\lambda \to e_{-a-\lambda b}$ is continuous, from $\mathbb{C}$ to $C^\infty(R^n)$, the $C^\infty(R^n)$-valued integral

$$F = \int_\Gamma e_{-a-\lambda b}\, d\lambda \tag{6}$$

is well defined. Evaluation at any $t \in R^n$ is a continuous linear functional on $C^\infty(R^n)$. It therefore commutes with the integral sign. Hence

$$F(t) = \int_\Gamma e_{-a-\lambda b}(t)\, d\lambda = \int_\Gamma e^{-ia\cdot t} e^{-i(b\cdot t)\lambda}\, d\lambda = 0.$$

Thus $F = 0$, and (6) gives

$$0 = u(F) = \int_\Gamma u(e_{-a-\lambda b})\, d\lambda = \int_\Gamma g(\lambda)\, d\lambda.$$

By Morera's theorem, g is entire.

The proof of part (*a*) will be completed by proving (2). Choose an auxiliary function h on the real line, infinitely differentiable, such that $h(s) = 1$ when $s < 1$ and $h(s) = 0$ when $s > 2$, and associate with each $z \in \mathcal{C}^n$ $(z \neq 0)$ the function

$$\phi_z(t) = e^{-iz\cdot t}h(|t||z| - r|z|) \qquad (t \in R^n). \tag{7}$$

Then $\phi_z \in \mathscr{D}(R^n)$. Since the support of u is in rB and $h(|t||z| - r|z|) = 1$ if $|t| \leq |z|^{-1} + r$, comparison of (1) and (7) shows that

$$f(z) = u(\phi_z). \tag{8}$$

Since u has order N, there is a $\gamma_0 < \infty$ such that $|u(\phi)| \leq \gamma_0\|\phi\|_N$ for all $\phi \in \mathscr{D}(R^n)$, where $\|\phi\|_N$ is as in (1) of Section 6.2; see (*d*) of Theorem 6.24. Hence (8) gives

$$|f(z)| \leq \gamma_0\|\phi_z\|_N. \tag{9}$$

On the support of ϕ_z, $|t| \leq r + 2/|z|$, so that

$$|e^{-iz\cdot t}| = e^{y\cdot t} \leq e^{2+r|\text{Im}\, z|}. \tag{10}$$

If we now apply the Leibniz formula to the product (7) and use (10), (9) implies (2).

This completes the proof of part (*a*).

(*b*) Since f now satisfies (2), we have

$$|f(x)| \leq \gamma(1 + |x|)^N \qquad (x \in R^n). \tag{11}$$

The restriction of f to R^n is therefore in $\mathscr{S}'_n$ and is the Fourier transform of some tempered distribution u.

Pick a function $h \in \mathscr{D}(R^n)$, with support in B, such that $\int h = 1$, define $h_\varepsilon(t) = \varepsilon^{-n}h(t/\varepsilon)$, for $\varepsilon > 0$, and put

$$f_\varepsilon(z) = f(z)\hat{h}_\varepsilon(z) \qquad (z \in \mathcal{C}^n), \tag{12}$$

where $\hat{h}_\varepsilon$ now denotes the entire function whose restriction to R^n is the Fourier transform of h_ε. Statement (*a*) of Theorem 7.22, applied to h_ε, leads to the conclusion that f_ε satisfies (2) of Theorem 7.22 with $r + \varepsilon$ in place of r. Therefore (*b*) of Theorem 7.22 implies that $f_\varepsilon = \hat{\phi}_\varepsilon$ for some $\phi_\varepsilon \in \mathscr{D}(R^n)$ whose support lies in $(r + \varepsilon)B$.

Consider some $\psi \in \mathscr{S}_n$ such that the support of $\hat{\psi}$ does not intersect rB. Then $\hat{\psi}\phi_\varepsilon = 0$ for all sufficiently small $\varepsilon > 0$. Since

$f\psi \in L^1(R^n)$ and $\hat{h}_\varepsilon(x) = \hat{h}(\varepsilon x) \to 1$ boundedly on R^n, we conclude that

$$u(\hat{\psi}) = \hat{u}(\psi) = \int f\psi \, dm_n = \lim_{\varepsilon \to 0} \int f_\varepsilon \psi \, dm_n$$

$$= \lim_{\varepsilon \to 0} \int \hat{\phi}_\varepsilon \psi \, dm_n = \lim_{\varepsilon \to 0} \int \hat{\psi} \phi_\varepsilon \, dm_n = 0.$$

Hence u has its support in rB.

Now we see that $z \to u(e_{-z})$ is an entire function, and since (1) holds for $z \in R^n$ (by the choice of u), Lemma 7.21 completes the proof of (*b*). ////

Sobolev's Lemma

If Ω is a proper open subset of R^n, no Fourier transform has been defined for functions whose domain is Ω or for distributions in Ω. Nevertheless, Fourier transform techniques can sometimes be used to attack local problems. Theorem 7.25, known as Sobolev's lemma, is an example of this.

7.24 Definitions A complex measurable function f, defined in an open set $\Omega \subset R^n$, is said to be *locally* L^2 *in* Ω if $\int_K |f|^2 \, dm_n < \infty$ for every compact $K \subset \Omega$.

Similarly, a distribution $u \in \mathscr{D}'(\Omega)$ is locally L^2 if there is a function g, locally L^2 in Ω, such that $u(\phi) = \int_\Omega g\phi \, dm_n$ for every $\phi \in \mathscr{D}(\Omega)$. To say that a function f has a distribution derivative $D^\alpha f$ which is locally L^2 refers to the *distribution* $D^\alpha f$ and means, explicitly, that there is a function g, locally L^2, such that

$$\int_\Omega g\phi \, dm_n = (-1)^{|\alpha|} \int_\Omega f D^\alpha \phi \, dm_n$$

for every $\phi \in \mathscr{D}(\Omega)$. A priori, this says nothing about the existence of $D^\alpha f$ in the classical sense, in terms of limits of quotients.

On the other hand, the class $C^{(p)}(\Omega)$ consists, for each nonnegative integer p, of those complex functions f in Ω whose derivatives $D^\alpha f$ exist in the classical sense, for each multi-index α with $|\alpha| \le p$, and are continuous functions in Ω.

We shall write D_i^k for the differential operator $(\partial/\partial x_i)^k$.

7.25 Theorem *Suppose n, p, r are integers, $n > 0$, $p \ge 0$, and*

$$r > p + \frac{n}{2}. \tag{1}$$

Suppose f is a function in an open set $\Omega \subset R^n$ whose distribution derivatives $D_i^k f$ are locally L^2 in Ω, for $1 \le i \le n$, $0 \le k \le r$.

Then there is a function $f_0 \in C^{(p)}(\Omega)$ such that $f_0(x) = f(x)$ for almost every $x \in \Omega$.

Note that the hypothesis involves no mixed derivatives, i.e., no terms like $D_1 D_2 f$. The conclusion is that f can be "corrected" so as to be in $C^{(p)}(\Omega)$, by redefining it on a set of measure 0.

Note also, as a corollary, that if *all* distribution derivatives of f are locally L^2 in Ω, then $f_0 \in C^\infty(\Omega)$.

PROOF. By hypothesis, there are functions g_{ik}, locally L^2 in Ω, that satisfy

$$\int_\Omega g_{ik}\,\phi\, dm_n = (-1)^k \int_\Omega f D_i^k \phi\, dm_n \qquad [\phi \in \mathscr{D}(\Omega)], \tag{2}$$

for $1 \le i \le n$, $0 \le k \le r$.

Let ω be an open set whose closure K is a compact subset of Ω. Choose $\psi \in \mathscr{D}(\Omega)$ so that $\psi = 1$ on K, and define F on R^n by

$$F(x) = \begin{cases} \psi(x) f(x) & \text{if } x \in \Omega, \\ 0 & \text{if } x \notin \Omega. \end{cases}$$

Then $F \in (L^2 \cap L^1)(R^n)$.

In Ω, the Leibniz formula gives

$$D_i^r F = \sum_{s=0}^{r} \binom{r}{s} (D_i^{r-s}\psi)(D_i^s f) = \sum_{s=0}^{r} \binom{r}{s} (D_i^{r-s}\psi) g_{is}. \tag{3}$$

In the complement Ω_0 of the support of ψ, $D_i^r F = 0$. These two distributions coincide in $\Omega \cap \Omega_0$. Hence $D_i^r F$, originally defined as a distribution in R^n, is actually in $L^2(R^n)$, for $1 \le i \le n$, because the functions $(D_i^{r-s}\psi) g_{is}$ are in $L^2(\Omega)$. [Having compact support, $D_i^r F$ is therefore also in $L^1(R^n)$.]

The Plancherel theorem, applied to F and to $D_1^r F, \ldots, D_n^r F$, shows now that

$$\int_{R^n} |\hat{F}|^2\, dm_n < \infty \tag{4}$$

and

$$\int_{R^n} y_i^{2r} |\hat{F}(y)|^2\, dm_n(y) < \infty \qquad (1 \le i \le n). \tag{5}$$

Since

$$(1 + |y|)^{2r} < (2n+2)^r (1 + y_1^{2r} + \cdots + y_n^{2r}), \tag{6}$$

where $|y| = (y_1^2 + \cdots + y_n^2)^{1/2}$, (4) and (5) imply

$$\int_{R^n} (1 + |y|)^{2r} |\hat{F}(y)|^2 \, dm_n(y) < \infty. \tag{7}$$

If J denotes the integral (7), and if σ_n is the $(n-1)$-dimensional volume of the unit sphere in R^n, the Schwarz inequality gives

$$\left\{\int_{R^n} (1 + |y|)^p |\hat{F}(y)| \, dm_n(y)\right\}^2 \le J \int_{R^n} (1 + |y|)^{2p-2r} \, dm_n(y)$$

$$= J\sigma_n \int_0^\infty (1 + t)^{2p-2r} t^{n-1} \, dt < \infty,$$

since $2p - 2r + n - 1 < -1$. We have thus proved that

$$\int_{R^n} (1 + |y|)^p |\hat{F}(y)| \, dm_n(y) < \infty. \tag{8}$$

Define

$$F_\omega(x) = \int_{R^n} \hat{F}(y) e^{ix \cdot y} \, dm_n(y) \qquad (x \in R^n). \tag{9}$$

By (*c*) of the inversion theorem 7.7, $F_\omega = F$ a.e. on R^n. Moreover, (8) implies that $y^\alpha \hat{F}(y)$ is in L^1 whenever $|\alpha| \le p$. Iteration of the proof of (*c*) of Theorem 7.4 leads therefore to the conclusion

$$F_\omega \in C^{(p)}(R^n). \tag{10}$$

Our given function f coincides with F in ω. Hence $f = F_\omega$ a.e. in ω.

If ω' is another set like ω, the preceding proof gives a function $F_{\omega'} \in C^{(p)}(R^n)$, which coincides with f a.e. in ω'. Hence $F_{\omega'} = F_\omega$ in $\omega' \cap \omega$. The desired function f_0 can therefore be defined in Ω by setting $f_0(x) = F_\omega(x)$ if $x \in \omega$. ////

Exercises

1. Suppose A is an invertible linear operator on R^n, $f \in L^1(R^n)$, and $g(x) = f(Ax)$. Express $\hat{g}$ in terms of $\hat{f}$. This generalizes (*d*) of Theorem 7.2.
2. Is the topology of $\mathscr{S}_n$ induced by some invariant metric which turns the Fourier transform into an isometry of $\mathscr{S}_n$ onto $\mathscr{S}_n$?
3. Suppose $f(x) = e^x$, $g(x) = e^x \cos(e^x)$, on the real line. Show that g is a tempered distribution but that f is not.
4. By Exercise 3 there exist distributions in R^n which are not tempered. Such distributions are continuous linear functionals on $\mathscr{D}(R^n)$ which have no continuous

linear extension to $\mathscr{S}_n$. Explain why this does not contradict the Hahn-Banach theorem.

5. (*a*) Construct a sequence in $\mathscr{D}(R^n)$ which converges to 0 in the topology of $\mathscr{S}_n$ but not in that of $\mathscr{D}(R^n)$.
   (*b*) Construct a sequence of polynomials which converges in the topology of $\mathscr{D}'(R^1)$ but not in that of $\mathscr{S}'_1$.
6. Prove that the operations listed in Theorem 7.13 are *continuous* mappings of $\mathscr{S}'_n$ into $\mathscr{S}'_n$.
7. If $u \in \mathscr{S}'_n$, prove that
$$(\tau_x u)^\wedge = e_{-x}\hat{u} \qquad \text{and} \qquad (e_x u)^\wedge = \tau_x \hat{u}$$
for every $x \in R^n$.
8. Suppose $f \in L^1(R^n)$, $f \neq 0$, λ is a complex number, and $\hat{f} = \lambda f$. What can you say about λ?
9. Prove (*a*) of Theorem 7.8 directly (without using Fourier transforms).
10. The Fourier transform of a complex Borel measure μ on R^n is customarily defined to be the function $\hat{\mu}$ given by
$$\hat{\mu}(x) = \int_{R^n} e^{-ix\cdot t}\, d\mu(t) \qquad (x \in R^n).$$
Of course, μ is also a tempered distribution, and as such its Fourier transform was defined in Section 7.14. Show that these two definitions are consistent. Prove that each $\hat{\mu}$ is bounded and uniformly continuous.
11. Suppose $\Lambda: \mathscr{S}_n \to C(R^n)$ is continuous, linear, and $\tau_x \Lambda = \Lambda \tau_x$ for every $x \in R^n$. Does it follow that there exists $u \in \mathscr{S}'_n$ such that
$$\Lambda\phi = u * \phi$$
for every $\phi \in \mathscr{S}_n$?
12. If $\{h_j\}$ is an approximate identity, as in Definition 6.31, and $u \in \mathscr{S}'_n$, does it follow that $u * h_j \to u$ as $j \to \infty$, in the weak*-topology of $\mathscr{S}'_n$?
13. Suppose X and Y are complete metric spaces, A is dense in X, $f: A \to Y$ is uniformly continuous.
    (*a*) Prove that f has a unique continuous extension $F: X \to Y$.
    (*b*) If f is an isometry, prove that the same is true of F, and prove that $F(X)$ is closed in Y.
    (This was used in the proof of the Plancherel theorem; see also Exercise 19, Chapter 1.)
14. Suppose F is an entire function in $\mathcal{C}^n$, and suppose that to each $\varepsilon > 0$ there correspond an integer $N(\varepsilon)$ and a constant $\gamma(\varepsilon) < \infty$ such that
$$|F(z)| \leq \gamma(\varepsilon)(1 + |z|)^{N(\varepsilon)} e^{\varepsilon|\text{Im } z|} \qquad (x \in \mathcal{C}^n).$$
Prove that F is a polynomial.
15. Suppose f is an entire function in $\mathcal{C}^n$, N is a positive integer, $r \geq 0$, and
$$|f(z)| \leq (1 + |z|)^N e^{r|\text{Im } z|} \qquad \text{for all } z \in \mathcal{C}^n,$$
$$|f(x)| \leq 1 \qquad \text{for all } x \in R^n.$$

Prove that then

$$|f(z)| \le e^{r|\text{Im } z|} \qquad \text{for all } z \in \mathcal{C}^n.$$

Suggestion: Fix $z = x + iy \in \mathcal{C}^n$; define

$$g_s(\lambda) = (1 - is\lambda)^{-N-1} e^{ir|y|\lambda} f(x + \lambda y)$$

for $\lambda \in \mathcal{C}$, $s > 0$, and apply the maximum modulus theorem to a large semi-circular region in the upper half-plane to deduce that $|g_s(i)| < 1$. Let $s \to 0$.

16. In (*b*) of Theorem 7.23 it is not asserted that u has order N. The following example shows that this is not always true.

Let μ be the Borel probability measure on R^3 which is concentrated on the unit sphere S^2 and which is invariant under all rotations of S^2. Compute (by using spherical coordinates) that

$$\hat{\mu}(x) = \frac{\sin|x|}{|x|} \qquad (x \in R^3).$$

Put $u = D_1\mu$. Then

$$|\hat{u}(x)| = |x_1\hat{\mu}(x)| \le 1 \qquad (x \in R^3).$$

Deduce from Exercise 15 that

$$|u(e_{-z})| \le \gamma e^{|\text{Im } z|} \qquad (z \in \mathcal{C}^3)$$

although u is not a distribution of order 0. (Its order is 1.) Find an explicit formula for the entire function $u(e_{-z})$, $z \in \mathcal{C}^3$.

17. Suppose u is a distribution in R^n, with compact support K, whose Fourier transform $\hat{u}$ is a *bounded* function on R^n.

(*a*) Assume $n = 1$ or $n = 2$, and prove that $\psi u = 0$ for every $\psi \in C^\infty(R^n)$ that vanishes on K.

(*b*) Assume $n = 2$, and assume that there is a real polynomial P, in two variables, that vanishes on K. Prove that $Pu = 0$ and that $\hat{u}$ therefore satisfies the partial differential equation $P(-D)\hat{u} = 0$. For example, when K is the unit circle, then

$$\hat{u} + \Delta\hat{u} = 0,$$

where $\Delta = \partial^2/\partial x_1^2 + \partial^2/\partial x_2^2$ is the Laplacian.

(*c*) Show, with the aid of Exercise 16 and the polynomial $1 - x_1^2 - x_2^2 - x_3^2$, that (*b*), hence also (*a*), becomes false with $n = 3$ in place of $n = 2$.

(*d*) Assume $n = 1$, $f \in L^1(R)$, $\hat{f} = 0$ on K, and $\hat{f}$ satisfies a Lipschitz condition of order $\frac{1}{2}$, that is, $|\hat{f}(t) - \hat{f}(s)| \le C|t - s|^{1/2}$. Prove that then

$$\int_{-\infty}^{\infty} f(x)\hat{u}(x)\,dx = 0.$$

Suggestion: For any n, let H_ε be the set of all points outside K whose distance from K is less than $\varepsilon > 0$. Let $\{h_\varepsilon\}$ be an approximate identity, as in the proof of (*b*) of Theorem 7.23, use the Plancherel theorem to obtain

$$\|u * h_\varepsilon\|_2 \le \|\hat{u}\|_\infty \varepsilon^{-n/2}\|h_1\|_2,$$

and show that therefore

$$|u(\phi)| \leq \|\hat{u}\|_\infty \|h_1\|_2 \liminf_{\varepsilon \to 0} \left\{ \varepsilon^{-n} \int_{H_\varepsilon} |\phi|^2 \, dm_n \right\}^{1/2}$$

for any $\phi \in \mathscr{D}(R^n)$ that vanishes on K.

This yields (*a*). A slight modification yields (*d*); (*b*) follows from (*a*).

18. Was it necessary to introduce the function ψ into the proof of Theorem 7.25? Could the proof have been simplified by setting $F(x) = f(x)$ on K, $F(x) = 0$ off K?

19. Show that the hypotheses of Theorem 7.25 imply that $D^\alpha f$ is locally L^2 for every multi-index α with $|\alpha| \leq r$.

20. Let $f \in L^2(R^2)$ be the continuous function whose Fourier transform is

$$\hat{f}(y) = (1 + |y|)^{-4} \{\log(2 + |y|)\}^{-1} \qquad (y \in R^2).$$

Since $|y|^3 \hat{f}(y)$ is in $L^2(R^2)$, Theorem 7.25 implies that $f \in C^{(1)}(R^2)$. Show that the stronger conclusion $f \in C^{(2)}(R^2)$ is false, by proving that

$$\frac{f(h, 0) + f(-h, 0) - 2f(0, 0)}{h^2} \to -\infty \text{ as } h \to 0.$$

This shows that $>$ cannot be replaced by $\geq$ in (1) of Theorem 7.25.

21. Suppose u is a distribution in R^n whose first derivatives $D_1 u, \ldots, D_n u$ are functions in $L^2(R^n)$. Prove that u is also a function and that u is locally L^2. (Show that "locally" cannot be omitted in the conclusion.) *Hint:* u is in fact the sum of an L^2-function and an entire function.

When $n = 1$, show that u is actually a continuous function. Show that this stronger conclusion is false when $n = 2$. For example, the gradient of the function

$$u(re^{i\theta}) = \log\log\left(2 + \frac{1}{r}\right)$$

is in $L^2(R^2)$. See Exercise 11, Chapter 8, for the same result under weaker hypotheses.

22. Periodic distributions, or distributions on a torus T^n, have *Fourier series* whose theory is somewhat simpler than that of Fourier transforms. This is mainly due to the compactness of T^n: Every distribution on T^n has compact support. In particular, *tempered* distributions are nothing special.

Prove the various assertions made in the following basic outline:

$$T^n = \{(e^{ix_1}, \ldots, e^{ix_n}) : x_j \text{ real}\}.$$

Functions ϕ on T^n can be identified with functions $\tilde{\phi}$ on R^n that are 2π-periodic in each variable, by setting

$$\tilde{\phi}(x_1, \ldots, x_n) = \phi(e^{ix_1}, \ldots, e^{ix_n}).$$

Z^n is the set (or additive group) of n-tuples $k = (k_1, \ldots, k_n)$ of integers k_j. For $k \in Z^n$, the function e_k is defined on T^n by

$$e_k(e^{ix_1}, \ldots, e^{ix_n}) = e^{ik \cdot x} = \exp\{i(k_1 x_1 + \cdots + k_n x_n)\}.$$

σ_n is the Haar measure of T^n. If $\phi \in L^1(\sigma_n)$, the Fourier coefficients of ϕ are

$$\hat{\phi}(k) = \int_{T^n} e_{-k}\phi \, d\sigma_n \qquad (k \in Z^n).$$

$\mathscr{D}(T^n)$ is the space of all functions ϕ on T^n such that $\tilde{\phi} \in C^\infty(R^n)$. If $\phi \in \mathscr{D}(T^n)$ then

$$\left\{ \sum_{k \in Z^n} (1 + k \cdot k)^N |\hat{\phi}(k)|^2 \right\}^{1/2} < \infty$$

for $N = 0, 1, 2, \ldots$. These norms define a Fréchet space topology on $\mathscr{D}(T^n)$, which coincides with the one given by the norms

$$\max_{|\alpha| \le N} \sup_{x \in R^n} |(D^\alpha \tilde{\phi})(x)| \qquad (N = 0, 1, 2, \ldots).$$

$\mathscr{D}'(T^n)$ is the space of all continuous linear functionals on $\mathscr{D}(T^n)$; its members are the *distributions* on T^n. The Fourier coefficients of any $u \in \mathscr{D}'(T^n)$ are defined by

$$\hat{u}(k) = u(e_{-k}) \qquad (k \in Z^n).$$

To each $u \in \mathscr{D}'(T^n)$ correspond an N and a C such that

$$|\hat{u}(k)| \le C(1 + |k|)^N \qquad (k \in Z^n).$$

Conversely, if g is a complex function on Z^n that satisfies $|g(k)| \le C(1 + |k|)^N$ for some C and N, then $g = \hat{u}$ for some $u \in \mathscr{D}'(T^n)$.

There is thus a linear one-to-one correspondence between distributions on T^n, on one hand, and functions of polynomial growth on Z^n, on the other.

If $E_1 \subset E_2 \subset E_3 \subset \cdots$ are finite sets whose union is Z^n, and if $u \in \mathscr{D}'(T^n)$, the "partial sums"

$$\sum_{k \in E_j} \hat{u}(k) e_k$$

converge to u as $j \to \infty$, in the weak*-topology of $\mathscr{D}'(T^n)$.

The convolution $u * v$ of $u \in \mathscr{D}'(T^n)$ and $v \in \mathscr{D}'(T^n)$ is most easily defined as having Fourier coefficients $\hat{u}(k)\hat{v}(k)$. The analogues of Theorems 6.30 and 6.37 are true; the proofs are much simpler.

23. Modify the proof of Theorem 7.25 so that Fourier series are used in place of Fourier transforms, by replacing F by a suitable periodic function.

24. Put $c = (2/\pi)^{1/2}$. For $j = 1, 2, 3, \ldots$, define g_j on the real line by

$$g(t) = \begin{cases} c/t & \text{if } 1/j < |t| < j \\ 0 & \text{otherwise.} \end{cases}$$

Prove that $\{\hat{g}_j\}$ is a uniformly bounded sequence of functions which converge pointwise, as $j \to \infty$. If $f \in L^2(R^1)$, it follows that $f * g_j$ converges, in the L^2 metric, to a function $Hf \in L^2$. This is the *Hilbert transform* of f; formally,

$$(Hf)(x) = \frac{1}{\pi} \int_{-\infty}^{\infty} \frac{f(t)}{x - t} \, dt.$$

(The integral exists, in the principal value sense, for almost every x, but this is not so easy to prove; if f satisfies a Lipschitz condition of order 1, for instance, the proof is trivial.) Prove that

$$\|Hf\|_2 = \|f\|_2 \qquad \text{and} \qquad H(Hf) = -f,$$

for every $f \in L^2(R^1)$. Thus H is an L^2-isometry, of period 4.

Is it true that $Hf \in \mathscr{S}_1$ if $f \in \mathscr{S}_1$?

CHAPTER 8

APPLICATIONS TO DIFFERENTIAL EQUATIONS

Fundamental Solutions

8.1 Introduction We shall be concerned with linear partial differential equations with constant coefficients. These are equations of the form

(1) $$P(D)u = v$$

where P is a nonconstant polynomial in n variables (with complex coefficients), $P(D)$ is the corresponding differential operator (see Section 7.1), v is a given function or distribution, and the function (or distribution) u is a *solution* of (1).

A distribution $E \in \mathscr{D}'(R^n)$ is said to be a *fundamental solution* of the operator $P(D)$ if it satisfies (1) with $v = \delta$, the Dirac measure:

(2) $$P(D)E = \delta.$$

The basic result (Theorem 8.5, due to Malgrange and Ehrenpreis) that will be proved here is that *such fundamental solutions always exist.*

Suppose we have an E that satisfies (2), suppose v has compact support, and put

(3) $$u = E * v.$$

Then u is a solution of (1), because

$$P(D)(E * v) = (P(D)E) * v = \delta * v = v, \tag{4}$$

by Theorems 6.35 and 6.37.

The existence of a fundamental solution thus leads to a general existence theorem for the equation (1); note also that every solution of (1) differs from $E * v$ by a solution of the homogeneous equation $P(D)u = 0$. Moreover, (3) gives some additional information about u. For instance, if $v \in \mathscr{D}(R^n)$, then $u \in C^\infty(R^n)$.

It may of course happen that the convolution $E * v$ exists for certain v whose support is not compact. This raises the problem of finding E so that its behavior at infinity is well under control. The best possible result would of course be to find an E with compact support. But this can never be done. If it could, $\hat{E}$ would be an entire function, and (2) would imply $P\hat{E} = 1$. But the product of an entire function and a polynomial cannot be 1 unless both are constant.

However, the equation $P\hat{E} = 1$ can sometimes be used to find E, namely, when $1/P$ is a tempered distribution; in this case, the Fourier transform of $1/P$ furnishes a fundamental solution which is a tempered distribution. For examples of this, see Exercises 5 to 9.

Another related question concerns the existence of solutions of (1) with compact support if the support of v is compact. The answer (given in Theorem 8.4) shows very clearly that it is not enough to study P on R^n in problems of this sort but that the behavior of P in the complex space $\mathbb{C}^n$ is highly significant.

8.2 Notations T^n is the torus that consists of all points

$$w = (e^{i\theta_1}, \ldots, e^{i\theta_n}) \tag{1}$$

in $\mathbb{C}^n$, where $\theta_1, \ldots, \theta_n$ are real; σ_n is the Haar measure of T^n, that is, Lebesgue measure divided by $(2\pi)^n$.

A polynomial in $\mathbb{C}^n$, *of degree N*, is a function

$$P(z) = \sum_{|\alpha| \le N} c(\alpha) z^\alpha \qquad (z \in \mathbb{C}^n), \tag{2}$$

where α ranges over multi-indices and $c(\alpha) \in \mathbb{C}$. If (2) holds and if $c(\alpha) \neq 0$ for at least one α with $|\alpha| = N$, P is said to have *exact degree N*.

8.3 Lemma *If P is a polynomial in $\mathbb{C}^n$, of exact degree N, then there is a constant $A < \infty$, depending only on P, such that*

$$|f(z)| \le Ar^{-N} \int_{T^n} |(fP)(z + rw)| \, d\sigma_n(w) \tag{1}$$

for every entire function f in $\mathbb{C}^n$, for every $z \in \mathbb{C}^n$, and for every $r > 0$.

PROOF. Assume first that F is an entire function of one complex variable and that

$$Q(\lambda) = c \prod_{i=1}^{N} (\lambda + a_i) \qquad (\lambda \in \mathcal{C}). \tag{2}$$

Put $Q_0(\lambda) = c \prod (1 + \bar{a}_i \lambda)$. Then $cF(0) = (FQ_0)(0)$. Since $|Q_0| = |Q|$ on the unit circle, it follows that

$$|cF(0)| \le \frac{1}{2\pi} \int_{-\pi}^{\pi} |(FQ)(e^{i\theta})| \, d\theta. \tag{3}$$

The given polynomial P can be written in the form $P = P_0 + P_1 + \cdots + P_N$, where each P_j is a homogeneous polynomial of degree j. Define A by

$$\frac{1}{A} = \int_{T^n} |P_N| \, d\sigma_n. \tag{4}$$

This integral is positive, since P has *exact* degree N. [See part (*b*) of Exercise 1.] If $z \in \mathcal{C}^n$ and $w \in T^n$, define

$$F(\lambda) = f(z + r\lambda w), \qquad Q(\lambda) = P(z + r\lambda w) \qquad (\lambda \in \mathcal{C}). \tag{5}$$

The leading coefficient of Q is $r^N P_N(w)$. Hence (3) implies

$$r^N |P_N(w)| |f(z)| \le \frac{1}{2\pi} \int_{-\pi}^{\pi} |(fP)(z + re^{i\theta} w)| \, d\theta. \tag{6}$$

If we integrate (6) with respect to σ_n, we get

$$|f(z)| \le Ar^{-N} \cdot \frac{1}{2\pi} \int_{-\pi}^{\pi} d\theta \int_{T^n} |(fP)(z + re^{i\theta} w)| \, d\sigma_n(w). \tag{7}$$

The measure σ_n is invariant under the change of variables $w \to e^{i\theta} w$. The inner integral in (7) is therefore independent of θ. This gives (1). ////

8.4 Theorem *Suppose P is a polynomial in n variables, $v \in \mathscr{D}'(R^n)$, and v has compact support. Then the equation*

$$P(D)u = v \tag{1}$$

has a solution with compact support if and only if there is an entire function g in $\mathcal{C}^n$ such that

$$Pg = \hat{v}. \tag{2}$$

When this condition is satisfied, (1) has a unique solution u with compact support; the support of this u lies in the convex hull of the support of v.

PROOF. If (1) has a solution u with compact support, (*a*) of Theorem 7.23 shows that (2) holds with $g = \hat{u}$.

Conversely, suppose (2) holds for some entire g. Choose $r > 0$ so that v has its support in $rB = \{x \in R^n : |x| \le r\}$. By Lemma 8.3, (2) implies

(3) $$|g(z)| \le A \int_{T^n} |\hat{v}(z + w)|\, d\sigma_n(w) \qquad (z \in \mathcal{C}^n).$$

By (*a*) of Theorem 7.23, there exist N and γ such that

(4) $$|\hat{v}(z + w)| \le \gamma(1 + |z + w|)^N \exp\{r|\operatorname{Im}(z + w)|\}.$$

There are constants c_1 and c_2 that satisfy

(5) $$1 + |z + w| \le c_1(1 + |z|)$$

and

(6) $$|\operatorname{Im}(z + w)| \le c_2 + |\operatorname{Im} z|$$

for all $z \in \mathcal{C}^n$ and all $w \in T^n$. It follows from these inequalities that

(7) $$|g(z)| \le B(1 + |z|)^N \exp\{r|\operatorname{Im} z|\} \qquad (z \in \mathcal{C}^n),$$

where B is another constant (depending on γ, A, N, c_1, c_2, and r). By (7) and (*b*) of Theorem 7.23, $g = \hat{u}$ for some distribution u with support in rB. Hence (2) becomes $P\hat{u} = \hat{v}$, which is equivalent to (1).

The uniqueness of u is obvious, since there is at most one entire function $\hat{u}$ that satisfies $P\hat{u} = \hat{v}$.

The preceding argument showed that the support S_u of u lies in every closed ball centered at the origin that contains the support S_v of v. Since (1) implies

(8) $$P(D)(\tau_x u) = \tau_x v \qquad (x \in R^n),$$

the same statement is true of $x + S_u$ and $x + S_v$. Consequently, S_u lies in the intersection of all closed balls (centered anywhere in R^n) that contain S_v. Since this intersection is the convex hull of S_v, the proof is complete. ////

8.5 Theorem *If P is a polynomial in $\mathcal{C}$, of exact degree N, and if $r > 0$, then the differential operator $P(D)$ has a fundamental solution E that satisfies*

(1) $$|E(\psi)| \le Ar^{-N} \int_{T^n} d\sigma_n(w) \int_{R^n} |\hat{\psi}(t + rw)|\, dm_n(t)$$

for every $\psi \in \mathcal{D}(R^n)$.

Here A is the constant that appears in Lemma 8.3. The main point of the theorem is the *existence* of a fundamental solution, rather than the estimate (1) which arises from the proof.

PROOF. Fix $r > 0$, and define

$$\|\psi\| = \int_{T^n} d\sigma_n(w) \int_{R^n} |\hat{\psi}(t + rw)| \, dm_n(t). \tag{2}$$

In preparation for the main part of the proof, let us first show that

$$\lim_{j\to\infty} \|\psi_j\| = 0 \qquad \text{if } \psi_j \to 0 \text{ in } \mathscr{D}(R^n). \tag{3}$$

Note that $\hat{\psi}(t + w) = (e_{-w}\psi)^\wedge(t)$ if $t \in R^n$ and $w \in \mathcal{C}^n$. Hence

$$\|\psi\| = \int_{T^n} d\sigma_n(w) \int_{R^n} |(e_{-rw}\psi)^\wedge| \, dm_n. \tag{4}$$

If $\psi_j \to 0$ in $\mathscr{D}(R^n)$, all ψ_j have their supports in some compact set K. The functions e_{rw} ($w \in T^n$) are uniformly bounded on K. It follows from the Leibniz formula that

$$\|D^\alpha(e_{-rw}\psi_j)\|_\infty \le C(K, \alpha) \max_{\beta\le\alpha} \|D^\beta\psi_j\|_\infty. \tag{5}$$

The right side of (5) tends to 0, for every α. Hence, given $\varepsilon > 0$, there exists j_0 such that

$$\|(I - \Delta)^n(e_{-rw}\psi_j)\|_2 < \varepsilon \qquad (j > j_0, w \in T^n), \tag{6}$$

where $\Delta = D_1^2 + \cdots + D_n^2$ is the Laplacian. By the Plancherel theorem, (6) is the same as

$$\int_{R^n} |(1 + |t|^2)^n \hat{\psi}_j(t + rw)|^2 \, dm_n(t) < \varepsilon^2, \tag{7}$$

from which it follows, by the Schwarz inequality and (2), that $\|\psi_j\| < C\varepsilon$ for all $j > j_0$, where

$$C^2 = \int_{R^n} (1 + |t|^2)^{-2n} \, dm_n(t) < \infty. \tag{8}$$

This proves (3).

Suppose now that $\phi \in \mathscr{D}(R^n)$ and that

$$\psi = P(D)\phi. \tag{9}$$

Then $\hat{\psi} = P\hat{\phi}$, $\hat{\phi}$ and $\hat{\psi}$ are entire, hence ψ determines ϕ. In particular, $\phi(0)$ is a linear functional of ψ, defined on the range of $P(D)$. The crux of the proof consists in showing that this functional is continuous, i.e., that there is a distribution $u \in \mathscr{D}'(R^n)$ that satisfies

$$u(P(D)\phi) = \phi(0) \qquad (\phi \in \mathscr{D}(R^n)), \tag{10}$$

because then the distribution $E = \check{u}$ satisfies

$$\begin{aligned}(P(D)E)(\phi) = E(P(-D)\phi) &= u((P(-D)\phi)^\vee) \\ &= u(P(D)\check{\phi}) = \check{\phi}(0) = \phi(0) = \delta(\phi),\end{aligned}$$

so that $P(D)E = \delta$, as desired.

Lemma 8.3, applied to $P\hat{\phi} = \hat{\psi}$, yields

$$|\hat{\phi}(t)| \le Ar^{-N} \int_{T^n} |\hat{\psi}(t + rw)| \, d\sigma_n(w) \qquad (t \in R^n). \tag{11}$$

By the inversion theorem, $\phi(0) = \int_{R^n} \hat{\phi} \, dm_n$. Thus (11), (2), and (9) give

$$|\phi(0)| \le Ar^{-N} \|P(D)\phi\| \qquad (\phi \in \mathscr{D}(R^n)). \tag{12}$$

Let Y be the subspace of $\mathscr{D}(R^n)$ that consists of the functions $P(D)\phi$, $\phi \in \mathscr{D}(R^n)$. By (12), the Hahn-Banach theorem 3.3 shows that the linear functional that is defined on Y by $P(D)\phi \to \phi(0)$ extends to a linear functional u on $\mathscr{D}(R^n)$ that satisfies (10) as well as

$$|u(\psi)| \le Ar^{-N} \|\psi\| \qquad (\psi \in \mathscr{D}(R^n)). \tag{13}$$

By (3), $u \in \mathscr{D}'(R^n)$. This completes the proof. ////

Elliptic Equations

8.6 Introduction If u is a twice continuously differentiable function in some open set $\Omega \subset R^2$ that satisfies the Laplace equation

$$\frac{\partial^2 u}{\partial x^2} + \frac{\partial^2 u}{\partial y^2} = 0, \tag{1}$$

then it is very well known that u is actually in $C^\infty(\Omega)$, simply because every real harmonic function in Ω is (locally) the real part of a holomorphic function. Any theorem of this type—one which asserts that every solution of a certain differential equation has stronger smoothness properties than is a priori evident—is called a *regularity theorem.*

We shall give a proof of a rather general regularity theorem for elliptic partial differential equations. The term "elliptic" will be defined presently. It may be of interest to see, first of all, that the equation

$$\frac{\partial^2 u}{\partial x \, \partial y} = 0 \tag{2}$$

behaves quite differently from (1), since it is satisfied by every function u of the form $u(x, y) = f(y)$, where f is any differentiable function. In fact, if (2) is interpreted to mean

(3) $$\frac{\partial}{\partial y}\left(\frac{\partial u}{\partial x}\right) = 0,$$

then f can be a perfectly arbitrary function.

8.7 Definitions Suppose Ω is open in R^n, N is a positive integer, $f_\alpha \in C^\infty(\Omega)$ for every multi-index α with $|\alpha| \le N$, and at least one f_α with $|\alpha| = N$ is not identically 0. These data determine a linear differential operator

(1) $$L = \sum_{|\alpha| \le N} f_\alpha D_\alpha$$

which acts on distributions $u \in \mathscr{D}'(\Omega)$ by

(2) $$Lu = \sum_{|\alpha| \le N} f_\alpha D_\alpha u.$$

The *order* of L is N. The operator

(3) $$\sum_{|\alpha| = N} f_\alpha D_\alpha$$

is the *principal part* of L. The *characteristic polynomial* of L is

(4) $$p(x, y) = \sum_{|\alpha| = N} f_\alpha(x) y^\alpha \qquad (x \in \Omega,\ y \in R^n).$$

This is a homogeneous polynomial of degree N in the variables $y = (y_1, \ldots, y_n)$, with coefficients in $C^\infty(\Omega)$.

The operator L is said to be *elliptic* if $p(x, y) \ne 0$ for every $x \in \Omega$ and for every $y \in R^n$, except, of course, for $y = 0$. Note that ellipticity is defined in terms of the principal part of L; the lower-order terms that appear in (1) play no role.

For example, the characteristic polynomial of the Laplacian

(5) $$\Delta = \frac{\partial^2}{\partial x_1^2} + \cdots + \frac{\partial^2}{\partial x_n^2}$$

is $p(x, y) = -(y_1^2 + \cdots + y_n^2)$, so that Δ is elliptic.

On the other hand, if $L = \partial^2/\partial x_1\, \partial x_2$, then $p(x, y) = -y_1 y_2$, and L is not elliptic.

The main result that we are aiming at (Theorem 8.12) involves some special spaces of tempered distributions, which we now describe.

8.8 Sobolev spaces Associate to each real number s a positive measure μ_s on R^n by setting

(1) $$d\mu_s(y) = (1 + |y|^2)^s\, dm_n(y).$$

If $f \in L^2(\mu_s)$, that is, if $\int |f|^2 \, d\mu_s < \infty$, then f is a tempered distribution [Example (c) of Section 7.12]; hence f is the Fourier transform of a tempered distribution u. The vector space of all u so obtained will be denoted by H^s; equipped with the norm

$$\|u\|_s = \left(\int_{R^n} |\hat{u}|^2 \, d\mu_s \right)^{1/2} \tag{2}$$

H^s is clearly isometrically isomorphic to $L^2(\mu_s)$.

These spaces H^s are called *Sobolev spaces.* The dimension n will be fixed throughout, and no reference to it will be made in the notation.

By the Plancherel theorem, $H^0 = L^2$.

It is obvious that $H^s \subset H^t$ if $t < s$. The union X of all spaces H^s is therefore a vector space. A linear operator $\Lambda: X \to X$ is said to have *order* t if the restriction of Λ to each H^s is a continuous mapping of H^s into H^{s-t}; note that t need not be an integer and that every operator of order t also has order t' if $t' > t$.

Here are the properties of the Sobolev spaces that will be needed.

8.9 Theorem

(a) *Every distribution with compact support lies in some H^s.*

(b) *If $-\infty < t < \infty$, the mapping $u \to v$ given by*

$$\hat{v}(y) = (1 + |y|^2)^{t/2}\hat{u}(y) \qquad (y \in R^n)$$

is a linear isometry of H^s onto H^{s-t} and is therefore an operator of order t whose inverse has order $-t$.

(c) *If $b \in L^\infty(R^n)$, the mapping $u \to v$ given by $\hat{v} = b\hat{u}$ is an operator of order* 0.

(d) *For every multi-index α, D_α is an operator of order $|\alpha|$.*

(e) *If $f \in \mathscr{S}_n$, then $u \to fu$ is an operator of order* 0.

PROOF. If $u \in \mathscr{D}'(R^n)$ has compact support, (a) of Theorem 7.23 shows that

$$|\hat{u}(y)| \le C(1 + |y|)^N \qquad (y \in R^n), \tag{1}$$

for some constants C and N. Hence $u \in H^s$ if $s < -N - n/2$. This proves part (a); (b) and (c) are obvious. The relation

$$|(D_\alpha u)^\wedge(y)| = |y^\alpha| |\hat{u}(y)| \le (1 + |y|^2)^{|\alpha|/2} |\hat{u}(y)|$$

implies

$$\|D_\alpha u\|_{s-|\alpha|} \le \|u\|_s, \tag{2}$$

so that (d) holds.

The proof of (*e*) depends on the inequality

$$(3) \qquad (1 + |x + y|^2)^s \leq 2^{|s|}(1 + |x|^2)^s(1 + |y|^2)^{|s|},$$

valid for $x \in R^n$, $y \in R^n$, $-\infty < s < \infty$. The case $s = 1$ of (3) is obvious. From it the case $s = -1$ is obtained by replacing x by $x - y$ and then y by $-y$. The general case of (3) is obtained from these two by raising everything to the power $|s|$. It follows from (3) that

$$(4) \qquad \int_{R^n} |h(x - y)|^2 \, d\mu_s(x) \leq 2^{|s|}(1 + |y|^2)^{|s|} \int_{R^n} |h|^2 \, d\mu_s$$

for every measurable function h on R^n.

Now suppose $u \in H^s$, $f \in \mathscr{S}_n$, $t > |s| + n/2$. Since $\hat{f} \in \mathscr{S}_n$, $\|f\|_t < \infty$. Put $\gamma = \mu_{|s|-t}(R^n)$. Then $\gamma < \infty$. Define $F = |\hat{u}| * |\hat{f}|$. By Theorem 7.19,

$$(5) \qquad |(fu)^\wedge| = |\hat{u} * \hat{f}| \leq |\hat{u}| * |\hat{f}| = F.$$

By the Schwarz inequality,

$$(6) \qquad |F(x)|^2 \leq \int_{R^n} |\hat{f}(y)|^2 \, d\mu_t(y) \int_{R^n} |\hat{u}(x - y)|^2 \, d\mu_{-t}(y)$$

for every $x \in R^n$. Integrate (6) over R^n, with respect to μ_s. By (4), the result is

$$(7) \qquad \int_{R^n} |F|^2 \, d\mu_s \leq 2^{|s|}\gamma \|f\|_t^2 \|u\|_s^2.$$

It follows from (5) and (7) that

$$(8) \qquad \|fu\|_s \leq (2^{|s|}\gamma)^{1/2} \|f\|_t \|u\|_s.$$

This proves (*e*). ////

8.10 Definition Let Ω be open in R^n. A distribution $u \in \mathscr{D}'(\Omega)$ is said to be *locally* H^s if there corresponds to each point $x \in \Omega$ a distribution $v \in H^s$ such that $u = v$ in some neighborhood ω of x. (See Section 6.19.)

8.11 Theorem *If $u \in \mathscr{D}'(\Omega)$ and $-\infty < s < \infty$, the following two statements are equivalent:*

(*a*) *u is locally H^s.*

(*b*) *$\psi u \in H^s$ for every $\psi \in \mathscr{D}(\Omega)$.*

Moreover, if s is a nonnegative integer, (a) and (b) are equivalent to

(*c*) *$D_\alpha u$ is locally L^2 for every α with $|\alpha| \leq s$.*

Statement (*b*) may need some clarification, since u acts only on test functions whose supports lie in Ω. However, ψu is the functional that assigns to each $\phi \in \mathscr{D}(R^n)$ the number

$$(\psi u)(\phi) = u(\psi\phi).$$

Note that $\psi\phi \in \mathscr{D}(\Omega)$, so that $u(\psi\phi)$ is defined.

PROOF. Assume u is locally H^s. Let K be the support of some $\psi \in \mathscr{D}(\Omega)$. Since K is compact, there are finitely many open sets $\omega_i \subset \Omega$, whose union covers K, and in which u coincides with some $v_i \in H^s$. There exist functions $\psi_i \in \mathscr{D}(\omega_i)$ such that $\sum \psi_i = 1$ on K. If $\phi \in \mathscr{D}(R^n)$ it follows that

$$u(\psi\phi) = \sum u(\psi_i \psi\phi) = \sum v_i(\psi_i \psi\phi),$$

since $\psi_i \psi\phi \in \mathscr{D}(\omega_i)$. Thus $\psi u = \sum \psi_i \psi v_i$. By (*e*) of Theorem 8.9, $\psi_i \psi v_i \in H^s$ for each i. Thus $\psi u \in H^s$, and (*a*) implies (*b*).

If (*b*) holds, if $x \in \Omega$, and if $\psi \in \mathscr{D}(\Omega)$ is 1 in a neighborhood ω of x, then $u = \psi u$ in ω, and $\psi u \in H^s$ by assumption. Thus (*b*) implies (*a*).

Assume again that (*b*) holds. If $\psi \in \mathscr{D}(\Omega)$, then $\psi u \in H^s$, hence $D_\alpha(\psi u) \in H^{s-|\alpha|}$, by (*d*) of Theorem 8.9. If $|\alpha| \le s$, then

$$H^{s-|\alpha|} \subset H^0 = L^2(R^n).$$

Thus $D_\alpha(\psi u) \in L^2(R^n)$. Taking $\psi = 1$ in some neighborhood of a point $x \in \Omega$ shows that $D_\alpha u$ is locally L^2 in Ω. Thus (*b*) implies (*c*).

Finally, assume $D_\alpha u$ is locally L^2 for every α with $|\alpha| \le s$. Fix $\psi \in \mathscr{D}(\Omega)$. The Leibniz formula shows that $D_\alpha(\psi u) \in L^2(R^n)$ if $|\alpha| \le s$. Hence

$$\int_{R^n} |y^\alpha|^2 |(\psi u)^\wedge(y)|^2 \, dm_n(y) < \infty \qquad (|\alpha| \le s). \tag{1}$$

If s is a nonnegative integer, (1) holds with the monomials $y_1^s, \ldots, y_n^s$ in place of y^α. It follows, as in the proof of Theorem 7.25, that

$$\int_{R^n} (1 + |y|^2)^s |(\psi u)^\wedge(y)|^2 \, dm_n(y) < \infty. \tag{2}$$

Thus $\psi u \in H^s$, (*c*) implies (*b*), and the proof is complete. ////

8.12 Theorem *Assume Ω is an open set in R^n, and*

(*a*) *$L = \sum f_\alpha D_\alpha$ is a linear elliptic differential operator in Ω, of order $N \ge 1$, with coefficients $f_\alpha \in C^\infty(\Omega)$,*

(*b*) *for each α with $|\alpha| = N$, f_α is a constant,*

(c) *u and v are distributions in Ω that satisfy*

$$Lu = v, \tag{1}$$

and v is locally H^s.

Then u is locally H^{s+N}

Corollary. *If L satisfies (a) and (b) and if $v \in C^\infty(\Omega)$, then every solution u of (1) belongs to $C^\infty(\Omega)$. In particular, every solution of the homogeneous equation $Lu = 0$ is in $C^\infty(\Omega)$.*

For if $v \in C^\infty(\Omega)$, then $\psi v \in \mathscr{D}(R^n)$ for every $\psi \in \mathscr{D}(\Omega)$; hence v is locally H^s for every s, and the theorem implies that u is locally H^s for every s; it follows from Theorems 8.11 and 7.25 that $u \in C^\infty(\Omega)$.

Assumption (*b*) can be dropped from the theorem, but its presence makes the proof considerably easier.

PROOF. Fix a point $x \in \Omega$, let $B_0 \subset \Omega$ be a closed ball with center at x, and let $\phi_0 \in \mathscr{D}(\Omega)$ be 1 on some open set containing B_0. By (*a*) of Theorem 8.9, $\phi_0 u \in H^t$ for some t. Since H^t becomes larger as t decreases, we may assume that $t = s + N - k$, where k is a positive integer. Choose closed balls

$$B_0 \supset B_1 \supset \cdots \supset B_k,$$

each centered at x, and each properly contained in the preceding one. Choose $\phi_1, \ldots, \phi_k \in \mathscr{D}(\Omega)$ so that $\phi_i = 1$ on some open set containing B_i, and $\phi_i = 0$ off B_{i-1}. Since $\phi_0 u \in H^t$, the following "bootstrap" proposition implies that

$$\phi_1 u \in H^{t+1}, \ldots, \phi_k u \in H^{t+k}.$$

It therefore leads to the conclusion that u is locally H^{s+N}, because $t + k = s + N$ and $\phi_k = 1$ on B_k. ////

Proposition. *If, in addition to the hypotheses of Theorem 8.12, $\psi u \in H^t$ for some $t \leq s + N - 1$ and for some $\psi \in \mathscr{D}(\Omega)$ which is 1 on an open set containing the support of a function $\phi \in \mathscr{D}(\Omega)$, then $\phi u \in H^{t+1}$.*

PROOF. We begin by showing that

$$L(\phi u) \in H^{t-N+1}. \tag{2}$$

Consider the distribution

$$\Lambda = L(\phi u) - \phi L u = L(\phi u) - \phi v. \tag{3}$$

Since its support lies in the support of ϕ, u can be replaced by ψu in (3), without changing Λ:

$$\Lambda = L(\phi\psi u) - \phi L(\psi u) = \sum_{|\alpha| \le N} f_\alpha \cdot [D_\alpha(\phi\psi u) - \phi D_\alpha(\psi u)]. \tag{4}$$

If the Leibniz formula is applied to $D_\alpha(\phi \cdot \psi u)$, one sees that the derivatives of order N of ψu cancel in (4). Therefore Λ is a linear combination [with coefficients in $\mathscr{D}(R^n)$] of derivatives of ψu, of orders at most $N - 1$. Since $\psi u \in H^t$, parts (*d*) and (*e*) of Theorem 8.9 imply that $\Lambda \in H^{t-N+1}$. By Theorem 8.11, $\phi v \in H^s$, and since $t - N + 1 \le s$, we have $\phi v \in H^{t-N+1}$. Now (2) follows from (3).

Since L is elliptic, its characteristic polynomial

$$p(y) = \sum_{|\alpha| = N} f_\alpha y^\alpha \qquad (y \in R^n) \tag{5}$$

has no zero in R^n, except at $y = 0$. Define functions

$$q(y) = |y|^{-N} p(y), \qquad r(y) = (1 + |y|^N) q(y), \tag{6}$$

for $y \in R^n$, $y \ne 0$, and define operators Q, R, S on the union of the Sobolev spaces by

$$(Qw)^\wedge = q\hat{w}, \qquad (Rw)^\wedge = r\hat{w} \tag{7}$$

and

$$S = \sum_{|\alpha| < N} \psi f_\alpha D_\alpha. \tag{8}$$

Since p is a homogeneous polynomial of degree N, $q(\lambda y) = q(y)$ if $\lambda > 0$, and since p vanishes only at the origin, the compactness of the unit sphere in R^n implies that both q and $1/q$ are bounded functions. It follows from (*c*) of Theorem 8.9 that *both Q and Q^{-1} are operators of order* 0.

Since both $(1 + |y|^2)^{-N/2}(1 + |y|^N)$ and its reciprocal are bounded functions on R^n, it follows from the preceding paragraph, combined with (*b*) and (*c*) of Theorem 8.9, that *R is an operator of order N whose inverse R^{-1} has order $-N$.*

Since $\psi f_\alpha \in \mathscr{D}(R^n)$ it follows from (*d*) and (*e*) of Theorem 8.9 that *S is an operator of order $N - 1$.*

Since $p = r - q$, and since p is assumed to have constant coefficients f_α, we have

$$\left(\sum_{|\alpha| = N} f_\alpha D_\alpha w\right)^\wedge = p\hat{w} = (r - q)\hat{w} = (Rw - Qw)^\wedge \tag{9}$$

if w lies in some Sobolev space. Hence

$$(R - Q + S)(\phi u) = L(\phi u). \tag{10}$$

By (2), $L(\phi u) \in H^{t-N+1}$.

Since $\psi u \in H^t$ and $\phi\psi = \phi$, (*e*) of Theorem 8.9 implies that $\phi u = \phi\psi u \in H^t$. Hence

$$(Q - S)(\phi u) \in H^{t-N+1}, \tag{11}$$

because Q has order 0 and S has order $N - 1 \geq 0$. It now follows from (10) that

$$R(\phi u) \in H^{t-N+1}, \tag{12}$$

and since R^{-1} has order $-N$, we finally conclude that $\phi u \in H^{t+1}$. ////

8.13 Example Suppose L is an *elliptic* differential operator in R^n, with *constant* coefficients, and E is a fundamental solution of L. In the complement of the origin, the equation $LE = \delta$ reduces to $LE = 0$. Theorem 8.12 implies therefore that, except at the origin, E is an infinitely differentiable function. The nature of the singularity of E at the origin depends, of course, on L.

8.14 Example The origin in R^2 is the only zero of the polynomial $p(y) = y_1 + iy_2$. If Ω is open in R^2, and if $u \in \mathscr{D}'(\Omega)$ is a distribution solution of the Cauchy-Riemann equation

$$\left(\frac{\partial}{\partial x_1} + i\frac{\partial}{\partial x_2}\right)u = 0,$$

Theorem 8.12 implies that $u \in C^\infty(\Omega)$. It follows that u is a holomorphic function of $z = x_1 + ix_2$ in Ω. In other words, *every holomorphic distribution is a holomorphic function.*

Exercises

1. The following simple properties of holomorphic functions of several variables were tacitly used in this chapter. Prove them.
   (*a*) If f is entire in $\mathcal{C}^n$, if $w \in \mathcal{C}^n$, and if $\phi(\lambda) = f(\lambda w)$, then ϕ is an entire function of one complex variable.
   (*b*) If P is a polynomial in $\mathcal{C}^n$ and if
   $$\int_{T^n} |P| \, d\sigma_n = 0$$
   then P is identically 0. *Hint:* Compute $\int_{T^n} |P|^2 \, d\sigma_n$.
   (*c*) If P is a polynomial (not identically 0) and g is an entire function in $\mathcal{C}^n$, then there is at most one entire function f that satisfies $Pf = g$.

   Find generalizations of these three properties.
2. Prove the statement about convex hulls made in the last sentence of the proof of Theorem 8.4.

3. Find a fundamental solution for the operator $\partial^2/\partial x_1\,\partial x_2$ in R^2. (There is one that is the characteristic function of a certain subset of R^2.)
4. Show that the equation

$$\frac{\partial^2 u}{\partial x_1^2} - \frac{\partial^2 u}{\partial x_2^2} = 0$$

is satisfied (in the distribution sense) by every locally integrable function u of the form

$$u(x_1, x_2) = f(x_1 + x_2) \qquad \text{or} \qquad u(x_1, x_2) = f(x_1 - x_2)$$

and that even classical solutions (i.e., twice continuously differentiable functions) need not be in C^∞. Note the contrast between this and the Laplace equation.
5. For $x \in R^3$, define $f(x) = (1 + |x|^2)^{-1}$. Show that $f \in L^2(R^3)$ and that $\hat{f}$ is a fundamental solution of the operator $I - \Delta$ in R^3. Find $\hat{f}$, by direct computation and also by the following reasoning:
(a) Since f is a radial function (i.e., one that depends only on the distance from the origin) the same is true of $\hat{f}$; see Exercise 1 of Chapter 7.
(b) Away from the origin, $(I - \Delta)\hat{f} = 0$, and $\hat{f} \in C^\infty$.
(c) If $F(|y|) = \hat{f}(y)$, (b) implies that F satisfies an ordinary differential equation in $(0, \infty)$ that can easily be solved explicitly.
Ans. $\hat{f}(y) = (\pi/2)^{1/2}|y|^{-1} \exp(-|y|)$.
Do the same with R^n in place of R^3; you will meet Bessel functions.
6. For $0 < \lambda < n$ and $x \in R^n$, define

$$K_\lambda(x) = |x|^{-\lambda}.$$

Show that

(a) $$\hat{K}_\lambda(y) = c(n, \lambda)K_{n-\lambda}(y) \qquad (y \in R^n),$$

where

$$c(n, \lambda) = \frac{2^{n/2-\lambda}\Gamma[(n-\lambda)/2]}{\Gamma(\lambda/2)}.$$

Suggestion: If $n < 2\lambda < 2n$, K_λ is the sum of an L^1-function and an L^2-function. For these λ, Equation (a) can be deduced from the homogeneity condition

$$K_\lambda(tx) = t^{-\lambda}K_\lambda(x) \qquad (x \in R^n,\ t > 0).$$

The case $0 < 2\lambda < n$ follows from the inversion theorem (for tempered distributions). A passage to the limit gives the case $2\lambda = n$. The constants $c(n, \lambda)$ can be computed from $\int f\hat{\phi} = \int \hat{f}\phi$, with $\phi(x) = \exp(-|x|^2/2)$.
7. Take $n \geq 3$ and $\lambda = 2$ in Exercise 6, and deduce that $-c(n, 2)K_{n-2}$ is a fundamental solution of the Laplacian Δ in R^n. For example, if v has compact support in R^3, show that a solution of $\Delta u = v$ is given by

$$u(x) = -\frac{1}{4\pi}\int_{R^3} |x - y|^{-1} v(y)\, dy.$$

8. Identify R^2 and $\mathcal{C}$ (so that $z = x_1 + ix_2$); put

$$\partial = \frac{\partial}{\partial x_1} - i\frac{\partial}{\partial x_2}, \qquad \bar{\partial} = \frac{\partial}{\partial x_1} + i\frac{\partial}{\partial x_2}.$$

Show that the Fourier transform of $1/z$ (regarded as a tempered distribution) is $-i/z$. Show that this result is equivalent to the Cauchy formula

$$\phi(z) = \int_{R^2} (\bar{\partial}\phi)(w)\frac{dm_2(w)}{w - z} \qquad [\phi \in \mathscr{D}(R^2)].$$

Since $\partial \log |w| = 1/w$ and $\Delta = \partial\bar{\partial}$, deduce that

$$\phi(z) = \int_{R^2} (\Delta\phi)(w) \log |w - z| \, dm_2(w) \qquad [\phi \in \mathscr{D}(R^2)].$$

Thus $\log |z|$ is a fundamental solution of the Laplacian in R^2.

9. Use Exercise 6 to compute that

$$\lim_{\varepsilon \to 0} [\varepsilon^{-1} - b - \hat{K}_{2-\varepsilon}(y)] = \log |y| \qquad (y \in R^2),$$

where b is a certain constant. Show that this leads to another proof of the last statement in Exercise 8.

10. Suppose $P(D) = D^2 + aD + bI$. (We are now in the case $n = 1$.) Let f and g be solutions of $P(D)u = 0$ which satisfy

$$f(0) = g(0) \qquad \text{and} \qquad f'(0) - g'(0) = 1.$$

Define

$$G(x) = \begin{cases} f(x) & \text{if } x \le 0, \\ g(x) & \text{if } x > 0, \end{cases}$$

and put

$$\Lambda\phi = -\int_{-\infty}^{\infty} \phi(x)G(x)\, dx \qquad [\phi \in \mathscr{D}(R)].$$

Prove that Λ is a fundamental solution of $P(D)$.

11. Suppose u is a distribution in R^n whose first derivatives $D_1 u, \ldots, D_n u$ are locally L^2. Prove that u is then locally L^2. *Hint:* If $\psi \in \mathscr{D}(R^n)$ is 1 in a neighborhood of the origin and if $\Delta E = \delta$, then $\Delta(\psi E) - \delta \in \mathscr{D}(R^n)$. Hence

$$u - \sum_{i=1}^{n} (D_i u) * D_i(\psi E)$$

is in $C^\infty(R^n)$. Each $D_i(\psi E)$ is an L^1-function with compact support.

12. Suppose u is a distribution in R^n whose Laplacian Δu is a continuous function. Prove that u is then a continuous function. *Hint:* As in Exercise 11,

$$u - (\psi E) * (\Delta u) \in C^\infty(R^n).$$

13. Prove analogues of Exercises 11 and 12, with R^n replaced by an arbitrary open set Ω.

14. Show, under the hypotheses of Exercise 12, that

(*a*) $\partial^2 u/\partial x_1^2$ is locally L^2, but

(*b*) $\partial^2 u/\partial x_1^2$ need not be a continuous function.

Outline of (*b*) for periodic distributions in R^2 (Exercise 22, Chapter 7): If $g \in C(T^2)$ has Fourier coefficients $\hat{g}(m, n)$ and if f is defined by

$$\hat{f}(m, n) = (1 + m^2 + n^2)^{-1}\hat{g}(m, n),$$

then $f \in C(T^2)$ and $\Delta f = f - g \in C(T^2)$, since $\sum |\hat{f}(m, n)| < \infty$. The Fourier coefficients of $\partial^2 f/\partial x_1^2$ are $-m^2\hat{f}(m, n)$. If $\partial^2 f/\partial x_1^2$ were continuous for *every* $g \in C(T^2)$, then $(\partial^2 f/\partial x_1^2)(0, 0)$ would be a continuous linear functional of g. Hence there would be a complex Borel measure μ on T^2, with Fourier coefficients

$$\hat{\mu}(m, n) = \frac{m^2}{1 + m^2 + n^2}.$$

The next exercise shows that no such measure exists.

15. If μ is a complex Borel measure on T^2, and if

$$\gamma(A, B) = \frac{1}{(2A + 1)(2B + 1)} \sum_{n=-A}^{A} \sum_{m=-B}^{B} \hat{\mu}(m, n),$$

prove that

$$\lim_{A\to\infty}\left[\lim_{B\to\infty} \gamma(A, B)\right] = \lim_{B\to\infty}\left[\lim_{A\to\infty} \gamma(A, B)\right].$$

Suggestion: If $D_A(t) = (2A + 1)^{-1} \sum_{-A}^{A} e^{int}$, then $D_A(x) = 1$ if $x = 0$, $D_A(x) \to 0$ otherwise, and

$$\gamma(A, B) = \int_{T^2} D_A(x) D_B(y)\, d\mu(x, y).$$

Conclude that each of the two iterated limits exists and that both are equal to $\mu(\{0, 0\})$.

If μ were as in Exercise 14, one of the iterated limits would be 1, the other 0.

16. Suppose L is an elliptic linear operator in some open set $\Omega \subset R^n$, and suppose that the order of L is *odd*.

(*a*) Prove that then $n = 1$ or $n = 2$.

(*b*) If $n = 2$, prove that the coefficients of the characteristic polynomial of L cannot all be real.

In view of (*a*), the Cauchy-Riemann operator is not a very typical example of an elliptic operator.

CHAPTER 9

TAUBERIAN THEORY

Wiener's Theorem

9.1 Introduction A tauberian theorem is one in which the asymptotic behavior of a sequence or of a function is deduced from the behavior of some of its averages. Tauberian theorems are often converses of fairly obvious results, but usually these converses depend on some additional assumption, called a *tauberian condition*. To see an example of this, consider the following three properties of a sequence of complex numbers $s_n = a_0 + \cdots + a_n$.

(*a*) $\lim_{n\to\infty} s_n = s.$

(*b*) If $f(r) = \sum_0^\infty a_n r^n$, $0 < r < 1$, then $\lim_{r\to 1} f(r) = s.$

(*c*) $\lim_{n\to\infty} na_n = 0.$

Since $f(r) = (1 - r) \sum s_n r^n$ and $(1 - r) \sum r^n = 1$, $f(r)$ is, for each $r \in (0, 1)$, an average of the sequence $\{s_n\}$. It is extremely easy to prove that (*a*) implies (*b*). The converse is not true, but (*b*) and (*c*) together imply (*a*); this is also quite easy and was proved by Tauber. The tauberian condition (*c*) can be replaced by the weaker assumption that $\{na_n\}$ is bounded (Littlewood). It is remarkable how much more difficult this weakening of (c) makes the proof.

Wiener's tauberian theorem deals with bounded measurable functions, originally on the real line. If $\phi \in L^\infty(R)$ and if $\phi(x) \to 0$ as $x \to +\infty$, then it is almost trivial that $(K * \phi)(x) \to 0$ as $x \to +\infty$ for every $K \in L^1(R)$. The convolutions $K * \phi$ may be regarded as averages of ϕ, at least when $\int K = 1$. Wiener's converse [(a) of Theorem 9.7] states that if $(K * \phi)(x) \to 0$ for *one* $K \in L^1(R)$ and if the Fourier transform of this K vanishes at no point of R, then $(f * \phi)(x) \to 0$ for *every* $f \in L^1(R)$; the stronger conclusion that $\phi(x) \to 0$ need not hold under these hypotheses, but it does hold if a slight additional condition (slow oscillation) is imposed on ϕ [(b) of Theorem 9.7].

The unexpected tauberian condition—the nonvanishing of $\hat{K}$—enters the proof in the following manner: If $(K * \phi)(x) \to 0$, the same is true if K is replaced by any of its translates, hence also if K is replaced by any finite linear combination g of translates of K. When $\hat{K}$ has no zero, it turns out that the set of these functions g is dense in L^1 (Theorem 9.5). One is thus led to the study of translation-invariant subspaces of L^1.

9.2 Lemma *Suppose $f \in L^1(R^n)$, $t \in R^n$, and $\varepsilon > 0$. Then there exists $h \in L^1(R^n)$, with $\|h\|_1 < \varepsilon$, such that*

(1) $$\hat{h}(s) = \hat{f}(t) - \hat{f}(s)$$

for all s in some neighborhood of t.

The lemma states that f is approximated, in the L^1-norm, by a function $f + h$ whose Fourier transform is constant in a neighborhood of the point t.

PROOF. Choose $g \in L^1(R^n)$ so that $\hat{g} = 1$ in some neighborhood of the origin. For $\lambda > 0$, put

(2) $$g_\lambda(x) = e^{it \cdot x}\lambda^{-n}g(x/\lambda) \qquad (x \in R^n)$$

and define

(3) $$h_\lambda(x) = \hat{f}(t)g_\lambda(x) - (f * g_\lambda)(x).$$

Since $\hat{g}_\lambda(s) = 1$ in some neighborhood V_λ of t, (3) shows that (1) holds for $s \in V_\lambda$, with h_λ in place of h. Next,

(4) $$h_\lambda(x) = \int_{R^n} f(y)[e^{-it \cdot y}g_\lambda(x) - g_\lambda(x - y)]\, dm_n(y).$$

The absolute value of the expression in brackets is

(5) $$|\lambda^{-n}g(\lambda^{-1}x) - \lambda^{-n}g(\lambda^{-1}(x - y))|.$$

It follows that

$$\|h_\lambda\|_1 \le \int_{R^n} |f(y)|\, dm_n(y) \int_{R^n} |g(\xi) - g(\xi - \lambda^{-1}y)|\, dm_n(\xi), \tag{6}$$

by the change of variables $x = \lambda\xi$. The inner integral in (6) is at most $2\|g\|_1$, and it tends to 0 for every $y \in R^n$, as $\lambda \to \infty$. Hence $\|h_\lambda\|_1 \to 0$, as $\lambda \to \infty$, by the dominated convergence theorem. ////

9.3 Theorem *If $\phi \in L^\infty(R^n)$, Y is a subspace of $L^1(R^n)$, and*

$$f * \phi = 0 \text{ for every } f \in Y, \tag{1}$$

then the set

$$Z(Y) = \bigcap_{f \in Y} \{s \in R^n : \hat{f}(s) = 0\} \tag{2}$$

contains the support of the tempered distribution $\hat{\phi}$.

PROOF. Fix a point t in the complement of $Z(Y)$. Then $\hat{f}(t) = 1$ for a certain $f \in Y$. Lemma 9.2 furnishes $h \in L^1(R^n)$, with $\|h\|_1 < 1$, such that $\hat{h}(s) = 1 - \hat{f}(s)$ in some neighborhood V of t.

To prove the theorem, it suffices to show that $\hat{\phi} = 0$ in V, or, equivalently, that $\hat{\phi}(\hat{\psi}) = 0$ for every $\psi \in \mathscr{S}_n$ whose Fourier transform $\hat{\psi}$ has its support in V. Since

$$\hat{\phi}(\hat{\psi}) = \phi(\check{\psi}) = (\phi * \psi)(0), \tag{3}$$

it suffices to show that $\phi * \psi = 0$.

Fix such a ψ. Put $g_0 = \psi$, $g_m = h * g_{m-1}$ for $m \ge 1$. Then $\|g_m\|_1 \le \|h\|_1^m \|\psi\|_1$, and since $\|h\|_1 < 1$, the function $G = \sum g_m$ is in $L^1(R^n)$. Since $\hat{h}(s) = 1 - \hat{f}(s)$ on the support of $\hat{\psi}$, we have

$$(1 - \hat{h}(s))\hat{\psi}(s) = \hat{\psi}(s)\hat{f}(s) \qquad (s \in R^n), \tag{4}$$

or

$$\hat{\psi} = \sum_{m=0}^{\infty} \hat{h}^m \hat{\psi} \hat{f} = \hat{G}\hat{f}. \tag{5}$$

Thus $\psi = G * f$, and (1) implies

$$\psi * \phi = G * f * \phi = 0. \tag{6}$$

////

9.4 Wiener's theorem *If Y is a closed translation-invariant subspace of $L^1(R^n)$ and if $Z(Y)$ is empty, then $Y = L^1(R^n)$.*

PROOF. To say that Y is translation-invariant means that $\tau_x f \in Y$ if $f \in Y$ and $x \in R^n$. If $\phi \in L^\infty(R^n)$ is such that $\int f\check{\phi} = 0$ for every $f \in Y$, the translation-invariance of Y implies that $f * \phi = 0$ for every $f \in Y$. By Theorem 9.3, the support of the distribution $\hat{\phi}$ is therefore empty, hence $\hat{\phi} = 0$ (Theorem 6.24), and since the Fourier transform maps $\mathscr{S}'_n$ to $\mathscr{S}'_n$ in a one-to-one fashion (Theorem 7.15), it follows that $\phi = 0$ as a distribution. Hence ϕ is the zero element of $L^\infty(R^n)$.

Thus $Y^\perp = \{0\}$. By the Hahn-Banach theorem, this implies that $Y = L^1(R^n)$. ////

9.5 Theorem *Suppose $K \in L^1(R^n)$ and Y is the smallest closed translation-invariant subspace of $L^1(R^n)$ that contains K. Then $Y = L^1(R^n)$ if and only if $\hat{K}(t) \neq 0$ for every $t \in R^n$.*

PROOF. Note that $Z(Y) = \{t \in R^n: \hat{K}(t) = 0\}$. The theorem thus asserts that $Y = L^1(R^n)$ if and only if $Z(Y)$ is empty. One-half of this is Theorem 9.4; the other half is trivial. ////

9.6 Definition A function $\phi \in L^\infty(R^n)$ is said to be *slowly oscillating* if to every $\varepsilon > 0$ correspond an $A < \infty$ and a $\delta > 0$ such that

(1) $$|\phi(x) - \phi(y)| < \varepsilon \qquad \text{if } |x| > A,\ |y| > A,\ |x - y| < \delta.$$

If $n = 1$, one can also define what it means for ϕ to be *slowly oscillating at* $+\infty$: the requirement (1) is replaced by

(2) $$|\phi(x) - \phi(y)| < \varepsilon \qquad \text{if } x > A,\ y > A,\ |x - y| < \delta.$$

The same definition can of course be made at $-\infty$.

Note that every uniformly continuous bounded function is slowly oscillating but that some slowly oscillating functions are not continuous.

We now come to Wiener's tauberian theorem; part (*b*) was added by Pitt.

9.7 Theorem

(*a*) *Suppose $\phi \in L^\infty(R^n)$, $K \in L^1(R^n)$, $\hat{K}(t) \neq 0$ for every $t \in R^n$, and*

(1) $$\lim_{|x| \to \infty} (K * \phi)(x) = a\hat{K}(0).$$

Then

(2) $$\lim_{|x| \to \infty} (f * \phi)(x) = a\hat{f}(0),$$

for every $f \in L^1(R^n)$.

(*b*) *If, in addition, ϕ is slowly oscillating, then*

(3) $$\lim_{|x| \to \infty} \phi(x) = a.$$

PROOF. Put $\psi(x) = \phi(x) - a$. Let Y be the set of all $f \in L^1(R^n)$ for which

(4) $$\lim_{|x| \to \infty} (f * \psi)(x) = 0.$$

It is clear that Y is a vector space. Also, Y is closed. To see this, suppose $f_i \in Y$, $\|f - f_i\|_1 \to 0$. Since

(5) $$\|f * \psi - f_i * \psi\|_\infty \le \|f - f_i\|_1 \|\psi\|_\infty,$$

$f_i * \psi \to f * \psi$ uniformly on R^n; hence (4) holds. Since

(6) $$((\tau_y f) * \psi)(x) = (\tau_y(f * \psi))(x) = (f * \psi)(x - y),$$

Y is translation-invariant. Finally, $K \in Y$, by (1), since $K * a = a\hat{K}(0)$.

Theorem 9.5 now applies and shows that $Y = L^1(R^n)$. Thus every $f \in L^1(R^n)$ satisfies (4), which is the same as (2). This proves part (*a*).

If ϕ is slowly oscillating and if $\varepsilon > 0$, choose A and δ as in Definition 9.6, and choose $f \in L^1(R^n)$ so that $f \ge 0$, $\hat{f}(0) = 1$, and $f(x) = 0$ if $|x| \ge \delta$. By (2),

(7) $$\lim_{|x| \to \infty} (f * \phi)(x) = a.$$

Also,

(8) $$\phi(x) - (f * \phi)(x) = \int_{|y| < \delta} [\phi(x) - \phi(x - y)] f(y)\, dm_n(y).$$

If $|x| > A + \delta$, our choice of A, δ, and f shows that

(9) $$|\phi(x) - (f * \phi)(x)| < \varepsilon.$$

Now (3) follows from (7) and (9).

This completes the proof. ////

9.8 Remark If $n = 1$, Theorem 9.7 can be modified in an obvious fashion, by writing $x \to +\infty$ in place of $|x| \to \infty$ wherever the latter occurs and by assuming in (*b*) that ϕ is slowly oscillating at $+\infty$. The proof remains unchanged.

The Prime Number Theorem

9.9 Introduction For any positive number x, $\pi(x)$ denotes the number of primes p that satisfy $p \le x$. The *prime number theorem* is the statement that

(1) $$\lim_{x \to \infty} \frac{\pi(x) \log x}{x} = 1.$$

We shall prove this by means of a tauberian theorem due to Ingham, based on that of Wiener. The idea is to replace the rather irregular function π by a function F whose asymptotic behavior is very easily established and to use the tauberian theorem to draw a conclusion about π from knowledge of F.

9.10 Preparation The letter p will now always denote a prime; m and n will be positive integers; x will be a positive number; $[x]$ is the integer that satisfies $x - 1 < [x] \le x$; the symbol $d \mid n$ means that d and n/d are positive integers. Define

(1) $$\Lambda(n) = \begin{cases} \log p & \text{if } n = p, p^2, p^3, \ldots, \\ 0 & \text{otherwise,} \end{cases}$$

(2) $$\psi(x) = \sum_{n \le x} \Lambda(n),$$

(3) $$F(x) = \sum_{m=1}^{\infty} \psi\left(\frac{x}{m}\right).$$

The following properties of ψ and F will be used:

(4) $$\frac{\psi(x)}{x} \le \frac{\pi(x) \log x}{x} < \frac{1}{\log x} + \frac{\psi(x) \log x}{x \log (x/\log^2 x)}$$

if $x > e$, and

(5) $$F(x) = x \log x - x + b(x) \log x,$$

where $b(x)$ remains bounded as $x \to \infty$.

By (4), the prime number theorem is a consequence of the relation

(6) $$\lim_{x \to \infty} \frac{\psi(x)}{x} = 1,$$

which will be proved from (3) and (5) by a tauberian theorem.

PROOF OF (4). $[\log x/\log p]$ is the number of powers of p that do not exceed x. Hence

$$\psi(x) = \sum_{p \le x} \left[\frac{\log x}{\log p}\right] \log p \le \sum_{p \le x} \log x = \pi(x) \log x.$$

This gives the first inequality in (4). If $1 < y < x$, then

$$\pi(x) - \pi(y) = \sum_{y < p \le x} 1 \le \sum_{y < p \le x} \frac{\log p}{\log y} \le \frac{\psi(x)}{\log y}.$$

Hence $\pi(x) < y + \psi(x)/\log y$. With $y = x/\log^2 x$, this gives the second half of (4). ////

PROOF OF (5). If $n > 1$, then

$$F(n) - F(n-1) = \sum_{m=1}^{\infty} \left\{ \psi\left(\frac{n}{m}\right) - \psi\left(\frac{n-1}{m}\right) \right\}.$$

The mth summand is 0 except when n/m is an integer, in which case it is $\Lambda(n/m)$. Hence

$$F(n) - F(n-1) = \sum_{m|n} \Lambda\left(\frac{n}{m}\right) = \sum_{d|n} \Lambda(d) = \log n.$$

The last equality depends on the factorization of n into a product of powers of distinct primes. Since $F(1) = 0$, we have computed that

(7) $$F(n) = \sum_{m=1}^{n} \log m = \log (n!) \qquad (n = 1, 2, 3, \ldots),$$

which suggests comparison of $F(x)$ with the integral

(8) $$J(x) = \int_1^x \log t \, dt = x \log x - x + 1.$$

If $n \le x \le n + 1$ then

(9) $$J(n) < F(n) \le F(x) \le F(n+1) < J(n+2)$$

so that

(10) $$|F(x) - J(x)| < 2 \log (x + 2).$$

Now (5) follows from (8) and (10). ////

9.11 The Riemann zeta function As is the custom in analytic number theory, complex variables will now be written in the form $s = \sigma + it$. In the half-plane $\sigma > 1$, the zeta function is defined by the series

(1) $$\zeta(s) = \sum_{n=1}^{\infty} n^{-s}$$

Since $|n^{-s}| = n^{-\sigma}$, the series converges uniformly on every compact subset of this half-plane, and ζ is holomorphic there.

A simple computation gives

$$s \int_1^{N+1} [x] x^{-1-s} \, dx = s \sum_{n=1}^{N} n \int_n^{n+1} x^{-1-s} \, dx = \sum_{n=1}^{N} n^{-s} - N(N+1)^{-s}.$$

When $\sigma > 1$, $N(N+1)^{-s} \to 0$ as $N \to \infty$. Hence

(2) $$\zeta(s) = s \int_1^{\infty} [x] x^{-1-s} \, dx \qquad (\sigma > 1).$$

If $b(x) = [x] - x$, it follows from (2) that

$$\zeta(s) = \frac{s}{s-1} + s\int_1^{\infty} b(x)x^{-1-s}\,dx \qquad (\sigma > 1). \tag{3}$$

Since b is bounded, the last integral defines a holomorphic function in the half-plane $\sigma > 0$. Thus (3) furnishes an analytic continuation of ζ to $\sigma > 0$, which is holomorphic except for a simple pole at $s = 1$, with residue 1. The most important property we shall need is that ζ has no zeros on the line $\sigma = 1$:

$$\zeta(1 + it) \neq 0 \qquad (-\infty < t < \infty). \tag{4}$$

The proof of (4) depends on the identity

$$\zeta(s) = \prod_p (1 - p^{-s})^{-1} \qquad (\sigma > 1). \tag{5}$$

Since $(1 - p^{-s})^{-1} = 1 + p^{-s} + p^{-2s} + \cdots$, the fact that the product (5) equals the series (1) is an immediate consequence of the fact that every positive integer has a unique factorization into a product of powers of primes. Since $\sum p^{-\sigma} < \infty$ if $\sigma > 1$, (5) shows that $\zeta(s) \neq 0$ if $\sigma > 1$ and that

$$\log \zeta(s) = \sum_p \sum_{m=1}^{\infty} m^{-1} p^{-ms} \qquad (\sigma > 1). \tag{6}$$

Fix a real $t \neq 0$. If $\sigma > 1$, (6) implies that

$$\begin{aligned} &\log |\zeta^3(\sigma)\zeta^4(\sigma + it)\zeta(\sigma + 2it)| \\ &\qquad = \sum_{p,\,m} m^{-1}p^{-m\sigma} \operatorname{Re}\{3 + 4p^{-imt} + p^{-2imt}\} \geq 0, \end{aligned} \tag{7}$$

because $\operatorname{Re}(6 + 8e^{i\theta} + 2e^{2i\theta}) = (e^{i\theta/2} + e^{-i\theta/2})^4 \geq 0$ for all real θ. Hence

$$|(\sigma - 1)\zeta(\sigma)|^3 \left|\frac{\zeta(\sigma + it)}{\sigma - 1}\right|^4 |\zeta(\sigma + 2it)| \geq \frac{1}{\sigma - 1}. \tag{8}$$

If $\zeta(1 + it)$ were 0, the left side of (8) would converge to a limit, namely, $|\zeta'(1 + it)|^4 |\zeta(1 + 2it)|$, as σ decreases to 1. Since the right side of (8) tends to infinity, this is impossible, and (4) is proved.

9.12 Ingham's tauberian theorem *Suppose g is a real nondecreasing function on $(0, \infty)$, $g(x) = 0$ if $x < 1$,*

$$G(x) = \sum_{n=1}^{\infty} g\left(\frac{x}{n}\right) \qquad (0 < x < \infty), \tag{1}$$

and

$$G(x) = ax \log x + bx + x\varepsilon(x), \tag{2}$$

where a, b are constants and $\varepsilon(x) \to 0$ as $x \to \infty$. Then

(3) $$\lim_{x\to\infty} x^{-1}g(x) = a.$$

If g is the function ${}^*\psi$ defined in Section 9.10, Ingham's theorem implies, in view of Equations (3) and (5) of Section 9.10, that (6) of Section 9.10 holds, and this, as we saw there, gives the prime number theorem.

PROOF. We first show that $x^{-1}g(x)$ is bounded. Since g is nondecreasing,

$$g(x) - g\left(\frac{x}{2}\right) \leq \sum_{n=1}^{\infty} (-1)^{n+1} g\left(\frac{x}{n}\right) = G(x) - 2G\left(\frac{x}{2}\right)$$

$$= x\left\{a \log 2 + \varepsilon(x) - \varepsilon\left(\frac{x}{2}\right)\right\} < Ax,$$

where A is some constant. Since

$$g(x) = g(x) - g\left(\frac{x}{2}\right) + g\left(\frac{x}{2}\right) - g\left(\frac{x}{4}\right) + \cdots,$$

it follows that

(4) $$g(x) < A\left(x + \frac{x}{2} + \frac{x}{4} + \cdots\right) = 2Ax.$$

We now make a change of variables that will enable us to use Fourier transforms in a familiar setting. For $-\infty < x < \infty$, define

(5) $$h(x) = g(e^x), \qquad H(x) = \sum_{n=1}^{\infty} h(x - \log n).$$

Then $h(x) = 0$ if $x < 0$, $H(x) = G(e^x)$; hence (2) becomes

(6) $$H(x) = e^x(ax + b + \varepsilon_1(x))$$

where $\varepsilon_1(x) \to 0$ as $x \to \infty$. If

(7) $$\phi(x) = e^{-x}h(x) \qquad (-\infty < x < \infty),$$

then ϕ is bounded, by (4). We have to prove that

(8) $$\lim_{x\to\infty} \phi(x) = a.$$

Put $k(x) = [e^x]e^{-x}$, let λ be a positive irrational number, and define

(9) $$K(x) = 2k(x) - k(x-1) - k(x-\lambda) \qquad (-\infty < x < \infty).$$

Then $K \in L^1(-\infty, \infty)$; in fact, $e^x K(x)$ is bounded. (See Exercise 8.) If $s = \sigma + it$, $\sigma > 0$, then formula (2) of Section 9.11 shows that

$$\int_{-\infty}^{\infty} k(x)e^{-xs}\, dx = \int_0^{\infty} [e^x]e^{-x(s+1)}\, dx = \int_1^{\infty} [y]y^{-2-s}\, dy = \frac{\zeta(1+s)}{1+s}.$$

Repeat this with $k(x-1)$ and $k(x-\lambda)$ in place of $k(x)$, use (9), and then let $\sigma \to 0$. The result is

$$\int_{-\infty}^{\infty} K(x)e^{-itx}\, dx = (2 - e^{-it} - e^{-i\lambda t})\frac{\zeta(1+it)}{1+it}. \tag{10}$$

Since $\zeta(1+it) \neq 0$ and since λ is irrational, $\hat{K}(t) \neq 0$ if $t \neq 0$. Since ζ has a pole with residue 1 at $s = 1$, the right side of (10) tends to $1 + \lambda$ as $t \to 0$. Thus $\hat{K}(0) \neq 0$.

To apply Wiener's theorem, we have to estimate $K * \phi$. To do this, put $u(x) = [e^x]$, let v be the characteristic function of $[0, \infty)$, and let μ be the measure that assigns mass 1 to each point of the set $\{\log n\colon\ n = 1, 2, 3, \ldots\}$ and whose support is this set. By (5), $H = h * \mu$. Also, $u = v * \mu$. Hence

$$(h * u)(x) = (h * v * \mu)(x) = (H * v)(x) = \int_0^x H(y)\, dy. \tag{11}$$

(Note that we now take convolutions with respect to Lebesgue measure, not with respect to the normalized measure m_1.) Since

$$(\phi * k)(x) = \int_{-\infty}^{\infty} e^{y-x}h(x-y)[e^y]e^{-y}\, dy = e^{-x}(h * u)(x),$$

(6) and (11) imply that

$$(\phi * k)(x) = e^{-x}\int_0^x H(y)\, dy = ax + b - a + \varepsilon_2(x), \tag{12}$$

where $\varepsilon_2(x) \to 0$ as $x \to \infty$. By (12) and (9),

$$\lim_{x\to\infty} (K * \phi)(x) = (1+\lambda)a = a\int_{-\infty}^{\infty} K(y)\, dy. \tag{13}$$

Therefore Wiener's theorem 9.7 (see also Remark 9.8) implies that

$$\lim_{x\to\infty} (f * \phi)(x) = a\int_{-\infty}^{\infty} f(y)\, dy \tag{14}$$

for *every* $f \in L^1(-\infty, \infty)$.

Let f_1 and f_2 be nonnegative functions whose integral is 1 and whose supports lie in $[0, \varepsilon]$ and $[-\varepsilon, 0]$, respectively. By (7), $e^x\phi(x)$ is nondecreasing. Thus $\phi(y) \le e^{\varepsilon}\phi(x)$ if $x - \varepsilon \le y \le x$, and $\phi(y) \ge e^{-\varepsilon}\phi(x)$ if $x \le y \le x + \varepsilon$. Consequently,

$$e^{-\varepsilon}(f_1 * \phi)(x) \le \phi(x) \le e^{\varepsilon}(f_2 * \phi)(x). \tag{15}$$

It follows from (14) and (15) that the upper and lower limits of $\phi(x)$, as $x \to \infty$, lie between $ae^{-\varepsilon}$ and ae^{ε}. Since $\varepsilon > 0$ was arbitrary, (8) holds, and the proof is complete. ////

The Renewal Equation

As another application of Wiener's tauberian theorem we shall now give a brief discussion of the behavior of bounded solutions ϕ of the integral equation

$$\phi(x) - \int_{-\infty}^{\infty} \phi(x - t)\, d\mu(t) = f(x)$$

which occurs in probability theory. Here μ is a given Borel probability measure, f is a given function, and ϕ is assumed to be a bounded Borel function, so that the integral exists for *every* $x \in R$. The equation can be written in the form

$$\phi - \phi * \mu = f,$$

for brevity.

We begin with a uniqueness theorem.

9.13 Theorem *If μ is a Borel probability measure on R whose support does not lie in any cyclic subgroup of R, and if ϕ is a bounded Borel function that satisfies the homogeneous equation*

$$\phi(x) - (\phi * \mu)(x) = 0 \tag{1}$$

for every $x \in R$, then there is a constant A such that $\phi(x) = A$ except possibly in a set of Lebesgue measure 0.

PROOF. Since μ is a probability measure, $\hat{\mu}(0) = 1$. Suppose that $\hat{\mu}(t) = 1$ for some $t \ne 0$. Since

$$\hat{\mu}(t) = \int_{-\infty}^{\infty} e^{-ixt}\, d\mu(x), \tag{2}$$

it follows that μ must be concentrated on the set of all x at which $e^{-ixt} = 1$, that is, on the set of all integral multiples of $2\pi/t$. But this is ruled out by the hypothesis of the theorem.

If $\sigma = \delta - \mu$, where δ is the Dirac measure, then $\hat{\sigma} = 1 - \hat{\mu}$. Hence $\hat{\sigma}(t) = 0$ if and only if $t = 0$, and (1) can be written in the form

$$\phi * \sigma = 0. \tag{3}$$

Put $g(x) = \exp(-x^2)$; put $K = g * \sigma$. Then $K \in L^1$, $\hat{K}(t) = 0$ only if $t = 0$, and (3) shows that $K * \phi = 0$. By Theorem 9.3 (with the one-dimensional space generated by K in place of Y) the distribution $\hat{\phi}$ has its support in $\{0\}$. Hence $\hat{\phi}$ is a finite linear combination of δ and its derivatives (Theorem 6.25), so that ϕ is a polynomial, in the distribution sense. Since nonconstant polynomials are not bounded on R, and since ϕ is assumed to be bounded, we have reached the desired conclusion. ////

9.14 Convolutions of measures If μ and λ are complex Borel measures on R^n, then

$$f \to \int_{R^n} \int_{R^n} f(x + y)\, d\mu(x)\, d\lambda(y) \tag{1}$$

is a bounded linear functional on $C_0(R^n)$, the space of all continuous functions on R^n that vanish at infinity. By the Riesz representation theorem, there is a unique Borel measure $\mu * \lambda$ on R^n that satisfies

$$\int_{R^n} f\, d(\mu * \lambda) = \int_{R^n} \int_{R^n} f(x + y)\, d\mu(x)\, d\lambda(y) \qquad [f \in C_0(R^n)]. \tag{2}$$

A standard approximation argument shows that (2) then holds also for every bounded Borel function f. In particular, we see that

$$(\mu * \lambda)^\wedge = \hat{\mu}\hat{\lambda}. \tag{3}$$

Two other consequences of (2) will be used in the next theorem. One is the almost obvious inequality

$$\|\mu * \lambda\| \le \|\mu\|\, \|\lambda\|, \tag{4}$$

where the norm denotes total variation. The other is the fact that $\mu * \lambda$ is absolutely continuous (relative to Lebesgue measure m_n) if this is true of μ; for in that case,

$$\int_{R^n} f(x + y)\, d\mu(x) = 0 \tag{5}$$

for every $y \in R^n$, if f is the characteristic function of a Borel set E with $m_n(E) = 0$, and (2) shows that $(\mu * \lambda)(E) = 0$.

Recall that every complex Borel measure μ has a unique Lebesgue decomposition

$$\mu = \mu_a + \mu_s, \tag{6}$$

where μ_a is absolutely continuous relative to m_n and μ_s is singular.

The next theorem is due to Karlin.

9.15 Theorem *Suppose μ is a Borel probability measure on R, such that*

(1) $$\mu_a \neq 0,$$

(2) $$\int_{-\infty}^{\infty} |x| \, d\mu(x) < \infty,$$

(3) $$M = \int_{-\infty}^{\infty} x \, d\mu(x) \neq 0.$$

Suppose that $f \in L^1(R)$, that $f(x) \to 0$ as $x \to \pm\infty$, and that ϕ is a bounded function that satisfies

(4) $$\phi(x) - (\phi * \mu)(x) = f(x) \qquad (-\infty < x < \infty).$$

Then the limits

(5) $$\phi(\infty) = \lim_{x \to \infty} \phi(x), \qquad \phi(-\infty) = \lim_{x \to -\infty} \phi(x)$$

exist, and

(6) $$\phi(\infty) - \phi(-\infty) = \frac{1}{M} \int_{-\infty}^{\infty} f(y) \, dy.$$

PROOF. Put $\sigma = \delta - \mu$, as in the proof of Theorem 9.13. Define

(7) $$K(x) = \sigma((-\infty, x)) = \begin{cases} -\mu((-\infty, x)) & \text{if } x \le 0, \\ \mu([x, \infty)) & \text{if } x > 0. \end{cases}$$

The assumption (2) guarantees that $K \in L^1(R)$. A straightforward computation, whose details we omit, shows that

(8) $$\int_{-\infty}^{\infty} K(x) e^{-ixt} \, dx = \begin{cases} \hat{\sigma}(t)/it & \text{if } t \neq 0 \\ M & \text{if } t = 0 \end{cases}$$

and that

(9) $$\int_r^s f(x) \, dx = (K * \phi)(s) - (K * \phi)(r) \qquad (-\infty < r < s < \infty),$$

since $f + \phi * \sigma$.

By (1), μ is not singular. The argument used at the beginning of the proof of Theorem 9.13 shows therefore that $\hat{\sigma}(t) \neq 0$ if $t \neq 0$. Hence (8) and (3) imply that $\hat{K}$ has no zero in R.

Since $f \in L^1(R)$, (9) implies that $K * \phi$ has limits at $\pm\infty$, whose difference is $\int_{-\infty}^{\infty} f$.

We shall show that ϕ is slowly oscillating. Once this is done, (5) and (6) follow from the properties of K and $K * \phi$ that we just proved, by Pitt's theorem (*b*) of 9.7.

Repeated substitution of $\phi = f + \phi * \mu$ into its right-hand side gives

(10) $$\phi = f + f * \mu + \cdots + f * \mu^{n-1} + \phi * \mu^n$$
$$= f_n + g_n + h_n \qquad (n = 2, 3, 4, \ldots),$$

where $\mu^1 = \mu$, $\mu^n = \mu * \mu^{n-1}$, $f_n = f + \cdots + f * \mu^{n-1}$, and

(11) $$g_n = \phi * (\mu^n)_a, \qquad h_n = \phi * (\mu^n)_s.$$

For each n, $f_n(x) \to 0$ as $x \to \pm\infty$, and g_n is uniformly continuous. Hence $f_n + g_n$ is slowly oscillating. Since the total variations satisfy

(12) $$\|(\mu^n)_s\| \le \|(\mu_s)^n\| \le \|\mu_s\|^n,$$

we have

(13) $$|h_n(x)| \le \|\phi\| \cdot \|\mu_s\|^n \qquad (-\infty < x < \infty),$$

where $\|\phi\|$ is the supremum of $|\phi|$ on R. By (1), $\|\mu_s\| < 1$. Hence $h_n \to 0$, uniformly on R. Consequently, ϕ is the uniform limit of the slowly oscillating functions $f_n + g_n$. This implies that ϕ is slowly oscillating, and completes the proof. ////

Exercises

1. Prove the theorem of Tauber stated in Section 9.1.
2. Suppose $\phi \in L^\infty(R^n)$ and the support of the distribution $\hat{\phi}$ consists of k distinct points $s_1, \ldots, s_k$. Construct suitable functions $\psi_1, \ldots, \psi_k$ such that $(\phi * \psi_j)^\wedge$ has the singleton $\{s_j\}$ as support, and conclude that ϕ is a trigonometric polynomial, namely,
$$\phi(x) = a_1 e^{is_1 \cdot x} + \cdots + a_k e^{is_k \cdot x} \qquad \text{(a.e.)}.$$
(The case $k = 1$ is done in the proof of Theorem 9.13.)
3. Suppose Y is a closed translation-invariant subspace of $L^1(R^n)$ such that $Z(Y)$ consists of k distinct points. (The notation is as in Theorem 9.3.) Use Exercise 2 to prove that Y has codimension k in $L^1(R^n)$, and conclude from this that Y consists of exactly those $f \in L^1(R^n)$ whose Fourier transforms are 0 at every point of $Z(Y)$.
4. Prove the following analogue of (*a*) of Theorem 9.7: If $\phi \in L^\infty(R^n)$, and if to every $t \in R^n$ corresponds a function $K_t \in L^1(R^n)$ such that $\hat{K}_t(t) \neq 0$ and $(K_t * \phi)(x) \to 0$ as $|x| \to \infty$, then $(f * \phi)(x) \to 0$ as $|x| \to \infty$, for every $f \in L^1(R^n)$.
5. Assume $K \in L^1(R^n)$ and $\hat{K}$ has at least one zero in R^n. Show that then there exists $\phi \in L^\infty(R^n)$ such that $(K * \phi)(x) = 0$ for every $x \in R^n$, although ϕ does not satisfy the conclusion of (*a*) of Theorem 9.7.
6. If $\phi(x) = \sin(x^2)$, $-\infty < x < \infty$, show that
$$\lim_{|x| \to \infty} (f * \phi)(x) = 0$$

for every $f \in L^1(R)$, although the conclusion of (*b*) of Theorem 9.7 does not hold.

7. For $\alpha > 0$, let f_α be the characteristic function of the interval $[0, \alpha]$. Define f_β in the same way; put $g = f_\alpha + f_\beta$. Prove that the set of all finite linear combinations of translates of g is dense in $L^1(R)$ if and only if β/α is irrational.
8. If $\alpha > 0$ and $\alpha x = 1$, prove that

$$1 - \alpha < \alpha[x] \leq 1,$$

and deduce from this that $e^x K(x)$ is bounded, as asserted in the proof of Theorem 9.12.

9. Let Q denote the set of all rational numbers. Let μ be a probability measure on R that is concentrated on Q, and let ϕ be the characteristic function of Q. Show that $\phi(x) = (\phi * \mu)(x)$ for every $x \in R$, although ϕ is not constant. (Compare with Theorem 9.13.) What other sets could be used in place of Q to achieve the same effect?
10. Special cases of the following facts were used in Theorem 9.15. Prove them.
    (*a*) If $\phi \in L^\infty(R^n)$ and $k \in L^1(R^n)$, then $k * \phi$ is uniformly continuous.
    (*b*) If $\{\phi_j\}$ is a sequence of slowly oscillating functions on R^n that converges uniformly to a function ϕ, then ϕ is slowly oscillating.
    (*c*) If μ and λ are complex Borel measures on R^n, then

$$\|(\mu * \lambda)_s\| \leq \|\mu_s\| \, \|\lambda_s\|.$$

11. Put $\psi(x) = \cos(|x|^{1/3})$ and define

$$f(x) = \psi(x) - \frac{1}{2}\int_{-1}^{1} \psi(x - y)\, dy \qquad (-\infty < x < \infty).$$

Prove that $f \in (L^1 \cap C_0)(R)$ but that no bounded solution of the equation

$$\phi(x) - \frac{1}{2}\int_{-1}^{1} \phi(x - y)\, dy = f(x)$$

has limits at $+\infty$ or at $-\infty$. (This illustrates the relevance of the condition $M \neq 0$ in Theorem 9.15.)

12. Let μ be a probability measure concentrated on the integers. Prove that every function ϕ on R which is periodic with period 1 satisfies $\phi - \phi * \mu = 0$. (This is relevant to Theorems 9.13 and 9.15.)
13. Assume $\phi \in L^\infty(0, \infty)$,

$$\int_0^\infty |K(x)| \frac{dx}{x} < \infty, \qquad \int_0^\infty |H(x)| \frac{dx}{x} < \infty,$$

$$\int_0^\infty K(x) x^{-it} \frac{dx}{x} \neq 0 \qquad \text{for } -\infty < t < \infty,$$

and

$$\lim_{x \to \infty} \int_0^\infty K\left(\frac{x}{u}\right) \phi(u) \frac{du}{u} = 0.$$

Prove that

$$\lim_{x\to\infty} \int_0^\infty H\left(\frac{x}{u}\right)\phi(u)\,\frac{du}{u} = 0.$$

This is an analogue of (*a*) of Theorem 9.7. How would "slowly oscillating" have to be defined to obtain the corresponding analog of (*b*) of Theorem 9.7?

14. Complete the details in the following outline of Wiener's proof of Littlewood's theorem. Assume $|na_n| \le 1$, $f(r) = \sum_0^\infty a_n r^n$, and $f(r) \to 0$ as $r \to 1$. If $s_n = a_0 + \cdots + a_n$, it is to be proved that $s_n \to 0$ as $n \to \infty$.
(*a*) $|s_n - f(1 - 1/n)| < 2$. Hence $\{s_n\}$ is bounded.
(*b*) If $\phi(x) = s_n$ on $[n, n+1)$ and $0 < x < y$, then

$$|\phi(y) - \phi(x)| \le \frac{y + 1 - x}{x}.$$

(*c*) $\int_0^\infty xe^{-xt}\phi(t)\,dt = f(e^{-x}) \to 0$ as $x \to 0$. Hence

$$\lim_{x\to\infty} \int_0^\infty K\left(\frac{x}{u}\right)\phi(u)\,\frac{du}{u} = 0$$

if

$$K(x) = \left(\frac{1}{x}\right)\exp\left(-\frac{1}{x}\right).$$

(*d*) $\int_0^\infty K(x)x^{-it}\,\frac{dx}{x} = \Gamma(1 + it) \ne 0$ if t is real.
(*e*) Put $H(x) = 1/(\varepsilon x)$ if $(1+\varepsilon)^{-1} < x < 1$, $H(x) = 0$ otherwise. Conclude that

$$\lim_{x\to\infty} \frac{1}{\varepsilon x}\int_x^{(1+\varepsilon)x} \phi(y)\,dy = 0.$$

(*f*) By (*b*) and (*e*), $\lim_{x\to\infty} \phi(x) = 0$.

Note: If $na_n \to 0$ is assumed to hold, then a modification of step (*a*) is all that is needed for the proof.

15. Let Y be a closed subspace of $L^1(R^n)$. Prove that Y is translation-invariant if and only if $f * g \in Y$ whenever $f \in Y$ and $g \in L^1(R^n)$.
The closed translation invariant subspaces of $L^1(R^n)$ are thus exactly the same as the closed ideals in the convolution algebra $L^1(R^n)$.

PART III

BANACH ALGEBRAS AND SPECTRAL THEORY

CHAPTER 10

BANACH ALGEBRAS

Introduction

10.1 Definition A *complex algebra* is a vector space A over the complex field $\mathcal{C}$ in which a multiplication is defined that satisfies

$$x(yz) = (xy)z, \tag{1}$$

$$(x + y)z = xz + yz, \qquad x(y + z) = xy + xz, \tag{2}$$

and

$$\alpha(xy) = (\alpha x)y = x(\alpha y) \tag{3}$$

for all x, y, and z in A and for all scalars α.

If, in addition, A is a *Banach space* with respect to a norm that satisfies the *multiplicative inequality*

$$\|xy\| \leq \|x\| \|y\| \qquad (x \in A, y \in A) \tag{4}$$

and if A contains a *unit element* e such that

$$xe = ex = x \qquad (x \in A) \tag{5}$$

and

$$\|e\| = 1, \tag{6}$$

then A is called a *Banach algebra*.

Note that we have not required that A be commutative, i.e., that

$xy = yx$ for all x and y in A, and we shall not do so except when explicitly stated.

It is clear that there is at most one $e \in A$ that satisfies (5), for if e' also satisfies (5), then $e' = e'e = e$.

The presence of a unit is very often omitted from the definition of a Banach algebra. However, when there is a unit it makes sense to talk about inverses, so that the spectrum of an element of A can be defined in a more natural way than is otherwise possible. This leads to a more intuitive development of the basic theory. Moreover, the resulting loss of generality is small, because many naturally occurring Banach algebras have a unit, and because the others can be supplied with one in the following canonical fashion.

Suppose A satisfies conditions (1) to (4), but A has no unit element. Let A_1 consist of all ordered pairs (x, α), where $x \in A$ and $\alpha \in \mathbb{C}$. Define the vector space operations in A_1 componentwise, define multiplication in A_1 by

$$(x, \alpha)(y, \beta) = (xy + \alpha y + \beta x, \alpha\beta), \tag{7}$$

and define

$$\|(x, \alpha)\| = \|x\| + |\alpha|, \qquad e = (0, 1). \tag{8}$$

Then A_1 satisfies properties (1) to (6), and the mapping $x \to (x, 0)$ is an isometric isomorphism of A onto a subspace of A_1 (in fact, onto a closed two-sided ideal of A_1) whose codimension is 1. If x is identified with $(x, 0)$, then A_1 is simply A plus the one-dimensional vector space generated by e. See Examples 10.3(*d*) and 11.13(*e*).

The inequality (4) makes multiplication a *continuous* operation in A. This means that if $x_n \to x$ and $y_n \to y$ then $x_n y_n \to xy$, which follows from the identity

$$x_n y_n - xy = (x_n - x)y_n + x(y_n - y). \tag{9}$$

In particular, multiplication is *left-continuous* and *right-continuous*:

$$x_n y \to xy \qquad \text{and} \qquad xy_n \to xy \tag{10}$$

if $x_n \to x$ and $y_n \to y$.

It is interesting that (4) can be replaced by the (apparently) weaker requirement (10) and that (6) can be dropped without enlarging the class of algebras under consideration.

10.2 Theorem *Assume that A is a Banach space as well as a complex algebra with unit element $e \neq 0$, in which multiplication is left-continuous and right-continuous. Then there is a norm on A which induces the same topology as the given one and which makes A into a Banach algebra.*

(The assumption $e \neq 0$ rules out the uninteresting case $A = \{0\}$.)

PROOF. Assign to each $x \in A$ the left-multiplication operator M_x defined by

$$(1) \qquad M_x(z) = xz \qquad (z \in A).$$

Let $\tilde{A}$ be the set of all M_x. Since right multiplication is assumed to be continuous, $\tilde{A} \subset \mathscr{B}(A)$, the Banach space of all bounded linear operators on A.

It is clear that $x \to M_x$ is linear. The associative law implies that $M_{xy} = M_x M_y$. If $x \in A$, then

$$(2) \qquad \|x\| = \|xe\| = \|M_x e\| \leq \|M_x\| \, \|e\|.$$

These facts can be summarized by saying that $x \to M_x$ is an isomorphism of A onto the algebra $\tilde{A}$, whose inverse is continuous. Since

$$(3) \qquad \|M_x M_y\| \leq \|M_x\| \, \|M_y\| \qquad \text{and} \qquad \|M_e\| = \|I\| = 1,$$

$\tilde{A}$ is a Banach algebra, *provided* it is complete, i.e., provided it is a *closed* subspace of $\mathscr{B}(A)$, relative to the topology given by the operator norm. (See Theorem 4.1.) Once this is done, the open mapping theorem implies that $x \to M_x$ is also continuous. Hence $\|x\|$ and $\|M_x\|$ are equivalent norms on A.

Suppose $T \in \mathscr{B}(A)$, $T_i \in \tilde{A}$, and $T_i \to T$ in the topology of $\mathscr{B}(A)$. If T_i is left multiplication by $x_i \in A$, then

$$(4) \qquad T_i(y) = x_i y = (x_i e)y = T_i(e)y.$$

As $i \to \infty$, the first term in (4) tends to $T(y)$, and $T_i(e) \to T(e)$. Since multiplication is assumed to be left-continuous in A, it follows that the last term of (4) tends to $T(e)y$. Put $x = T(e)$. Then

$$(5) \qquad T(y) = T(e)y = xy = M_x(y) \qquad (y \in A),$$

so that $T = M_x \in \tilde{A}$, and $\tilde{A}$ is closed. ////

The theorem says, in particular, that, in the presence of completeness, left continuity plus right continuity implies "joint" continuity. Exercise 6 shows that this may fail in normed linear algebras that are not complete.

10.3 Examples (*a*) Let $C(K)$ be the Banach space of all complex continuous functions on a nonempty compact Hausdorff space K, with the supremum norm. Define multiplication in the usual way: $(fg)(p) = f(p)g(p)$. This makes $C(K)$ into a commutative Banach algebra; the constant function 1 is the unit element.

If K is a finite set, consisting of, say, n points, then $C(K)$ is simply $\mathcal{C}^n$, with coordinatewise multiplication.

In particular, when $n = 1$, we obtain the simplest Banach algebra, namely $\mathbb{C}$, with the absolute value as norm.

(*b*) Let X be a Banach space. Then $\mathscr{B}(X)$, the algebra of all bounded linear operators on X, is a Banach algebra, with respect to the usual operator norm. The identity operator I is its unit element. If $\dim X = n < \infty$, then $\mathscr{B}(X)$ is (isomorphic to) the algebra of all complex n-by-n matrices. If $\dim X > 1$, then $\mathscr{B}(X)$ is not commutative. (The trivial space $X = \{0\}$ must be excluded.)

Every closed subalgebra of $\mathscr{B}(X)$ that contains I is also a Banach algebra. The proof of Theorem 10.2 shows, in fact, that *every* Banach algebra is isomorphic to one of these.

(*c*) If K is a nonempty compact subset of $\mathbb{C}$, or of $\mathbb{C}^n$, and if A is the subalgebra of $C(K)$ that consists of those $f \in C(K)$ that are holomorphic in the interior of K, then A is complete (relative to the supremum norm) and is therefore a Banach algebra.

When K is the closed unit disc in $\mathbb{C}$, then A is called the *disc algebra.*

(*d*) $L^1(R^n)$, with *convolution* as multiplication, satisfies all requirements of Definition 10.1, except that it lacks a unit. One can adjoin one by the abstract procedure outlined in Section 10.1 or one can do it more concretely by enlarging $L^1(R^n)$ to the algebra of all complex Borel measures μ on R^n of the form

$$d\mu = f\,dm_n + \lambda\,d\delta$$

where $f \in L^1(R^n)$, δ is the Dirac measure on R^n, and λ is a scalar.

(*e*) Let $M(R^n)$ be the algebra of all complex Borel measures on R^n, with convolution as multiplication, normed by the total variation. This is a commutative Banach algebra, with unit δ, which contains (*d*) as a closed subalgebra.

10.4 Remarks There are several reasons for restricting our attention to Banach algebras over the *complex* field, although real Banach algebras (whose definition should be obvious) have also been studied.

One reason is that certain elementary facts about holomorphic functions play an important role in the foundations of the subject. This may be observed in Theorems 10.9 and 10.13 and becomes even more obvious in the symbolic calculus.

Another reason—one whose implications are not quite so obvious—is that $\mathbb{C}$ has a natural nontrivial involution (see Definition 11.14), namely, conjugation, and that many of the deeper properties of certain types of Banach algebras depend on the presence of an involution. (For the same reason, the theory of complex Hilbert spaces is richer than that of real ones.)

At one point (Theorem 10.34) a topological difference between $\mathbb{C}$ and R will even play a role.

Among the important mappings from one Banach algebra into another are the *homomorphisms*. These are linear mappings h that are also *multiplicative*:

$$h(xy) = h(x)h(y).$$

Of particular interest is the case in which the range is the simplest of all Banach algebras, namely, $\mathbb{C}$ itself. Many of the significant features of the commutative theory depend crucially on a sufficient supply of homomorphisms onto $\mathbb{C}$.

Complex Homomorphisms

10.5 Definition Suppose A is a complex algebra and ϕ is a linear functional on A which is not identically 0. If

$$\phi(xy) = \phi(x)\phi(y) \tag{1}$$

for all $x \in A$ and $y \in A$, then ϕ is called a *complex homomorphism* on A.

(The exclusion of $\phi \equiv 0$ is, of course, just a matter of convenience.)

An element $x \in A$ is said to be *invertible* if it has an *inverse* in A, that is, if there exists an element $x^{-1} \in A$ such that

$$x^{-1}x = xx^{-1} = e, \tag{2}$$

where e is the unit element of A.

Note that no $x \in A$ has more than one inverse, for if $yx = e = xz$ then

$$y = ye = y(xz) = (yx)z = ez = z.$$

10.6 Proposition *If ϕ is a complex homomorphism on a complex algebra A with unit e, then $\phi(e) = 1$, and $\phi(x) \neq 0$ for every invertible $x \in A$.*

PROOF. For some $y \in A$, $\phi(y) \neq 0$. Since

$$\phi(y) = \phi(ye) = \phi(y)\phi(e),$$

it follows that $\phi(e) = 1$. If x is invertible, then

$$\phi(x)\phi(x^{-1}) = \phi(xx^{-1}) = \phi(e) = 1,$$

so that $\phi(x) \neq 0$. ////

Parts (*a*) and (*c*) of the following theorem are perhaps the most widely used facts in the theory of Banach algebras; in particular, (*c*) implies that all complex homomorphisms of Banach algebras are continuous.

10.7 Theorem *Suppose A is a Banach algebra, $x \in A$, $\|x\| < 1$. Then* (*a*) $e - x$ *is invertible,*

(*b*) $\|(e-x)^{-1} - e - x\| \le \dfrac{\|x\|^2}{1 - \|x\|}$,

(*c*) $|\phi(x)| < 1$ *for every complex homomorphism* ϕ *on* A.

PROOF. Since $\|x^n\| \le \|x\|^n$ and $\|x\| < 1$, the elements

(1) $$s_n = e + x + x^2 + \cdots + x^n$$

form a Cauchy sequence in A. Since A is complete, there exists $s \in A$ such that $s_n \to s$. Since $x^n \to 0$ and

(2) $$s_n \cdot (e - x) = e - x^{n+1} = (e - x) \cdot s_n,$$

the continuity of multiplication implies that s is the inverse of $e - x$. Next, (1) shows that

$$\|s - e - x\| = \|x^2 + x^3 + \cdots\| \le \sum_{n=2}^{\infty} \|x\|^n = \frac{\|x\|^2}{1 - \|x\|}.$$

Finally, suppose $\lambda \in \mathcal{C}$, $|\lambda| \ge 1$. By (*a*), $e - \lambda^{-1}x$ is invertible. By Proposition 10.6,

$$1 - \lambda^{-1}\phi(x) = \phi(e - \lambda^{-1}x) \ne 0.$$

Hence $\phi(x) \ne \lambda$. This completes the proof. ////

We now interrupt the main line of development and insert a theorem which shows, for Banach algebras, that Proposition 10.6 actually characterizes the complex homomorphisms among the linear functionals. This striking result has apparently found no interesting applications as yet.

10.8 Lemma *Suppose* f *is an entire function of one complex variable,* $f(0) = 1, f'(0) = 0$, *and*

(1) $$0 < |f(\lambda)| \le e^{|\lambda|} \qquad (\lambda \in \mathcal{C}).$$

Then $f(\lambda) = 1$ *for all* $\lambda \in \mathcal{C}$.

PROOF. Since f has no zero, there is an entire function g such that $f = \exp\{g\}$, $g(0) = g'(0) = 0$, and $\operatorname{Re}[g(\lambda)] \le |\lambda|$. This inequality implies

(2) $$|g(\lambda)| \le |2r - g(\lambda)| \qquad (|\lambda| \le r).$$

The function

(3) $$h_r(\lambda) = \frac{r^2 g(\lambda)}{\lambda^2[2r - g(\lambda)]}$$

is holomorphic in $\{\lambda: |\lambda| < 2r\}$, and $|h_r(\lambda)| \le 1$ if $|\lambda| = r$. By the maximum modulus theorem,

$$|h_r(\lambda)| \le 1 \qquad (|\lambda| \le r). \tag{4}$$

Fix λ and let $r \to \infty$. Then (3) and (4) imply that $g(\lambda) = 0$. ////

10.9 Theorem (Gleason, Kahane, Zelazko) *If ϕ is a linear functional on a Banach algebra A, such that $\phi(e) = 1$ and $\phi(x) \ne 0$ for every invertible $x \in A$, then*

$$\phi(xy) = \phi(x)\phi(y) \qquad (x \in A,\ y \in A). \tag{1}$$

Note that the continuity of ϕ is not part of the hypothesis.

PROOF. Let N be the null space of ϕ. If $x \in A$ and $y \in A$, the assumption $\phi(e) = 1$ shows that

$$x = a + \phi(x)e, \qquad y + b + \phi(y)e, \tag{2}$$

where $a \in N$, $b \in N$. If ϕ is applied to the product of the equations (2), one obtains

$$\phi(xy) = \phi(ab) + \phi(x)\phi(y). \tag{3}$$

The desired conclusion (1) is therefore equivalent to the assertion that

$$ab \in N \qquad \text{if } a \in N \text{ and } b \in N. \tag{4}$$

Suppose we had proved a special case of (4), namely,

$$a^2 \in N \qquad \text{if } a \in N. \tag{5}$$

Then (3), with $x = y$, implies

$$\phi(x^2) = [\phi(x)]^2 \qquad (x \in A). \tag{6}$$

Replacement of x by $x + y$ in (6) results in

$$\phi(xy + yx) = 2\phi(x)\phi(y) \qquad (x \in A,\ y \in A). \tag{7}$$

Hence

$$xy + yx \in N \qquad \text{if } x \in N,\ y \in A. \tag{8}$$

Consider the identity

$$(xy - yx)^2 + (xy + yx)^2 = 2[x(yxy) + (yxy)x]. \tag{9}$$

If $x \in N$, the right side of (9) is in N, by (8), and so is $(xy + yx)^2$, by (8) and (6). Hence $(xy - yx)^2$ is in N, and another application of (6) yields

$$xy - yz \in N \qquad \text{if } x \in N,\ y \in A. \tag{10}$$

Addition of (8) and (10) gives (4), hence (1).

Thus (5) implies (1), for purely algebraic reasons. The proof of (5) uses analytic methods.

By hypothesis, N contains no invertible element of A. Thus $\|e - x\| \geq 1$ for every $x \in N$, by (*a*) of Theorem 10.7. Hence

$$\|\lambda e - x\| \geq |\lambda| = |\phi(\lambda e - x)| \qquad (x \in N, \lambda \in \mathcal{C}). \tag{11}$$

We conclude that ϕ is a *continuous* linear functional on A, of norm 1.

To prove (5), fix $a \in N$, assume $\|a\| = 1$ without loss of generality, and define

$$f(\lambda) = \sum_{n=0}^{\infty} \frac{\phi(a^n)}{n!} \lambda^n \qquad (\lambda \in \mathcal{C}). \tag{12}$$

Since $|\phi(a^n)| \leq \|a^n\| \leq \|a\|^n = 1$, f is entire and satisfies $|f(\lambda)| \leq \exp|\lambda|$ for all $\lambda \in \mathcal{C}$. Also, $f(0) = \phi(e) = 1$, and $f'(0) = \phi(a) = 0$.

If we can prove that $f(\lambda) \neq 0$ for every $\lambda \in \mathcal{C}$, Lemma 10.8 will imply that $f''(0) = 0$; hence $\phi(a^2) = 0$, which proves (5).

The series

$$E(\lambda) = \sum_{n=0}^{\infty} \frac{\lambda^n}{n!} a^n \tag{13}$$

converges in the norm of A, for every $\lambda \in \mathcal{C}$. The continuity of ϕ shows that

$$f(\lambda) = \phi(E(\lambda)) \quad (\lambda \in \mathcal{C}). \tag{14}$$

The functional equation $E(\lambda + \mu) = E(\lambda)E(\mu)$ follows from (13) exactly as in the scalar case. In particular,

$$E(\lambda)E(-\lambda) = E(0) = e \qquad (\lambda \in \mathcal{C}). \tag{15}$$

Hence $E(\lambda)$ is an invertible element of A, for every $\lambda \in \mathcal{C}$. This implies, by hypothesis, that $\phi(E(\lambda)) \neq 0$, and therefore $f(\lambda) \neq 0$, by (14). This completes the proof. ////

Basic Properties of Spectra

10.10 Definitions Let A be a Banach algebra; let $G = G(A)$ be the set of all invertible elements of A. If $x \in G$ and $y \in G$, then $y^{-1}x$ is the inverse of $x^{-1}y$; thus $x^{-1}y \in G$, and G is a *group*.

If $x \in A$, the *spectrum* $\sigma(x)$ of x is the set of all complex numbers λ such that $\lambda e - x$ is *not* invertible. The complement of $\sigma(x)$ is the *resolvent set* of x; it consists of all $\lambda \in \mathcal{C}$ for which $(\lambda e - x)^{-1}$ exists.

The *spectral radius* of x is the number

$$\rho(x) = \sup\{|\lambda| : \lambda \in \sigma(x)\}. \tag{1}$$

It is the radius of the smallest closed circular disc in $\mathcal{C}$, with center at 0, which contains $\sigma(x)$. Of course, (1) makes no sense if $\sigma(x)$ is empty. But this never happens, as we shall see.

10.11 Theorem *Suppose A is a Banach algebra, $x \in G(A)$, $h \in A$, and $\|h\| < \frac{1}{2}\|x^{-1}\|^{-1}$. Then $x + h \in G(A)$, and*

$$\|(x+h)^{-1} - x^{-1} + x^{-1}hx^{-1}\| \le 2\|x^{-1}\|^3\|h\|^2. \tag{1}$$

PROOF. Since $x + h = x(e + x^{-1}h)$ and $\|x^{-1}h\| < \frac{1}{2}$, Theorem 10.7 implies that $x + h \in G(A)$ and that the norm of the right member of the identity

$$(x+h)^{-1} - x^{-1} + x^{-1}hx^{-1} = [(e + x^{-1}h)^{-1} - e + x^{-1}h]x^{-1}$$

is at most $2\|x^{-1}h\|^2\|x^{-1}\|$. ////

10.12 Theorem *If A is a Banach algebra, then $G(A)$ is an open subset of A, and the mapping $x \to x^{-1}$ is a homeomorphism of $G(A)$ onto $G(A)$.*

PROOF. That $G(A)$ is open and that $x \to x^{-1}$ is continuous follows from Theorem 10.11. Since $x \to x^{-1}$ maps $G(A)$ onto $G(A)$ and since it is its own inverse, it is a homeomorphism. ////

10.13 Theorem *If A is a Banach algebra and $x \in A$, then*

(a) *the spectrum $\sigma(x)$ of x is compact and nonempty, and*
(b) *the spectral radius $\rho(x)$ of x satisfies*

$$\rho(x) = \lim_{n\to\infty} \|x^n\|^{1/n} = \inf_{n\ge 1} \|x^n\|^{1/n}. \tag{1}$$

Note that the existence of the limit in (1) is part of the conclusion and that the inequality

$$\rho(x) \le \|x\| \tag{2}$$

is contained in the *spectral radius formula* (1).

PROOF. If $|\lambda| > \|x\|$ then $e - \lambda^{-1}x$ lies in $G(A)$, by Theorem 10.7, and so does $\lambda e - x$. Thus $\lambda \notin \sigma(x)$. This proves (2). In particular, $\sigma(x)$ is a bounded set.

To prove that $\sigma(x)$ is closed, define $g: \mathcal{C} \to A$ by $g(\lambda) = \lambda e - x$. Then g is continuous, and the complement Ω of $\sigma(x)$ is $g^{-1}(G(A))$, which is open, by Theorem 10.12. Thus $\sigma(x)$ is compact.

Now define $f: \Omega \to G(A)$ by

$$f(\lambda) = (\lambda e - x)^{-1} \qquad (\lambda \in \Omega). \tag{3}$$

Replace x by $\lambda e - x$ and h by $(\mu - \lambda)e$ in Theorem 10.11. If $\lambda \in \Omega$ and μ is sufficiently close to λ, the result of this substitution is

$$\|f(\mu) - f(\lambda) + (\mu - \lambda)f^2(\lambda)\| \le 2\|f(\lambda)\|^3 |\mu - \lambda|^2, \tag{4}$$

so that

$$\lim_{\mu \to \lambda} \frac{f(\mu) - f(\lambda)}{\mu - \lambda} = -f^2(\lambda) \qquad (\lambda \in \Omega). \tag{5}$$

Thus f is a strongly holomorphic A-valued function in Ω.

If $|\lambda| > \|x\|$, the argument used in Theorem 10.7 shows that

$$f(\lambda) = \sum_{n=0}^{\infty} \lambda^{-n-1}x^n = \lambda^{-1}e + \lambda^{-2}x + \cdots. \tag{6}$$

This series converges uniformly on every circle Γ_r with center at 0 and radius $r > \|x\|$. By Theorem 3.29, term-by-term integration is therefore legitimate. Hence

$$x^n = \frac{1}{2\pi i}\int_{\Gamma_r} \lambda^n f(\lambda)\, d\lambda \qquad (r > \|x\|,\ n = 0, 1, 2, \ldots). \tag{7}$$

If $\sigma(x)$ were empty, Ω would be $\mathcal{C}$, and the Cauchy theorem 3.31 would imply that all integrals in (7) are 0. But when $n = 0$, the left-hand side of (7) is $e \ne 0$. This contradiction shows that $\sigma(x)$ is not empty.

Since Ω contains all λ with $|\lambda| > \rho(x)$, an application of (3) of the Cauchy theorem 3.31 shows that the condition $r > \|x\|$ can be replaced in (7) by $r > \rho(x)$. If

$$M(r) = \max_{\theta} \|f(re^{i\theta})\| \qquad (r > \rho(x)), \tag{8}$$

the continuity of f implies that $M(r) < \infty$. Since (7) now gives

$$\|x^n\| \le r^{n+1}M(r), \tag{9}$$

we obtain

$$\limsup_{n\to\infty} \|x^n\|^{1/n} \le r \qquad (r > \rho(x)) \tag{10}$$

so that

$$\limsup_{n\to\infty} \|x^n\|^{1/n} \le \rho(x). \tag{11}$$

On the other hand, if $\lambda \in \sigma(x)$, the factorization

$$\lambda^n e - x^n = (\lambda e - x)(\lambda^{n-1}e + \cdots + x^{n-1}) \tag{12}$$

shows that $\lambda^n e - x^n$ is not invertible. Thus $\lambda^n \in \sigma(x^n)$. By (2), $|\lambda^n| \le \|x^n\|$ for $n = 1, 2, 3, \ldots$. Hence

$$\rho(x) \le \inf_{n \ge 1} \|x^n\|^{1/n}, \tag{13}$$

and (1) is an immediate consequence of (11) and (13). ////

The nonemptiness of $\sigma(x)$ leads to an easy characterization of those Banach algebras that are division algebras.

10.14 Theorem (Gelfand-Mazur) *If A is a Banach algebra in which every nonzero element is invertible, then A is (isometrically isomorphic to) the complex field.*

PROOF. If $x \in A$ and $\lambda_1 \ne \lambda_2$, then at most one of the elements $\lambda_1 e - x$ and $\lambda_2 e - x$ is 0; hence at least one of them is invertible. Since $\sigma(x)$ is not empty, it follows that $\sigma(x)$ consists of exactly one point, say $\lambda(x)$, for each $x \in A$. Since $\lambda(x)e - x$ is not invertible, it is 0. Hence $x = \lambda(x)e$. The mapping $x \to \lambda(x)$ is therefore an isomorphism of A onto $\mathbb{C}$, which is also an isometry, since $|\lambda(x)| = \|\lambda(x)e\| = \|x\|$ for every $x \in A$. ////

Theorems 10.13 and 10.14 are among the key results of this chapter. Much of the content of Chapters 11 to 13 is independent of the remainder of Chapter 10.

10.15 Remarks (*a*) Whether an element of A is or is not invertible in A is a purely algebraic property. The spectrum and the spectral radius of an $x \in A$ are thus defined in terms of the algebraic structure of A, regardless of any metric (or topological) considerations. On the other hand, $\lim \|x^n\|^{1/n}$ depends obviously on metric properties of A. This is one of the remarkable features of the spectral radius formula: It asserts the equality of certain quantities which arise in entirely different ways.

(*b*) Our algebra A may be a subalgebra of a larger Banach algebra B, and it may then very well happen that some $x \in A$ is not invertible in A but is invertible in B. The spectrum of x depends therefore on the algebra. The inclusion $\sigma_A(x) \supset \sigma_B(x)$ holds (the notation is self-explanatory); the two spectra can be different. The spectral radius is, however, unaffected by the passage from A to B, since the spectral radius formula expresses it in terms of metric properties of powers of x, and these are independent of anything that happens outside A.

Theorem 10.18 will describe the relation between $\sigma_A(x)$ and $\sigma_B(x)$ in greater detail.

10.16 Lemma *Suppose V and W are open sets in some topological space X, $V \subset W$, and W contains no boundary point of V. Then V is a union of components of W.*

Recall that a *component* of W is, by definition, a maximal connected subset of W.

PROOF. Let Ω be a component of W that intersects V. Let U be the complement of $\bar{V}$. Since W contains no boundary point of V, Ω is the union of the two disjoint open sets $\Omega \cap V$ and $\Omega \cap U$. Since Ω is connected, $\Omega \cap U$ is empty. Thus $\Omega \subset V$. ////

10.17 Lemma *Suppose A is a Banach algebra, $x_n \in G(A)$ for $n = 1, 2, 3, \ldots$, x is a boundary point of $G(A)$, and $x_n \to x$ as $n \to \infty$. Then $\|x_n^{-1}\| \to \infty$ as $n \to \infty$.*

PROOF. If the conclusion is false, there exists $M < \infty$ such that $\|x_n^{-1}\| < M$ for infinitely many n. For one of these, $\|x_n - x\| < 1/M$. For this n,

$$\|e - x_n^{-1}x\| = \|x_n^{-1}(x_n - x)\| < 1,$$

so that $x_n^{-1}x \in G(A)$. Since $x = x_n(x_n^{-1}x)$ and $G(A)$ is a group, it follows that $x \in G(A)$. This contradicts the hypothesis, since $G(A)$ is open. ////

10.18 Theorem

(*a*) *If A is a closed subalgebra of a Banach algebra B, and if A contains the unit element of B, then $G(A)$ is a union of components of $A \cap G(B)$.*

(*b*) *Under these conditions, if $x \in A$, then $\sigma_A(x)$ is the union of $\sigma_B(x)$ and a (possibly empty) collection of bounded components of the complement of $\sigma_B(x)$. In particular, the boundary of $\sigma_A(x)$ lies in $\sigma_B(x)$.*

PROOF. (*a*) Every member of A that has an inverse in A has the same inverse in B. Thus $G(A) \subset G(B)$. Both $G(A)$ and $A \cap G(B)$ are open subsets of A. By Lemma 10.16, it is sufficient to prove that $G(B)$ contains no boundary point y of $G(A)$.

Any such y is the limit of a sequence $\{x_n\}$ in $G(A)$. By Lemma 10.17, $\|x_n^{-1}\| \to \infty$. If y were in $G(B)$, the continuity of inversion in $G(B)$ (Theorem 10.12) would force x_n^{-1} to converge to y^{-1}. In particular $\{\|x_n^{-1}\|\}$ would be bounded. Hence $y \notin G(B)$, and (*a*) is proved.

(*b*) Let Ω_A and Ω_B be the complements of $\sigma_A(x)$ and of $\sigma_B(x)$, relative to $\mathbb{C}$. The inclusion $\Omega_A \subset \Omega_B$ is obvious, since $\lambda \in \Omega_A$ if and only if $\lambda e - x \in G(A)$. Let λ_0 be a boundary point of Ω_A. Then

$\lambda_0 e - x$ is a boundary point of $G(A)$. By (*a*), $\lambda_0 e - x \notin G(B)$. Hence $\lambda_0 \notin \Omega_B$. Lemma 10.16 implies now that Ω_A is the union of certain components of Ω_B. The other components of Ω_B are therefore subsets of $\sigma_A(x)$. This proves (*b*). ////

Corollaries. *Suppose* $x \in A \subset B$.

(*a*) *If* $\sigma_B(x)$ *does not separate* $\mathbb{C}$, *i.e., if its complement* Ω_B *is connected, then* $\sigma_A(x) = \sigma_B(x)$.

(*b*) *If* $\sigma_A(x)$ *is larger than* $\sigma_B(x)$, *then* $\sigma_A(x)$ *is obtained from* $\sigma_B(x)$ *by "filling in some holes" in* $\sigma_B(x)$.

(*c*) *If* $\sigma_A(x)$ *has empty interior, then* $\sigma_A(x) = \sigma_B(x)$.

The most important application of this corollary occurs when $\sigma_B(x)$ contains only real numbers.

As another application of Lemma 10.17 we now prove a theorem whose conclusion is the same as that of the Gelfand-Mazur theorem, although its consequences are not nearly so important.

10.19 Theorem *If A is a Banach algebra and if there exists* $M < \infty$ *such that*

(1) $$\|x\| \, \|y\| \le M\|xy\| \qquad (x \in A, y \in A),$$

then A is (isometrically isomorphic to) $\mathbb{C}$.

PROOF. Let y be a boundary point of $G(A)$. Then $y = \lim y_n$ for some sequence $\{y_n\}$ in $G(A)$. By Lemma 10.17, $\|y_n^{-1}\| \to \infty$. By hypothesis,

(2) $$\|y_n\| \, \|y_n^{-1}\| \le M\|e\| \qquad (n = 1, 2, 3, \ldots).$$

Hence $\|y_n\| \to 0$ and therefore $y = 0$.

If $x \in A$, each boundary point λ of $\sigma(x)$ gives rise to a boundary point $\lambda e - x$ of $G(A)$. Thus $x = \lambda e$. In other words, $A = \{\lambda e: \lambda \in \mathbb{C}\}$. ////

It is natural to ask whether the spectra of two elements x and y of A are close together, in some suitably defined sense, if x and y are close to each other. The next theorem gives a very simple answer.

10.20 Theorem *Suppose A is a Banach algebra,* $x \in A$, Ω *is an open set in* $\mathbb{C}$, *and* $\sigma(x) \subset \Omega$. *Then there exists* $\delta > 0$ *such that* $\sigma(x + y) \subset \Omega$ *for every* $y \in A$ *with* $\|y\| < \delta$.

PROOF. Since $\|(\lambda e - x)^{-1}\|$ is a continuous function of λ in the complement of $\sigma(x)$, and since this norm tends to 0 as $\lambda \to \infty$, there is a number $M < \infty$ such that

$$\|(\lambda e - x)^{-1}\| < M$$

for all λ outside Ω. If $y \in A$, $\|y\| < 1/M$, and $\lambda \notin \Omega$, it follows that

$$\lambda e - (x + y) = (\lambda e - x)[e - (\lambda e - x)^{-1} y]$$

is invertible in A, since $\|(\lambda e - x)^{-1} y\| < 1$; hence $\lambda \notin \sigma(x + y)$. This gives the desired conclusion, with $\delta = 1/M$. ////

Symbolic Calculus

10.21 Introduction If x is an element of a Banach algebra A and if $f(\lambda) = \alpha_0 + \cdots + \alpha_n \lambda^n$ is a polynomial with complex coefficients α_i, there can be no doubt about the meaning of the symbol $f(x)$; it obviously denotes the element of A defined by

$$f(x) = \alpha_0 e + \alpha_1 x + \cdots + \alpha_n x^n.$$

The question arises whether $f(x)$ can be defined in a meaningful way for other functions f. We have already encountered some examples of this. For instance, during the proof of Theorem 10.9 we came very close to defining the exponential function in A. In fact, if $f(\lambda) = \sum \alpha_k \lambda^k$ is any entire function in $\mathbb{C}$, it is natural to define $f(x) \in A$ by $f(x) = \sum \alpha_k x^k$; this series always converges. Another example is given by the meromorphic functions

$$f(\lambda) = \frac{1}{\alpha - \lambda}.$$

In this case, the natural definition of $f(x)$ is

$$f(x) = (\alpha e - x)^{-1}$$

which makes sense for all x whose spectrum does not contain α.

One is thus led to the conjecture that $f(x)$ should be definable, within A, whenever f is holomorphic in an open set that contains $\sigma(x)$. This turns out to be correct and can be accomplished by a version of the Cauchy formula that converts complex functions defined in open subsets of $\mathbb{C}$ to A-valued ones defined in certain open subsets of A. (Just as in classical analysis, the Cauchy formula is a much more adaptable tool than the power series representation.) Moreover, the entities $f(x)$ so defined (see Definition 10.26) turn out to have interesting properties. The most important of these are summarized in Theorems 10.27 to 10.29.

In certain algebras one can go further. For instance, if x is a bounded normal operator on a Hilbert space H, the symbol $f(x)$ can be interpreted as a bounded normal operator on H when f is any continuous complex function on $\sigma(x)$, and even when f is any complex bounded Borel function on $\sigma(x)$. In Chapter 12 we shall see how this leads to an efficient proof of a very general form of the spectral theorem.

10.22 Integration of A-valued functions If A is a Banach algebra and f is a continuous A-valued function on some compact Hausdorff space Q on which a complex Borel measure μ is defined, then $\int f\,d\mu$ exists and has all the properties that were discussed in Chapter 3, simply because A is a Banach space. However, an additional property can be added to these and will be used in the sequel, namely: *If $x \in A$, then*

$$x \int_Q f\,d\mu = \int_Q xf(p)\,d\mu(p) \tag{1}$$

and

$$\left(\int_Q f\,d\mu\right)x = \int_Q f(p)x\,d\mu(p) \tag{2}$$

To prove (1), let M_x be left multiplication by x, as in the proof of Theorem 10.2, and let Λ be a bounded linear functional on A. Then ΛM_x is a bounded linear functional. Definition 3.26 implies therefore that

$$\Lambda M_x \int_Q f\,d\mu = \int_Q (\Lambda M_x f)\,d\mu = \Lambda \int_Q (M_x f)\,d\mu,$$

for every Λ, so that

$$M_x \int_Q f\,d\mu = \int_Q (M_x f)\,d\mu,$$

which is just another way of writing (1). To prove (2), interpret M_x to be right multiplication by x.

10.23 Contours Suppose K is a compact subset of an open $\Omega \subset \mathcal{C}$, and Γ is a collection of finitely many oriented line intervals $\gamma_1, \ldots, \gamma_n$ in Ω, none of which intersects K. In this situation, integration over Γ is defined by

$$\int_\Gamma \phi(\lambda)\,d\lambda = \sum_{j=1}^{n} \int_{\gamma_j} \phi(\lambda)\,d\lambda. \tag{1}$$

It is well known that Γ can be so chosen that

$$\text{Ind}_\Gamma(\zeta) = \frac{1}{2\pi i}\int_\Gamma \frac{d\lambda}{\lambda - \zeta} = \begin{cases} 1 & \text{if } \zeta \in K \\ 0 & \text{if } \zeta \notin \Omega \end{cases} \tag{2}$$

and that the Cauchy formula

$$f(\zeta) = \frac{1}{2\pi i} \int_\Gamma (\lambda - \zeta)^{-1} f(\lambda)\, d\lambda \tag{3}$$

then holds for every holomorphic function f in Ω and for every $\zeta \in K$. See, for instance, Theorem 13.5 of [23].

We shall describe the situation (2) briefly by saying that *the contour* Γ *surrounds* K *in* Ω.

Note that neither K nor Ω nor the union of the intervals γ_i has been assumed to be connected.

10.24 Lemma *Suppose A is a Banach algebra, $x \in A$, $\alpha \in \mathbb{C}$, $\alpha \notin \sigma(x)$, Ω is the complement of α in $\mathbb{C}$, and Γ surrounds $\sigma(x)$ in Ω. Then*

$$\frac{1}{2\pi i} \int_\Gamma (\alpha - \lambda)^n (\lambda e - x)^{-1}\, d\lambda = (\alpha e - x)^n \qquad (n = 0, \pm 1, \pm 2, \ldots). \tag{1}$$

PROOF. Denote the integral by y_n. When $\lambda \notin \sigma(x)$, then

$$(\lambda e - x)^{-1} = (\alpha e - x)^{-1} + (\alpha - \lambda)(\alpha e - x)^{-1}(\lambda e - x)^{-1}$$

By Section 10.22, y_n is therefore the sum of

$$(\alpha e - x)^{-1} \cdot \frac{1}{2\pi i} \int_\Gamma (\alpha - \lambda)^n\, d\lambda = 0, \tag{2}$$

since $\mathrm{Ind}_\Gamma(\alpha) = 0$, and

$$(\alpha e - x)^{-1} \cdot \frac{1}{2\pi i} \int_\Gamma (\alpha - \lambda)^{n+1} (\lambda e - x)^{-1}\, d\lambda. \tag{3}$$

Hence

$$(\alpha e - x) y_n = y_{n+1} \qquad (n = 0, \pm 1, \pm 2, \ldots). \tag{4}$$

This recursion formula shows that (1) follows from the case $n = 0$. We thus have to prove that

$$\frac{1}{2\pi i} \int_\Gamma (\lambda e - x)^{-1}\, d\lambda = e. \tag{5}$$

Let Γ_r be a positively oriented circle, centered at 0, with radius $r > \|x\|$. On Γ_r, $(\lambda e - x)^{-1} = \sum \lambda^{-n-1} x^n$. Termwise integration of this series gives (5), with Γ_r in place of Γ. Since the integrand in (5) is a holomorphic A-valued function in the complement of $\sigma(x)$ (see the proof of Theorem 10.13), and since

$$\mathrm{Ind}_{\Gamma_r}(\zeta) = 1 = \mathrm{Ind}_\Gamma(\zeta) \tag{6}$$

for every $\zeta \in \sigma(x)$, the Cauchy theorem 3.31 shows that the integral (5) is unaffected if Γ is replaced by Γ_r. This completes the proof. ////

10.25 Theorem *Suppose*

$$R(\lambda) = P(\lambda) + \sum_{m,k} c_{m,k}(\lambda - \alpha_m)^{-k} \tag{1}$$

is a rational function with poles at the points α_m. *[P is a polynomial, and the sum in* (1) *has only finitely many terms.] If* $x \in A$ *and if* $\sigma(x)$ *contains no pole of R, define*

$$R(x) = P(x) + \sum_{m,k} c_{m,k}(x - \alpha_m e)^{-k}. \tag{2}$$

If Ω *is an open set in* $\mathbb{C}$ *that contains* $\sigma(x)$ *and in which R is holomorphic, and if* Γ *surrounds* $\sigma(x)$ *in* Ω, *then*

$$R(x) = \frac{1}{2\pi i}\int_\Gamma R(\lambda)(\lambda e - x)^{-1}\,d\lambda. \tag{3}$$

PROOF. Apply Lemma 10.24. ////

Note that (2) is certainly the most natural definition of a rational function of $x \in A$. The conclusion (3) shows that the Cauchy formula achieves the same result. This motivates the following definition.

10.26 Definition Suppose A is a Banach algebra, Ω is an open set in $\mathbb{C}$, and $H(\Omega)$ is the algebra of all complex holomorphic functions in Ω. By Theorem 10.20,

$$A_\Omega = \{x \in A : \sigma(x) \subset \Omega\} \tag{1}$$

is an open subset of A.

We define $\tilde{H}(A_\Omega)$ to be the set of all A-valued functions $\tilde{f}$, with domain A_Ω, that arise from an $f \in H(\Omega)$ by the formula

$$\tilde{f}(x) = \frac{1}{2\pi i}\int_\Gamma f(\lambda)(\lambda e - x)^{-1}\,d\lambda, \tag{2}$$

where Γ is any contour that surrounds $\sigma(x)$ in Ω.

This definition calls for some comments.

(*a*) Since Γ stays away from $\sigma(x)$ and since inversion is continuous in A, the integrand is continuous in (2), so that the integral exists and defines $\tilde{f}(x)$ as an element of A.

(*b*) The integrand is actually a holomorphic A-valued function in the complement of $\sigma(x)$. (This was observed in the proof of Theorem 10.13. See

Exercise 3.) The Cauchy theorem 3.31 implies therefore that $\tilde{f}(x)$ *is independent of the choice of* Γ, provided only that Γ surrounds $\sigma(x)$ in Ω.

(*c*) If $x = \alpha e$ and $\alpha \in \Omega$, (2) becomes

$$\tilde{f}(\alpha e) = f(\alpha)e. \tag{3}$$

Note that $\alpha e \in A_\Omega$ if and only if $\alpha \in \Omega$. If we identity $\lambda \in \mathcal{C}$ with $\lambda e \in A$, every $f \in H(\Omega)$ may be regarded as mapping a certain subset of A_Ω (namely, the intersection of A_Ω with the one-dimensional subspace of A generated by e) into A, and then (3) shows that $\tilde{f}$ may be regarded as an *extension* of f.

In most treatments of this topic, $f(x)$ is written in place of our $\tilde{f}(x)$. The notation $\tilde{f}$ is used here because it avoids certain ambiguities that might cause misunderstandings.

(*d*) If S is any set and A is any algebra, the collection of all A-valued functions on S is an algebra, if scalar multiplication, addition, and multiplication are defined pointwise. For instance, if u and v map S into A, then

$$(uv)(s) = u(s)v(s) \qquad (s \in S).$$

This will be applied to A-valued functions defined in A_Ω.

10.27 Theorem *Suppose A, $H(\Omega)$, and $\tilde{H}(A_\Omega)$ are as in Definition* 10.26. *Then $\tilde{H}(A_\Omega)$ is a complex algebra. The mapping $f \to \tilde{f}$ is an algebra isomorphism of $H(\Omega)$ onto $\tilde{H}(A_\Omega)$ which is continuous in the following sense:*

If $f_n \in H(\Omega)$ $(n = 1, 2, 3, \ldots)$ and $f_n \to f$ uniformly on compact subsets of Ω, then

$$\tilde{f}(x) = \lim_{n\to\infty} \tilde{f}_n(x) \qquad (x \in A_\Omega). \tag{1}$$

If $u(\lambda) = \lambda$ and $v(\lambda) = 1$ in Ω, then $\tilde{u}(x) = x$ and $\tilde{v}(x) = e$ for every $x \in A_\Omega$.

PROOF. The last sentence follows from Theorem 10.25. The integral representation (2) in Section 10.26 makes it obvious that $f \to \tilde{f}$ is linear. If $\tilde{f} = 0$, then

$$f(\alpha)e = \tilde{f}(\alpha e) = 0 \qquad (\alpha \in \Omega), \tag{2}$$

so that $f = 0$. Thus $f \to \tilde{f}$ is one-to-one.

The asserted continuity follows directly from the integral (2) in Section 10.26, since $\|(\lambda e - x)^{-1}\|$ is bounded on Γ. (Use the same Γ for all f_n, and apply Theorem 3.29.)

It remains to be proved that $f \to \tilde{f}$ is multiplicative. Explicitly, if $f \in H(\Omega)$, $g \in H(\Omega)$, and $h(\lambda) = f(\lambda)g(\lambda)$ for all $\lambda \in \Omega$, it has to be shown that

$$\tilde{h}(x) = \tilde{f}(x)\tilde{g}(x) \qquad (x \in A_\Omega). \tag{3}$$

If f and g are rational functions without poles in Ω, and if $h = fg$, then $h(x) = f(x)g(x)$, and since Theorem 10.25 asserts that $R(x) = \tilde{R}(x)$, (3) holds. In the general case, Runge's theorem (Th. 13.9 of [23]) allows us to approximate f and g by rational functions f_n and g_n, uniformly on compact subsets of Ω. Then $f_n g_n$ converges to h in the same manner, and (3) follows from the continuity of the mapping $f \to \tilde{f}$. ////

Since $H(\Omega)$ is obviously a commutative algebra, Theorem 10.27 implies that $\tilde{H}(A_\Omega)$ is also commutative. This may be surprising, because $\tilde{f}(x)$ and $\tilde{f}(y)$ need not commute. However, $\tilde{f}(x)$ and $\tilde{g}(x)$ do commute in A, for every $x \in A_\Omega$. Hence $\tilde{f}\tilde{g} = \tilde{g}\tilde{f}$, by Definition 10.26 (*d*).

10.28 Theorem *Suppose $x \in A_\Omega$ and $f \in H(\Omega)$.*

(*a*) *$\tilde{f}(x)$ is invertible in A if and only if $f(\lambda) \neq 0$ for every $\lambda \in \sigma(x)$.*
(*b*) $\sigma(\tilde{f}(x)) = f(\sigma(x))$.

Part (*b*) is called the *spectral mapping theorem.*

PROOF. (*a*) If f has no zero on $\sigma(x)$, then $g = 1/f$ is holomorphic in an open set Ω_1 such that $\sigma(x) \subset \Omega_1 \subset \Omega$. Since $fg = 1$ in Ω_1, Theorem 10.27 (with Ω_1 in place of Ω) shows that $\tilde{f}(x)\tilde{g}(x) = e$, and thus $\tilde{f}(x)$ is invertible. Conversely, if $f(\alpha) = 0$ for some $\alpha \in \sigma(x)$ then there exists $h \in H(\Omega)$ such that

$$(1) \qquad (\lambda - \alpha)h(\lambda) = f(\lambda) \qquad (\lambda \in \Omega),$$

which implies

$$(2) \qquad (x - \alpha e)\tilde{h}(x) = \tilde{f}(x) = \tilde{h}(x)(x - \alpha e),$$

by Theorem 10.27. Since $x - \alpha e$ is not invertible in A, neither is $\tilde{f}(x)$, by (2).

(*b*) Fix $\beta \in \mathbb{C}$. By definition, $\beta \in \sigma(\tilde{f}(x))$ if and only if $\tilde{f}(x) - \beta e$ is not invertible in A. By (*a*), applied to $f - \beta$ is place of f, this happens if and only if $f - \beta$ has a zero in $\sigma(x)$, that is, if and only if $\beta \in f(\sigma(x))$. ////

The spectral mapping theorem makes it possible to include composition of functions among the operations of the symbolic calculus.

10.29 Theorem *Suppose $x \in A_\Omega$, $f \in H(\Omega)$, Ω_1 is an open set containing $f(\sigma(x))$, $g \in H(\Omega_1)$, and $h(\lambda) = g(f(\lambda))$ in Ω_0, the set of all $\lambda \in \Omega$ with $f(\lambda) \in \Omega_1$.*

Then $\tilde{f}(x) \in A_{\Omega_1}$ *and* $\tilde{h}(x) = \tilde{g}(\tilde{f}(x))$.

Briefly, $\tilde{h} = \tilde{g} \circ \tilde{f}$ if $h = g \circ f$.

PROOF. By (*b*) of Theorem 10.28, $\sigma(\tilde{f}(x)) \subset \Omega_1$, and therefore $\tilde{g}(\tilde{f}(x))$ is defined.

Fix a contour Γ_1 that surrounds $f(\sigma(x))$ in Ω_1. There is an open set W, with $\sigma(x) \subset W \subset \Omega_0$, so small that

$$\text{(1)} \qquad \operatorname{Ind}_{\Gamma_1}(f(\lambda)) = 1 \qquad (\lambda \in W).$$

Fix a contour Γ_0 that surrounds $\sigma(x)$ in W. If $\zeta \in \Gamma_1$, then $1/(\zeta - f) \in H(W)$. Hence Theorem 10.27, with W in place of Ω, shows that

$$\text{(2)} \quad [\zeta e - \tilde{f}(x)]^{-1} = \frac{1}{2\pi i}\int_{\Gamma_0} [\zeta - f(\lambda)]^{-1}(\lambda e - x)^{-1}\, d\lambda \qquad (\zeta \in \Gamma_1).$$

Since Γ_1 surrounds $\sigma(\tilde{f}(x))$ in Ω_1, (1) and (2) imply

$$\begin{aligned}
\tilde{g}(\tilde{f}(x)) &= \frac{1}{2\pi i}\int_{\Gamma_1} g(\zeta)[\zeta e - \tilde{f}(x)]^{-1}\, d\zeta \\
&= \frac{1}{2\pi i}\int_{\Gamma_0} \frac{1}{2\pi i}\int_{\Gamma_1} g(\zeta)[\zeta - f(\lambda)]^{-1}\, d\zeta\, (\lambda e - x)^{-1}\, d\lambda \\
&= \frac{1}{2\pi i}\int_{\Gamma_0} g(f(\lambda))(\lambda e - x)^{-1}\, d\lambda = \frac{1}{2\pi i}\int_{\Gamma_0} h(\lambda)(\lambda e - x)^{-1}\, d\lambda = \tilde{h}(x).
\end{aligned}$$

////

We shall now give some applications of this symbolic calculus. The first one deals with the existence of roots and logarithms. To say that an element $x \in A$ has an *n*th *root* in A means that $x = y^n$ for some $y \in A$. If $x = \exp(y)$ for some $y \in A$, then y is a *logarithm* of x.

Note that $\exp(y) = \sum_0^\infty y^n/n!$ but that the exponential function can also be defined by contour integration, as in Definition 10.26. The continuity assertion of Theorem 10.27 shows that these definitions coincide (as they do for every entire function).

10.30 Theorem *Suppose A is a Banach algebra, $x \in A$, and the spectrum $\sigma(x)$ of x does not separate 0 from ∞. Then*

(*a*) *x has roots of all orders in A,*
(*b*) *x has a logarithm in A, and*
(*c*) *if $\varepsilon > 0$, there is a polynomial P such that $\|x^{-1} - P(x)\| < \varepsilon$.*

Moreover, if $\sigma(x)$ lies in the positive real axis, the roots in (a) can be chosen so as to satisfy the same condition.

PROOF. By hypothesis, 0 lies in the unbounded component of the complement of $\sigma(x)$. Hence there is a function f, holomorphic in a simply connected open set $\Omega \supset \sigma(x)$, which satisfies

$$\exp(f(\lambda)) = \lambda.$$

It follows from Theorem 10.29 that

$$\exp(\tilde{f}(x)) = x,$$

so that $y = \tilde{f}(x)$ is a logarithm of x. If $0 < \lambda < \infty$ for every $\lambda \in \sigma(x)$, f can be chosen so as to be real on $\sigma(x)$, so that $\sigma(y)$ lies in the real axis, by the spectral mapping theorem. If $z = \exp(y/n)$, then $z^n = x$, and another application of the spectral mapping theorem shows that $\sigma(z) \subset (0, \infty)$ if $\sigma(y) \subset (-\infty, \infty)$. This proves (*a*) and (*b*); of course (*a*) could have been proved directly, without passing through (*b*).

To prove (*c*), note that $1/\lambda$ can be approximated by polynomials, uniformly on some open set containing $\sigma(x)$ (Runge's theorem), and use the continuity assertion of Theorem 10.27. ////

These results are not quite trivial even when A is a finite-dimensional algebra. For example, it is a special case of (*b*) that a complex n-by-n matrix M is the exponential of some matrix if and only if 0 is not an eigenvalue of M, that is, if and only if M is invertible. To deduce this from (*b*), let A be the algebra of all complex n-by-n matrices (or the algebra of all bounded linear operators on $\mathcal{C}^n$).

If some $x \in A$ satisfies a polynomial identity, i.e., if $P(x) = 0$ for some polynomial P, then $\tilde{f}(x)$ can always be calculated as a polynomial in x, without using the Cauchy integral as in Definition 10.26. If A is finite-dimensional, then this remark applies to every $x \in A$. Here are the details:

10.31 Theorem *Let $P(\lambda) = (\lambda - \alpha_1)^{m_1} \cdots (\lambda - \alpha_s)^{m_s}$ be a polynomial of degree $n = m_1 + \cdots + m_s$ and let Ω be an open set in $\mathcal{C}$ which contains the zeros $\alpha_1, \ldots, \alpha_s$ of P.*

If A is a Banach algebra, $x \in A$, and $P(x) = 0$, then

(*a*) *$\sigma(x) \subset \{\alpha_1, \ldots, \alpha_s\}$ and*

(*b*) *to every $f \in H(\Omega)$ corresponds a polynomial Q, of degree $< n$, and a function $g \in H(\Omega)$, so that*

$$f(\lambda) - Q(\lambda) = P(\lambda)g(\lambda) \qquad (\lambda \in \Omega) \tag{1}$$

and

$$\tilde{f}(x) = Q(x). \tag{2}$$

PROOF. By the spectral mapping theorem,

$$P(\sigma(x)) = \sigma(P(x)) = \sigma(0) = \{0\}. \tag{3}$$

This proves (*a*).

If all multiplicities m_i are 1, Q can be obtained with the aid of the Lagrange interpolation formula

$$Q(\lambda) = \sum_{i=1}^{n} \frac{f(\alpha_i)P(\lambda)}{P'(\alpha_i)(\lambda - \alpha_i)}. \tag{4}$$

This gives $Q(\alpha_i) = f(\alpha_i)$ $(1 \le i \le n)$; hence $(f - Q)/P$ is holomorphic in Ω.

In the general case, the Laurent series of f/P about the points $\alpha_1, \ldots, \alpha_s$ give constants c_{ik} so that

$$g(\lambda) = \frac{f(\lambda)}{P(\lambda)} - \sum_{i=1}^{s} \sum_{k=1}^{m_i} \frac{c_{ik}}{(\lambda - \alpha_i)^k} \tag{5}$$

is holomorphic in Ω.

This proves (1), and now (2) is a consequence of Theorem 10.27, because (1) implies

$$\tilde{f}(x) = Q(x) + P(x)\tilde{g}(x), \tag{6}$$

and $P(x) = 0$. ////

10.32 Definition Let $\mathscr{B}(X)$ be the Banach algebra of all bounded linear operators on the Banach space X. The *point spectrum* $\sigma_p(T)$ of an operator $T \in \mathscr{B}(X)$ is the set of all eigenvalues of T. Thus $\lambda \in \sigma_p(T)$ if and only if the null space $\mathscr{N}(T - \lambda I)$ of $T - \lambda I$ has positive dimension.

When $A = \mathscr{B}(X)$. the spectral mapping theorem can be refined in the following way.

10.33 Theorem *Suppose $T \in \mathscr{B}(X)$, Ω is open in $\mathcal{C}$, $\sigma(T) \subset \Omega$, and $f \in H(\Omega)$.*

(*a*) *If $x \in X$, $\alpha \in \Omega$, and $Tx = \alpha x$, then $\tilde{f}(T)x = f(\alpha)x$.*

(*b*) $f(\sigma_p(T)) \subset \sigma_p(\tilde{f}(T))$.

(*c*) *If $\alpha \in \sigma_p(\tilde{f}(T))$ and $f - \alpha$ does not vanish identically in any component of Ω, then $\alpha \in f(\sigma_p(T))$.*

(*d*) *If f is not constant in any component of Ω, then $f(\sigma_p(T)) = \sigma_p(\tilde{f}(T))$.*

Part (*a*) states that every eigenvector of T, with eigenvalue α, is also an eigenvector of $\tilde{f}(T)$, with eigenvalue $f(\alpha)$.

PROOF. (*a*) If $x = 0$ there is nothing to be proved. Assume $x \neq 0$ and $Tx = \alpha x$. Then $\alpha \in \sigma(T)$, and there exists $g \in H(\Omega)$ such that

$$f(\lambda) - f(\alpha) = g(\lambda)(\lambda - \alpha). \tag{1}$$

By Theorem 10.27, (1) implies

$$\tilde{f}(T) - f(\alpha)I = \tilde{g}(T)(T - \alpha I). \tag{2}$$

Since $(T - \alpha I)x = 0$, (2) proves (*a*).

Thus $f(\alpha)$ is an eigenvalue of $\tilde{f}(T)$ whenever α is an eigenvalue of T. It follows that (*a*) implies (*b*).

Under the hypotheses of (*c*),

$$\alpha \in \sigma_p(\tilde{f}(T)) \subset \sigma(\tilde{f}(T)) = f(\sigma(T)), \tag{3}$$

so that

$$f^{-1}(\alpha) \cap \sigma(T) \neq \varnothing. \tag{4}$$

Moreover, the set (4) is finite, because $\sigma(T)$ is a compact subset of Ω and $f - \alpha$ does not vanish identically in any component of Ω. Let $\zeta_1, \ldots, \zeta_n$ be the zeros of $f - \alpha$ in $\sigma(T)$, counted according to their multiplicities. Then

$$f(\lambda) - \alpha = g(\lambda)(\lambda - \zeta_1) \cdots (\lambda - \zeta_n), \tag{5}$$

where $g \in H(\Omega)$ and g has no zero on $\sigma(T)$, so that

$$\tilde{f}(T) - \alpha I = \tilde{g}(T)(T - \zeta_1 I) \cdots (T - \zeta_n I). \tag{6}$$

By (*a*) of Theorem 10.28, $\tilde{g}(T)$ is invertible in $\mathscr{B}(X)$. Since α is an eigenvalue of $\tilde{f}(T)$, $\tilde{f}(T) - \alpha I$ is not one-to-one on X. Hence (6) implies that at least one of the operators $T - \zeta_i I$ must fail to be one-to-one. The corresponding ζ_i is in $\sigma_P(T)$, and since $f(\zeta_i) = \alpha$ the proof of (*c*) is complete.

Finally, (*d*) is an immediate consequence of (*b*) and (*c*). ////

The Group of Invertible Elements

We shall now take a closer look at the structure of $G = G(A)$, the multiplicative group of all invertible elements of a Banach algebra A.

G_1 will denote the component of G that contains e, the identity element of G. Sometimes G_1 is called the *principal component* of G. By the definition of *component*, G_1 is the union of all connected subsets of G that contain e.

The group G contains the set

$$\exp(A) = \{\exp(x) : x \in A\},$$

the range of the exponential function in A, simply because $\exp(-x)$ is the inverse of $\exp(x)$. In fact, the power series definition of $\exp(x)$ yields the functional equation

$$\exp(x+y) = \exp(x)\exp(y),$$

provided that $xy = yx$; also, $\exp(0) = e$.

Note also that G is a topological group (see Section 5.12) since multiplication and inversion are continuous in G.

10.34 Theorem

(a) *G_1 is an open normal subgroup of G.*
(b) *G_1 is the group generated by* $\exp(A)$.
(c) *If A is commutative, then $G_1 = \exp(A)$.*
(d) *If A is commutative, the quotient group G/G_1 contains no element of finite order (except for the identity).*

PROOF. (*a*) Theorem 10.11 shows that every $x \in G_1$ is the center of an open ball $U \subset G$. Since U intersects G_1 and U is connected, $U \subset G_1$. Therefore G_1 is open.

If $x \in G_1$ then $x^{-1}G_1$ is a connected subset of G which contains $x^{-1}x = e$. Hence $x^{-1}G_1 \subset G_1$, for every $x \in G_1$. This proves that G_1 is a subgroup of G. Also, $y^{-1}G_1y$ is homeomorphic to G_1, hence connected, for every $y \in G$, and contains e. Thus $y^{-1}G_1y \subset G_1$. By definition, this says that G_1 is a normal subgroup of G.

(*b*) Let Γ be the group generated by $\exp(A)$. For $n = 1, 2, 3, \ldots$, let E_n be the set of all products of n members of $\exp(A)$. Since $y^{-1} \in \exp(A)$ whenever $y \in \exp(A)$, Γ is the union of the sets E_n. Since the product of any two connected sets is connected, induction shows that each E_n is connected. Each E_n contains e, and so $E_n \subset G_1$. Hence Γ is a subgroup of G_1.

Next, $\exp(A)$ has nonempty interior, relative to G (see Theorem 10.30); hence so has Γ. Since Γ is a group and since multiplication by any $x \in G$ is a homeomorphism of G onto G, Γ is open.

Each coset of Γ in G_1 is therefore open, and so is any union of these cosets. Since Γ is the complement of a union of its cosets, Γ is closed, relative to G_1.

Thus Γ is an open and closed subset of G_1. Since G_1 is connected, $\Gamma = G_1$.

(*c*) If A is commutative, the functional equation satisfied by exp shows that $\exp(A)$ is a group. Hence (*b*) implies (*c*).

(*d*) We have to prove the following proposition:

If A is commutative, if $x \in G$, and if $x^n \in G_1$ for some positive integer n, then $x \in G_1$.

Under these conditions, $x^n = \exp(a)$ for some $a \in A$, by (*c*). Put $y = \exp(n^{-1}a)$ and $z = xy^{-1}$. Since $y \in G_1$, it suffices to prove that $z \in G_1$.

The commutativity of A shows that

$$z^n = x^n y^{-n} = \exp(a)\exp(-a) = e.$$

Therefore $\sigma(z)$ does not separate 0 from ∞ (it consists of at most n points, lying on the unit circle), and this implies, by Theorem 10.30, that $z = \exp(w)$ for some $w \in A$. Put

$$f(\lambda) = \exp(\lambda w).$$

Then $f: \mathcal{C} \to G$ is continuous, $f(0) = e \in G_1$, hence $f(\mathcal{C}) \subset G_1$. In particular, $z = f(1) \in G_1$. ////

Theorem 12.38 will show that $\exp(A)$ is not always a group.

Lomonosov's Invariant Subspace Theorem

An *invariant subspace* of an operator $T \in \mathscr{B}(X)$ is, by definition, a closed subspace M of X such that $M \neq \{0\}$, $M \neq X$, and $Tx \in M$ for every $x \in M$; or, briefly, $T(M) \subset M$.

The question arises (and was asked more than half a century ago) whether it is true, for every complex Banach space X, that every $T \in \mathscr{B}(X)$ has an invariant subspace. In recent years some counterexamples have been constructed in some nonreflexive spaces, and even in ℓ^1. (References are given in Appendix B.) Positive results have been found for certain classes of operators on a Hilbert space (in particular, for normal operators; see Chapter 12), but even there the general question is still open.

Lomonosov's proof of the following striking theorem used Schauder's fixed point theorem to produce an eigenvalue (namely, 1). T. M. Hilden observed that this could also be done by an appeal to the spectral radius formula; the resulting proof is a slight simplification of the original one.

10.35 Theorem *Suppose that X is an infinite-dimensional complex Banach space and that $T \in \mathscr{B}(X)$ is compact, $T \neq 0$.*

Then there is a closed subspace M of X such that $M \neq \{0\}$, $M \neq X$, and

$$S(M) \subset M \tag{1}$$

for every $S \in \mathscr{B}(X)$ that commutes with T.

Observe, as a corollary, that every $S \in \mathscr{B}(X)$ which commutes with some nonzero compact operator has an invariant subspace!

PROOF. Let us introduce the notations

$$\Gamma = \{S \in \mathscr{B}(X): ST = TS\} \tag{2}$$

and, for each $y \in X$,

$$\Gamma(y) = \{Sy: S \in \Gamma\}. \tag{3}$$

It is easy to see that Γ is a subalgebra of $\mathscr{B}(X)$ and that $\Gamma(y)$ is therefore a subspace of X which contains y. Thus $\Gamma(y) \neq \{0\}$ if $y \neq 0$. Moreover,

$$S(\Gamma(y)) \subset \Gamma(y) \tag{4}$$

for all $y \in X$ and $S \in \Gamma$, simply because Γ is closed under multiplication. Thus (1) holds for every $\Gamma(y)$, hence also for its closure.

If the conclusion of the theorem is false, *it follows that $\Gamma(y)$ is dense in X for every $y \neq 0$.*

Let us assume this.

Pick $x_0 \in X$ so that $Tx_0 \neq 0$. Then $x_0 \neq 0$, and the continuity of T shows that there is an open ball B, centered at x_0, so small that

$$\|Tx\| \geq \tfrac{1}{2}\|Tx_0\| \qquad \text{and} \qquad \|x\| \geq \tfrac{1}{2}\|x_0\| \tag{5}$$

for every $x \in B$. Our assumption about $\Gamma(y)$ implies that every $y \neq 0$ has a neighborhood W which is mapped into the open set B by some $S \in \Gamma$. Since T is a compact operator, $K = \overline{T(B)}$ is a compact set. By (5), $0 \notin K$. Therefore there are open sets $W_1, \ldots, W_n$, whose union covers K, such that $S_i(W_i) \subset B$ for some $S_i \in \Gamma$, $1 \leq i \leq n$. Put

$$\mu = \max\{\|S_1\|, \ldots, \|S_n\|\}. \tag{6}$$

Starting with x_0, Tx_0 lies in K, hence in some W_{i_1}, and $S_{i_1}Tx_0 \in B$. Therefore $TS_{i_1}Tx_0$ lies in K, hence in some W_{i_2}, and $S_{i_2}TS_{i_1}TX_0$ is back in B. Continuing this ping-pong game, we obtain vectors

$$x_N = S_{i_N}T \cdots S_{i_1}Tx_0 = S_{i_N} \cdots S_{i_1}T^N x_0 \tag{7}$$

in B. Hence

$$\tfrac{1}{2}\|x_0\| \leq \|x_N\| \leq \mu^N \|T^N\| \, \|x_0\| \qquad (N = 1, 2, 3, \ldots), \tag{8}$$

and this gives information about the spectral radius of T, namely,

$$\rho(T) = \lim_{N\to\infty} \|T^N\|^{1/N} \geq \frac{1}{\mu} > 0. \tag{9}$$

We now invoke Theorem 4.25. Since $\rho(T) > 0$, T has an eigenvalue $\lambda \neq 0$. The corresponding eigenspace

$$M_\lambda = \{x \in X: Tx = \lambda x\} \tag{10}$$

is finite-dimensional; hence $M_\lambda \neq X$. If $S \in \Gamma$ and $x \in M_\lambda$, then

$$T(Sx) = S(Tx) = S(\lambda x) = \lambda Sx, \tag{11}$$

so that $Sx \in M_\lambda$. This says that $S(M_\lambda) \subset M_\lambda$.

Thus M_λ satisfies the conclusion of the theorem, even though we assumed that this conclusion failed. ////

Exercises

Throughout this set of exercises, A denotes a Banach algebra and $x, y, \ldots$ denote elements of A, unless the contrary is stated.

1. Use the identity $(xy)^n = x(yx)^{n-1}y$ to prove that xy and yx always have the same spectral radius.
2. (*a*) If x and xy are invertible in A, prove that y is invertible.
   (*b*) If xy and yx are invertible in A, prove that x and y are invertible. (The commutative case of this was tacitly used in the proofs of Theorems 10.13 and 10.28.)
   (*c*) Show that it is possible to have $xy = e$ but $yx \neq e$. For example, consider the right and left shifts S_R and S_L, defined on some Banach space of functions f on the nonnegative integers by
   $$(S_R f)(n) = f(n-1) \qquad \text{if } n \geq 1,$$
   $$(S_R f)(0) = 0,$$
   $$(S_L f)(n) = f(n+1) \qquad \text{for all } n \geq 0.$$
   (*d*) If $xy = e$ and $yx = z \neq e$, show that z is a nontrivial idempotent. (This means that $z^2 = z$, $z \neq 0$, $z \neq e$.)
3. Prove that every finite-dimensional A is isomorphic to an algebra of matrices. *Hint:* The proof of Theorem 10.2 shows that every A is isomorphic to a subalgebra of $\mathscr{B}(A)$. Conclude that $xy = e$ implies $yx = e$ if $\dim A < \infty$.
4. (*a*) Prove that $e - yx$ is invertible in A if $e - xy$ is invertible. *Suggestion:* Put $z = (e - xy)^{-1}$, write z as a geometric series (assume $\|x\| < 1$, $\|y\| < 1$), and use the identity stated in Exercise 1 to obtain a finite formula for $(e - yx)^{-1}$ in terms of x, y, z. Then show that this formula works without any restrictions on $\|x\|$ or $\|y\|$.
   (*b*) If $\lambda \in \mathbb{C}$, $\lambda \neq 0$, and $\lambda \in \sigma(xy)$, prove that $\lambda \in \sigma(yx)$. Thus $\sigma(xy) \cup \{0\} = \sigma(yx) \cup \{0\}$. Show, however, that $\sigma(xy)$ is not always equal to $\sigma(yx)$.

5. Let A_0 and A_1 be the algebras of all complex 2-by-2 matrices of the form

$$\begin{pmatrix} \alpha & 0 \\ 0 & \beta \end{pmatrix}, \quad \begin{pmatrix} \alpha & \beta \\ 0 & \alpha \end{pmatrix}.$$

Prove that every two-dimensional complex algebra A with unit e is isomorphic to one of these and that A_0 is not isomorphic to A_1. *Hint:* Show that A has a basis $\{e, a\}$ in which $a^2 = \lambda e$ for some $\lambda \in \mathbb{C}$. Distinguish between the cases $\lambda = 0$, $\lambda \neq 0$. Show that there exists a three-dimensional noncommutative Banach algebra.

6. Let A be the algebra of all complex functions f on $\{1, 2, 3, \ldots\}$ which are 0 except at finitely many points, with pointwise addition and multiplication and norm

$$\|f\| = \sum_{k=1}^{\infty} k^{-2} |f(k)|.$$

Show that multiplication is left-continuous (hence also right-continuous, since A is commutative) but not jointly continuous. (Adjunction of a unit element, as suggested in Section 10.1, would make no difference.) Show, in fact, that there is a sequence $\{f_n\}$ in A so that $\|f_n\| \to 0$ but $\|f_n^2\| \to \infty$, as $n \to \infty$.

7. Let $C^2 = C^2([0, 1])$ be the space of all complex functions on $[0, 1]$ whose second derivative is continuous. Choose $a > 0$, $b > 0$, and define

$$\|f\| = \|f\|_\infty + a\|f'\|_\infty + b\|f''\|_\infty.$$

Show that this makes C^2 into a Banach space, for every choice of a, b, but that the Banach algebra axioms hold if and only if $2b \le a^2$. (For necessity, consider x and x^2.)

8. What happens if the process of adjoining a unit (described in Section 10.1) is applied to an algebra A which already has a unit? Clearly, the result cannot be an algebra A_1 with two units. Explain.

9. Suppose that Ω is open in $\mathbb{C}$ and that $f: \Omega \to A$ and $\varphi: \Omega \to \mathbb{C}$ are holomorphic. Prove that $\varphi f: \Omega \to A$ is holomorphic. [This was used in the proof of Theorem 10.13, with $\varphi(\lambda) = \lambda^n$.]

10. Another proof that $\sigma(x) \neq \varnothing$ can be based on Liouville's theorem 3.32 and the fact that $(\lambda e - x)^{-1} \to 0$ as $\lambda \to \infty$. Complete the details.

11. Call $x \in A$ a *topological divisor of zero* if there is a sequence $\{y_n\}$ in A, with $\|y_n\| = 1$, such that

$$\lim_{n\to\infty} xy_n = 0 = \lim_{n\to\infty} y_n x.$$

(*a*) Prove that every boundary point x of the set of all invertible elements of A is a topological divisor of zero. *Hint:* Take $y_n = x_n^{-1}/\|x_n^{-1}\|$, where $x_n \to x$.

(*b*) In which Banach algebras is 0 the only topological divisor of 0?

12. Find the spectrum of the operator $T \in \mathscr{B}(\ell^2)$ given by

$$T(x_1, x_2, x_3, x_4, \ldots) = (-x_2, x_1, -x_4, x_3, \ldots).$$

13. Suppose $K = \{\lambda \in \mathbb{C}: 1 \le |\lambda| \le 2\}$. Put $f(\lambda) = \lambda$. Let A be the smallest closed subalgebra of $C(K)$ that contains 1 and f. Let B be the smallest closed subalgebra of $C(K)$ that contains f and $1/f$. Describe the spectra $\sigma_A(f)$ and $\sigma_B(f)$.

14. (*a*) Fubini's theorem was applied to vector-valued integrals in the proof of Theorem 10.29. Justify this.

(*b*) Construct a proof of Theorem 10.29 that uses no contour integrals, as follows: Prove the theorem first for polynomials g, then for rational functions g with no poles in Ω_1, and obtain the general case from Runge's theorem.

15. Suppose X is a Banach space, $T \in \mathscr{B}(X)$ is compact, and $\|T^n\| \geq 1$ for all $n \geq 1$. Prove that the point spectrum of T is not empty.

16. Let $X = C([0, 1])$ and define $T \in \mathscr{B}(X)$ by

$$(Tx)(s) = \int_0^s x(u)\, du \qquad (0 \leq s \leq 1).$$

Show that $\sigma_p(T) = \varnothing$. Hence $f(\sigma_p(T)) = \varnothing$ for every f, but if $f \equiv 0$, then $\tilde{f}(T) = 0$; hence

$$\sigma_p(\tilde{f}(T)) = \sigma_p(0) = \{0\} \neq \varnothing.$$

The extra hypotheses in (*c*) and (*d*) of Theorem 10.33 are thus needed.

17. Suppose that the spectrum of some $x \in A$ is not connected. Prove that A contains a nontrivial idempotent z. (This is defined in Exercise 2.)

Prove also that $A = A_0 \oplus A_1$, where

$$A_0 = \{x\colon zx = 0\}, \qquad A_1 = \{x\colon zx = x\}.$$

18. Suppose Ω is open in $\mathbb{C}$, α is an isolated boundary point of Ω, $f\colon \Omega \to X$ is a holomorphic X-valued function (where X is some complex Banach space), and there is a smallest nonnegative integer n such that

$$|\lambda - \alpha|^n \|f(\lambda)\|$$

is bounded as $\lambda \to \alpha$. If $n > 0$, then f is said to have a pole of order n at α.

(*a*) Suppose $x \in A$ and $(\lambda e - x)^{-1}$ has a pole at every point of $\sigma(x)$. [Note that this can happen only when $\sigma(x)$ is a finite set.] Prove that there is a nontrivial polynomial P such that $P(x) = 0$.

(*b*) As a special case of (*a*), assume $\sigma(x) = \{0\}$ and $(\lambda e - x)^{-1}$ has a pole of order n at 0. Prove that $x^n = 0$.

19. Let S_R be the right shift, acting on ℓ^2, as in Exercise 2. Let $\{c_n\}$ be a sequence of complex numbers such that $c_n \neq 0$ but $c_n \to 0$ as $n \to \infty$. Define $M \in \mathscr{B}(\ell^2)$ by

$$(Mf)(n) = c_n f(n) \qquad (n \geq 0),$$

and define $T \in \mathscr{B}(\ell^2)$ by $T = MS_R$.

(*a*) Compute $\|T^m\|$, for $m = 1, 2, 3, \ldots$.

(*b*) Show that $\sigma(T) = \{0\}$.

(*c*) Show that T has no eigenvalue. (Its point spectrum is therefore empty, although its spectrum consists of a single point!)

(*d*) Show that $(\lambda I - T)^{-1}$ does not have a pole at 0.

(*e*) Show that T is a compact operator.

20. Suppose $x \in A$, $x_n \in A$, and $\lim x_n = x$. Suppose Ω is an open set in $\mathbb{C}$ that contains a component of $\sigma(x)$. Prove that $\sigma(x_n)$ intersects Ω for all sufficiently large n. (This strengthens Theorem 10.20.) *Hint*: If $\sigma(x) \subset \Omega \cup \Omega_0$, where Ω_0 is an open set disjoint from Ω, consider the function f that is 1 in Ω, 0 in Ω_0.

21. Let C_R be the algebra of all *real* continuous functions on $[0, 1]$, with the supremum norm. This satisfies all requirements of a Banach algebra, except that the scalars are now real.

(*a*) If $\phi(f) = \int_0^1 f(t)\,dt$, then $\phi(1) = 1$, and $\phi(f) \neq 0$ if f is invertible in C_R, but ϕ is not multiplicative.

(*b*) If G and G_1 are defined in C_R as in Theorem 10.34, show that G/G_1 is a group of order 2.

The analogues of Theorem 10.9 and (*d*) of Theorem 10.34 are thus false for real scalars. Exactly where would the proof of (*d*) of Theorem 10.34 break down?

22. Suppose $A = C(T)$, the algebra of all continuous complex functions on the unit circle T, with the supremum norm. Show that two invertible members of $C(T)$ are in the same coset of G_1 if and only if they are homotopic mappings of T into the set of all nonzero complex numbers. Deduce from this that G/G_1 is isomorphic to the additive group of the integers. (The notation is as in Theorem 10.34.)

23. Suppose $A = M(R)$, the convolution algebra of all complex Borel measures on the real line; see (*e*) of Example 10.3. Supply the details in the following proof that G/G_1 is uncountable: If $\alpha \in R$, let δ_α be the unit mass concentrated at α. Assume $\delta_\alpha \in G_1$. Then $\delta_\alpha = \exp(\mu_\alpha)$ for some $\mu_\alpha \in M(R)$; hence, for $-\infty < t < \infty$,

$$-i\alpha t = \hat{\mu}_\alpha(t) + 2k\pi i,$$

where k is an integer. Since $\hat{\mu}_\alpha$ is a bounded function, $\alpha = 0$. Thus δ_0 is the only δ_α in G_1. No coset of G_1 in G contains therefore more than one δ_α.

24. (*a*) Prove that A is commutative if there is a constant $M < \infty$ such that $\|xy\| \le M\|yx\|$ for all x and y in A. *Hint*: $\|w^{-1}yw\| \le M\|y\|$ if w is invertible in A. Replace w by $\exp(\lambda x)$, where $x \in A$ and $\lambda \in \mathbb{C}$. Continue as in Theorem 12.16.

(*b*) Prove that A is commutative if $\|x^2\| = \|x\|^2$ for every $x \in A$. *Hint*: Show that $\|x\| = \rho(x)$. Use Exercise 1 to deduce that $\|w^{-1}yw\| = \|y\|$. Continue as in (*a*).

25. As regards the invariant subspace problem, described in the introduction to Theorem 10.35, explain why the problem is

(*a*) trivial in $\mathbb{C}^n$,

(*b*) different in R^n,

(*c*) uninteresting if X is not separable.

How should Lomonosov's theorem be reformulated when $X = \mathbb{C}^n$?

26. Let S_R be the right shift on ℓ^2, as in Exercise 2. Prove that 0 is the only *compact* $T \in \mathscr{B}(\ell^2)$ that commutes with S_R. *Hint*: If $T \neq 0$ then

$$\|T(S_R^N x) - T(S_R^M x)\|$$

does not tend to 0 when $N - M \to \infty$.

CHAPTER 11

COMMUTATIVE BANACH ALGEBRAS

This chapter deals primarily with the Gelfand theory of commutative Banach algebras, although some of the results of this theory will be applied to noncommutative situations. The terminology of the preceding chapter will be used without change. In particular, Banach algebras will not be assumed to be commutative unless this is explicitly stated, but the presence of a unit will be assumed without special mention, as will the fact that the scalar field is $\mathbb{C}$.

Ideals and Homomorphisms

11.1 Definition A subset J of a commutative complex algebra A is said to be an *ideal* if

(*a*) J is a subspace of A (in the vector space sense), and

(*b*) $xy \in J$ whenever $x \in A$ and $y \in J$.

If $J \neq A$, J is a *proper* ideal. *Maximal ideals* are proper ideals which are not contained in any larger proper ideal.

11.2 Proposition

(*a*) *No proper ideal of A contains any invertible element of A.*

(*b*) *If J is an ideal in a commutative Banach algebra A, then its closure $\bar{J}$ is also an ideal.*

The proofs are so simple that they are left as an exercise.

11.3 Theorem

(a) *If A is a commutative complex algebra with unit, then every proper ideal of A is contained in a maximal ideal of A.*

(b) *If A is a commutative Banach algebra, then every maximal idea of A is closed.*

PROOF. (a) Let J be a proper ideal of A. Let $\mathscr{P}$ be the collection of all proper ideals of A that contain J. Partially order $\mathscr{P}$ by set inclusion, let $\mathscr{Q}$ be a maximal totally ordered subcollection of $\mathscr{P}$ (the existence of $\mathscr{Q}$ is assured by Hausdorff's maximality theorem), and let M be the union of all members of $\mathscr{Q}$. Being the union of a *totally* ordered collection of ideals, M is an ideal. Obviously $J \subset M$, and $M \neq A$ since no member of $\mathscr{P}$ contains the unit of A. The maximality of $\mathscr{Q}$ implies that M is a maximal ideal of A.

(b) Suppose M is a maximal ideal in A. Since M contains no invertible element of A and since the set of all invertible elements is open, $\bar{M}$ contains no invertible element either. Thus $\bar{M}$ is a proper ideal of A, and the maximality of M shows therefore that $M = \bar{M}$. ////

11.4 Homomorphisms and quotient algebras If A and B are commutative Banach algebras and ϕ is a homomorphism of A into B (see Section 10.4) then the null space or *kernel* of ϕ is obviously an ideal in A, which is closed if ϕ is continuous.

Conversely, suppose J is a proper *closed* ideal in A and $\pi: A \to A/J$ is the quotient map, as in Definition 1.40. Then A/J is a Banach space, with respect to the quotient norm (Theorem 1.41). *We will show that A/J is actually a Banach algebra and that π is a homomorphism.*

If $x' - x \in J$ and $y' - y \in J$, the identity

$$x'y' - xy = (x' - x)y' + x(y' - y) \tag{1}$$

shows that $x'y' - xy \in J$; hence $\pi(x'y') = \pi(xy)$. Multiplication can therefore be unambiguously defined in A/J by

$$\pi(x)\pi(y) = \pi(xy) \qquad (x \in A,\ y \in A). \tag{2}$$

It is then easily verified that A/J is a complex algebra and that π is a homomorphism. Since $\|\pi(x)\| \le \|x\|$, by the definition of the quotient norm, π is continuous.

Suppose $x_i \in A$ $(i = 1, 2)$ and $\delta > 0$. Then

$$\|x_i + y_i\| \le \|\pi(x_i)\| + \delta \qquad (i = 1, 2) \tag{3}$$

for some $y_i \in J$, by the definition of the quotient norm. Since

$$(x_1 + y_1)(x_2 + y_2) \in x_1 x_2 + J,$$

we have

$$\|\pi(x_1 x_2)\| \le \|(x_1 + y_1)(x_2 + y_2)\| \le \|x_1 + y_1\| \|x_2 + y_2\|, \tag{4}$$

so that (3) implies the multiplicative inequality

$$\|\pi(x_1)\pi(x_2)\| \le \|\pi(x_1)\| \; \|\pi(x_2)\|. \tag{5}$$

Finally, if e is the unit element of A, then (2) shows that $\pi(e)$ is the unit of A/J, and since $\pi(e) \neq 0$, (5) shows that $\|\pi(e)\| \ge 1 = \|e\|$. Since $\|\pi(x)\| \le \|x\|$ for every $x \in A$, $\|\pi(e)\| = 1$. This completes the proof.

Part (*a*) of the next theorem is one of the key facts of the whole theory. The set Δ that appears in it will later be given a compact Hausdorff topology (Theorem 11.9). The study of commutative Banach algebras will then to a large extent be reduced to the study of more familiar (and more special) objects, namely, algebras of continuous complex functions on Δ, with pointwise addition and multiplication. However, Theorem 11.5 has interesting concrete consequences even without the introduction of this topology. Sections 11.6 and 11.7 illustrate this point.

11.5 Theorem *Let A be a commutative Banach algebra, and let Δ be the set of all complex homomorphisms of A.*

(*a*) *Every maximal ideal of A is the kernel of some $h \in \Delta$.*

(*b*) *If $h \in \Delta$, the kernel of h is a maximal ideal of A.*

(*c*) *An element $x \in A$ is invertible in A if and only if $h(x) \neq 0$ for every $h \in \Delta$.*

(*d*) *An element $x \in A$ is invertible in A if and only if x lies in no proper ideal of A.*

(*e*) *$\lambda \in \sigma(x)$ if and only if $h(x) = \lambda$ for some $h \in \Delta$.*

PROOF. (*a*) Let M be a maximal ideal of A. Then M is closed (Theorem 11.3) and A/M is therefore a Banach algebra. Choose $x \in A$, $x \notin M$, and put

$$J = \{ax + y : a \in A, y \in M\}. \tag{1}$$

Then J is an ideal in A which is larger than M, since $x \in J$. (Take $a = e$, $y = 0$.) Thus $J = A$, and $ax + y = e$ for some $a \in A$, $y \in M$. If $\pi: A \to A/M$ is the quotient map, it follows that $\pi(a)\pi(x) = \pi(e)$. Every nonzero element $\pi(x)$ of the Banach algebra A/M is therefore invertible in A/M. By the Gelfand-Mazur theorem, there is an isomorphism j of A/M onto $\mathbb{C}$. Put $h = j \circ \pi$. Then $h \in \Delta$, and M is the null space of h.

(*b*) If $h \in \Delta$, then $h^{-1}(0)$ is an ideal in A which is maximal because it has codimension 1.

(*c*) If x is invertible in A and $h \in \Delta$, then

$$h(x)h(x^{-1}) = h(xx^{-1}) = h(e) = 1,$$

so that $h(x) \neq 0$. If x is not invertible, then the set $\{ax: a \in A\}$ does not contain e, hence is a proper ideal which lies in a maximal one (Theorem 11.3) and which is therefore annihilated by some $h \in \Delta$, because of (*a*).

(*d*) No invertible element lies in any proper ideal. The converse was proved in the proof of (*c*).

(*e*) Apply (*c*) to $\lambda e - x$ in place of x. ////

Our first application concerns functions on R^n that are sums of absolutely convergent trigonometric series. The notation is as in Exercise 22 of Chapter 7.

11.6 Wiener's lemma *Suppose f is a function on R^n, and*

$$f(x) = \sum a_m e^{im \cdot x}, \qquad \sum |a_m| < \infty, \tag{1}$$

where both sums are extended over all $m \in Z^n$. If $f(x) \neq 0$ for every $x \in R^n$, then

$$\frac{1}{f(x)} = \sum b_m e^{im \cdot x} \qquad \text{with} \quad \sum |b_m| < \infty. \tag{2}$$

PROOF. Let A be the set of functions of the form (1), normed by $\|f\| = \sum |a_m|$. One checks easily that A is a commutative Banach algebra, with respect to pointwise multiplication. Its unit is the constant function 1. For each x, the evaluation $f \to f(x)$ is a complex homomorphism of A. The assumption about the given function f is that no evaluation annihilates it. If we can prove that A has no other complex homomorphisms, (*c*) of Theorem 11.5 will imply that f is invertible in A, which is exactly the desired conclusion.

For $r = 1, \ldots, n$, put $g_r(x) = \exp(ix_r)$, where x_r is the rth coordinate of x. Then g_r and $1/g_r$ are in A and have norm 1. If $h \in \Delta$, it follows from (*c*) of Theorem 10.7 that

$$|h(g_r)| \leq 1 \qquad \text{and} \qquad \left|\frac{1}{h(g_r)}\right| = \left|h\left(\frac{1}{g_r}\right)\right| \leq 1.$$

Hence there are real numbers y_r such that

$$h(g_r) = \exp(iy_r) = g_r(y) \qquad (1 \leq r \leq n), \tag{3}$$

where $y = (y_1, \ldots, y_n)$. If P is a trigonometric polynomial (which means, by definition, that P is a finite linear combination of products of integral powers of the functions g_r and $1/g_r$), then (3) implies

$$h(P) = P(y), \tag{4}$$

because h is linear and multiplicative. Since h is continuous on A (Theorem 10.7) and since the set of all trigonometric polynomials is dense in A (as is obvious from the definition of the norm), (4) implies that $h(f) = f(y)$ for every $f \in A$. Thus h is evaluation at y, and the proof is complete. ////

This lemma was used (with $n = 1$) in the original proof of the tauberian theorem 9.7. To see the connection, let us reinterpret the lemma. Regard Z^n as being embedded in R^n in the obvious way. The given coefficients a_m define then a measure μ on R^n, concentrated on Z^n, which assigns mass a_m to each $m \in Z^n$. Consider the problem of finding a complex measure σ, concentrated on Z^n, such that the convolution $\mu * \sigma$ is the Dirac measure δ. Wiener's lemma states that this problem can be solved if (and trivially only if) the Fourier transform of μ has no zero on R^n; this is precisely the tauberian hypothesis in Theorem 9.7.

For our next application, let U^n be the set of all points $z = (z_1, \ldots, z_n)$ in $\mathcal{C}^n$ such that $|z_i| < 1$ for $1 \le i \le n$. In other words, this *polydisc* U^n is the cartesian product of n copies of the open unit disc U in $\mathcal{C}$. We define $A(U^n)$ to be the set of all functions f that are holomorphic in U^n (see Definition 7.20) and that are continuous on its closure $\bar{U}^n$.

11.7 Theorem *Suppose $f_1, \ldots, f_k \in A(U^n)$, and suppose that to each $z \in \bar{U}^n$ there corresponds at least one i such that $f_i(z) \neq 0$. Then there exist functions $\phi_1, \ldots, \phi_k \in A(U^n)$ such that*

$$f_1(z)\phi_1(z) + \cdots + f_k(z)\phi_k(z) = 1 \qquad (z \in \bar{U}^n). \tag{1}$$

PROOF. $A = A(U^n)$ is a commutative Banach algebra, with pointwise multiplication and the supremum norm. Let J be the set of all sums $\sum f_i \phi_i$, with $\phi_i \in A$. Then J is an ideal. If the conclusion is false, then $J \neq A$; hence J lies in some maximal ideal of A (Theorem 11.3), and some $h \in \Delta$ annihilates J, by (*a*) of Theorem 11.5.

For $1 \le r \le n$, put $g_r(z) = z_r$. Then $\|g_r\| = 1$; hence $h(g_r) = w_r$, with $|w_r| \le 1$. Put $w = (w_1, \ldots, w_n)$. Then $w \in \bar{U}^n$, and $h(g_r) = g_r(w)$. It follows that $h(P) = P(w)$ for every polynomial P, since h is a homomorphism. The polynomials are dense in $A(U^n)$ (Exercise 4). Hence $h(f) = f(w)$ for every $f \in A$, by essentially the same argument that was used in the proof of Theorem 11.6.

Since h annihilates J, $f_i(w) = 0$ for $1 \le i \le k$. This contradicts the hypothesis. ////

Gelfand Transforms

11.8 Definitions Let Δ be the set of all complex homomorphisms of a commutative Banach algebra A. The formula

$$\hat{x}(h) = h(x) \qquad (h \in \Delta) \tag{1}$$

assigns to each $x \in A$ a function $\hat{x}: \Delta \to \mathcal{C}$; we call $\hat{x}$ the *Gelfand transform* of x.

Let $\hat{A}$ be the set of all $\hat{x}$, for $x \in A$. The *Gelfand topology* of Δ is the weak topology induced by $\hat{A}$, that is, the weakest topology that makes every $\hat{x}$ continuous. Then obviously $\hat{A} \subset C(\Delta)$, the algebra of all complex continuous functions on Δ.

Since there is a one-to-one correspondence between the maximal ideals of A and the members of Δ (Theorem 11.5), Δ, equipped with its Gelfand topology, is usually called the *maximal ideal space* of A.

The term "Gelfand transform" is also applied to the *mapping* $x \to \hat{x}$ of A onto $\hat{A}$.

The *radical* of A, denoted by rad A, is the intersection of all maximal ideals of A. If rad $A = \{0\}$, A is called *semisimple*.

11.9 Theorem *Let Δ be the maximal ideal space of a commutative Banach algebra A.*

(a) *Δ is a compact Hausdorff space.*

(b) *The Gelfand transform is a homomorphism of A onto a subalgebra $\hat{A}$ of $C(\Delta)$, whose kernel is* rad A. *The Gelfand transform is therefore on isomorphism if and only if A is semisimple.*

(c) *For each $x \in A$, the range of $\hat{x}$ is the spectrum $\sigma(x)$. Hence*

$$\|\hat{x}\|_\infty = \rho(x) \le \|x\|,$$

where $\|\hat{x}\|_\infty$ is the maximum of $|\hat{x}(h)|$ on Δ, and $x \in$ rad A *if and only if $\rho(x) = 0$.*

PROOF. We first prove (*b*) and (*c*). Suppose $x \in A$, $y \in A$, $\alpha \in \mathcal{C}$, $h \in \Delta$. Then

$$(\alpha x)^\wedge(h) = h(\alpha x) = \alpha h(x) = (\alpha\hat{x})(h),$$

$$(x + y)^\wedge(h) = h(x + y) = h(x) + h(y) = \hat{x}(h) + \hat{y}(h) = (\hat{x} + \hat{y})(h),$$

and

$$(xy)^\wedge(h) = h(xy) = h(x)h(y) = \hat{x}(h)\hat{y}(h) = (\hat{x}\hat{y})(h).$$

Thus $x \to \hat{x}$ is a homomorphism. Its kernel consists of those $x \in A$ which satisfy $h(x) = 0$ for every $h \in \Delta$; by Theorem 11.5, this is the intersection of all maximal ideals of A, that is, rad A.

To say that λ is in the range of $\hat{x}$ means that $\lambda = \hat{x}(h) = h(x)$ for some $h \in \Delta$. By (*e*) of Theorem 11.5, this happens if and only if $\lambda \in \sigma(x)$. This proves (*b*) and (*c*).

To prove (*a*), let A^* be the dual space of A (regarded as a Banach space), and let K be the norm-closed unit ball of A^*. By the Banach-Alaoglu theorem, K is weak*-compact. By (*c*) of Theorem 10.7, $\Delta \subset K$. The Gelfand topology of Δ is evidently the restriction to Δ of the weak*-topology of A^*. It is therefore enough to show that Δ is a weak*-closed subset of A^*.

Let Λ_0 be in the weak*-closure of Δ. We have to show that

$$\Lambda_0(xy) = \Lambda_0 x \Lambda_0 y \qquad (x \in A, y \in A) \tag{1}$$

and

$$\Lambda_0 e = 1. \tag{2}$$

[Note that (2) is necessary; otherwise Λ_0 would be the zero homomorphism, which is not in Δ.]

Fix $x \in A$, $y \in A$, $\varepsilon > 0$. Put

$$W = \{\Lambda \in A^*: |\Lambda z_i - \Lambda_0 z_i| < \varepsilon \text{ for } 1 \leq i \leq 4\}, \tag{3}$$

where $z_1 = e$, $z_2 = x$, $z_3 = y$, $z_4 = xy$. Then W is a weak*-neighborhood of Λ_0 which therefore contains an $h \in \Delta$. For this h,

$$|1 - \Lambda_0 e| = |h(e) - \Lambda_0 e| < \varepsilon, \tag{4}$$

which gives (2), and

$$\begin{aligned}\Lambda_0(xy) - \Lambda_0 x \Lambda_0 y &= [\Lambda_0(xy) - h(xy)] + [h(x)h(y) - \Lambda_0 x \Lambda_0 y] \\ &= [\Lambda_0(xy) - h(xy)] + [h(y) - \Lambda_0 y]h(x) \\ &\qquad + [h(x) - \Lambda_0 x]\Lambda_0 y,\end{aligned}$$

which gives

$$|\Lambda_0(xy) - \Lambda_0 x \Lambda_0 y| < (1 + \|x\| + |\Lambda_0 y|)\varepsilon. \tag{5}$$

Since (5) implies (1), the proof is complete. ////

Semisimple algebras have an important property which was earlier proved for $\mathcal{C}$:

11.10 Theorem *If $\psi: A \to B$ is a homomorphism of a Banach algebra A into a semisimple commutative Banach algebra B, then ψ is continuous.*

PROOF. Suppose $x_n \to x$ in A and $\psi(x_n) \to y$ in B. By the closed graph theorem, it is enough to show that $y = \psi(x)$.

Pick some homomorphism $h: B \to \mathcal{C}$. Then $\varphi = h \circ \psi$ is a homomorphism of A into $\mathcal{C}$. Theorem 10.7 shows that h and φ are continuous. Hence

$$h(y) = \lim h(\psi(x_n)) = \lim \varphi(x_n) = \varphi(x) = h(\psi(x)),$$

for every $h \in \Delta_B$. Thus $y - \psi(x) \in \operatorname{rad} B$. Since $\operatorname{rad} B = \{0\}$, $y = \psi(x)$. ////

Corollary. *Every isomorphism between two semisimple commutative Banach algebras is a homeomorphism.*

In particular, this is true of every automorphism of a semisimple commutative Banach algebra. The topology of such an algebra is therefore completely determined by its algebraic structure.

In Theorem 11.9, the algebra $\hat{A}$ may or may not be closed in $C(\Delta)$, with respect to the supremum norm. Which of these cases occurs can be decided by comparing $\|x^2\|$ with $\|x\|^2$, for all $x \in A$. Recall that $\|x^2\| \le \|x\|^2$ is always true.

11.11 Lemma *If A is a commutative Banach algebra and*

$$(1) \qquad r = \inf \frac{\|x^2\|}{\|x\|^2}, \qquad s = \inf \frac{\|\hat{x}\|_\infty}{\|x\|} \qquad (x \in A,\ x \neq 0),$$

then $s^2 \le r \le s$.

PROOF. Since $\|\hat{x}\|_\infty \ge s\|x\|$,

$$(2) \qquad \|x^2\| \ge \|\hat{x}^2\|_\infty = \|\hat{x}\|_\infty^2 \ge s^2\|x\|^2$$

for every $x \in A$. Thus $s^2 \le r$.

Since $\|x^2\| \ge r\|x\|^2$ for every $x \in A$, induction on n shows that

$$(3) \qquad \|x^m\| \ge r^{m-1}\|x\|^m \qquad (m = 2^n,\ n = 1, 2, 3, \ldots).$$

Take mth roots in (3) and let $m \to \infty$. By the spectral radius formula and (c) of Theorem 11.9,

$$(4) \qquad \|\hat{x}\|_\infty = \rho(x) \ge r\|x\| \qquad (x \in A).$$

Hence $r \le s$. ////

11.12 Theorem *Suppose A is a commutative Banach algebra.*

(a) *The Gelfand transform is an isometry (that is, $\|x\| = \|\hat{x}\|_\infty$ for every $x \in A$) if and only if $\|x^2\| = \|x\|^2$ for every $x \in A$.*

(b) *A is semisimple and $\hat{A}$ is closed in $C(\Delta)$ if and only if there exists $K < \infty$ such that $\|x\|^2 \le K\|x^2\|$ for every $x \in A$.*

PROOF. (*a*) In the terminology of Lemma 11.11, the Gelfand transform is an isometry if and only if $s = 1$, which happens (by the lemma) if and only if $r = 1$.

(*b*) The existence of K is equivalent to $r > 0$, hence to $s > 0$, by the lemma. If $s > 0$, then $x \to \hat{x}$ is one-to-one and has a continuous inverse, so that $\hat{A}$ is complete (hence closed) in $C(\Delta)$. Conversely, if $x \to \hat{x}$ is one-to-one and if $\hat{A}$ is closed in $C(\Delta)$, the open mapping theorem implies that $s > 0$. ////

11.13 Examples In some cases, the maximal ideal space of a given commutative Banach algebra can easily be described explicitly. In others, extreme pathologies occur. We shall now give some examples to illustrate this.

(*a*) Let X be a compact Hausdorff space, put $A = C(X)$, with the supremum norm. For each $x \in X, f \to f(x)$ is a complex homomorphism h_x. Since $C(X)$ separates points on X (Urysohn's lemma), $x \neq y$ implies $h_x \neq h_y$. Thus $x \to h_x$ embeds X in Δ.

We claim that each $h \in \Delta$ is an h_x. If this is false, there is a maximal ideal M in $C(X)$ which contains, for each $p \in X$, a function f with $f(p) \neq 0$. The compactness of X implies then that M contains finitely many functions $f_1, \ldots, f_n$ such that at least one of them is $\neq 0$ at each point of X. Put

$$g = f_1 \bar{f}_1 + \cdots + f_n \bar{f}_n.$$

Then $g \in M$, since M is an ideal; $g > 0$ at every point of X; hence g is invertible in $C(X)$. But proper ideals contain no invertible elements.

Thus $x \leftrightarrow h_x$ is one-to-one correspondence between X and Δ and can be used to identify Δ with X. This identification is also correct in terms of the two topologies that are involved: The Gelfand topology γ of X is the weak topology induced by $C(X)$ and is therefore weaker than τ, the original one, but γ is a Hausdorff topology; hence $\gamma = \tau$. [See (*a*) of Section 3.8.]

Summing up, X "is" the maximal ideal space of $C(X)$, and the Gelfand transform is the identity mapping on $C(X)$.

(*b*) Let A be the algebra of all absolutely convergent trigonometric series, as in Section 11.6. We found there that the complex homomorphisms are the evaluations at points of R^n. Since the members of A are 2π-periodic in each variable, Δ is the torus T^n obtained from R^n by the mapping

$$(x_1, \ldots, x_n) \to (e^{ix_1}, \ldots, e^{ix_n}).$$

This is an example in which $\hat{A}$ is dense in $C(\Delta)$, although $\hat{A} \neq C(\Delta)$.

(*c*) In the same way, the proof of Theorem 11.7 contains the result that $\bar{U}^n$ is the maximal ideal space of $A(U^n)$. The argument used at the end of (*a*) shows that the natural topology of $\bar{U}^n$ is the same as the Gelfand topology induced by $A(\bar{U}^n)$; the same remark applies to (*b*).

(*d*) The preceding example has interesting generalizations. Let A now be a commutative Banach algebra with a *finite set of generators*, say $x_1, \ldots, x_n$. This means that $x_i \in A$ $(1 \le i \le n)$ and that the set of all polynomials in $x_1, \ldots, x_n$ is dense in A. Define

$$\phi(h) = (\hat{x}_1(h), \ldots, \hat{x}_n(h)) \qquad (h \in \Delta). \tag{1}$$

Then ϕ is a homeomorphism of Δ onto a compact set $K \subset \mathcal{C}^n$. Indeed, ϕ is continuous since $\hat{A} \subset C(\Delta)$. If $\phi(h_1) = \phi(h_2)$, then $h_1(x_i) = h_2(x_i)$ for all i; hence $h_1(x) = h_2(x)$ whenever x is a polynomial in $x_1, \ldots, x_n$, and since these polynomials are dense in A, $h_1 = h_2$. Thus ϕ is one-to-one.

We can now transfer $\hat{A}$ from Δ to K and may thus regard K as the maximal ideal space of A. To make this precise, define

$$\psi(x) = \hat{x} \circ \phi^{-1} \qquad (x \in A). \tag{2}$$

Then ψ is a homomorphism (an isomorphism if A is semisimple) of A onto a subalgebra $\psi(A)$ of $C(K)$. One verifies easily that

$$\psi(x_i)(z) = z_i \qquad \text{if } z = (z_1, \ldots, z_n) \in K, \tag{3}$$

and therefore

$$\psi(P(x_1, \ldots, x_n))(z) = P(z) \qquad (z \in K) \tag{4}$$

for every polynomial P in n variables.

It follows that every member of $\psi(A)$ is a uniform limit of polynomials, on K.

The sets $K \subset \mathcal{C}^n$ which arise in this fashion as maximal ideal spaces have a property known as *polynomial convexity:*

If $w \in \mathcal{C}^n$ and $w \notin K$, there exists a polynomial P such that $|P(z)| \le 1$ for every $z \in K$, but $|P(w)| > 1$.

To prove this, assume there is no such polynomial. The norm-decreasing property of the Gelfand transform implies then that

$$|P(w)| \le \|P(x_1, \ldots, x_n)\| \tag{5}$$

for every polynomial P; the norm is that of A. Since $\{x_1, \ldots, x_n\}$ is a set of generators of A, it follows from (5) that there is an $h \in \Delta$ such that $\phi(h) = w$. But then $w \in K$, and we have a contradiction.

The compact polynomially convex subsets of $\mathcal{C}$ are simply those whose complement is connected; this is an easy consequence of Runge's theorem. In $\mathcal{C}^n$, the structure of the polynomially convex sets is by no means fully understood.

(*e*) Our next example shows that the Gelfand transform is a generalization of the Fourier transform, at least in the L^1-context.

Let A be $L^1(R^n)$ with a unit attached, as described in (*d*) of Section 10.3. The members of A are of the form $f + \alpha\delta$, where $f \in L^1(R^n)$, $\alpha \in \mathcal{C}$, and δ is the Dirac measure on R^n; multiplication in A is convolution:

$$(f + \alpha\delta) * (g + \beta\delta) = (f * g + \beta f + \alpha g) + \alpha\beta\delta.$$

For each $t \in R^n$, the formula

$$h_t(f + \alpha\delta) = \hat{f}(t) + \alpha \tag{6}$$

defines a complex homomorphism of A; here $\hat{f}$ is the Fourier transform of f. In addition,

$$h_\infty(f + \alpha\delta) = \alpha \tag{7}$$

also defines a complex homomorphism. There are no others. (A proof will be sketched presently.) Thus Δ, as a set, is $R^n \cup \{\infty\}$. Give Δ the topology of the one-point compactification of R^n. Since $\hat{f}(t) \to 0$ as $|t| \to \infty$, for every $f \in L^1(R^n)$, it follows from (6) and (7) that $\hat{A} \subset C(\Delta)$. Since $\hat{A}$ separates points on Δ, the weak topology induced on Δ by $\hat{A}$ is the same as the one that we just chose.

It remains to be proved that every $h \in \Delta$ is of the form (6) or (7). If $h(f) = 0$ for every $f \in L^1(R^n)$, then $h = h_\infty$. Assume $h(f) \neq 0$ for some $f \in L^1(R^n)$. Then $h(f) = \int f\beta\, dm_n$, for some $\beta \in L^\infty(R^n)$. Since $h(f * g) = h(f)h(g)$, one can prove that β coincides almost everywhere with a continuous function b which satisfies

$$b(x + y) = b(x)b(y) \qquad (x, y \in R^n). \tag{8}$$

Finally, every bounded solution of (8) is of the form

$$b(x) = e^{-ix \cdot t} \qquad (x \in R^n) \tag{9}$$

for some $t \in R^n$. Thus $h(f) = \hat{f}(t)$, and h has the form (6).

For $n = 1$, the details that complete the preceding sketch may be found in Sec. 9.22 of [23]. The case $n > 1$ is quite similar.

(*f*) Our final example is $L^\infty(m)$. Here m is Lebesgue measure on the unit interval $[0, 1]$, and $L^\infty(m)$ is the usual Banach space of equivalence classes (modulo sets of measure 0) of complex bounded measurable functions on $[0, 1]$, normed by the essential supremum. Under pointwise multiplication, this is obviously a commutative Banach algebra.

If $f \in L^\infty(m)$ and G_f is the union of all open sets $G \subset \mathcal{C}$ with $m(f^{-1}(G)) = 0$, then the complement of G_f (called the *essential range* of f) is easily seen to coincide with the spectrum $\sigma(f)$ of f, hence with the range of its Gelfand transform $\hat{f}$. It follows that $\hat{f}$ is real if f is real. Hence $L^\infty(m)^\wedge$ is closed under complex conjugation. By the Stone-Weierstrass theorem, $L^\infty(m)^\wedge$ is therefore dense in $C(\Delta)$, where Δ is the maximal ideal space of $L^\infty(m)$. It also follows that $f \to \hat{f}$ is an isometry, so that $L^\infty(m)^\wedge$ is closed in $C(\Delta)$.

We conclude that $f \to \hat{f}$ is an isometry of $L^\infty(m)$ onto $C(\Delta)$.

Next, $\hat{f} \to \int f\,dm$ is a bounded linear functional on $C(\Delta)$. By the Riesz representation theorem, there is therefore a regular Borel probability measure μ on Δ that satisfies

$$\int_\Delta \hat{f}\,d\mu = \int_0^1 f\,dm \tag{10}$$

for every $f \in L^\infty(m)$. This measure is related to the topology of Δ in the following way:

(i) *$\mu(V) > 0$ if V is open and nonempty.*

(ii) *To every bounded Borel function φ on Δ corresponds an $\hat{f} \in C(\Delta)$ such that $\hat{f} = \varphi$ a.e. $[\mu]$.*

(iii) *If V is open, so is $\bar{V}$.*

(iv) *If E is a Borel set in Δ, then*

$$\mu(E^0) = \mu(E) = \mu(\bar{E}). \tag{11}$$

If V is as in (*i*), Urysohn's lemma implies that there is an $\hat{f} \in C(\Delta)$, $\hat{f} \geq 0$, such that $\hat{f} = 0$ outside V and $\hat{f}(p) = 1$ at some $p \in V$. Hence f is not the zero element of $L^\infty(m)$, and the integrals (10) are positive. This gives (*i*).

In (*ii*), assume $|\varphi| \leq 1$. Since $C(\Delta)$ is dense in $L^2(\mu)$ (recall that μ is a regular Borel probability measure), there are functions $\hat{f}_n \in C(\Delta)$ such that $\int |\hat{f}_n - \varphi|^2\,d\mu \to 0$, and which can be so adjusted that $|\hat{f}_n| \leq 1$. Then $\|f_n\|_\infty \leq 1$, and (10) implies that $\{f_n\}$ is a Cauchy sequence in $L^2(m)$. Hence there is an $f \in L^\infty(m)$ such that

$$\int_\Delta |\hat{f}_n - \hat{f}|^2\,d\mu = \int_0^1 |f_n - f|^2\,dm \to 0 \tag{12}$$

as $n \to \infty$. Thus $\varphi = \hat{f}$ a.e. $[\mu]$.

Next, let V be open and let W be the complement of $\bar{V}$. By (*ii*) there is an $\hat{f} \in C(\Delta)$ such that $\hat{f} = 1$ a.e. $[\mu]$ off V. The set on which $\hat{f}$ is neither 1 nor 0 is open and has μ-measure 0, hence is empty, by (*i*). The same reasoning shows that the sets $V \cap \{\hat{f} \neq 1\}$ and $W \cap \{\hat{f} \neq 0\}$ are empty. Hence $\hat{f} = 1$ on $\bar{V}$, $\hat{f} = 0$ on W.

This proves (*iii*), and shows also that $\mu(\bar{V}) = \mu(V)$. Taking complements, we see that $\mu(K^0) = \mu(K)$ for every compact $K \subset \Delta$.

If E is a Borel set in Δ, and $\varepsilon > 0$, then there is a compact K and an open V such that $K \subset E \subset V$ and $\mu(V) < \mu(K) + \varepsilon$. Hence

$$\mu(\bar{E}) \leq \mu(\bar{V}) = \mu(V) < \mu(K) + \varepsilon = \mu(K^0) + \varepsilon \leq \mu(E^0) + \varepsilon,$$

and this proves (*iv*).

It is an easy consequence of (*iii*) that disjoint open sets have disjoint closures.

If we define two bounded Borel functions φ and ψ to be equivalent, provided $\mu\{\varphi \neq \psi\} = 0$, then (*ii*) asserts that every equivalence class contains one continuous function—and only one, by (*i*). Hence (with an obvious interpretation) $L^\infty(\mu) = C(\Delta)$.

Property (*iv*) asserts, among other things, that of two disjoint Borel sets in Δ, at most one can be dense in Δ, even though no point of Δ is isolated (Exercise 18).

We conclude with an application to measure theory. If E and F are measurable sets, let us say that F *almost contains* E if F contains E except for a set of measure 0, that is, if $m(E - F) = 0$.

The union of an uncountable collection of measurable sets is not always measurable. However, the following is true:

If $\{E_\alpha\}$ is an arbitrary collection of measurable sets in $[0, 1]$, *there is a measurable set $E \subset [0, 1]$ with the following two properties:*

(*i*) *E almost contains every E_α.*

(*ii*) *If F is measurable and F almost contains every E_α, then F almost contains E.*

Thus E is the *least upper bound* of $\{E_\alpha\}$. The existence of E implies that the Boolean algebra of measurable sets (modulo sets of measure 0) is *complete.*

With the machinery now at our disposal, the proof is very simple. Let f_α be the characteristic function of E_α. Its Gelfand transform $\hat{f}_\alpha$ is then the characteristic function of an open (and closed) set $\Omega_\alpha \subset \Delta$. Let Ω be the union of all these Ω_α. Then Ω is open, so is its closure $\bar{\Omega}$, and there exists $f \in L^\infty(m)$ such that $\hat{f}$ is the characteristic function of $\bar{\Omega}$. The desired set E is the set of all $x \in [0, 1]$ at which $f(x) = 1$.

Involutions

11.14 Definition A mapping $x \to x^*$ of a complex (not necessarily commutative) algebra A into A is called an *involution* on A if it has the following four properties, for all $x \in A$, $y \in A$, and $\lambda \in \mathbb{C}$:

(1) $$(x + y)^* = x^* + y^*.$$

(2) $$(\lambda x)^* = \bar{\lambda} x^*.$$

(3) $$(xy)^* = y^* x^*.$$

(4) $$x^{**} = x.$$

In other words, an involution is a conjugate-linear antiautomorphism of period 2.

Any $x \in A$ for which $x^* = x$ is called *hermitian*, or *self-adjoint*.

For example, $f \to \bar{f}$ is an involution on $C(X)$. The one that we will be most concerned with later is the passage from an operator on a Hilbert space to its adjoint.

11.15 Theorem *If A is a Banach algebra with an involution, and if $x \in A$, then*

(*a*) *$x + x^*$, $i(x - x^*)$, and xx^* are hermitian,*

(*b*) *x has a unique representation $x = u + iv$, with $u \in A$, $v \in A$, and both u and v hermitian,*

(*c*) *the unit e is hermitian,*

(*d*) *x is invertible in A if and only if x^* is invertible, in which case $(x^*)^{-1} = (x^{-1})^*$, and*

(*e*) *$\lambda \in \sigma(x)$ if and only if $\bar{\lambda} \in \sigma(x^*)$.*

PROOF. Statement (*a*) is obvious. If $2u = x + x^*$, $2v = i(x^* - x)$, then $x = u + iv$ is a representation as in (*b*). Suppose $x = u' + iv'$ is another one. Put $w = v' - v$. Then both w and iw are hermitian, so that

$$iw = (iw)^* = -iw^* = -iw.$$

Hence $w = 0$, and the uniqueness follows.

Since $e^* = ee^*$; (*a*) implies (*c*); (*d*) follows from (*c*) and $(xy)^* = y^*x^*$. Finally, (*e*) follows if (*d*) is applied to $\lambda e - x$ in place of x. ////

11.16 Theorem *If the Banach algebra A is commutative and semisimple, then every involution on A is continuous.*

PROOF. Let h be a complex homomorphism of A, and define $\phi(x) = \bar{h}(x^*)$. Properties (1) to (3) of Definition 11.14 show that ϕ is a complex homomorphism. Hence ϕ is continuous. Suppose $x_n \to x$ and $x_n^* \to y$ in A. Then

$$\bar{h}(x^*) = \phi(x) = \lim \phi(x_n) = \lim \bar{h}(x_n^*) = \bar{h}(y).$$

Since A is semisimple, $y = x^*$. Hence $x \to x^*$ is continuous, by the closed graph theorem. ////

11.17 Definition A Banach algebra A with an involution $x \to x^*$ that satisfies

$$\|xx^*\| = \|x\|^2 \tag{1}$$

for every $x \in A$ is called a B^*-algebra.

Note that $\|x\|^2 = \|xx^*\| \le \|x\|\|x^*\|$ implies $\|x\| \le \|x^*\|$, hence also

$$\|x^*\| \le \|x^{**}\| = \|x\|.$$

Thus

$$\|x^*\| = \|x\| \tag{2}$$

in every B^*-algebra. It also follows that

$$\|xx^*\| = \|x\|\|x^*\|. \tag{3}$$

Conversely, (2) and (3) obviously imply (1).

The following theorem is the key to the proof of the spectral theorem that will be given in Chapter 12.

11.18 Theorem (Gelfand-Naimark) *Suppose A is a commutative B^*-algebra, with maximal ideal space Δ. The Gelfand transform is then an isometric isomorphism of A onto $C(\Delta)$, which has the additional property that*

$$h(x^*) = \overline{h(x)} \qquad (x \in A, h \in \Delta), \tag{1}$$

or, equivalently, that

$$(x^*)^\wedge = \overline{\hat{x}} \qquad (x \in A). \tag{2}$$

In particular, x is hermitian if and only if $\hat{x}$ is a real-valued function.

The interpretation of (2) is that the Gelfand transform carries the given involution on A to the natural involution on $C(\Delta)$, which is conjugation. Isomorphisms that preserve involutions in this manner are often called $*$-isomorphisms.

PROOF. Assume first that $u \in A$, $u = u^*$, $h \in \Delta$. We have to prove that $h(u)$ is real. Put $z = u + ite$, for real t. If $h(u) = \alpha + i\beta$, with α and β real, then

$$h(z) = \alpha + i(\beta + t), \qquad zz^* = u^2 + t^2 e,$$

so that

$$\alpha^2 + (\beta + t)^2 = |h(z)|^2 \le \|z\|^2 = \|zz^*\| \le \|u\|^2 + t^2,$$

or

$$\alpha^2 + \beta^2 + 2\beta t \le \|u\|^2 \qquad (-\infty < t < \infty). \tag{3}$$

By (3), $\beta = 0$; hence $h(u)$ is real.

If $x \in A$, then $x = u + iv$, with $u = u^*$, $v = v^*$. Hence $x^* = u - iv$. Since $\hat{u}$ and $\hat{v}$ are real, (2) is proved.

Thus $\hat{A}$ is closed under complex conjugation. By the Stone-Weierstrass theorem, $\hat{A}$ is therefore dense in $C(\Delta)$.

If $x \in A$ and $y = xx^*$, then $y = y^*$ so that $\|y^2\| = \|y\|^2$. It follows, by induction on n, that $\|y^m\| = \|y\|^m$ for $m = 2^n$. Hence $\|\hat{y}\|_\infty = \|y\|$, by the spectral radius formula and (c) of Theorem 11.9. Since $y = xx^*$, (2) implies that $\hat{y} = |\hat{x}|^2$. Hence

$$\|\hat{x}\|_\infty^2 = \|\hat{y}\|_\infty = \|y\| = \|xx^*\| = \|x\|^2,$$

or $\|\hat{x}\|_\infty = \|x\|$. Thus $x \to \hat{x}$ is an isometry. Hence $\hat{A}$ is closed in $C(\Delta)$. Since $\hat{A}$ is also dense in $C(\Delta)$, we conclude that $\hat{A} = C(\Delta)$. This completes the proof. ////

The next theorem is a special case of the one just proved. We shall state it in a form that involves the *inverse* of the Gelfand transform, in order to make contact with the symbolic calculus.

11.19 Theorem *If A is a commutative B^*-algebra which contains an element x such that the polynomials in x and x^* are dense in A, then the formula*

$$(\Psi f)^\wedge = f \circ \hat{x} \tag{1}$$

defines an isometric isomorphism Ψ of $C(\sigma(x))$ onto A which satisfies

$$\Psi \bar{f} = (\Psi f)^* \tag{2}$$

for every $f \in C(\sigma(x))$. Moreover, if $f(\lambda) = \lambda$ on $\sigma(x)$, then $\Psi f = x$.

PROOF. Let Δ be the maximal ideal space of A. Then $\hat{x}$ is a continuous function on Δ whose range is $\sigma(x)$. Suppose $h_1 \in \Delta$, $h_2 \in \Delta$, and $\hat{x}(h_1) = \hat{x}(h_2)$, that is, $h_1(x) = h_2(x)$. Theorem 11.18 implies then that $h_1(x^*) = h_2(x^*)$. If P is any polynomial in two variables, it follows that

$$h_1(P(x, x^*)) = h_2(P(x, x^*)),$$

since h_1 and h_2 are homomorphisms. By hypothesis, elements of the form $P(x, x^*)$ are dense in A. The continuity of h_1 and h_2 implies therefore that $h_1(y) = h_2(y)$ for every $y \in A$. Hence $h_1 = h_2$. We have proved that $\hat{x}$ is one-to-one. Since Δ is compact, it follows that $\hat{x}$ is a *homeomorphism* of Δ onto $\sigma(x)$.

The mapping $f \to f \circ \hat{x}$ is therefore an isometric isomorphism of $C(\sigma(x))$ onto $C(\Delta)$ which also preserves complex conjugation.

Each $f \circ \hat{x}$ is thus (by Theorem 11.18) the Gelfand transform of a unique element of A which we denote by Ψf and which satisfies $\|\Psi f\| = \|f\|_\infty$. Assertion (2) comes from (2) of Theorem 11.18. If $f(\lambda) = \lambda$, then $f \circ \hat{x} = \hat{x}$, so that (1) gives $\Psi f = x$. ////

Remark. In the situation described by Theorem 11.19, it makes perfectly good sense to write $f(x)$ for the element of A whose Gelfand transform is $f \circ \hat{x}$. This notation is indeed frequently used. It extends the symbolic calculus (for these particular algebras) to arbitrary continuous functions on the spectrum of x, whether they are holomorphic or not.

The existence of square roots is often of special interest, and in algebras with involution one may ask under what conditions hermitian elements have hermitian square roots.

11.20 Theorem *Suppose A is a commutative Banach algebra with an involution, $x \in A$, $x = x^*$, and $\sigma(x)$ contains no real λ with $\lambda \leq 0$. Then there exists $y \in A$ with $y = y^*$ and $y^2 = x$.*

Note that the given involution is not assumed to be continuous. This will give us an opportunity to use the radical of A. We shall see later, in Theorem 11.26, that commutativity can be dropped from the hypotheses. This will be used in the proof of Theorem 11.31.

PROOF. Let Ω be the complement (in $\mathbb{C}$) of the set of all nonpositive real numbers. There exists $f \in H(\Omega)$ such that $f^2(\lambda) = \lambda$, and $f(1) = 1$. Since $\sigma(x) \subset \Omega$, we can define $y \in A$ by

$$y = \tilde{f}(x), \tag{1}$$

as in Definition 10.26. Then $y^2 = x$, by Theorem 10.27. We will prove that $y^* = y$.

Since Ω is simply connected, Runge's theorem furnishes polynomials P_n that converge to f, uniformly on compact subsets of Ω. Define Q_n by

$$2Q_n(\lambda) = P_n(\lambda) + \overline{P_n(\bar{\lambda})}. \tag{2}$$

Since $f(\bar{\lambda}) = \overline{f(\lambda)}$, the polynomials Q_n converge to f in the same manner. Define

$$y_n = Q_n(x) \quad (n = 1, 2, 3, \ldots). \tag{3}$$

By (2), the polynomials Q_n have *real* coefficients. Since $x = x^*$, it follows that $y_n = y_n^*$. By Theorem 10.27,

$$y = \lim_{n \to \infty} y_n, \tag{4}$$

since $Q_n \to f$, so that $Q_n(x) \to \tilde{f}(x)$. If the involution were assumed to be continuous, the set of hermitian elements would be closed, and $y^* = y$ would follow directly from (4).

Let R be the radical of A. Let $\pi: A \to A/R$ be the quotient map.

If $\pi(x) = \pi(y)$ and $z = x - y$, then $z \in R$; hence $z^* \in R$ because $\rho(z^*) = \rho(z) = 0$ (see Theorem 11.15), and therefore $\pi(x^*) = \pi(y^*)$. This shows that the formula

$$(5) \qquad [\pi(a)]^* = \pi(a^*) \qquad (a \in A)$$

defines an involution in A/R. If $a \in A$ is hermitian, so is $\pi(a)$. Since π is continuous, $\pi(y_n) \to \pi(y)$. Since A/R is isomorphic to $\hat{A}$ (Theorem 11.9), A/R is semisimple, and therefore every involution in A/R is continuous (Theorem 11.16). It follows that $\pi(y)$ is hermitian. Hence $\pi(y) = \pi(y^*)$.

We conclude that $y^* - y$ lies in the radical of A.

By Theorem 11.15, $y = u + iv$, where $u = u^*$ and $v = v^*$. We just proved that $v \in R$. Since $x = y^2$. we have

$$(6) \qquad x = u^2 - v^2 + 2iuv.$$

Let h be any complex homomorphism of A. Since $v \in R$, $h(v) = 0$. Hence $h(x) = [h(u)]^2$. By hypothesis, $0 \notin \sigma(x)$. Thus $h(x) \neq 0$; hence $h(u) \neq 0$. By Theorem 11.5, u is invertible in A. Since $x = x^*$, (6) implies that $uv = 0$. Since $v = u^{-1}(uv)$, we conclude that $v = 0$. This completes the proof. ////

Remark. If $\sigma(x) \subset (0, \infty)$, then also $\sigma(y) \subset (0, \infty)$. This follows from (1) (the definition of y) and the spectral mapping theorem.

Applications to Noncommutative Algebras

Noncommutative algebras always contain commutative ones. Their presence can sometimes be exploited to extend certain theorems from the commutative situation to the noncommutative one. On a trivial level, we have already done this: In the elementary discussion of spectra, our attention was usually fixed on one element $x \in A$; the (closed) subalgebra A_0 of A that x generates is commutative, and much of the discussion took place within A_0. One possible difficulty was that x might have different spectra with respect to A and A_0. There is a simple construction (Theorem 11.22) that circumvents this. Another device (Theorem 11.25) can be used when A has an involution.

11.21 Centralizers If S is a subset of a Banach algebra A, the *centralizer* of S is the set

$$\Gamma(S) = \{x \in A: xs = sx \text{ for every } s \in S\}.$$

We say that S *commutes* if any two elements of S commute with each other. We shall use the following simple properties of centralizers.

(a) *$\Gamma(S)$ is a closed subalgebra of A.*

(b) *$S \subset \Gamma(\Gamma(S))$.*

(c) *If S commutes, then $\Gamma(\Gamma(S))$ commutes.*

Indeed, if x and y commute with every $s \in S$, so do λx, $x + y$, and xy; since multiplication is continuous in A, $\Gamma(S)$ is closed. This proves (a). Since every $s \in S$ commutes with every $x \in \Gamma(S)$, (b) holds. If S commutes, then $S \subset \Gamma(S)$, hence $\Gamma(S) \supset \Gamma(\Gamma(S))$, which proves (c), since $\Gamma(E)$ obviously commutes whenever $\Gamma(E) \subset E$.

11.22 Theorem *Suppose A is a Banach algebra, $S \subset A$, S commutes, and $B = \Gamma(\Gamma(S))$. Then B is a commutative Banach algebra, $S \subset B$, and $\sigma_B(x) = \sigma_A(x)$ for every $x \in B$.*

PROOF. Since $e \in B$, Section 11.21 shows that B is a commutative Banach algebra that contains S. Suppose $x \in B$ and x is invertible in A. We have to show that $x^{-1} \in B$. Since $x \in B$, $xy = yx$ for every $y \in \Gamma(S)$; hence $y = x^{-1}yx$, $yx^{-1} = x^{-1}y$. This says that $x^{-1} \in \Gamma(\Gamma(S)) = B$. ////

11.23 Theorem *Suppose A is a Banach algebra, $x \in A$, $y \in A$, and $xy = yx$. Then*

$$\sigma(x + y) \subset \sigma(x) + \sigma(y) \quad \textit{and} \quad \sigma(xy) \subset \sigma(x)\sigma(y).$$

PROOF. Put $S = \{x, y\}$; put $B = \Gamma(\Gamma(S))$. Then $x + y \in B$, $xy \in B$, and Theorem 11.22 shows that we have to prove that

$$\sigma_B(x + y) \subset \sigma_B(x) + \sigma_B(y) \quad \text{and} \quad \sigma_B(xy) \subset \sigma_B(x)\sigma_B(y).$$

Since B is commutative, $\sigma_B(z)$ is the range of the Gelfand transform $\hat{z}$, for every $z \in B$. (The Gelfand transforms are now functions on the maximal ideal space of B.) Since

$$(x + y)^\wedge = \hat{x} + \hat{y} \quad \text{and} \quad (xy)^\wedge = \hat{x}\hat{y},$$

we have the desired conclusion. ////

11.24 Definition Let A be an algebra with an involution. If $x \in A$ and $xx^* = x^*x$, then x is said to be *normal*. A set $S \subset A$ is said to be normal if S commutes and if $x^* \in S$ whenever $x \in S$.

11.25 Theorem *Suppose A is a Banach algebra with an involution, and B is a normal subset of A that is maximal with respect to being normal. Then*

(*a*) *B is a closed commutative subalgebra of A, and*
(*b*) $\sigma_B(x) = \sigma_A(x)$ *for every* $x \in B$.

Note that the involution is not assumed to be continuous but that B nevertheless turns out to be closed.

PROOF. We begin with a simple criterion for membership in B: *If* $x \in A$, *if* $xx^* = x^*x$, *and if* $xy = yx$ *for every* $y \in B$, *then* $x \in B$.

For if x satisfies these conditions, we also have $xy^* = y^*x$ for all $y \in B$, since B is normal, and therefore $x^*y = yx^*$. It follows that $B \cup \{x, x^*\}$ is normal. Hence $x \in B$, since B is maximal.

This criterion makes it clear that sums and products of members of B are in B. Thus B is a commutative algebra.

Suppose $x_n \in B$ and $x_n \to x$. Since $x_n y = yx_n$ for all $y \in B$, and multiplication is continuous, we have $xy = yx$ and therefore also

$$x^*y = (y^*x)^* = (xy^*)^* = yx^*.$$

In particular, $x^*x_n = x_n x^*$ for all n, which leads to $x^*x = xx^*$. Hence $x \in B$, by the above criterion. This proves that B is closed and completes (*a*).

Note also that $e \in B$. To prove (*b*), assume $x \in B$, $x^{-1} \in A$. Since x is normal, so is x^{-1}, and since x commutes with every $y \in B$, so does x^{-1}. Hence $x^{-1} \in B$. ////

Our first application of this is a generalization of Theorem 11.20:

11.26 Theorem *The word "commutative" may be dropped from the hypothesis of Theorem* 11.20.

PROOF. By Hausdorff's maximality theorem, the given hermitian (hence normal) $x \in A$ lies in some maximal normal set B. By Theorem 11.25 we can apply Theorem 11.20 with B in place of A. ////

Our next application of Theorem 11.25 will extend some consequences of Theorem 11.18 to arbitrary (not necessarily commutative) B^*-algebras.

11.27 Definition In a Banach algebra with involution, the statement "$x \geq 0$" means that $x = x^*$ and that $\sigma(x) \subset [0, \infty)$.

11.28 Theorem *Every B^*-algebra A has the following properties:*

(*a*) *Hermitian elements have real spectra.*
(*b*) *If $x \in A$ is normal, then $\rho(x) = \|x\|$.*
(*c*) *If $y \in A$, then $\rho(yy^*) = \|y\|^2$.*
(*d*) *If $u \in A$, $v \in A$, $u \geq 0$, and $v \geq 0$, then $u + v \geq 0$.*
(*e*) *If $y \in A$, then $yy^* \geq 0$.*
(*f*) *If $y \in A$, then $e + yy^*$ is invertible in A.*

PROOF. Every normal $x \in A$ lies in a maximal normal set $B \subset A$. By Theorems 11.18 and 11.25, B is a commutative B^*-algebra which is isometrically isomorphic to its Gelfand transform $\hat{B} = C(\Delta)$ and which has the property that

$$(1) \qquad \sigma(z) = \hat{z}(\Delta) \qquad (z \in B).$$

Here $\sigma(z)$ is the spectrum of z relative to A, Δ is the maximal ideal space of B, and $\hat{z}(\Delta)$ is the range of the Gelfand transform of z, regarded as an element of B.

If $x = x^*$, Theorem 11.18 shows that $\hat{x}$ is a real-valued function on Δ. Hence (1) implies (*a*).

For any normal x, (1) implies $\rho(x) = \|\hat{x}\|_\infty$. Also, $\|\hat{x}\|_\infty = \|x\|$, since B and $\hat{B}$ are isometric. This proves (*b*).

If $y \in A$, then yy^* is hermitian. Hence (*c*) follows from (*b*), since $\rho(yy^*) = \|yy^*\| = \|y\|^2$.

Suppose now that u and v are as in (*d*). Put $\alpha = \|u\|$, $\beta = \|v\|$, $w = u + v$, $\gamma = \alpha + \beta$. Then $\sigma(u) \subset [0, \alpha]$, so that

$$(2) \qquad \sigma(\alpha e - u) \subset [0, \alpha]$$

and (*b*) implies therefore that $\|\alpha e - u\| \leq \alpha$. For the same reason, $\|\beta e - v\| \leq \beta$. Hence

$$(3) \qquad \|\gamma e - w\| \leq \gamma.$$

Since $w = w^*$, (*a*) implies that $\sigma(\gamma e - w)$ is real. Thus

$$(4) \qquad \sigma(\gamma e - w) \subset [-\gamma, \gamma],$$

because of (3). But (4) implies that $\sigma(w) \subset [0, 2\gamma]$. Thus $w \geq 0$, and (*d*) is proved.

We turn to the proof of (*e*). Put $x = yy^*$. Then x is hermitian, and if B is chosen as in the first paragraph of this proof, then $\hat{x}$ is a real-valued function on Δ. By (1), we have to show that $\hat{x} \geq 0$ on Δ.

Since $\hat{B} = C(\Delta)$, there exists $z \in B$ such that

$$(5) \qquad \hat{z} = |\hat{x}| - \hat{x} \qquad \text{on } \Delta.$$

Then $z = z^*$, because $\hat{z}$ is real (Theorem 11.18). Put

$$(6) \qquad zy = w = u + iv,$$

where u and v are hermitian elements of A. Then

$$(7) \qquad ww^* = zyy^*z^* = zxz = z^2x$$

and therefore

$$(8) \qquad w^*w = 2u^2 + 2v^2 - ww^* = 2u^2 + 2v^2 - z^2x.$$

Since $u = u^*$, $\sigma(u)$ is real, by (*a*), hence $u^2 \geq 0$, by the spectral mapping theorem. Likewise $v^2 \geq 0$. By (5), $\hat{z}^2\hat{x} \leq 0$ on Δ. Since $z^2x \in B$, it follows from (1) that $-z^2x \geq 0$. Now (8) and (*d*) imply that $w^*w \geq 0$.

But $\sigma(ww^*) \subset \sigma(w^*w) \cup \{0\}$ (Exercise 4, Chapter 10). Hence $ww^* \geq 0$. By (7), this means that $\hat{z}^2\hat{x} \geq 0$ on Δ. By (5), this last inequality holds only when $\hat{x} = |\hat{x}|$. Thus $\hat{x} \geq 0$, and (*e*) is proved.

Finally, (*f*) is a corollary of (*e*). ////

Equality of spectra can now be proved in yet another situation, in which commutativity plays no role.

11.29 Theorem *Suppose A is a B^*-algebra, B is a closed subalgebra of A, $e \in B$, and $x^* \in B$ for every $x \in B$. Then $\sigma_A(x) = \sigma_B(x)$ for every $x \in B$.*

PROOF. Suppose $x \in B$ and x has an inverse in A. We have to show that $x^{-1} \in B$. Since x is invertible in A, so is x^*, hence also xx^*, and therefore $0 \notin \sigma_A(xx^*)$. By (*a*) of Theorem 11.28, $\sigma_A(xx^*) \subset (-\infty, \infty)$, so that $\sigma_A(xx^*)$ has a connected complement in $\mathbb{C}$. Theorem 10.18 shows now that $\sigma_B(xx^*) = \sigma_A(xx^*)$. Hence $(xx^*)^{-1} \in B$, and finally $x^{-1} = x^*(xx^*)^{-1} \in B$. ////

Positive Functionals

11.30 Definition A *positive functional* is a linear functional F on a Banach algebra A with an involution, that satisfies

$$F(xx^*) \geq 0$$

for every $x \in A$. Note that A is not assumed to be commutative and that continuity of F is not postulated. (The meaning of the term "positive" depends of course on the particular involution that is under consideration.)

11.31 Theorem *Every positive functional F on a Banach algebra A with involution has the following properties:*

(*a*) $F(x^*) = \overline{F(x)}$.

(*b*) $|F(xy^*)|^2 \le F(xx^*)F(yy^*)$.

(*c*) $|F(x)|^2 \le F(e)F(xx^*) \le F(e)^2\rho(xx^*)$.

(*d*) $|F(x)| \le F(e)\rho(x)$ *for every normal* $x \in A$.

(*e*) *F is a bounded linear functional on A. Moreover,* $\|F\| = F(e)$ *if A is commutative, and* $\|F\| \le \beta^{1/2}F(e)$ *if the involution satisfies* $\|x^*\| \le \beta\|x\|$ *for every* $x \in A$.

PROOF. If $x \in A$ and $y \in A$, put

$$(1)\qquad p = F(xx^*), \qquad q = F(yy^*), \qquad r = F(xy^*), \qquad s = F(yx^*).$$

Since $F[(x + \alpha y)(x^* + \bar{\alpha}y^*)] \ge 0$ for every $\alpha \in \mathcal{C}$,

$$(2)\qquad p + \bar{\alpha}r + \alpha s + |\alpha|^2 q \ge 0.$$

With $\alpha = 1$ and $\alpha = i$, (2) shows that $s = r$ and $i(s - r)$ are real. Hence $s = \bar{r}$. With $y = e$, this gives (*a*).

If $r = 0$, (*b*) is obvious. If $r \ne 0$, take $\alpha = tr/|r|$ in (2), where t is real. Then (2) becomes

$$(3)\qquad p + 2|r|t + qt^2 \ge 0 \qquad (-\infty < t < \infty),$$

so that $|r|^2 \le pq$. This proves (*b*).

Since $ee^* = e$, the first half of (*c*) is a special case of (*b*). For the second half, pick $t > \rho(xx^*)$. Then $\sigma(te - xx^*)$ lies in the open right half-plane. By Theorem 11.26, there exists $u \in A$, with $u = u^*$, such that $u^2 = te - xx^*$. Hence

$$(4)\qquad tF(e) - F(xx^*) = F(u^2) \ge 0.$$

It follows that

$$(5)\qquad F(xx^*) \le F(e)\rho(xx^*).$$

This completes part (*c*).

If x is normal, i.e., if $xx^* = x^*x$, Theorem 11.23 implies that $\sigma(xx^*) \subset \sigma(x)\sigma(x^*)$, so that

$$(6)\qquad \rho(xx^*) \le \rho(x)\rho(x^*) = \rho(x)^2.$$

Clearly, (*d*) follows from (6) and (*c*).

If A is commutative, then (*d*) holds for every $x \in A$, so that $\|F\| = F(e)$. If $\|x^*\| \le \beta\|x\|$, (*c*) implies $|F(x)| \le F(e)\beta^{1/2}\|x\|$, since $\rho(xx^*) \le \|x\|\|x^*\|$. This disposes of the special cases of part (*e*).

Before turning to the general case, we observe that $F(e) \ge 0$ and that $F(x) = 0$ for every $x \in A$ if $F(e) = 0$; this follows from (*c*). In the remainder of this proof we shall therefore assume, without loss of generality, that

$$F(e) = 1. \tag{7}$$

Let $\bar{H}$ be the closure of H, the set of all hermitian elements of A. Note that H and iH are *real* vector spaces and that $A = H + iH$, by Theorem 11.15. By (*d*), the restriction of F to H is a real-linear functional of norm 1, which therefore extends to a real-linear functional Φ on $\bar{H}$, also of norm 1. We claim that

$$\Phi(y) = 0 \qquad \text{if } y \in \bar{H} \cap i\bar{H}, \tag{8}$$

for if $y = \lim u_n = \lim (iv_n)$, where $u_n \in H$ and $v_n \in H$, then $u_n^2 \to y^2$, $v_n^2 \to -y^2$, so that (*c*) and (*d*) imply

$$|F(u_n)|^2 \le F(u_n^2) \le F(u_n^2 + v_n^2) \le \|u_n^2 + v_n^2\| \to 0. \tag{9}$$

Since $\Phi(y) = \lim F(u_n)$, (8) is proved.

By Theorem 5.20, there is a constant $\gamma < \infty$ such that every $x \in A$ has a representation

$$x = x_1 + ix_2, \quad x_1 \in \bar{H}, \quad x_2 \in \bar{H}, \qquad \|x_1\| + \|x_2\| \le \gamma\|x\|. \tag{10}$$

If $x = u + iv$, with $u \in H$, $v \in H$, then $x_1 - u$ and $x_2 - v$ lie in $\bar{H} \cap i\bar{H}$. Hence (8) yields

$$F(x) = F(u) + iF(v) = \Phi(x_1) + i\Phi(x_2), \tag{11}$$

so that

$$|F(x)| \le |\Phi(x_1)| + |\Phi(x_2)| \le \|x_1\| + \|x_2\| \le \gamma\|x\|. \tag{12}$$

This completes the proof. ////

Exercise 13 contains further information about part (*e*).

Examples of positive functionals—and a relation between them and positive measures—are furnished by the next theorem. It contains Bochner's classical theorem about positive-definite functions as a very special case. The identifications that lead from one to the other are indicated in Exercise 14.

11.32 Theorem *Suppose A is a commutative Banach algebra, with maximal ideal space Δ, and with an involution that is symmetric in the sense that*

$$h(x^*) = \overline{h(x)} \qquad (x \in A,\ h \in \Delta). \tag{1}$$

Let K be the set of all positive functionals F on A that satisfy $F(e) \le 1$.

Let M be the set of all positive regular Borel measures μ on Δ that satisfy $\mu(\Delta) \le 1$. Then the formula

$$(2) \qquad F(x) = \int_\Delta \hat{x}\, d\mu \qquad (x \in A)$$

establishes a one-to-one correspondence between the convex sets K and M, which carries extreme points to extreme points.

Consequently, the multiplicative linear functionals on A are precisely the extreme points of K.

PROOF. If $\mu \in M$ and F is defined by (2), then F is obviously linear, and $F(xx^*) = \int |\hat{x}|^2\, d\mu \ge 0$, because (1) implies that $(xx^*)^\wedge = |\hat{x}|^2$. Since $F(e) = \mu(\Delta)$, $F \in K$.

If $F \in K$, then F vanishes on the radical of A, by (*d*) of Theorem 11.31. Hence there is a functional $\hat{F}$ on $\hat{A}$ that satisfies $\hat{F}(\hat{x}) = F(x)$ for all $x \in A$. In fact,

$$(3) \qquad |\hat{F}(\hat{x})| = |F(x)| \le F(e)\rho(x) = F(e)\|\hat{x}\|_\infty \qquad (x \in A).$$

by (*d*) of Theorem 11.31. It follows that $\hat{F}$ is a linear functional of norm $F(e)$ on the subspace $\hat{A}$ of $C(\Delta)$. This extends to a functional on $C(\Delta)$, with the same norm, and now the Riesz representation theorem furnishes a regular Borel measure μ, with $\|\mu\| = F(e)$, that satisfies (2). Since

$$(4) \qquad \mu(\Delta) = \int_\Delta \hat{e}\, d\mu = F(e) = \|\mu\|,$$

we see that $\mu \ge 0$. Thus $\mu \in M$.

By (1), $\hat{A}$ satisfies the hypotheses of the Stone-Weierstrass theorem and is therefore dense in $C(\Delta)$. This implies that μ is uniquely determined by F.

One extreme point of M is 0; the others are unit masses concentrated at points $h \in \Delta$. Since every complex homomorphism of A has the form $x \to \hat{x}(h)$, for some $h \in \Delta$, the proof is complete. ////

We conclude by showing that the extreme points of K are multiplicative even if (1) is not satisfied.

11.33 Theorem *Let K be the set of all positive functionals F on a commutative Banach algebra A with an involution that satisfy $F(e) \le 1$. If $F \in K$, then each of the following three properties implies the other two:*

(*a*) *$F(xy) = F(x)F(y)$ for all x and $y \in A$.*

(*b*) $F(xx^*) = F(x)F(x^*)$ *for every* $x \in A$.

(*c*) F *is an extreme point of* K.

PROOF. It is trivial that (*a*) implies (*b*). Suppose (*b*) holds. With $x = e$, (*b*) shows that $F(e) = F(e)^2$, and so $F(e) = 0$ or $F(e) = 1$. When $F(e) = 0$, then $F = 0$, by (*c*) of Theorem 11.31, and so F is an extreme point of K. Assume $F(e) = 1$, and $2F = F_1 + F_2$, $F_1 \in K$, $F_2 \in K$. We have to show that $F_1 = F$. Clearly, $F_1(e) = 1 = F(e)$. If $x \in A$ is such that $F(x) = 0$, then

$$|F_1(x)|^2 \leq F_1(xx^*) \leq 2F(xx^*) = 2F(x)F(x^*) = 0, \tag{1}$$

by (*b*) and Theorem 11.31. Thus F_1 coincides with F on the null space of F and at e. It follows that $F_1 = F$. Hence (*b*) implies (*c*).

To show that (*c*) implies (*a*), let F be an extreme point of K. Either $F(e) = 0$, in which case there is nothing to prove, or $F(e) = 1$. We shall first prove a special case of (*a*), namely,

$$F(xx^*y) = F(xx^*)F(y) \qquad (x \in A,\ y \in A). \tag{2}$$

Choose x so that $\|xx^*\| < 1$. By Theorem 11.20, there exists $z \in A$, $z = z^*$, such that $z^2 = e - xx^*$. Define

$$\Phi(y) = F(xx^*y) \qquad (y \in A). \tag{3}$$

Then

$$\Phi(yy^*) = F(xx^*yy^*) = F[(xy)(xy)^*] \geq 0, \tag{4}$$

and also

$$(F - \Phi)(yy^*) = F[(e - xx^*)yy^*] = F(z^2yy^*) = F[(yz)(yz)^*] \geq 0. \tag{5}$$

Since

$$0 \leq \Phi(e) = F(xx^*) \leq F(e)\|xx^*\| < 1, \tag{6}$$

(4) and (5) show that both Φ and $F - \Phi$ are in K. If $\Phi(e) = 0$, then $\Phi = 0$. If $\Phi(e) > 0$, (6) shows that

$$F = \Phi(e) \cdot \frac{\Phi}{\Phi(e)} + (F - \Phi)(e) \cdot \frac{F - \Phi}{F(e) - \Phi(e)}, \tag{7}$$

a convex combination of members of K. Since F is extreme, we conclude that

$$\Phi = \Phi(e)F. \tag{8}$$

Now (2) follows from (8) and (3).

Finally, the passage from (2) to (*a*) is accomplished by any of the following identities, which are satisfied by every involution:

If $n = 3, 4, 5, \ldots$, *if* $\omega = \exp(2\pi i/n)$, *if* $x \in A$, *and if* $z_p = e + \omega^{-p}x$, *then*

$$x = \frac{1}{n}\sum_{p=1}^{n} \omega^p z_p z_p^*. \tag{9}$$

The proof of (9) is a straightforward computation which uses the fact that

$$\sum_{p=1}^{n} \omega^p = \sum_{p=1}^{n} \omega^{2p} = 0. \tag{10}$$ ////

Exercises

1. Prove Proposition 11.2.
2. State and prove an analogue of Wiener's lemma 11.6 for power series that converge absolutely on the closed unit disc.
3. If X is a compact Hausdorff space, show that there is a natural one-to-one correspondence between closed subsets of X and closed ideals of $C(X)$.
4. Prove that the polynomials are dense in the polydisc algebra $A(U^n)$. (See Theorem 11.7.) *Suggestion:* If $f \in A(U^n)$, $0 < r < 1$, and f_r is defined by $f_r(z) = f(rz)$, then f_r is the sum of an absolutely (hence uniformly) convergent multiple power series on $\bar{U}^n$.
5. Suppose A is a commutative Banach algebra, $x \in A$, and f is holomorphic in some open set $\Omega \subset \mathcal{C}$ that contains the range of $\hat{x}$. Prove that there exists $y \in A$ such that $\hat{y} = f \circ \hat{x}$, that is, such that $h(y) = f(h(x))$ for every complex homomorphism h of A. Prove that y is uniquely determined by x and f if A is semisimple.
6. Suppose A and B are commutative Banach algebras, B is semisimple, $\psi: A \to B$ is a homomorphism whose range is dense in B, and $\alpha: \Delta_B \to \Delta_A$ is defined by

   $$(\alpha h)(x) = h(\psi(x)) \qquad (x \in A, h \in \Delta_B).$$

   Prove that α is a homeomorphism of Δ_B onto a compact subset of Δ_A. [The fact that $\psi(A)$ is dense in B implies that α is one-to-one and that the topology of Δ_B is the weak topology induced by the Gelfand transforms of the elements $\psi(x)$, for $x \in A$.]

   Let A be the disc algebra, let $B = C(K)$, where K is an arc in the unit disc, and let ψ be the restriction mapping of A into B. This example shows that $\alpha(\Delta_B)$ may be a proper subset of Δ_A, even if ψ is one-to-one.

   Find an example in which $\psi(A) = B$ but $\alpha(\Delta_B) \neq \Delta_A$.
7. In Example 11.13(*b*) it was asserted that $\hat{A} \neq C(\Delta)$. Find several proofs of this.
8. Which properties of Lebesgue measure are used in Example 11.13(*f*)? Can Lebesgue measure be replaced by any positive measure, without changing any of the results?

   Supply the details for the last paragraph in Example 11.13(*f*).
9. Let C' be the algebra of all continuously differentiable complex functions on the unit interval $[0, 1]$, with pointwise multiplication, normed by

$$\|f\| = \|f\|_\infty + \|f'\|_\infty.$$

(a) Show that C' is a semisimple commutative Banach algebra. Find its maximal ideal space.

(b) Fix p, $0 \le p \le 1$; let J be the set of all $f \in C'$ for which $f(p) = f'(p) = 0$. Show that J is a closed ideal in C' and that C'/J is a two-dimensional algebra which has a one-dimensional radical. (This gives an example of a semisimple algebra with a quotient algebra that is not semisimple.) To which of the two algebras described in Exercise 5 of Chapter 10 is C'/J isomorphic?

10. Let A be the disc algebra. Associate to each $f \in A$ a function $f^* \in A$ by the formula

$$f^*(z) = \overline{f(\bar{z})}.$$

Then $f \to f^*$ is an involution on A.

(a) Does this involution turn A into a B^*-algebra?

(b) Does $\sigma(ff^*)$ always lie in the real axis?

(c) Which complex homomorphisms of A are positive functionals, with respect to this involution?

(d) If μ is a positive finite Borel measure on $[-1, 1]$, then

$$f \to \int_{-1}^{1} f(t)\, d\mu(t)$$

is a positive functional on A. Are there any others?

11. Show that commuting idempotents have distance ≥ 1. Explicitly, if $x^2 = x$, $y^2 = y$, $xy = yx$ for some x and y in a Banach algebra, then either $x = y$ or $\|x - y\| \ge 1$. Show that this may fail if $xy \ne yx$.

12. If $xy = yx$ for some x and y in a Banach algebra, prove that

$$\rho(xy) \le \rho(x)\rho(y) \qquad \text{and} \qquad \rho(x + y) \le \rho(x) + \rho(y).$$

13. Let t be a large positive number, and define a norm on $\mathbb{C}^2$ by

$$\|w\| = |w_1| + t|w_2| \qquad \text{if } w = (w_1, w_2).$$

Let A be the algebra of all complex 2-by-2 matrices, with the corresponding operator norm:

$$\|y\| = \max\{\|y(w)\| : \|w\| = 1\} \qquad (y \in A).$$

For $y \in A$, let y^* be the conjugate transpose of y. Consider a fixed $x \in A$, namely,

$$x = \begin{pmatrix} 0 & t^2 \\ 1 & 0 \end{pmatrix}.$$

Prove the following statements.

(a) $\|x(w)\| = t\|w\|$; hence $\|x\| = t$.

(b) $\sigma(x) = \{t, -t\} = \sigma(x^*)$.

(c) $\sigma(xx^*) = (1, t^4\} = \sigma(x^*x)$.

(d) $\sigma(x + x^*) = \{1 + t^2, -1 - t^2\}$.

(e) Therefore commutativity is required in Theorem 11.23 and in Exercise 12.
(f) If $F(y)$ is the sum of the four entries in y, for $y \in A$, then F is a positive functional on A.
(g) The equality $\|F\| = F(e)$ [see (e) of Theorem 11.31] does not hold, because $F(e) = 2$ and $F(x) = 1 + t^2$, so that $\|F\| > t$.
(h) If K is the set of all positive functionals f on A that satisfy $f(e) \le 1$ (as in Theorem 11.33), then K has many extreme points, although 0 is the only multiplicative linear functional on A. Commutativity is therefore required in the implication (c) → (a) of Theorem 11.33.

14. A complex function ϕ, defined on R^n, is said to be *positive-definite* if

$$\sum_{i,j=1}^{r} c_i \bar{c}_j \phi(x_i - x_j) \ge 0$$

for every choice of $x_1, \ldots, x_r$, in R^n and for every choice of complex numbers $c_1, \ldots, c_r$.

(a) Show that $|\phi(x)| \le \phi(0)$ for every $x \in R^n$.
(b) Show that the Fourier transform of every finite positive Borel measure on R^n is positive-definite.
(c) Complete the following outline of the converse of (b) (Bochner's theorem): *If ϕ is continuous and positive-definite, then ϕ is the Fourier transform of a finite positive Borel measure.*

Let A be the convolution algebra $L^1(R^n)$, with a unit attached, as described in (d) of Section 10.3 and (e) of Section 11.13. Define $\tilde{f}(x) = \overline{f(-x)}$. Show that

$$f + \alpha\delta \to \tilde{f} + \bar{\alpha}\delta$$

is an involution on A and that

$$f + \alpha\delta \to \int_{R^n} f\phi \, dm_n + \alpha\phi(0)$$

is a positive functional on A. By Theorem 11.32 and (e) of Section 11.13, there is a positive measure μ on the one-point compactification Δ of R^n, such that

$$\int_{R^n} f\phi \, dm_n + \alpha\phi(0) = \int_{\Delta} (\hat{f} + \alpha) \, d\mu.$$

If σ is the restriction of μ to R^n, it follows that

$$\int_{R^n} f\phi \, dm_n = \int_{R^n} \hat{f} \, d\sigma$$

for every $f \in L^1(R^n)$. Hence $\phi = \hat{\sigma}$. (Actually, μ is already concentrated on R^n, so that $\sigma = \mu$.)

(d) Let P be the set of all continuous positive-definite functions ϕ on R^n that satisfy $\phi(0) \le 1$. Find all extreme points of this convex set.

15. Let Δ be the maximal ideal space of a commutative Banach algebra A. Call a closed set $\beta \subset \Delta$ an *A-boundary* if the maximum of $|\hat{x}|$ on Δ equals its maximum on β, for every $x \in A$. (Trivially, Δ is an A-boundary.)

Prove that the intersection ∂_A of all A-boundaries is an A-boundary.

∂_A is called the *Shilov boundary* of A. The terminology is suggested by the maximum modulus property of holomorphic functions. For instance, when A is the disc algebra, then ∂_A is the unit circle, which is the topological boundary of Δ, the closed unit disc.

Outline of proof: Show first that there is an A-boundary β_0 which is minimal in the sense that no proper subset of β_0 is an A-boundary. (Partially order the collection of A-boundaries by set inclusion, etc.) Then pick $h_0 \in \beta_0$, pick $x_1, \ldots, x_n \in A$ with $\hat{x}_i(h_0) = 0$, and put

$$V = \{h \in \Delta : |\hat{x}_i(h)| < 1 \text{ for } 1 \le i \le n\}.$$

Since β_0 is minimal, there exists $x \in A$ with $\|\hat{x}\|_\infty = 1$ and $|\hat{x}(h)| < 1$ on $\beta_0 - V$. If $y = x^m$ and m is sufficiently large, then $|\hat{x}_i \hat{y}| < 1$ on β_0, for all i. Hence $\|\hat{x}_i \hat{y}\|_\infty < 1$. Conclude from this first that $|\hat{y}(h)| = \|\hat{y}\|_\infty$ only in V, hence that V intersects every A-boundary β, and finally that $h_0 \in \beta$. Thus $\beta_0 \subset \beta$, and $\beta_0 = \partial_A$.

16. Suppose A is a Banach algebra, m is an integer, $m \ge 2$, $K < \infty$, and

$$\|x\|^m \le K\|x^m\|$$

for every $x \in A$. Show that there exist constants $K_n < \infty$, for $n = 1, 2, 3, \ldots$, such that

$$\|x\|^n \le K_n\|x^n\| \qquad (x \in A).$$

(This extends Theorem 11.12.)

17. Suppose $\{\omega_n\}$ $(-\infty < n < \infty)$ are positive numbers such that $\omega_0 = 1$ and

$$\omega_{m+n} \le \omega_m \omega_n$$

for all integers m and n. Let $A = A\{\omega_n\}$ be the set of all complex functions f on the integers for which the norm

$$\|f\| = \sum_{-\infty}^{\infty} |f(n)|\omega_n$$

is finite. Define multiplication in A by

$$(f * g)(n) = \sum_{k=-\infty}^{\infty} f(n-k)g(k).$$

(*a*) Show that each $A\{\omega_n\}$ is a commutative Banach algebra.

(*b*) Show that $R_+ = \lim_{n\to\infty} (\omega_n)^{1/n}$ exists and is finite, by showing that $R_+ = \inf_{n\ge 0} (\omega_n)^{1/n}$.

(*c*) Show similarly that $R_- = \lim_{n\to\infty} (\omega_{-n})^{-1/n}$ exists and that $R_- \le R_+$.

(*d*) Put $\Delta = \{\lambda \in \mathbb{C} : R_- \le |\lambda| \le R_+\}$. Show that Δ can be identified with the maximal ideal space of $A\{\omega_n\}$ and that the Gelfand transforms are absolutely convergent Laurent series on Δ.

(*e*) Consider the following choices for $\{\omega_n\}$:

(*i*) $\omega_n = 1$.

(*ii*) $\omega_n = 2^n$.

(*iii*) $\omega_n = 2^n$ if $n \geq 0$, $\omega_n = 1$ if $n < 0$.

(*iv*) $\omega_n = 1 + 2n^2$.

(*v*) $\omega_n = 1 + 2n^2$ if $n \geq 0$, $\omega_n = 1$ if $n < 0$.

For which of these is Δ a circle? For which choices is $A\{\omega_n\}$ self-adjoint, in the sense that $\hat{A}$ is closed under complex conjugation?

(*f*) Is $A\{\omega_n\}$ always semisimple?

(*g*) Is there an $A\{\omega_n\}$, with Δ the unit circle, such that $\hat{A}$ consists entirely of infinitely differentiable functions?

18. Let Δ be the maximal ideal space of $L^\infty(m)$, as in Section 11.13. Show that

(*a*) Δ has no isolated point, and

(*b*) Δ contains no convergent sequence of distinct points. *Hint:* If $p_1, p_2, p_3, \ldots,$ are distinct points in Δ, none of which is a limit point of the others, and if $\{w_i\}$ is any bounded sequence of numbers, then there exist pairwise disjoint open sets V_i in Δ such that $p_i \in V_i$, and there exists a function $\varphi \in C(\Delta)$ such that $\varphi = w_i$ on V_i.

19. Let $L^\infty(m)$ be as above and prove: If $f_n \in L^\infty(m)$ and $f_n \to 0$ in the weak topology of $L^\infty(m)$, then $\int_0^1 |f_n|^p \, dm \to 0$ for every $p \in (0, \infty)$. Show, by constructing an example, that the converse is false.

20. Prove the following partial converse of Theorem 11.31: If F is a bounded linear functional on a B^*-algebra A, and $\|F\| = F(0) = 1$, then F is positive.

Suggestion: Choose $x \in A$, $\|x\| \leq 1$, put $F(xx^*) = \alpha + \beta i$,

$$y_t = xx^* - (\tfrac{1}{2} + it)e,$$

for $-\infty < t < \infty$. Use Theorem 11.28 to show that $\sigma(xx^*) \subset [0, 1]$ and that therefore

$$|F(y_t)| \leq \|y_t\| = \rho(y_t) \leq |\tfrac{1}{2} + it|.$$

Proceed as in Lemma 5.26.

21. In $\mathcal{C}^2$, let K_1 consist of all points $(e^{i\theta}, e^{-i\theta})$, and let K_2 consist of the points $(e^{i\theta}, e^{i\theta})$, $0 \leq \theta \leq 2\pi$. Of these circles, show that K_1 is polynomially convex but K_2 is not. How about $K_3 = \{(\cos\theta, \sin\theta) : 0 \leq \theta \leq 2\pi\}$?

22. Show that a 3-by-3 matrix M commutes with

$$\begin{bmatrix} 0 & 0 & 1 \\ 0 & 0 & 0 \\ 0 & 0 & 0 \end{bmatrix} \quad \text{if and only if} \quad M = \begin{bmatrix} a & x & y \\ 0 & z & w \\ 0 & 0 & a \end{bmatrix}.$$

Deduce from this that centralizers (see Section 11.21) need not be commutative.

CHAPTER 12

BOUNDED OPERATORS ON A HILBERT SPACE

Basic Facts

12.1 Definitions A complex vector space H is called an *inner product space* (or *unitary space*) if to each ordered pair of vectors x and y in H is associated a complex number (x, y), called the *inner product* or *scalar product* of x and y, such that the following rules hold:

(*a*) $(y, x) = \overline{(x, y)}$. (The bar denotes complex conjugation.)
(*b*) $(x + y, z) = (x, z) + (y, z)$.
(*c*) $(\alpha x, y) = \alpha(x, y)$ if $x \in H$, $y \in H$, $\alpha \in \mathcal{C}$.
(*d*) $(x, x) \geq 0$ for all $x \in H$.
(*e*) $(x, x) = 0$ only if $x = 0$.

For fixed y, (x, y) is therefore a linear function of x. For fixed x, it is a conjugate-linear function of y. Such functions of two variables are sometimes called *sesquilinear*.

If $(x, y) = 0$, x is said to be *orthogonal* to y, and the notation $x \perp y$ is sometimes used. Since $(x, y) = 0$ implies $(y, x) = 0$, the relation $\perp$ is symmetric. If $E \subset H$ and $F \subset H$, the notation $E \perp F$ means that $x \perp y$ whenever $x \in E$ and $y \in F$. Also, $E^\perp$ is the set of all $y \in H$ that are orthogonal to every $x \in E$.

Every inner product space can be normed by defining

$$\|x\| = (x, x)^{1/2}.$$

Theorem 12.2 implies this. If the resulting normed space is complete, it is called a *Hilbert space.*

12.2 Theorem *If $x \in H$ and $y \in H$, where H is an inner product space, then*

$$(1) \qquad |(x, y)| \le \|x\| \|y\|$$

and

$$(2) \qquad \|x + y\| \le \|x\| + \|y\|.$$

Moreover

$$(3) \qquad \|y\| \le \|\lambda x + y\| \qquad \textit{for every } \lambda \in \mathbb{C}$$

if and only if $x \perp y$.

PROOF. Put $\alpha = (x, y)$. A simple computation gives

$$(4) \qquad 0 \le \|\lambda x + y\|^2 = |\lambda|^2 \|x\|^2 + 2 \operatorname{Re}(\alpha\lambda) + \|y\|^2.$$

Hence (3) holds if $\alpha = 0$. If $x = 0$, (1) and (3) are obvious. If $x \neq 0$, take $\lambda = -\bar{\alpha}/\|x\|^2$. With this λ, (4) becomes

$$(5) \qquad 0 \le \|\lambda x + y\|^2 = \|y\|^2 - \frac{|\alpha|^2}{\|x\|^2}.$$

This proves (1) and shows that (3) is false when $\alpha \neq 0$. By squaring both sides of (2), one sees that (2) is a consequence of (1). ////

Note: Unless the contrary is explicitly stated, the letter H will from now on denote a Hilbert space.

12.3 Theorem *Every nonempty closed convex set $E \subset H$ contains a unique x of minimal norm.*

PROOF. The parallelogram law

$$(1) \qquad \|x + y\|^2 + \|x - y\|^2 = 2\|x\|^2 + 2\|y\|^2 \qquad (x \in H, y \in H)$$

follows directly from the definition $\|x\|^2 = (x, x)$. Put

$$(2) \qquad d = \inf \{\|x\| : x \in E\}.$$

Choose $x_n \in E$ so that $\|x_n\| \to d$. Since $\frac{1}{2}(x_n + x_m) \in E$, $\|x_n + x_m\|^2 \geq 4d^2$. If x and y are replaced by x_n and x_m in (1), the right side of (1) tends to $4d^2$. Hence (1) implies that $\{x_n\}$ is a Cauchy sequence in H, which therefore converges to some $x \in E$, with $\|x\| = d$.

If $y \in E$ and $\|y\| = d$, the sequence $\{x, y, x, y, \ldots\}$ must converge, as we just saw. Hence $y = x$. ////

12.4 Theorem *If M is a closed subspace of H, then*

$$H = M \oplus M^{\perp}$$

The conclusion is, more explicitly, that M and $M^{\perp}$ are closed subspaces of H whose intersection is $\{0\}$ and whose sum is H. The space $M^{\perp}$ is called the *orthogonal complement* of M.

PROOF. If $E \subset H$, the linearity of (x, y) as a function of x shows that $E^{\perp}$ is a subspace of H, and the Schwarz inequality (1) of Theorem 12.2 implies then that $E^{\perp}$ is closed.

If $x \in M$ and $x \in M^{\perp}$, then $(x, x) = 0$; hence $x = 0$. Thus $M \cap M^{\perp} = \{0\}$.

If $x \in H$, apply Theorem 12.3 to the set $x - M$ to conclude that there exists $x_1 \in M$ that minimizes $\|x - x_1\|$. Put $x_2 = x - x_1$. Then $\|x_2\| \leq \|x_2 + y\|$ for all $y \in M$. Hence $x_2 \in M^{\perp}$, by Theorem 12.2. Since $x = x_1 + x_2$, we have shown that $M + M^{\perp} = H$. ////

Corollary. *If M is a closed subspace of H, then*

$$(M^{\perp})^{\perp} = M.$$

PROOF. The inclusion $M \subset (M^{\perp})^{\perp}$ is obvious. Since

$$M \oplus M^{\perp} = H = M^{\perp} \oplus (M^{\perp})^{\perp},$$

M cannot be a proper subspace of $(M^{\perp})^{\perp}$. ////

We now describe the dual space H^* of H.

12.5 Theorem *There is a conjugate-linear isometry $y \to \Lambda$ of H onto H^*, given by*

(1) $$\Lambda x = (x, y) \qquad (x \in H).$$

PROOF. If $y \in H$ and Λ is defined by (1), the Schwarz inequality (1) of Theorem 12.2 shows that $\Lambda \in H^*$ and that $\|\Lambda\| \leq \|y\|$. Since

(2) $$\|y\|^2 = (y, y) = \Lambda y \leq \|\Lambda\| \|y\|,$$

it follows that $\|\Lambda\| = \|y\|$.

It remains to be shown that every $\Lambda \in H^*$ has the form (1).

If $\Lambda = 0$, take $y = 0$. If $\Lambda \neq 0$, let $\mathscr{N}(\Lambda)$ be the null space of Λ. By Theorem 12.4 there exists $z \in \mathscr{N}(\Lambda)^\perp$, $z \neq 0$. Since

$$(3) \qquad (\Lambda x)z - (\Lambda z)x \in \mathscr{N}(\Lambda) \qquad (x \in H),$$

it follows that $(\Lambda x)(z, z) - (\Lambda z)(x, z) = 0$. Hence (1) holds with $y = (z, z)^{-1}\overline{(\Lambda z)}z$. ////

12.6 Theorem *If $\{x_n\}$ is a sequence of pairwise orthogonal vectors in H, then each of the following three statements implies the other two.*

(*a*) $\sum_{n=1}^{\infty} x_n$ *converges, in the norm topology of H.*

(*b*) $\sum_{n=1}^{\infty} \|x_n\|^2 < \infty$.

(*c*) $\sum_{n=1}^{\infty} (x_n, y)$ *converges, for every $y \in H$.*

Thus strong convergence (*a*) and weak convergence (*c*) are equivalent for series of orthogonal vectors.

PROOF. Since $(x_i, x_j) = 0$ if $i \neq j$, the equality

$$(1) \qquad \|x_n + \cdots + x_m\|^2 = \|x_n\|^2 + \cdots + \|x_m\|^2$$

holds whenever $n \le m$. Hence (*b*) implies that the partial sums of $\sum x_n$ form a Cauchy sequence in H. Since H is complete, (*b*) implies (*a*). The Schwarz inequality shows that (*a*) implies (*c*). Finally, assume that (*c*) holds. Define $\Lambda_n \in H^*$ by

$$(2) \qquad \Lambda_n y = \sum_{i=1}^{n} (y, x_i) \qquad (y \in H, n = 1, 2, 3, \ldots).$$

By (*c*), $\{\Lambda_n y\}$ converges for every $y \in H$; hence $\{\|\Lambda_n\|\}$ is bounded, by the Banach-Steinhaus theorem. But

$$(3) \qquad \|\Lambda_n\| = \|x_1 + \cdots + x_n\| = \{\|x_1\|^2 + \cdots + \|x_n\|^2\}^{1/2}.$$

Hence (*c*) implies (*b*). ////

Bounded Operators

In conformity with notations used earlier, $\mathscr{B}(H)$ will now denote the Banach algebra of all bounded linear operators T on a Hilbert space $H \neq \{0\}$, normed by

$$\|T\| = \sup \{\|Tx\|: x \in H, \|x\| \le 1\}.$$

We shall see that $\mathscr{B}(H)$ has an involution which makes it into a B^*-algebra.

We begin with a simple but useful uniqueness theorem.

12.7 Theorem *If $T \in \mathscr{B}(H)$ and if $(Tx, x) = 0$ for every $x \in H$, then $T = 0$.*

PROOF. Since $(T(x + y), x + y) = 0$, we see that

(1) $$(Tx, y) + (Ty, x) = 0 \qquad (x \in H, y \in H).$$

If y is replaced by iy in (1), the result is

(2) $$-i(Tx, y) + i(Ty, x) = 0 \qquad (x \in H, y \in H).$$

Multiply (2) by i and add to (1), to obtain

(3) $$(Tx, y) = 0 \qquad (x \in H, y \in H).$$

With $y = Tx$, (3) gives $\|Tx\|^2 = 0$. Hence $Tx = 0$. ////

Corollary. *If $S \in \mathscr{B}(H)$, $T \in \mathscr{B}(H)$, and*

$$(Sx, x) = (Tx, x)$$

for every $x \in H$, then $S = T$.

PROOF. Apply the theorem to $S - T$. ////

Note that Theorem 12.7 would fail if the scalar field were R. To see this, consider rotations in R^2.

12.8 Theorem *If $f: H \times H \to \mathbb{C}$ is sesquilinear and bounded, in the sense that*

(1) $$M = \sup \{|f(x, y)| : \|x\| = \|y\| = 1\} < \infty,$$

then there exists a unique $S \in \mathscr{B}(H)$ that satisfies

(2) $$f(x, y) = (x, Sy) \qquad (x \in H, y \in H).$$

Moreover, $\|S\| = M$.

PROOF. Since $|f(x, y)| \le M\|x\| \|y\|$, the mapping

$$x \to f(x, y)$$

is, for each $y \in H$, a bounded linear functional on H, of norm at most $M\|y\|$. It now follows from Theorem 12.5 that to each $y \in H$ corresponds a unique element $Sy \in H$ such that (2) holds; also, $\|Sy\| \le M\|y\|$. It is clear that $S: H \to H$ is additive. If $\alpha \in \mathcal{C}$, then

$$(x, S(\alpha y)) = f(x, \alpha y) = \bar{\alpha} f(x, y) = \bar{\alpha}(x, Sy) = (x, \alpha Sy)$$

for all x and y in H. It follows that S is linear. Hence $S \in \mathscr{B}(H)$, and $\|S\| \le M$.

But we also have

$$|f(x, y)| = |(x, Sy)| \le \|x\| \|Sy\| \le \|x\| \|S\| \|y\|,$$

which gives the opposite inequality $M \le \|S\|$. ////

12.9 Adjoints If $T \in \mathscr{B}(H)$, then (Tx, y) is linear in x, conjugate-linear in y, and bounded. Theorem 12.8 shows therefore that there exists a unique $T^* \in \mathscr{B}(H)$ for which

(1) $$(Tx, y) = (x, T^*y) \qquad (x \in H, y \in H)$$

and also that

(2) $$\|T^*\| = \|T\|.$$

We claim that $T \to T^*$ is an involution on $\mathscr{B}(H)$, that is, that the following four properties hold:

(3) $$(T + S)^* = T^* + S^*.$$

(4) $$(\alpha T)^* = \bar{\alpha} T^*.$$

(5) $$(ST)^* = T^*S^*.$$

(6) $$T^{**} = T.$$

Of these, (3) is obvious. The computations

$$(\alpha Tx, y) = \alpha(Tx, y) = \alpha(x, T^*y) = (x, \bar{\alpha}T^*y),$$
$$(STx, y) = (Tx, S^*y) = (x, T^*S^*y),$$
$$(Tx, y) = \overline{(T^*y, x)} = \overline{(y, T^{**}x)} = (T^{**}x, y)$$

give (4), (5), and (6). Since

$$\|Tx\|^2 = (Tx, Tx) = (T^*Tx, x) \le \|T^*T\| \|x\|^2$$

for every $x \in H$, we have $\|T\|^2 \le \|T^*T\|$. On the other hand, (2) gives

$$\|T^*T\| \le \|T^*\| \|T\| = \|T\|^2.$$

Hence the equality

(7) $$\|T^*T\| = \|T\|^2$$

holds for every $T \in \mathscr{B}(H)$.

We have thus proved that $\mathscr{B}(H)$ is a B^*-algebra, relative to the involution $T \to T^*$ defined by (1).

Note: In the preceding setting, T^* is sometimes called the *Hilbert space adjoint* of T, to distinguish it from the Banach space adjoint that was discussed in Chapter 4. The only difference between the two is that in the Hilbert space setting $T \to T^*$ is conjugate-linear instead of linear. This is due to the conjugate-linear nature of the isometry described in Theorem 12.5. If T^* were regarded as an operator on H^* rather than on H, we would be exactly in the situation of Chapter 4.

12.10 Theorem *If $T \in \mathscr{B}(H)$, then*

$$\mathscr{N}(T^*) = \mathscr{R}(T)^\perp \quad \text{and} \quad \mathscr{N}(T) = \mathscr{R}(T^*)^\perp.$$

We recall that $\mathscr{N}(T)$ and $\mathscr{R}(T)$ denote the null space and range of T, respectively.

PROOF. Each of the following four statements is clearly equivalent to the one that follows and/or precedes it.

(1) $$T^*y = 0.$$

(2) $$(x, T^*y) = 0 \text{ for every } x \in H.$$

(3) $$(Tx, y) = 0 \text{ for every } x \in H.$$

(4) $$y \in \mathscr{R}(T)^\perp.$$

Thus $\mathscr{N}(T^*) = \mathscr{R}(T)^\perp$. Since $T^{**} = T$, the second assertion follows from the first if T is replaced by T^*. ////

12.11 Definition An operator $T \in \mathscr{B}(H)$ is said to be

(*a*) *normal* if $TT^* = T^*T$,
(*b*) *self-adjoint* (or *hermitian*) if $T^* = T$,
(*c*) *unitary* if $T^*T = I = TT^*$, where I is the identity operator on H,
(*d*) a *projection* if $T^2 = T$.

It is clear that self-adjoint operators and unitary operators are normal. Most of the theorems obtained in this chapter will be about normal operators.

This algebraic requirement, namely, that T should commute with its adjoint, has remarkably strong analytic and geometric consequences.

12.12 Theorem *An operator $T \in \mathscr{B}(H)$ is normal if and only if*

(1) $$\|Tx\| = \|T^*x\|$$

for every $x \in H$. Normal operators T have the following properties:

(a) $\mathscr{N}(T) = \mathscr{N}(T^*)$.
(b) $\mathscr{R}(T)$ *is dense in H if and only if T is one-to-one.*
(c) *T is invertible if and only if there exists* $\delta > 0$ *such that* $\|Tx\| \geq \delta\|x\|$ *for every* $x \in H$.
(d) *If* $Tx = \alpha x$ *for some* $x \in H$, $\alpha \in \mathbb{C}$, *then* $T^*x = \bar{\alpha}x$.
(e) *If α and β are distinct eigenvalues of T, then the corresponding eigenspaces are orthogonal to each other.*

PROOF. The equalities

$$\|Tx\|^2 = (Tx, Tx) = (T^*Tx, x),$$
$$\|T^*x\|^2 = (T^*x, T^*x) = (TT^*x, x),$$

combined with the corollary to Theorem 12.7, prove the first statement, and (a) is an immediate consequence. Since $\mathscr{R}(T)^\perp = \mathscr{N}(T^*)$, (a) implies (b). If there is a $\delta > 0$ as in (c), then $\mathscr{R}(T)$ is closed, by Theorem 1.26, and is dense, by (b); hence $\mathscr{R}(T) = H$ and T is invertible. The converse follows from the open mapping theorem. To obtain (d), apply (a) to $T - \alpha I$ in place of T. Finally, if $Tx = \alpha x$ and $Ty = \beta y$, then (d) shows that

$$\alpha(x, y) = (\alpha x, y) = (Tx, y) = (x, T^*y) = (x, \bar{\beta}y) = \beta(x, y).$$

Since $\alpha \neq \beta$, we conclude that $x \perp y$. ////

12.13 Theorem *If* $U \in \mathscr{B}(H)$, *the following three statements are equivalent.*

(a) *U is unitary.*
(b) $\mathscr{R}(U) = H$ *and* $(Ux, Uy) = (x, y)$ *for all* $x \in H$, $y \in H$.
(c) $\mathscr{R}(U) = H$ *and* $\|Ux\| = \|x\|$ *for every* $x \in H$.

PROOF. If U is unitary, then $\mathscr{R}(U) = H$ because $UU^* = I$. Also, $U^*U = I$, so that

$$(Ux, Uy) = (x, U^*Uy) = (x, y).$$

Thus (a) implies (b). It is obvious that (b) implies (c). If (c) holds, then

$$(U^*Ux, x) = (Ux, Ux) = \|Ux\|^2 = \|x\|^2 = (x, x)$$

for every $x \in H$, so that $U^*U = I$. But (c) implies also that U is a linear isometry of H onto H, so that U is invertible in $\mathscr{B}(H)$. Since $U^*U = I$, $U^{-1} = U^*$, and therefore U is unitary. ////

Note: The equivalence of (*a*) and (*b*) shows that the unitary operators are precisely those linear isomorphisms of H that also preserve the inner product. They are therefore the *Hilbert space automorphisms*.

The equivalence of (*b*) and (*c*) is also a corollary of Exercise 2.

The preceding proof shows that an operator $T \in \mathscr{B}(H)$ is an isometry (i.e., satisfies $\|Tx\| = \|x\|$ for every $x \in H$) if and only if $T^*T = I$. This is one half of what is needed to be unitary, but it is not enough. For example, let T be the right shift S_R on ℓ^2 (see Exercise 2, Chapter 10) whose adjoint is easily seen to be S_L.

12.14 Theorem *Each of the following four properties of a projection $P \in \mathscr{B}(H)$ implies the other three:*

(*a*) *P is self-adjoint.*
(*b*) *P is normal.*
(*c*) *$\mathscr{R}(P) = \mathscr{N}(P)^\perp$*
(*d*) *$(Px, x) = \|Px\|^2$ for every $x \in H$.*

Moreover, two self-adjoint projections P and Q have $\mathscr{R}(P) \perp \mathscr{R}(Q)$ if and only if $PQ = 0$.

Property (*c*) is usually expressed by saying that P is an *orthogonal* projection.

PROOF. It is trivial that (*a*) implies (*b*). Statement (*a*) of Theorem 12.12 shows that $\mathscr{N}(P) = \mathscr{R}(P)^\perp$ if P is normal; since P is a projection, $\mathscr{R}(P) = \mathscr{N}(I - P)$, so that $\mathscr{R}(P)$ is closed. It now follows from the corollary to Theorem 12.4 that (*b*) implies (*c*).

If (*c*) holds, every $x \in H$ has the form $x = y + z$, with $y \perp z$, $Py = 0$, $Pz = z$. Hence $Px = z$, and $(Px, x) = (z, z)$. This proves (*d*).

Finally, assume (*d*) holds. Then

$$\|Px\|^2 = (Px, x) = (x, P^*x) = (P^*x, x).$$

The last equality holds because $\|Px\|^2$ is real and $(x, P^*x) = \|Px\|^2$. Thus $(Px, x) = (P^*x, x)$, for every $x \in H$, so that $P = P^*$, by Theorem 12.7. Hence (*d*) implies (*a*).

The identity $(Px, Qy) = (x, PQy)$ proves the last assertion. ////

12.15 Theorem
(*a*) *If U is unitary and $\lambda \in \sigma(U)$, then $|\lambda| = 1$.*
(*b*) *If S is self-adjoint and $\lambda \in \sigma(S)$, then λ is a real number.*

PROOF. (*a*) Theorem 12.13 shows that $\|U\| = 1$, and therefore $|\lambda| \le 1$ if $\lambda \in \sigma(U)$. On the other hand, if $|\lambda| < 1$, then $\|\lambda U^*\| < 1$; hence

$$\lambda I - U = -U(I - \lambda U^*)$$

is invertible in $\mathscr{B}(H)$ (Theorem 10.7), and therefore $\lambda \notin \sigma(U)$.

(*b*) Suppose $S = S^*$, $\lambda = \alpha + i\beta \in \sigma(S)$. Put $S_\lambda = S - \lambda I$. A simple calculation gives

$$\|S_\lambda x\|^2 = \|Sx - \alpha x\|^2 + \beta^2\|x\|^2,$$

so that $\|S_\lambda x\| \ge |\beta|\|x\|$. If $\beta \ne 0$, it follows that S_λ is invertible [by (*c*) of Theorem 12.12], and thus $\lambda \notin \sigma(S)$. ////

A Commutativity Theorem

Let x and y be commuting elements in some Banach algebra with an involution. It is then obvious that x^* and y^* commute, simply because $x^*y^* = (yx)^*$. Does it follow that x commutes with y^*? Of course, the answer is negative whenever x is not normal and $y = x$. But it can be negative even when both x and y are normal (Exercise 28). It is therefore an interesting fact that the answer is affirmative (if x is normal) in $\mathscr{B}(H)$, relative to the involution furnished by the Hilbert space adjoint:

If $N \in \mathscr{B}(H)$ is normal, if $T \in \mathscr{B}(H)$, and if $NT = TN$, then $N^*T = TN^*$.

In fact, a more general result is true:

12.16 Theorem (Fuglede-Putnam-Rosenblum) *Assume that M, N, $T \in \mathscr{B}(H)$, M and N are normal, and*

(1) $$MT = TN.$$

Then $M^*T = TN^*$.

PROOF. Suppose first that $S \in \mathscr{B}(H)$. Put $V = S - S^*$, and define

(2) $$Q = \exp(V) = \sum_{n=0}^{\infty} \left(\frac{1}{n!}\right) V^n.$$

Then $V^* = -V$, and therefore

(3) $$Q^* = \exp(V^*) = \exp(-V) = Q^{-1}.$$

Hence Q is unitary. The consequence we need is that

(4) $$\|\exp(S - S^*)\| = 1 \qquad \text{for every } S \in \mathscr{B}(H).$$

If (1) holds, then $M^kT = TN^k$ for $k = 1, 2, 3, \ldots$, by induction. Hence

(5) $$\exp(M)T = T\exp(N),$$

or

(6) $$T = \exp(-M)T\exp(N).$$

Put $U_1 = \exp(M^* - M)$, $U_2 = \exp(N - N^*)$. Since M and N are normal, it follows from (6) that

(7) $$\exp(M^*)T\exp(-N^*) = U_1 T U_2.$$

By (4), $\|U_1\| = \|U_2\| = 1$, so that (7) implies

(8) $$\|\exp(M^*)T\exp(-N^*)\| \le \|T\|.$$

We now define

(9) $$f(\lambda) = \exp(\lambda M^*)T\exp(-\lambda N^*) \qquad (\lambda \in \mathbb{C}).$$

The hypotheses of the theorem hold with $\bar{\lambda}M$ and $\bar{\lambda}N$ in place of M and N. Therefore (8) implies that $\|f(\lambda)\| \le \|T\|$ for every $\lambda \in \mathbb{C}$. Thus f is a bounded entire $\mathscr{B}(H)$-valued function. By Liouville's theorem 3.32, $f(\lambda) = f(0) = T$, for every $\lambda \in \mathbb{C}$. Hence (9) becomes

(10) $$\exp(\lambda M^*)T = T\exp(\lambda N^*) \qquad (\lambda \in \mathbb{C}).$$

If we equate the coefficients of λ in (10), we obtain $M^*T = TN^*$. ////

Remark. Inspection of this proof shows that it used no properties of $\mathscr{B}(H)$ which are not shared by every B^*-algebra. This observation does not lead to a generalization of the theorem, however, because of Theorem 12.41.

Note that the hypotheses of Theorem 12.16 do *not* imply that $MT^* = T^*N$, even when M and N are self-adjoint and T is normal: If

$$M = \begin{bmatrix} 1 & 0 \\ 0 & -1 \end{bmatrix}, \quad N = \begin{bmatrix} 0 & 1 \\ 1 & 0 \end{bmatrix}, \quad T = \begin{bmatrix} 1 & 1 \\ -1 & 1 \end{bmatrix},$$

then $MT = TN$ but $MT^* \ne T^*N$.

Resolutions of the Identity

12.17 Definition Let $\mathfrak{M}$ be a σ-algebra in a set Ω, and let H be a Hilbert space. In this setting, a *resolution of the identity* (on $\mathfrak{M}$) is a mapping

$$E\colon \mathfrak{M} \to \mathscr{B}(H)$$

with the following properties:

(*a*) $E(\varnothing) = 0$, $E(\Omega) = I$.

(*b*) Each $E(\omega)$ is a self-adjoint projection.

(*c*) $E(\omega' \cap \omega'') = E(\omega')E(\omega'')$.

(*d*) If $\omega' \cap \omega'' = \varnothing$, then $E(\omega' \cup \omega'') = E(\omega') + E(\omega'')$.

(*e*) For every $x \in H$ and $y \in H$, the set function $E_{x,y}$ defined by

$$E_{x,y}(\omega) = (E(\omega)x, y)$$

is a complex measure on $\mathfrak{M}$.

When $\mathfrak{M}$ is the σ-algebra of all Borel sets on a compact or locally compact Hausdorff space, it is customary to add another requirement to (*e*): Each $E_{x,y}$ should be a *regular* Borel measure. (This is automatically satisfied on compact metric spaces, for instance. See [23].)

Here are some immediate consequences of these properties.

Since each $E(\omega)$ is a self-adjoint projection, we have

$$E_{x,x}(\omega) = (E(\omega)x, x) = \|E(\omega)x\|^2 \qquad (x \in H) \tag{1}$$

so that each $E_{x,x}$ is a positive measure on $\mathfrak{M}$ whose total variation is

$$\|E_{x,x}\| = E_{x,x}(\Omega) = \|x\|^2. \tag{2}$$

By (*c*), any two of the projections $E(\omega)$ commute with each other.

If $\omega' \cap \omega'' = \varnothing$, (*a*) and (*c*) show that the ranges of $E(\omega')$ and $E(\omega'')$ are orthogonal to each other (Theorem 12.14).

By (*d*), E is finitely additive. The question arises whether E is countably additive, i.e., whether the series

$$\sum_{n=1}^{\infty} E(\omega_n) \tag{3}$$

converges, *in the norm topology of* $\mathscr{B}(H)$, to $E(\omega)$, whenever ω is the union of the disjoint sets $\omega_n \in \mathfrak{M}$. Since the norm of any projection is either 0 or at least 1, the partial sums of the series (3) cannot form a Cauchy sequence, unless all but finitely many of the $E(\omega_n)$ are 0. Thus E is not countably additive, except in some trivial situations.

However, let $\{\omega_n\}$ be as above, and fix $x \in H$. Since $E(\omega_n)E(\omega_m) = 0$ when $n \neq m$, the vectors $E(\omega_n)x$ and $E(\omega_m)x$ are orthogonal to each other (Theorem 12.14). By (*e*),

$$\sum_{n=1}^{\infty} (E(\omega_n)x, y) = (E(\omega)x, y) \tag{4}$$

for every $y \in H$. It now follows from Theorem 12.6 that

$$\sum_{n=1}^{\infty} E(\omega_n)x = E(\omega)x. \tag{5}$$

The series (5) converges in the norm topology of H. We summarize the result just proved:

12.18 Proposition *If E is a resolution of the identity, and if $x \in H$, then*

$$\omega \to E(\omega)x$$

is a countably additive H-valued measure on $\mathfrak{M}$.

Moreover, sets of measure zero can be handled in the usual way:

12.19 Proposition *Suppose E is a resolution of the identity. If $\omega_n \in \mathfrak{M}$ and $E(\omega_n) = 0$ for $n = 1, 2, 3, \ldots$, and if $\omega = \bigcup_{n=1}^{\infty} \omega_n$, then $E(\omega) = 0$.*

PROOF. Since $E(\omega_n) = 0$, $E_{x,x}(\omega_n) = 0$ for every $x \in H$. Since $E_{x,x}$ is countably additive, it follows that $E_{x,x}(\omega) = 0$. But $\|E(\omega)x\|^2 = E_{x,x}(\omega)$. Hence $E(\omega) = 0$. ////

12.20 The algebra $L^{\infty}(E)$ Let E be a resolution of the identity on $\mathfrak{M}$, as above. Let f be a complex $\mathfrak{M}$-measurable function on Ω. There is a countable collection $\{D_i\}$ of open discs which forms a base for the topology of $\mathcal{C}$. Let V be the union of those D_i for which $E(f^{-1}(D_i)) = 0$. By Proposition 12.19, $E(f^{-1}(V)) = 0$. Also, V is the largest open subset of $\mathcal{C}$ with this property.

The *essential range* of f is, by definition, the complement of V. It is the smallest closed subset of $\mathcal{C}$ that contains $f(p)$ for almost all $p \in \Omega$, that is, for all $p \in \Omega$ except those that lie in some set $\omega \in \mathfrak{M}$ with $E(\omega) = 0$.

We say that f is *essentially bounded* if its essential range is bounded, hence compact. In that case, the largest value of $|\lambda|$, as λ runs through the essential range of f, is called the *essential supremum* $\|f\|_{\infty}$ of f.

Let B be the algebra of all bounded complex $\mathfrak{M}$-measurable functions on Ω; with the norm

$$\|f\| = \sup \{|f(p)|: p \in \Omega\},$$

one sees easily that B is a Banach algebra and that

$$N = \{f \in B: \|f\|_{\infty} = 0\}$$

is an ideal of B which is *closed*, by Proposition 12.19. Hence B/N is a Banach algebra, which we denote (in the usual manner) by $L^{\infty}(E)$.

The norm of any coset $[f] = f + N$ of $L^{\infty}(E)$ is then equal to $\|f\|_{\infty}$, and its spectrum $\sigma([f])$ is the essential range of f. As is usually done in measure theory, the distinction between f and its equivalence class $[f]$ will be ignored.

Our next concern will be the integration of functions with respect to the projection-valued measures described above. The resulting integrals $\int f\,dE$ turn out to be not only linear (as all good integrals ought to be) but also multiplicative!

12.21 Theorem *If E is a resolution of the identity, as above, then there exists an isometric*-isomorphism Ψ of the Banach algebra $L^\infty(E)$ onto a closed normal subalgebra A of $\mathscr{B}(H)$, which is related to E by the formula.*

(1) $$(\Psi(f)x, y) = \int_\Omega f\,dE_{x,y} \qquad (x, y \in H, f \in L^\infty(E)).$$

This justifies the notation

(2) $$\Psi(f) = \int_\Omega f\,dE.$$

Moreover,

(3) $$\|\Psi(f)x\|^2 = \int_\Omega |f|^2\,dE_{x,x} \qquad (x \in H, f \in L^\infty(E)),$$

and an operator $Q \in \mathscr{B}(H)$ commutes with every $E(\omega)$ if and only if Q commutes with every $\Psi(f)$.

Recall that a *normal* subalgebra A of $\mathscr{B}(H)$ is a commutative one which contains T^* for every $T \in A$. To say that Ψ is a *-isomorphism means that Ψ is one-to-one, linear, and multiplicative and that

(4) $$\Psi(\bar{f}) = \Psi(f)^* \qquad (f \in L^\infty(E)).$$

PROOF. To begin with, let $\{\omega_1, \ldots, \omega_n\}$ be a partition of Ω, with $\omega_i \in \mathfrak{M}$, and let s be a simple function, such that $s = \alpha_i$ on ω_i. Define $\Psi(s) \in \mathscr{B}(H)$ by

(5) $$\Psi(s) = \sum_{i=1}^{n} \alpha_i E(\omega_i).$$

Since each $E(\omega_i)$ is self-adjoint,

(6) $$\Psi(s)^* = \sum_{i=1}^{n} \bar{\alpha}_i E(\omega_i) = \Psi(\bar{s}).$$

If $\{\omega'_1, \ldots, \omega'_m\}$ is another partition of this kind, and if $t = \beta_j$ on ω'_j, then

$$\Psi(s)\Psi(t) = \sum_{i,j} \alpha_i \beta_j E(\omega_i)E(\omega'_j) = \sum_{i,j} \alpha_i \beta_j E(\omega_i \cap \omega'_j).$$

Since st is the simple function that equals $\alpha_i \beta_j$ on $\omega_i \cap \omega'_j$, it follows that

$$\Psi(s)\Psi(t) = \Psi(st). \tag{7}$$

An entirely analogous argument shows that

$$\Psi(\alpha s + \beta t) = \alpha\Psi(s) + \beta\Psi(t). \tag{8}$$

If $x \in H$ and $y \in H$, (5) leads to

$$(\Psi(s)x, y) = \sum_{i=1}^{n} \alpha_i (E(\omega_i)x, y) = \sum_{i=1}^{n} \alpha_i E_{x,y}(\omega_i) = \int_\Omega s \, dE_{x,y}. \tag{9}$$

By (6) and (7),

$$\Psi(s)^*\Psi(s) = \Psi(\bar{s})\Psi(s) = \Psi(\bar{s}s) = \Psi(|s|^2). \tag{10}$$

Hence (9) yields

$$\|\Psi(s)x\|^2 = (\Psi(s)^*\Psi(s)x, x) = (\Psi(|s|^2)x, x) = \int_\Omega |s|^2 \, dE_{x,x}, \tag{11}$$

so that

$$\|\Psi(s)x\| \le \|s\|_\infty \|x\|, \tag{12}$$

by formula (2) of Section 12.17. On the other hand, if $x \in \mathscr{R}(E(\omega_j))$, then

$$\Psi(s)x = \alpha_j E(\omega_j)x = \alpha_j x, \tag{13}$$

since the projections $E(\omega_i)$ have mutually orthogonal ranges. If j is chosen so that $|\alpha_j| = \|s\|_\infty$, it follows from (12) and (13) that

$$\|\Psi(s)\| = \|s\|_\infty. \tag{14}$$

Now suppose $f \in L^\infty(E)$. There is a sequence of simple measurable functions s_k that converges to f in the norm of $L^\infty(E)$. By (14), the corresponding operators $\Psi(s_k)$ form a Cauchy sequence in $\mathscr{B}(H)$ which is therefore norm-convergent to an operator that we call $\Psi(f)$; it is easy to see that $\Psi(f)$ does not depend on the particular choice of $\{s_k\}$. Obviously (14) leads to

$$\|\Psi(f)\| = \|f\|_\infty \qquad (f \in L^\infty(E)). \tag{15}$$

Now (1) follows from (9) (with s_k in place of s), since each $E_{x,y}$ is a finite measure; (2) and (3) follow from (6) and (11); and if bounded measurable functions f and g are approximated, in the norm of $L^\infty(E)$, by simple measurable functions s and t, we see that (7) and (8) hold with f and g in place of s and t.

Thus Ψ is an isometric isomorphism of $L^\infty(E)$ into $\mathscr{B}(H)$. Since $L^\infty(E)$ is complete, its image $A = \Psi(L^\infty(E))$ is closed in $\mathscr{B}(H)$, because of (15).

Finally, if Q commutes with every $E(\omega)$, then Q commutes with $\Psi(s)$ whenever s is simple, and therefore the approximation process used above shows that Q commutes with every member of A. ////

It is perhaps worth mentioning that the equality

$$\|f\|_\infty^2 = \sup\left\{\int_\Omega |f|^2\, dE_{x,x}\colon \|x\| \leq 1\right\} \tag{16}$$

holds for every $f \in L^\infty(E)$, because of (3) and (15).

The Spectral Theorem

The principal assertion of the spectral theorem is that every bounded normal operator T on a Hilbert space induces (in a canonical way) a resolution E of the identity on the Borel subsets of its spectrum $\sigma(T)$ and that T can be reconstructed from E by an integral of the type discussed in Theorem 12.21. A large part of the theory of normal operators depends on this fact.

It should perhaps be stated explicitly that the spectrum $\sigma(T)$ of an operator $T \in \mathscr{B}(H)$ will always refer to the full algebra $\mathscr{B}(H)$. In other words, $\lambda \in \sigma(T)$ if and only if $T - \lambda I$ has no inverse in $\mathscr{B}(H)$. Sometimes we shall also be concerned with closed subalgebras A of $\mathscr{B}(H)$ which have the additional property that $I \in A$ and $T^* \in A$ whenever $T \in A$. (Such algebras are sometimes called *-algebras.)

Let A be such an algebra, and suppose that $T \in A$ and $T^{-1} \in \mathscr{B}(H)$. Since TT^* is self-adjoint, $\sigma(TT^*)$ is a compact subset of the real line (Theorem 12.15), hence does not separate $\mathbb{C}$, and therefore $\sigma_A(TT^*) = \sigma(TT^*)$, by the corollary to Theorem 10.18. Since TT^* is invertible in $\mathscr{B}(H)$, this equality shows that $(TT^*)^{-1} \in A$, and therefore $T^{-1} = T^*(TT^*)^{-1}$ is also in A.

*Thus T has the same spectrum relative to all closed *-algebras in $\mathscr{B}(H)$ that contain T.*

Theorem 12.23 will be obtained as a special case of the following result, which deals with normal algebras of operators rather than with individual ones.

12.22 Theorem *If A is a closed normal subalgebra of $\mathscr{B}(H)$ which contains the identity operator I and if Δ is the maximal ideal space of A, then the following assertions are true:*

(*a*) *There exists a unique resolution E of the identity on the Borel subsets of Δ which satisfies*

(1)
$$T = \int_{\Delta} \hat{T}\, dE$$

for every $T \in A$, where $\hat{T}$ is the Gelfand transform of T.

(*b*) *The inverse of the Gelfand transform (i.e., the map that takes $\hat{T}$* back to *T) extends to an isometric *-isomorphism Φ of the algebra $L^\infty(E)$ onto a closed subalgebra B of $\mathscr{B}(H)$, $B \supset A$, given by*

(2)
$$\Phi f = \int_{\Delta} f\, dE \qquad (f \in L^\infty(E)).$$

Explicitly, Φ is linear and multiplicative and satisfies

(3)
$$\Phi \bar{f} = (\Phi f)^*, \quad \|\Phi f\| = \|f\|_\infty \qquad (f \in L^\infty(E)).$$

(*c*) *B is the closure [in the norm topology of $\mathscr{B}(H)$] of the set of all finite linear combinations of the projections $E(\omega)$.*

(*d*) *If $\omega \subset \Delta$ is open and nonempty, then $E(\omega) \neq 0$.*

(*e*) *An operator $S \in \mathscr{B}(H)$ commutes with every $T \in A$ if and only if S commutes with every projection $E(\omega)$.*

PROOF. Recall that (1) is an abbreviation for

(4)
$$(Tx, y) = \int_{\Delta} \hat{T}\, dE_{x,y} \qquad (x, y \in H,\ T \in A).$$

Since $\mathscr{B}(H)$ is a B^*-algebra (Section 12.9), our given algebra A is a commutative B^*-algebra. The Gelfand-Naimark theorem 11.18 asserts therefore that $T \to \hat{T}$ is an isometric *-isomorphism of A *onto* $C(\Delta)$.

This leads to an easy proof of the uniqueness of E. Suppose E satisfies (4). Since $\hat{T}$ ranges over all of $C(\Delta)$, the assumed regularity of the complex Borel measures $E_{x,y}$ shows that each $E_{x,y}$ is uniquely determined by (4); this follows from the uniqueness assertion that is part of the Riesz representation theorem ([23], Th. 6.19). Since, by definition, $(E(\omega)x, y) = E_{x,y}(\omega)$, each projection $E(\omega)$ is also uniquely determined by (4).

This uniqueness proof motivates the following proof of the existence of E. If $x \in H$ and $y \in H$, Theorem 11.18 shows that $\hat{T} \to (Tx, y)$ is a bounded linear functional on $C(\Delta)$, of norm $\le \|x\|\,\|y\|$, since $\|\hat{T}\|_\infty = \|T\|$. The Riesz representation theorem supplies us therefore with unique regular complex Borel measures $\mu_{x,y}$ on Δ such that

(5)
$$(Tx, y) = \int_{\Delta} \hat{T}\, d\mu_{x,y} \qquad (x, y \in H,\ T \in A).$$

For fixed T, the left side of (5) is a bounded sesquilinear functional on H, hence so is the right side, and it remains so if the continuous function $\hat{T}$ is replaced by an arbitrary bounded Borel function f. To each such f corresponds therefore an operator $\Phi f \in \mathscr{B}(H)$ (see Theorem 12.8) such that

$$((\Phi f)x, y) = \int_\Delta f \, d\mu_{x,y} \qquad (x, y \in H). \tag{6}$$

Comparison of (5) and (6) shows that $\Phi \hat{T} = T$. Thus Φ is an extension of the inverse of the Gelfand transform.

It is clear that Φ is linear.

Part of the Gelfand-Naimark theorem states that T is self-adjoint if and only if $\hat{T}$ is real-valued. For such T,

$$\int_\Delta \hat{T} \, d\mu_{x,y} = (Tx, y) = (x, Ty) = \overline{(Ty, x)} = \overline{\int_\Delta \hat{T} \, d\mu_{y,x}},$$

and this implies that $\mu_{y,x} = \overline{\mu_{x,y}}$. Hence

$$((\Phi \bar{f})x, y) = \int_\Delta \bar{f} \, d\mu_{x,y} = \overline{\int_\Delta f \, d\mu_{y,x}} = \overline{((\Phi f)y, x)} = (x, (\Phi f)y)$$

for all $x, y \in H$, so that

$$\Phi \bar{f} = (\Phi f)^* \tag{7}$$

Our next objective is the equality

$$\Phi(fg) = (\Phi f)(\Phi g) \tag{8}$$

for bounded Borel functions f, g on Δ. If $S \in A$ and $T \in A$, then $(ST)^\wedge = \hat{S}\hat{T}$; hence

$$\int_\Delta \hat{S}\hat{T} \, d\mu_{x,y} = (STx, y) = \int_\Delta \hat{S} \, d\mu_{Tx,y}.$$

This holds for every $\hat{S} \in C(\Delta)$; hence the two integrals are equal if $\hat{S}$ is replaced by any bounded Borel function f. Thus

$$\int_\Delta f\hat{T} \, d\mu_{x,y} = \int_\Delta f \, d\mu_{Tx,y} = ((\Phi f)Tx, y) = (Tx, z) = \int_\Delta \hat{T} \, d\mu_{x,z},$$

where we put $z = (\Phi f)^* y$. Again, the first and last integrals remain equal if $\hat{T}$ is replaced by g. This gives

$$(\Phi(fg)x, y) = \int_\Delta fg \, d\mu_{x,y} = \int_\Delta g \, d\mu_{x,z}$$

$$= ((\Phi g)x, z) = ((\Phi g)x, (\Phi f)^* y) = (\Phi(f)\Phi(g)x, y),$$

and (8) is proved.

We are finally ready to define E: If ω is a Borel subset of Δ, let χ_ω be its characteristic function, and put

$$E(\omega) = \Phi(\chi_\omega). \tag{9}$$

By (8), $E(\omega \cap \omega') = E(\omega)E(\omega')$. With $\omega' = \omega$, this shows that each $E(\omega)$ is a projection. Since Φf is self-adjoint when f is real, by (7), each $E(\omega)$ is self-adjoint. It is clear that $E(\varnothing) = \Phi(0) = 0$. That $E(\Delta) = I$ follows from (5) and (6). The finite additivity of E is a consequence of (6), and, for all $x, y \in H$,

$$E_{x,y}(\omega) = (E(\omega)x, y) = \int_\Delta \chi_\omega \, d\mu_{x,y} = \mu_{x,y}(\omega).$$

Thus (6) becomes (2). That $\|\Phi f\| = \|f\|_\infty$ follows now from Theorem 12.21.

This completes the proof of (*a*) and (*b*).

Part (*c*) is now clear because every $f \in L^\infty(E)$ is a uniform limit of simple functions (i.e., of functions with only finitely many values).

Suppose next that ω is open and $E(\omega) = 0$. If $T \in A$ and $\hat{T}$ has its support in ω, (1) implies that $T = 0$; hence $\hat{T} = 0$. Since $\hat{A} = C(\Delta)$. Urysohn's lemma implies now that $\omega = \varnothing$. This proves (*d*).

To prove (*e*), choose $S \in \mathscr{B}(H)$, $x \in H$, $y \in H$, and put $z = S^*y$. For any $T \in A$ and any Borel set $\omega \subset \Delta$ we then have

$$(STx, y) = (Tx, z) = \int_\Delta \hat{T} \, dE_{x,z}, \tag{10}$$

$$(TSx, y) = \int_\Delta \hat{T} \, dE_{Sx,y}, \tag{11}$$

$$(SE(\omega)x, y) = (E(\omega)x, z) = E_{x,z}(\omega), \tag{12}$$

$$(E(\omega)Sx, y) = E_{Sx,y}(\omega). \tag{13}$$

If $ST = TS$ for every $T \in A$, the measures in (10) and (11) are equal, so that $SE(\omega) = E(\omega)S$. The same argument establishes the converse. This completes the proof. ////

We now specialize this theorem to a single operator.

12.23 Theorem *If $T \in \mathscr{B}(H)$ and T is normal, then there exists a unique resolution of the identity E on the Borel subsets of $\sigma(T)$ which satisfies*

$$T = \int \lambda \, dE(\lambda). \tag{1}$$

Furthermore, every projection $E(\omega)$ commutes with every $S \in \mathscr{B}(H)$ which commutes with T.

We shall refer to this E as *the spectral decomposition of* T.

Sometimes it is convenient to think of E as being defined for all Borel sets in $\mathcal{C}$; to achieve this put $E(\omega) = 0$ if $\omega \cap \sigma(T) = \emptyset$.

PROOF. Let A be the smallest closed subalgebra of $\mathscr{B}(H)$ that contains I, T, and T^*. Since T is normal, Theorem 12.22 applies to A. By Theorem 11.19, the maximal ideal space of A can be identified with $\sigma(T)$ in such a way that $\hat{T}(\lambda) = \lambda$ for every $\lambda \in \sigma(T)$. The existence of E follows now from Theorem 12.22.

On the other hand, if E exists so that (1) holds, Theorem 12.21 shows that

$$p(T, T^*) = \int_{\sigma(T)} p(\lambda, \bar{\lambda})\, dE(\lambda), \tag{2}$$

where p is any polynomial in two variables (with complex coefficients). By the Stone-Weierstrass theorem, these polynomials are dense in $C(\sigma(T))$. The projections $E(\omega)$ are therefore uniquely determined by the integrals (2), hence by T, just as in the uniqueness proof in Theorem 12.22.

If $ST = TS$, then also $ST^* = T^*S$, by Theorem 12.16; hence S commutes with every member of A. By (*e*) of Theorem 12.22, $SE(\omega) = E(\omega)S$ for every Borel set $\omega \subset \sigma(T)$. ////

12.24 The symbolic calculus for normal operators If E is the spectral decomposition of a normal operator $T \in \mathscr{B}(H)$, and if f is a bounded Borel function on $\sigma(T)$, it is customary to denote the operator

$$\Psi(f) = \int_{\sigma(T)} f\, dE \tag{1}$$

by $f(T)$.

Using this notation, part of the content of Theorems 12.21 to 12.23 can be summarized as follows:

The mapping $f \to f(T)$ is a homomorphism of the algebra of all bounded Borel functions on $\sigma(T)$ into $\mathscr{B}(H)$, which carries the function 1 to I, which carries the identity function on $\sigma(T)$ to T, and which satisfies

$$\bar{f}(T) = f(T)^* \tag{2}$$

and

$$\| f(T) \| \le \sup \{ |f(\lambda)| : \lambda \in \sigma(T) \}. \tag{3}$$

If $f \in C(\sigma(T))$, then equality holds in (3), and therefore $f \to f(T)$ is an isomorphism on $C(\sigma(T))$ which satisfies

$$\|f(T)x\|^2 = \int_{\sigma(T)} |f|^2 \, dE_{x,x}.$$

If $f_n \to f$ uniformly, then $\|f_n(T) - f(T)\| \to 0$, as $n \to \infty$.

If $S \in \mathscr{B}(H)$ and $ST = TS$, then $Sf(T) = f(T)S$ for every bounded Borel function f.

Since the identity function can be uniformly approximated, on $\sigma(T)$, by simple Borel functions, it follows that T is a limit, in the norm topology of $\mathscr{B}(H)$, of finite linear combinations of projections $E(\omega)$.

The following proof contains our first application of this symbolic calculus.

12.25 Theorem *If $T \in \mathscr{B}(H)$ is normal, then*

$$\|T\| = \sup \{|(Tx, x)| : x \in H, \|x\| \leq 1\}.$$

PROOF. Choose $\varepsilon > 0$. It is clearly enough to show that

$$|(Tx_0, x_0)| > \|T\| - \varepsilon \tag{1}$$

for some $x_0 \in H$ with $\|x_0\| = 1$.

Since $\|T\| = \|\hat{T}\|_\infty = \rho(T)$ (Theorem 11.18), there exists $\lambda_0 \in \sigma(T)$ such that $|\lambda_0| = \|T\|$. Let ω be the set of all $\lambda \in \sigma(T)$ for which $|\lambda - \lambda_0| < \varepsilon$. If E is the spectral decomposition of T, then (*d*) of Theorem 12.22 implies that $E(\omega) \neq 0$. Therefore there exists $x_0 \in H$ with $\|x_0\| = 1$ and $E(\omega)x_0 = x_0$.

Define $f(\lambda) = \lambda - \lambda_0$ for $\lambda \in \omega$; put $f(\lambda) = 0$ for all other $\lambda \in \sigma(T)$. Then

$$f(T) = (T - \lambda_0 I)E(\omega),$$

so that

$$f(T)x_0 = Tx_0 - \lambda_0 x_0.$$

Hence

$$|(Tx_0, x_0) - \lambda_0| = |(f(T)x_0, x_0)| \leq \|f(T)\| \leq \varepsilon,$$

since $|f(\lambda)| < \varepsilon$ for all $\lambda \in \sigma(T)$. This implies (1), because $|\lambda_0| = \|T\|$.

////

To see that normality is needed here, let T be the linear operator on $\mathbb{C}^2$ (with basis $\{e_1, e_2\}$) given by $Te_1 = 0$, $Te_2 = e_1$. It has $\|T\| = 1$, but $|(Tx, x)| \leq \frac{1}{2}$ if $\|x\| \leq 1$.

Our next result contains a converse to Theorem 12.15.

12.26 Theorem *A normal $T \in \mathscr{B}(H)$ is*

(*a*) *self-adjoint if and only if $\sigma(T)$ lies in the real axis,*
(*b*) *unitary if and only if $\sigma(T)$ lies on the unit circle.*

PROOF. Choose A as in the proof of Theorem 12.23. Then $\hat{T}(\lambda) = \lambda$ and $(T^*)^\wedge(\lambda) = \bar{\lambda}$ on $\sigma(T)$. Hence $T = T^*$ if and only if $\lambda = \bar{\lambda}$ on $\sigma(T)$, and $TT^* = I$ if and only if $\lambda\bar{\lambda} = 1$ on $\sigma(T)$. ////

12.27 Invariant subspaces As in Theorem 10.35, a closed subspace M of H is an *invariant subspace* of a set $\sum \subset \mathscr{B}(H)$ if every $T \in \sum$ maps M into M. For example, every eigenspace of T is an invariant subspace of T. When $\dim H < \infty$, the spectral theorem implies that the eigenspaces of every normal operator T span H. [Sketch of proof: The characteristic function of each point in $\sigma(T)$ corresponds to a projection in H. The sum of these projections is $E(\sigma(T)) = I$.] If $\dim H = \infty$, it can happen that T has no eigenvalues (Exercise 20). But normal operators still have invariant subspaces that are nontrivial (that is, $\neq \{0\}$ and $\neq H$).

In fact, let A be a normal algebra, as in Theorem 12.22, and let E be its resolution of the identity, on the Borel subsets of Δ. If Δ consists of a single point, then A consists of the scalar multiples of I, and every subspace of H is invariant under A. Suppose that $\Delta = \omega \cup \omega'$, where ω and ω' are nonempty disjoint Borel sets. Let M and M' be the ranges of $E(\omega)$ and $E(\omega')$. Then $TE(\omega) = E(\omega)T$ for every $T \in A$. If $x \in M$, it follows that

$$Tx = TE(\omega)x = E(\omega)Tx,$$

so that $Tx \in M$. The same holds for M'.

Hence M and M' are invariant subspaces of A.

Moreover, $M' = M^\perp$, and $H = M \oplus M'$.

Decompositions of Δ into finitely many (or even countably many) disjoint Borel sets induce, in the same manner, decompositions of H into pairwise orthogonal invariant subspaces of A.

It is an open problem whether every (nonnormal) $T \in \mathscr{B}(H)$ has a nontrivial invariant subspace if H is an infinite-dimensional separable Hilbert space.

Eigenvalues of Normal Operators

If $T \in \mathscr{B}(H)$ is normal, its eigenvalues bear a simple relation to its spectral decomposition (Theorem 12.29). This will be derived from the following application of the symbolic calculus:

12.28 Theorem *Suppose $T \in \mathscr{B}(H)$ is normal and E is its spectral decomposition. If $f \in C(\sigma(T))$ and if $\omega_0 = f^{-1}(0)$, then*

$$\mathscr{N}(f(T)) = \mathscr{R}(E(\omega_0)). \tag{1}$$

PROOF. Put $g(\lambda) = 1$ on ω_0, $g(\lambda) = 0$ at all other points of $\sigma(T)$. Then $fg = 0$, so that $f(T)g(T) = 0$. Since $g(T) = E(\omega_0)$, it follows that

$$\mathscr{R}(E(\omega_0)) \subset \mathscr{N}(f(T)). \tag{2}$$

For each positive integer n, let ω_n be the set of all $\lambda \in \sigma(T)$ where $1/n \le |f(\lambda)| < 1/(n-1)$. The complement $\tilde{\omega}$ of ω_0, relative to $\sigma(T)$, is then the union of the disjoint Borel sets ω_n. Define

$$f_n(\lambda) = \begin{cases} 1/f(\lambda) & \text{on } \omega_n, \\ 0 & \text{elsewhere on } \sigma(T). \end{cases} \tag{3}$$

Each f_n is a bounded Borel function on $\sigma(T)$, and

$$f_n(T)f(T) = E(\omega_n) \qquad (n = 1, 2, 3, \ldots). \tag{4}$$

If $f(T)x = 0$, it follows that $E(\omega_n)x = 0$. The countable additivity of the mapping $\omega \to E(\omega)x$ (Proposition 12.18) shows therefore that $E(\tilde{\omega})x = 0$. But $E(\tilde{\omega}) + E(\omega_0) = I$. Hence $E(\omega_0)x = x$. We have now proved that

$$\mathscr{N}(f(T)) \subset \mathscr{R}(E(\omega_0)), \tag{5}$$

and (1) follows from (2) and (5). ////

12.29 Theorem *Suppose E is the spectral decomposition of a normal $T \in \mathscr{B}(H)$, $\lambda_0 \in \sigma(T)$, and $E_0 = E(\{\lambda_0\})$. Then*

(a) $\mathscr{N}(T - \lambda_0 I) = \mathscr{R}(E_0)$,

(b) *λ_0 is an eigenvalue of T if and only if $E_0 \neq 0$, and*

(c) *every isolated point of $\sigma(T)$ is an eigenvalue of T.*

(d) *Moreover, if $\sigma(T) = \{\lambda_1, \lambda_2, \lambda_3, \ldots\}$ is a countable set, then every $x \in H$ has a unique expansion of the form*

$$x = \sum_{i=1}^{\infty} x_i,$$

where $Tx_i = \lambda_i x_i$. Also, $x_i \perp x_j$ whenever $i \neq j$.

Statements (*b*) and (*c*) explain the term *point spectrum of T* for the set of all eigenvalues of T.

PROOF. Part (*a*) is an immediate corollary of Theorem 12.28, with $f(\lambda) = \lambda - \lambda_0$. It is clear that (*b*) follows from (*a*). If λ_0 is an isolated point of $\sigma(T)$, then $\{\lambda_0\}$ is a nonempty open subset of $\sigma(T)$; hence $E_0 \neq 0$, by (*d*) of Theorem 12.22. Therefore (*c*) follows from (*b*).

To prove (*d*), put $E_i = E(\{\lambda_i\})$, $i = 1, 2, 3, \ldots$. At limit points λ_i of $\sigma(T)$, E_i may or may not be 0. In any case, the projections E_i have pairwise orthogonal ranges. The countable additivity of $\omega \to E(\omega)x$ (Proposition 12.18) shows that

$$\sum_{i=1}^{\infty} E_i x = E(\sigma(T))x = x \qquad (x \in H).$$

The series converges, in the norm of H. This gives the desired representation of x, if $x_i = E_i x$. The uniqueness follows from the orthogonality of the vectors x_i, and $Tx_i = \lambda_i x_i$ follows from (*a*). ////

12.30 Theorem *A normal operator $T \in \mathscr{B}(H)$ is compact if and only if it satisfies the following two conditions:*

(*a*) *$\sigma(T)$ has no limit point except possibly* 0.

(*b*) *If $\lambda \neq 0$, then* $\dim \mathscr{N}(T - \lambda I) < \infty$.

PROOF. For the necessity, see (*d*) of Theorem 4.18, and Theorem 4.25.

To prove the sufficiency, assume (*a*) and (*b*) hold, let $\{\lambda_i\}$ be an enumeration of the nonzero points of $\sigma(T)$ such that $|\lambda_1| \geq |\lambda_2| \geq \cdots$, define $f_n(\lambda) = \lambda$ if $\lambda = \lambda_i$ and $i \leq n$, and put $f_n(\lambda) = 0$ at the other points of $\sigma(T)$. If $E_i = E(\{\lambda_i\})$, as in Theorem 12.29, then

$$f_n(T) = \lambda_1 E_1 + \cdots + \lambda_n E_n.$$

Since $\dim \mathscr{R}(E_i) = \dim \mathscr{N}(T - \lambda_i I) < \infty$, each $f_n(T)$ is a compact operator. Since $|\lambda - f_n(\lambda)| \leq |\lambda_n|$ for all $\lambda \in \sigma(T)$, we have

$$\|T - f_n(T)\| \leq |\lambda_n| \to 0 \quad \text{as} \quad n \to \infty.$$

It now follows from (*c*) of Theorem 4.18 that T is compact. ////

We have tacitly assumed that $\sigma(T)$ is infinite. If $\sigma(T)$ contains only n points different from 0, then $f_n(T) = T$ in the preceding proof, and Theorem 4.18 is not needed.

12.31 Theorem *Suppose $T \in \mathscr{B}(H)$ is normal and compact. Then*

(*a*) *T has an eigenvalue λ with $|\lambda| = \|T\|$,*

(*b*) *$f(T)$ is compact if $f \in C(\sigma(T))$ and $f(0) = 0$, but*

(*c*) *$f(T)$ is not compact if $f \in C(\sigma(T))$, $f(0) \neq 0$, and* $\dim H = \infty$.

PROOF. Since T is normal, Theorem 11.18 shows that there exists $\lambda \in \sigma(T)$ with $|\lambda| = \|T\|$. If $\|T\| > 0$, this λ is an isolated point of $\sigma(T)$ (Theorem 12.30), hence an eigenvalue of T (Theorem 12.29). If $\|T\| = 0$, (*a*) is obvious.

To prove (*b*), choose $\varepsilon > 0$, then $\delta > 0$ so that $|f(\lambda)| < \varepsilon$ if $|\lambda| \leq \delta$; let $\lambda_1, \ldots, \lambda_N$ be the points in $\sigma(T)$ for which $|\lambda_i| > \delta$; find polynomials Q_k $(1 \leq k \leq N)$ so that $Q_k(\lambda_k) = 1$, $Q_k(\lambda_j) = 0$ for $j \neq k$, $1 \leq j \leq N$; and define

$$P(\lambda) = \sum_{k=1}^{N} f(\lambda_k)(\lambda/\lambda_k)^M Q_k(\lambda),$$

where M is a positive integer, so large that $|P(\lambda)| < \varepsilon$ if $|\lambda| \leq \delta$. The polynomial P has λ as a factor. Hence $P(T)$ is a compact operator, by (*f*) of Theorem 4.18. Also, $P(\lambda_j) = f(\lambda_j)$ for $1 \leq j \leq N$. It follows that $|P(\lambda) - f(\lambda)| < 2\varepsilon$ for all $\lambda \in \sigma(T)$. Hence $\|P(T) - f(T)\| < 2\varepsilon$, and (*c*) of Theorem 4.18 implies that $f(T)$ is compact.

In the proof of (*c*), assume $f(0) = 1$, without loss of generality. Then (*b*), applied to $1 - f$, shows that the operator $S = I - f(T)$ is compact. Let B be the unit ball of H. Then

$$B \subset S(B) + f(T)(B).$$

If $f(T)$ were compact, it would follow that B lies in the sum of two compact sets; hence H would be locally compact, hence finite-dimensional, contrary to our hypothesis. ////

Positive Operators and Square Roots

12.32 Theorem *Suppose $T \in \mathscr{B}(H)$. Then*

(*a*) $(Tx, x) \geq 0$ *for every $x \in H$ if and only if*
(*b*) $T = T^*$ *and* $\sigma(T) \subset [0, \infty)$.

If $T \in \mathscr{B}(H)$ satisfies (*a*), we call T a *positive* operator and write $T \geq 0$. The theorem asserts that this terminology agrees with Definition 11.27.

PROOF. In general, (Tx, x) and (x, Tx) are complex conjugates of each other. But if (*a*) holds, then (Tx, x) is real, so that

$$(x, T^*x) = (Tx, x) = (x, Tx)$$

for every $x \in H$. By Theorem 12.7, $T = T^*$, and thus $\sigma(T)$ lies in the real axis (Theorem 12.15). If $\lambda > 0$, (*a*) implies that

$$\lambda\|x\|^2 = (\lambda x, x) \leq ((T + \lambda I)x, x) \leq \|(T + \lambda I)x\|\|x\|,$$

so that

$$\|(T + \lambda I)x\| \geq \lambda \|x\|.$$

By Theorem 12.12(*c*), $T + \lambda I$ is invertible in $\mathscr{B}(H)$, and $-\lambda$ is not in $\sigma(T)$. It follows that (*a*) implies (*b*).

Assume now that (*b*) holds, and let E be the spectral decomposition of T, so that

$$(Tx, x) = \int_{\sigma(T)} \lambda \, dE_{x,x}(\lambda) \qquad (x \in H).$$

Since each $E_{x,x}$ is a positive measure, and since $\lambda \geq 0$ on $\sigma(T)$, we have $(Tx, x) \geq 0$. Thus (*b*) implies (*a*). ////

12.33 Theorem *Every positive $T \in \mathscr{B}(H)$ has a unique positive square root $S \in \mathscr{B}(H)$. If T is invertible, so is S.*

PROOF. Let A be any closed normal subalgebra of $\mathscr{B}(H)$ that contains I and T, and let Δ be the maximal ideal space of A. By Theorem 11.18, $\hat{A} = C(\Delta)$. Since T satisfies condition (*b*) of Theorem 12.32, and since $\sigma(T) = \hat{T}(\Delta)$, we see that $\hat{T} \geq 0$. Since every nonnegative continuous function has a unique nonnegative continuous square root, it follows that there is a unique $S \in A$ that satisfies $S^2 = T$ and $\hat{S} \geq 0$; by Theorem 12.32, $\hat{S} \geq 0$ is equivalent to $S \geq 0$.

In particular, let A_0 be the smallest of these algebras A. Then there exists $S_0 \in A_0$ such that $S_0^2 = T$ and $S_0 \geq 0$. If $S \in \mathscr{B}(H)$ is any positive square root of T, let A be the smallest closed subalgebra of $\mathscr{B}(H)$ that contains I and S. Then $T \in A$, since $T = S^2$. Hence $A_0 \subset A$, so that $S_0 \in A$. The conclusion of the preceding paragraph shows now that $S = S_0$.

Finally, if T is invertible, then $S^{-1} = T^{-1}S$, since S commutes with $S^2 = T$. ////

12.34 Theorem *If $T \in \mathscr{B}(H)$, then the positive square root of T^*T is the only positive operator $P \in \mathscr{B}(H)$ that satisfies $\|Px\| = \|Tx\|$ for every $x \in H$.*

PROOF. Note first that

(1) $$(T^*Tx, x) = (Tx, Tx) = \|Tx\|^2 \geq 0 \qquad (x \in H),$$

so that $T^*T \geq 0$. (In the more abstract setting of Theorem 11.28 this was much harder to prove!)

Next, if $P \in \mathscr{B}(H)$ and $P = P^*$, then

(2) $$(P^2x, x) = (Px, Px) = \|Px\|^2 \qquad (x \in H).$$

By Theorem 12.7, it follows that $\|Px\| = \|Tx\|$ for every $x \in H$ if and only if $P^2 = T^*T$.

This completes the proof. ////

The fact that every complex number λ can be factored in the form $\lambda = \alpha|\lambda|$, where $|\alpha| = 1$, suggests the problem of trying to factor $T \in \mathscr{B}(H)$ in the form $T = UP$, with U unitary and $P \geq 0$. When this is possible, we call UP a *polar decomposition of* T.

Note that U, being unitary, is an isometry. Theorem 12.34 shows therefore that P is uniquely determined by T.

12.35 Theorem

(*a*) *If $T \in \mathscr{B}(H)$ is invertible, then T has a unique polar decomposition $T = UP$.*

(*b*) *If $T \in \mathscr{B}(H)$ is normal, then T has a polar decomposition $T = UP$ in which U and P commute with each other and with T.*

PROOF. (*a*) If T is invertible, so are T^* and T^*T, and Theorem 12.33 shows that the positive square root P of T^*T is also invertible. Put $U = TP^{-1}$. Then U is invertible, and

$$U^*U = P^{-1}T^*TP^{-1} = P^{-1}P^2P^{-1} = I,$$

so that U is unitary. Since P is invertible, it is obvious that TP^{-1} is the only possible choice for U.

(*b*) Put $p(\lambda) = |\lambda|$, $u(\lambda) = \lambda/|\lambda|$ if $\lambda \neq 0$, $u(0) = 1$. Then p and u are bounded Borel functions on $\sigma(T)$. Put $P = p(T)$, $U = u(T)$. Since $p \geq 0$, Theorem 12.32 shows that $P \geq 0$. Since $u\bar{u} = 1$, $UU^* = U^*U = I$. Since $\lambda = u(\lambda)p(\lambda)$, the relation $T = UP$ follows from the symbolic calculus. ////

Remark. It is not true that every $T \in \mathscr{B}(H)$ has a polar decomposition. (See Exercise 19.) However, if P is the positive square root of T^*T, then $\|Px\| = \|Tx\|$ for every $x \in H$; hence $Tx = Ty$ if $Px = Py$, by linearity. The formula

$$VPx = Tx$$

defines a linear isometry V of $\mathscr{R}(P)$ onto $\mathscr{R}(T)$, which has a continuous extension to a linear isometry of the closure of $\mathscr{R}(P)$ onto the closure of $\mathscr{R}(T)$.

If there is a linear isometry of $\mathscr{R}(P)^\perp$ onto $\mathscr{R}(T)^\perp$, then V can be extended to a unitary operator on H, and then T has a polar decomposition. This always happens when $\dim H < \infty$, since $\mathscr{R}(P)$ and $\mathscr{R}(T)$ have then the same codimension.

If V is extended to a member of $\mathscr{B}(H)$ by defining $Vy = 0$ for all $y \in \mathscr{R}(P)^\perp$, then V is called a *partial isometry*.

Every $T \in \mathscr{B}(H)$ thus has a factorization $T = VP$ in which P is positive and V is a partial isometry.

In (*a*), no two of T, U, P need to commute. For example,

$$\begin{pmatrix} 0 & 1 \\ 2 & 0 \end{pmatrix} = \begin{pmatrix} 0 & 1 \\ 1 & 0 \end{pmatrix} \begin{pmatrix} 2 & 0 \\ 0 & 1 \end{pmatrix}$$

In combination with Theorem 12.16, the polar decomposition leads to an interesting result concerning similarity of normal operators.

12.36 Theorem *Suppose $M, N, T \in \mathscr{B}(H)$, M and N are normal, T is invertible, and*

$$M = TNT^{-1}. \tag{1}$$

If $T = UP$ is the polar decomposition of T, then

$$M = UNU^{-1}. \tag{2}$$

Two operators M and N that satisfy (1) are usually called *similar*. If U is unitary and (2) holds, M and N are said to be *unitarily equivalent*. The theorem thus asserts that similar normal operators are actually unitarily equivalent.

PROOF. By (1), $MT = TN$. Hence $M^*T = TN^*$, by Theorem 12.16. Consequently,

$$T^*M = (M^*T)^* = (TN^*)^* = NT^*,$$

so that

$$NP^2 = NT^*T = T^*MT = T^*TN = P^2N,$$

since $P^2 = T^*T$. Hence N commutes with $f(P^2)$, for every $f \in C(\sigma(P^2))$. (See Section 12.24.) Since $P \geq 0$, $\sigma(P^2) \subset [0, \infty)$. If $f(\lambda) = \lambda^{1/2} \geq 0$ on $\sigma(P^2)$, it follows that $NP = PN$. Hence (1) yields

$$M = (UP)N(UP)^{-1} = UPNP^{-1}U^{-1} = UNU^{-1}. \qquad ////$$

The Group of Invertible Operators

Some features of the group of all invertible elements in a Banach algebra A were described at the end of Chapter 10. The following two theorems contain further information about this group, in the special case $A = \mathscr{B}(H)$.

12.37 Theorem *The group G of all invertible operators $T \in \mathscr{B}(H)$ is connected, and every $T \in G$ is the product of two exponentials.*

Here an *exponential* is, of course, any operator of the form $\exp(S)$ with $S \in \mathscr{B}(H)$.

PROOF. Let $T = UP$ be the polar decomposition of some $T \in G$. Recall that U is unitary and that P is positive and invertible. Since $\sigma(P) \subset (0, \infty)$, log is a continuous real function on $\sigma(P)$. It follows from the symbolic calculus that there is a self-adjoint $S \in \mathscr{B}(H)$ such that $P = \exp(S)$. Since U is unitary, $\sigma(U)$ lies on the unit circle, so that there is a real bounded Borel function f on $\sigma(U)$ that satisfies

$$\exp\{if(\lambda)\} = \lambda \qquad [\lambda \in \sigma(U)].$$

(Note that there may not exist any *continuous* f with this property!) Put $Q = f(U)$. Then $Q \in \mathscr{B}(H)$ is self-adjoint, and $U = \exp(iQ)$. Thus

$$T = UP = \exp(iQ)\exp(S).$$

From this it follows easily that G is connected, for if T_r is defined, for $0 \le r \le 1$, by

$$T_r = \exp(irQ)\exp(rS)$$

then $r \to T_r$ is a continuous mapping of the unit interval $[0, 1]$ into G, $T_0 = I$, and $T_1 = T$. This completes the proof. ////

It is now natural to ask whether every $T \in G$ is an exponential, rather than merely the product of two exponentials. In other words, is every product of two exponentials an exponential? The answer is affirmative if $\dim H < \infty$; in fact, it is affirmative in every finite-dimensional Banach algebra, as a consequence of Theorem 10.30. But in general the answer is negative, as we shall now see.

12.38 Theorem *Let D be a bounded open set in $\mathbb{C}$ such that the set*

(1) $$\Omega = \{\alpha \in \mathbb{C} : \alpha^2 \in D\}$$

is connected and such that 0 is not in the closure of D. Let H be the space of all holomorphic functions f in D that satisfy

(2) $$\int_D |f|^2 \, dm_2 < \infty$$

(where m_2 is Lebesgue measure in the plane), with inner product

(3) $$(f, g) = \int_D f\bar{g} \, dm_2 .$$

Then H is a Hilbert space. Define the multiplication operator $M \in \mathscr{B}(H)$ by

(4) $$(Mf)(z) = zf(z) \qquad (f \in H, z \in D).$$

Then M is invertible, but M has no square root in $\mathscr{B}(H)$.

Since every exponential has roots of all orders, it follows that M is not an exponential.

PROOF. It is clear that (3) defines an inner product that makes H a unitary space. We show now that H is complete. Let K be a compact subset of D, whose distance from the complement of D if δ. If $z \in K$, if Δ is the open circular disc with radius δ and center z, and if $f(\zeta) = \sum a_n(\zeta - z)^n$ for $\zeta \in \Delta$, a simple computation shows that

$$(5) \qquad \sum_{n=0}^{\infty} (n+1)^{-1} |a_n|^2 \delta^{2n+2} = \frac{1}{\pi} \int_{\Delta} |f|^2 \, dm_2 .$$

Since $f(z) = a_0$, it follows that

$$(6) \qquad |f(z)| \le \pi^{-1/2} \delta^{-1} \|f\| \qquad (z \in K, f \in H),$$

where $\|f\| = (f, f)^{1/2}$. Every Cauchy sequence in H converges therefore uniformly on compact subsets of D. From this it follows easily that H is complete. Hence H is a Hilbert space.

Since D is bounded, $M \in \mathscr{B}(H)$. Since $1/z$ is bounded in D, $M^{-1} \in \mathscr{B}(H)$.

Assume now, to reach a contradiction, that $M = Q^2$ for some $Q \in \mathscr{B}(H)$. Fix $\alpha \in \Omega$. Put $\lambda = \alpha^2$. Then $\lambda \in D$. Define

$$(7) \qquad M_\lambda = M - \lambda I, \qquad S = Q - \alpha I, \qquad T = Q + \alpha I,$$

so that

$$(8) \qquad ST = M_\lambda = TS.$$

Since we are dealing with holomorphic functions, the formula

$$(9) \qquad (M_\lambda g)(z) = (z - \lambda) g(z) \qquad (z \in D, \, g \in H)$$

shows that M_λ is one-to-one and that its range $\mathscr{R}(M_\lambda)$ consists of exactly those $f \in H$ that satisfy $f(\lambda) = 0$. Hence (6) shows that $\mathscr{R}(M_\lambda)$ is a *closed* subspace of H, *of codimension* 1.

Since M_λ is one-to-one, the first equation (8) shows that T is one-to-one; the second shows that S is one-to-one. Since $\mathscr{R}(M_\lambda) \neq H$, M_λ is not invertible in $\mathscr{B}(H)$. Hence at least one of S and T is not invertible. Suppose S is not invertible. Since $M_\lambda = ST$, $\mathscr{R}(M_\lambda) \subset \mathscr{R}(S)$, so that $\mathscr{R}(S)$ is either $\mathscr{R}(M_\lambda)$ or H. In the latter case, the open mapping theorem would imply that S is invertible. Hence S is a one-to-one mapping of H onto $\mathscr{R}(M_\lambda)$. But the equation $M_\lambda = ST$ shows that S maps $\mathscr{R}(T)$ onto $\mathscr{R}(M_\lambda)$. Hence $\mathscr{R}(T) = H$, and another application of the open mapping theorem shows that $T^{-1} \in \mathscr{B}(H)$.

We have now proved that one and only one of the operators S and T is invertible in $\mathscr{B}(H)$. Therefore exactly one of the numbers α and $-\alpha$ lies in $\sigma(Q)$, if $\alpha \in \Omega$. It follows that Ω is the union of two disjoint congruent sets, $\sigma(Q) \cap \Omega$ and $-\sigma(Q) \cap \Omega$, both of which are closed (relative to Ω) since $\sigma(Q)$ is compact. The assumption that $M = Q^2$ leads thus to the conclusion that Ω is not connected, which contradicts the hypothesis.

This completes the proof. ////

The simplest example of a region D that satisfies the hypothesis of Theorem 12.38 is a circular annulus with center at 0. In that case, a more conceptual proof can be given. See Exercise 40.

A Characterization of B^*-algebras

The fact that every $\mathscr{B}(H)$ is a B^*-algebra has been exploited throughout this chapter. We shall now establish a converse (Theorem 12.41) which asserts that every B^*-algebra (commutative or not) is isometrically $*$-isomorphic to some closed subalgebra of some $\mathscr{B}(H)$. The proof depends on the existence of a sufficiently large supply of positive functionals.

12.39 Theorem *If A is a B^*-algebra and if $z \in A$, then there exists a positive functional F on A such that*

(1) $$F(e) = 1 \qquad \textit{and} \qquad F(zz^*) = \|z\|^2.$$

PROOF. Set $zz^* = x_0$. By (e) of Theorem 11.28, $\sigma(x_0) \subset [0, \infty)$. Let Δ_0 be the maximal ideal space of the closed algebra $A_0 \subset A$ generated by e and x_0. Then $\hat{A}_0 = C(\Delta_0)$ and (by Theorem 11.19) $\hat{x}_0$ is a nonnegative real continuous function on Δ_0. It attains its maximum at some point $h \in \Delta_0$. Thus

(2) $$\hat{x}_0(h) = \|\hat{x}_0\|_\infty = \|x_0\| = \|z\|^2.$$

Define a linear functional f on A_0 by $f(x) = \hat{x}(h)$. Then

(3) $$f(e) = 1, \qquad f(zz^*) = \|z\|^2,$$

and $\|f\| = 1$, because $|f(x)| \le \|\hat{x}\|_\infty = \|x\|$ for all $x \in A_0$.

The Hahn-Banach theorem extends f to a linear functional F on A, with $\|F\| = 1$. We have to prove that $F(yy^*) \ge 0$ for every $y \in A$.

Fix $y \subset A$ and let Δ_1 be the maximal ideal space of the closed algebra $A_1 \subset A$ generated by e and yy^*. Then $\hat{A}_1 = C(\Delta_1)$. Use F to define a linear functional φ on $C(\Delta_1)$ by setting

(4) $$\varphi(\hat{x}) = F(x) \qquad (x \in A_1).$$

Then $\varphi(1) = F(e) = f(e) = 1$, $|\varphi(\hat{x})| \le \|x\| = \|\hat{x}\|_\infty$, hence $\|\varphi\| = 1$, and now Lemma 5.26 shows that $\varphi(\hat{x}) \ge 0$ for all $x \subset A$ for which $\hat{x} \ge 0$ on Δ_1. If $x_1 = yy^*$, we see, as at the start of this proof, that $\hat{x}_1 \ge 0$ on Δ_1. Thus $F(yy^*) = F(x_1) = \varphi(\hat{x}_1) \ge 0$, as needed. ////

12.40 Theorem *If A is a B^*-algebra and if $u \in A$, $u \neq 0$, there exists a Hilbert space H_u and there exists a homomorphism T_u of A into $\mathscr{B}(H_u)$ that satisfies $T_u(e) = I$,*

(1) $$T_u(x^*) = T_u(x)^* \qquad (x \in A),$$

(2) $$\|T_u(x)\| \le \|x\| \qquad (x \in A),$$

and $\|T_u(u)\| = \|u\|$.

PROOF. We regard u as fixed and omit the subscripts u. Fix a positive functional F on A that satisfies

(3) $$F(e) = 1 \qquad \text{and} \qquad F(u^*u) = \|u\|^2.$$

Such an F exists, by Theorem 12.39. Define

(4) $$Y = \{y \in A\colon F(xy) = 0 \text{ for every } x \in A\}.$$

Since F is continuous (Theorem 11.31), Y is a closed subspace of A. Denote cosets of Y, that is, elements of A/Y, by x':

(5) $$x' = x + Y \qquad (x \in A).$$

We claim that

(6) $$(a', b') = F(b^*a)$$

defines an inner product on A/Y.

To see that (a', b') is well defined by (6), i.e., that it is independent of the choice of representatives a and b, it is enough to show that $F(b^*a) = 0$ if at least one of a or b lies in Y. If $a \in Y$, $F(b^*a) = 0$ follows from (4). If $b \in Y$, then

(7) $$F(b^*a) = \overline{F(a^*b)} = 0,$$

by (*a*) of Theorem 11.31 and another application of (4). Thus (a', b') is well defined, it is linear in a', and conjugate-linear in b', and

(8) $$(a', a') = F(a^*a) \ge 0,$$

since F is a positive functional. If $(a', a') = 0$, then $F(a^*a) = 0$; hence $F(xa) = 0$ for every $x \in A$, by (*b*) of Theorem 11.31, so that $a \in Y$ and $a' = 0$.

A/Y is thus an inner product space, with norm $\|a'\| = F(a^*a)^{1/2}$. Its completion H is the Hilbert space that we are looking for. We define linear operators $T(x)$ on A/Y by

(9) $$T(x)a' = (xa)'.$$

Again, one checks easily that this definition is independent of the choice of $a \in a'$, for if $y \in Y$, (4) implies that $xy \in Y$. (Y is a left ideal in A.) It is obvious that $x \to T(x)$ is linear and that

$$(10) \qquad T(x_1)T(x_2) = T(x_1x_2) \qquad (x_1 \in A,\ x_2 \in A);$$

in particular, (9) shows that $T(e)$ is the identity operator on A/Y. We now claim that

$$(11) \qquad \|T(x)\| \le \|x\| \qquad (x \in A).$$

Once this is shown, the uniform continuity of the operators $T(x)$ enables us to extend them to bounded linear operators on H. Note that

$$(12) \qquad \|T(x)a'\|^2 = ((xa)', (xa)') = F(a^*x^*xa).$$

For fixed $a \in A$, define $G(x) = F(a^*xa)$. Then G is a positive functional on A, so that

$$(13) \qquad G(x^*x) \le G(e)\|x\|^2,$$

by (*d*) of Theorem 11.31. Thus

$$(14) \qquad \|T(x)a'\|^2 = G(x^*x) \le F(a^*a)\|x\|^2 = \|a'\|^2\|x\|^2,$$

which proves (11).

Next, the computation

$$\begin{aligned}(T(x^*)a', b') &= ((x^*a)', b') = F(b^*x^*a) = F((xb)^*a)\\ &= (a', (xb)') = (a', T(x)b') = (T(x)^*a', b')\end{aligned}$$

shows that $T(x^*)a' = T(x)^*a'$, for all $a' \in A/Y$. Since A/Y is dense in H, this proves (1).

Finally, (3) and (12) show that

$$(15) \qquad \|u\|^2 = F(u^*u) = \|T(u)e'\|^2 \le \|T(u)\|^2$$

since $\|e'\|^2 = F(e^*e) = F(e) = 1$. In conjunction with (11), (15) gives $\|T(u)\| = \|u\|$, and the proof is complete. ////

12.41 Theorem *If A is a B^*-algebra, there exists an isometric *-isomorphism of A onto a closed subalgebra of $\mathscr{B}(H)$, where H is a suitably chosen Hilbert space.*

PROOF. Let H be the "direct sum" of the Hilbert spaces H_u constructed in Theorem 12.40. Here is a precise description of H: Let $\pi_u(v)$ be the H_u-coordinate of an element v of the cartesian product of the spaces H_u. Then, by definition, $v \in H$ if and only if

$$(1) \qquad \sum_u \|\pi_u(v)\|^2 < \infty,$$

where $\|\pi_u(v)\|$ denotes the H_u-norm of $\pi_u(v)$. The convergence of (1) implies that at most countably many $\pi_u(v)$ are different from 0. The inner product in H is given by

$$(v', v'') = \sum_u (\pi_u(v'), \pi_u(v'')) \qquad (v', v'' \in H), \tag{2}$$

so that $\|v\|^2 = (v, v)$ is the left side of (1). We leave it as an exercise to verify that all Hilbert space axioms are now satisfied by H.

If $S_u \in \mathscr{B}(H_u)$, if $\|S_u\| \le M$ for all u, and if Sv is defined to be the vector whose coordinate in H_u is

$$\pi_u(Sv) = S_u \pi_u(v), \tag{3}$$

one verifies easily that $Sv \in H$ if $v \in H$, that $S \in \mathscr{B}(H)$, and that

$$\|S\| = \sup_u \|S_u\|. \tag{4}$$

We now associate with each $x \in A$ an operator $T(x) \in \mathscr{B}(H)$, by requiring that

$$\pi_u(T(x)v) = T_u(x)(\pi_u(v)), \tag{5}$$

where T_u is as in Theorem 12.40. Since

$$\|T_u(x)\| \le \|x\| = \|T_x(x)\|, \tag{6}$$

by Theorem 12.40, it follows from (4) that

$$\|T(x)\| = \sup_u \|T_u(x)\| = \|x\|. \tag{7}$$

That the mapping $x \to T(x)$ of A into $\mathscr{B}(H)$ has the other required properties follows from a coordinatewise application of Theorem 12.40. ////

An Ergodic Theorem

12.42 Definitions The term "ergodic" comes from statistical mechanics, where it is applied to systems in which "time average = space average" holds for certain quantities. To see a simple mathematical example, let μ be a probability measure on some σ-algebra $\mathscr{M}$ in a set Ω, let ψ map Ω into Ω, and define its iterates by $\psi^1 = \psi$, $\psi^n = \psi \circ \psi^{n-1}$ $(n = 2, 3, 4, \ldots)$. If we think of time as discrete, the "time average" of a function f on Ω, relative to the transformation ψ, is

$$\lim_{n\to\infty} \frac{1}{n}(f + f \circ \psi + \cdots + f \circ \psi^{n-1}) \tag{1}$$

when this limit exists in some sense.

The "space average" of an $f \in L^1(\mu)$ is simply $\int_\Omega f\, d\mu$.

We will be concerned with *measure-preserving* one-to-one maps ψ of Ω onto Ω. This means that $\psi(E)$ and $\psi^{-1}(E)$ are in $\mathscr{M}$ for every $E \in \mathscr{M}$ and that their measure is $\mu(E)$. It is then clear that

$$\int_\Omega (f \circ \psi)\, d\mu = \int_\Omega f\, d\mu \tag{2}$$

for every $f \in L^1(\mu)$.

If, moreover, $\psi(E) = E \in \mathscr{M}$ occurs only when $\mu(E) = 0$ or $\mu(E) = 1$, then ψ is said to be *ergodic*. In that case it is clear that every measurable function g for which $g \circ \psi = g$ a.e. $[\mu]$ is constant a.e. $[\mu]$.

We can now state von Neumann's *mean ergodic theorem*; it is so named because L^2-convergence used to be called "convergence in the mean."

12.43 Theorem *Let $(\Omega, \mathscr{M}, \mu)$ be as above. If $\psi: \Omega \to \Omega$ is one-to-one and measure-preserving, and $f \in L^2(\mu)$, then the averages*

$$A_n f = \frac{1}{n}(f + f \circ \psi + \cdots + f \circ \psi^{n-1})$$

converge, in the L^2-metric, to some $g \in L^2(\mu)$, as $n \to \infty$.

Moreover, $g \circ \psi = g$. Thus g is the constant $\int_\Omega f\, d\mu$ if ψ is ergodic.

It is clear that the second assertion follows from the first. The first one says, explicitly, that

$$\lim_{n \to \infty} \int_\Omega |g - A_n f|^2\, d\mu = 0.$$

The key to the proof is the observation that the map $f \to f \circ \psi$ is an isometry of $L^2(\mu)$ onto $L^2(\mu)$. It is thus a unitary operator on the Hilbert space $L^2(\mu)$. The following abstract reformulation of Theorem 12.43 is then an easy consequence of the spectral theorem.

12.44 Theorem *If $U \in \mathscr{B}(H)$ is unitary and $x \in H$, then the averages*

$$A_n x = \frac{1}{n}(x + Ux + \cdots + U^{n-1}x) \tag{1}$$

converge, in the norm topology of H, to some $y \in H$.

PROOF. Let E be the spectral decomposition of U. Define functions a_n and b on the unit circle by

$$a_n(\lambda) = \frac{1}{n}(1 + \lambda + \cdots + \lambda^{n-1}), \tag{2}$$

$b(1) = 1$, $b(\lambda) = 0$ for $\lambda \neq 1$.

Then $A_n x = a_n(U)x$. Set $y = b(U)x$. This gives

$$(3) \qquad \|y - A_n x\|^2 = \|b(U)x - a_n(U)x\|^2 = \int_{\sigma(U)} |b - a_n|^2 \, dE_{x,x}.$$

Since $|b - a_n| < 1$ on the unit circle, and $(b - a_n)(\lambda) \to 0$ pointwise, the dominated convergence theorem shows that

$$(4) \qquad \lim_{n \to \infty} \|y - A_n x\| = 0. \qquad ////$$

Exercises

Throughout these exercises, the letter H denotes a Hilbert space.

1. The completion of an inner product space is a Hilbert space. Make this statement more precise, and prove it. (See the proof of Theorem 12.40 for an application.)

2. Suppose N is a positive integer, $\alpha \in \mathcal{C}$, $\alpha^N = 1$, and $\alpha^2 \neq 1$. Prove that every Hilbert space inner product satisfies the identities

$$(x, y) = \frac{1}{N} \sum_{n=1}^{N} \|x + \alpha^n y\|^2 \alpha^n$$

and

$$(x, y) = \frac{1}{2\pi} \int_{-\pi}^{\pi} \|x + e^{i\theta} y\|^2 e^{i\theta} \, d\theta.$$

Generalize this: Which functions f and measures μ on a set Ω give rise to the identity

$$(x, y) = \int_{\Omega} \|x + f(p)y\|^2 \, d\mu(p)?$$

3. (*a*) Assume x_n and y_n are in the closed unit ball of H, and $(x_n, y_n) \to 1$ as $n \to \infty$. Prove that then $\|x_n - y_n\| \to 0$.

(*b*) Assume $x_n \in H$, $x_n \to x$ weakly, and $\|x_n\| \to \|x\|$. Prove that then $\|x_n - x\| \to 0$.

4. Let H^* be the dual space of H; define $\psi: H^* \to H$ by

$$y^*(x) = (x, \psi y^*) \qquad (x \in H, \, y^* \in H^*).$$

(See Theorem 12.5.) Prove that H^* is a Hilbert space, relative to the inner product

$$[x^*, y^*] = (\psi y^*, \psi x^*).$$

If $\phi: H^{**} \to H^*$ satisfies $z^{**}(y^*) = [y^*, \phi z^{**}]$ for all $y^* \in H^*$ and $z^{**} \in H^{**}$, prove that $\psi\phi$ is an isomorphism of H^{**} onto H whose existence implies that H is reflexive.

5. Suppose $\{u_n\}$ is a sequence of unit vectors in H (that is $\|u_n\| = 1$), and assume that

$$\Gamma^2 = \sum_{i \neq j} |(u_i, u_j)|^2 < \infty.$$

If $\{\alpha_i\}$ is any sequence of scalars, prove that

$$(1 - \Gamma) \sum_{i=m}^{n} |\alpha_i|^2 \leq \left\| \sum_{i=m}^{n} \alpha_i u_i \right\|^2 \leq (1 + \Gamma) \sum_{i=m}^{n} |\alpha_i|^2,$$

and deduce that the following three properties of $\{\alpha_i\}$ are equivalent to each other:

(a) $\sum_{i=1}^{\infty} |\alpha_i|^2 < \infty.$

(b) $\sum_{i=1}^{\infty} \alpha_i u_i$ converges, in the norm of H.

(c) $\sum_{i=1}^{\infty} \alpha_i (u_i, y)$ converges, for every $y \in H$.

This generalizes Theorem 12.6.

6. Suppose E is a resolution of the identity, as in Section 12.17, and prove that

$$|E_{x,y}(\omega)|^2 \leq E_{x,x}(\omega) E_{y,y}(\omega)$$

for all $x \in H$, $y \in H$, and $\omega \in \mathfrak{M}$.

7. Suppose $U \in \mathscr{B}(H)$ is unitary, and $\varepsilon > 0$. Prove that scalars $\alpha_0, \ldots, \alpha_n$ can be chosen so that

$$\|U^{-1} - \alpha_0 I - \alpha_1 U - \cdots - \alpha_n U^n\| < \varepsilon,$$

if $\sigma(U)$ is a proper subset of the unit circle, but that this norm is never less than 1 if $\sigma(U)$ covers the whole circle.

8. Prove Theorem 12.35 with PU in place of UP.

9. Suppose $T = UP$ is the polar decomposition of an invertible $T \in \mathscr{B}(H)$. Prove that T is normal if and only if $UP = PU$.

10. Prove that every normal invertible $T \in \mathscr{B}(H)$ is the exponential of some normal $S \in \mathscr{B}(H)$.

11. Suppose $N \in \mathscr{B}(H)$ is normal, and $T \in \mathscr{B}(H)$ is invertible. Prove that TNT^{-1} is normal if and only if N commutes with T^*T.

12. (a) Suppose $S \in \mathscr{B}(H)$, $T \in \mathscr{B}(H)$, S and T are normal, and $ST = TS$. Prove that $S + T$ and ST are normal.

    (b) If, in addition, $S \geq 0$ and $T \geq 0$ (see Theorem 12.32), prove that $S + T \geq 0$ and $ST \geq 0$.

    (c) Show, however, that there exist $S \geq 0$ and $T \geq 0$ such that ST is not even normal (of course, then $ST \neq TS$). In fact, such examples exist if $\dim H = 2$.

13. If $T \in \mathscr{B}(H)$ is normal, show that $T^* = UT$, for some unitary U. When is U unique?

14. Assume $T \in \mathscr{B}(H)$ and T^*T is a compact operator. Show that T is then compact.

15. Find a noncompact $T \in \mathscr{B}(H)$ such that $T^2 = 0$. Can such an operator be normal?

16. Suppose $T \in \mathscr{B}(H)$ is normal, and $\sigma(T)$ is a finite set. Deduce as much information about T from this as you can.

17. Show, under the hypotheses of (*d*) of Theorem 12.29, that the equation $Ty = x$ has a solution $y \in H$ if and only if

$$\sum_{i=1}^{\infty} |\lambda_i|^{-2} \|x_i\|^2 < \infty.$$

(If $\lambda_i = 0$ for one i, then x_i must be 0, for this i.)

18. The spectrum $\sigma(T)$ of $T \in \mathscr{B}(H)$ can be divided into three disjoint pieces:

The *point spectrum* $\sigma_p(T)$ consists of all $\lambda \in \mathcal{C}$ for which $T - \lambda I$ is not one-to-one.

The *continuous spectrum* $\sigma_c(T)$ consists of all $\lambda \in \mathcal{C}$ such that $T - \lambda I$ is a one-to-one mapping of H onto a dense proper subspace of H.

The *residual spectrum* $\sigma_r(T)$ consists of all other $\lambda \in \sigma(T)$.

(*a*) Prove that every normal $T \in \mathscr{B}(H)$ has empty residual spectrum.

(*b*) Prove that the point spectrum of a normal $T \in \mathscr{B}(H)$ is at most countable, if H is separable.

(*c*) Let S_R and S_L be the right and left shifts (as defined in Exercise 2 of Chapter 10), acting on the Hilbert space ℓ^2.

Prove that $(S_R)^* = S_L$ and that

$$\sigma_p(S_L) = \sigma_r(S_R) = \{\lambda : |\lambda| < 1\},$$
$$\sigma_c(S_L) = \sigma_c(S_R) = \{\lambda : |\lambda| = 1\},$$
$$\sigma_r(S_L) = \sigma_p(S_R) = \varnothing.$$

19. Let S_R and S_L be as above. Prove that neither S_R nor S_L has polar decompositions UP, with U unitary and $P \geq 0$.

20. Let μ be a positive measure on a measure space Ω, let $H = L^2(\mu)$, with the usual inner product

$$(f, g) = \int_\Omega f\bar{g}\, d\mu.$$

For $\phi \in L^\infty(\mu)$, define the multiplication operator M_ϕ by $M_\phi(f) = \phi f$. Then $M_\phi \in \mathscr{B}(H)$.

Under what conditions on ϕ does M_ϕ have eigenvalues? Give an example in which $\sigma(M_\phi) = \sigma_c(M_\phi)$. Show that every M_ϕ is normal. What is the relation between $\sigma(M_\phi)$ and the essential range of ϕ? Show that $\phi \to M_\phi$ is an isometric *-isomorphism of $L^\infty(\mu)$ onto a closed subalgebra A of $\mathscr{B}(H)$. (Certain pathological measures μ have to be excluded in order to make this last statement correct.) Is A a maximal commutative sub-algebra of $\mathscr{B}(H)$? *Hint:* If $T \in \mathscr{B}(H)$ and $TM_\phi = M_\phi T$ for all $\phi \in L^\infty(\mu)$, and if $\mu(\Omega) < \infty$, show that T is a multiplication by $T(1)$ and hence that $T \in A$.

21. Suppose $T \in \mathscr{B}(H)$ is normal, A is the closed subalgebra of $\mathscr{B}(H)$ generated by I, T, and T^*, and T can be approximated, in the norm topology of $\mathscr{B}(H)$, by finite linear combinations of projections *that belong to A*.

Under what (necessary and sufficient) conditions on $\sigma(T)$ does this happen?

22. Does every normal $T \in \mathscr{B}(H)$ have a square root in $\mathscr{B}(H)$? What can you say about the cardinality of the set of all square roots of T? Can it happen that two square roots of the same T do not commute? Can this happen when $T = I$?
23. Show that the Fourier transform $f \to \hat{f}$ is a unitary operator on $L^2(R^n)$. What is its spectrum? *Suggestion:* When $n = 1$, compute the Fourier transforms of

$$\exp\left(\frac{1}{2}x^2\right)\left(\frac{d}{dx}\right)^m \exp(-x^2) \qquad (m = 0, 1, 2, \ldots).$$

24. Show that any two infinite-dimensional separable Hilbert spaces are isometrically isomorphic (via countable orthonormal bases; see [23]). Show that the space H in Theorem 12.38 is separable. Show that the answer to the question that precedes Theorem 12.38 is therefore negative for every infinite-dimensional H, separable or not.
25. Suppose $T \in \mathscr{B}(H)$ is normal, f is a bounded Borel function on $\sigma(T)$, and $S = f(T)$. If E_T and E_S are the spectral decompositions of T and S, respectively, prove that

$$E_S(\omega) = E_T(f^{-1}(\omega))$$

for every Borel set $\omega \subset \sigma(S)$.

26. If $S \in \mathscr{B}(H)$ and $T \in \mathscr{B}(H)$, the notation $S \geq T$ means that $S - T \geq 0$, that is, that

$$(Sx, x) \geq (Tx, x)$$

for all $x \in H$. Prove the equivalence of the following four properties of a pair of self-adjoint projections P and Q:

(a) $P \geq Q$.
(b) $\mathscr{R}(P) \supset \mathscr{R}(Q)$.
(c) $PQ = Q$.
(d) $QP = Q$.

If E is a resolution of the identity, it follows that $E(\omega') \geq E(\omega'')$ if $\omega' \supset \omega''$.

27. Suppose * is an involution in a complex algebra A, q is an invertible element of A such that $q^* = q$ and $x^\#$ is defined by

$$x^\# = q^{-1}x^*q$$

for every $x \in A$. Show that $^\#$ is an involution in A.

28. Let A be the algebra of all complex 4-by-4 matrices. If $M = (m_{ij}) \in A$, let M^* be the conjugate transpose of M: $m_{ij}^* = \overline{m_{ji}}$. Put

$$Q = \begin{pmatrix} 0 & 0 & 0 & 1 \\ 0 & 0 & 1 & 0 \\ 0 & 1 & 0 & 0 \\ 1 & 0 & 0 & 0 \end{pmatrix} \qquad S = \begin{pmatrix} 0 & 0 & 0 & 0 \\ 1 & 0 & 0 & 0 \\ 0 & 0 & 0 & 0 \\ 0 & 0 & 0 & 0 \end{pmatrix} \qquad T = \begin{pmatrix} 0 & 0 & 0 & 0 \\ 0 & 0 & 0 & 0 \\ 0 & 0 & 0 & 0 \\ 0 & 0 & 0 & 1 \end{pmatrix}.$$

As in Exercise 27, define

$$M^\# = Q^{-1}M^*Q \qquad (M \in A).$$

(*a*) Show that S and T are normal, with respect to the involution $^\#$, that $ST = TS$, but that $ST^\# \neq T^\# S$.
(*b*) Show that $S + T$ is not $^\#$-normal.
(*c*) Compare $\|SS^\#\|$ with $\|S\|^2$.
(*d*) Compute the spectral radius $\rho(S + S^\#)$; show that it is different from $\|S + S^\#\|$.
(*e*) Define $V = (v_{ij}) \in A$ so that $v_{12} = v_{24} = i$, $v_{31} = v_{43} = -i$, $v_{ij} = 0$ otherwise. Compute $\sigma(VV^\#)$; it does not lie in $[0, \infty)$.

Part (*a*) shows that Theorem 12.16 fails for some involutions. Part (*b*) does the same for part (*a*) of Exercise 12; (*c*), (*d*), and (*e*) show that various parts of Theorem 11.28 fail for the involution $^\#$.

29. Let X be the vector space of all trigonometric polynomials on the real line: these are functions of the form

$$f(t) = c_1 e^{is_1 t} + \cdots + c_n e^{is_n t},$$

where $s_k \in R$ and $c_k \in \mathcal{C}$, for $1 \le k \le n$. Show that

$$(f, g) = \lim_{A \to \infty} \frac{1}{2A} \int_{-A}^{A} f(t)\overline{g(t)}\, dt$$

exists and defines an inner product on X, that

$$\|f\|^2 = (f, f) = |c_1|^2 + \cdots + |c_n|^2,$$

and that the completion of X is a nonseparable Hilbert space H. Show that H contains all *uniform* limits of trigonometric polynomials; these are the so-called "almost-periodic" functions on R.

30. Let H_w be an infinite-dimensional Hilbert space, with its weak topology. Prove that the inner product is a separately continuous function on $H_w \times H_w$ which is not jointly continuous.

31. Assume $T_n \in \mathscr{B}(H)$ for $n = 1, 2, 3, \ldots$, and

$$\lim_{n \to \infty} \|T_n x\| = 0$$

for every $x \in H$. Does it follow that

$$\lim_{n \to \infty} \|T_n^* x\| = 0$$

for every $x \in H$?

32. Let X be a *uniformly convex* Banach space. This means, by definition, that the assumptions

$$\|x_n\| \le 1, \qquad \|y_n\| \le 1, \qquad \|x_n + y_n\| \to 2$$

imply that $\|x_n - y_n\| \to 0$.

For example, every Hilbert space is uniformly convex.
(*a*) Prove that Theorem 12.3 holds in X.
(*b*) Assume $\|x_n\| = 1$, $\Lambda \in X^*$, $\|\Lambda\| = 1$, and $\Lambda x_n \to 1$. Prove that $\{x_n\}$ is a Cauchy sequence (in the norm-topology of X). *Hint:* Consider $\Lambda(x_n + x_m)$.
(*c*) Prove that every $\Lambda \in X^*$ attains its maximum on the closed unit ball of X.

(*d*) Assume that $x_n \to x$ weakly and $\|x_n\| \to \|x\|$. Prove that $\|x_n - x\| \to 0$. *Hint:* Reduce to the case $\|x_n\| = 1$. Consider $\Lambda(x_n + x)$, for a suitable Λ.

(*c*) Show that the preceding four properties fail in certain Banach spaces (for instance, in L^1, or in C). These are therefore not uniformly convex.

33. Prove the assertion about the case dim $H < \infty$ made in the remark that follows Theorem 12.35.

34. Find an operator $T \in \mathscr{B}(H)$, with $\sigma(T) = \{1\}$, which is neither unitary nor self-adjoint.

35. If S is self-adjoint and $U = \exp(iS)$, show that U is unitary. Deduce from this, and from the fact that $\sigma(U)$ lies on the unit circle, that $\sigma(S)$ lies on the real axis.

36. Show that $\mathscr{R}(T^*) = \mathscr{R}(T)$ if $T \in \mathscr{B}(H)$ is normal. *Hint:* Using Theorem 12.35, $T = T^*U^2$.

37. Define T on $H = L^2(0, 1)$ by $(Tf)(x) = xf(x)$. Show that T is self-adjoint and that $\mathscr{R}(T)$ is a dense proper subspace of H.

38. Find a nonnormal $T \in \mathscr{B}(H)$ such that

$$\|T\| = \sup \{|(Tx, x)| : x \in H, \|x\| \le 1\}.$$

(This shows that Theorem 12.25 has no converse.)

39. Show that T and T^* can have the same null space without being normal.

40. Let D be a circular annulus in $\mathcal{C}$ with center at 0. Define H and $M \in \mathscr{B}(H)$ as in Theorem 12.38. Prove that M has no square root in $\mathscr{B}(H)$ by completing the following outline: Assume $Q \in \mathscr{B}(H)$, $Q^2 = M$. Put $u(z) = 1$, $v(z) = z$, $h = Qu$. Since $QM = MQ$, induction shows that $Qv^n = hv^n$ for all integers n. It follows from the Laurent series expansion that $Qf = hf$ for every $f \in H$. This leads to $h^2 = v$, i.e., $h^2(z) = z$ for all $z \in D$, an impossibility.

Find the adjoint M^* of M. (Use Laurent series.)

CHAPTER 13

UNBOUNDED OPERATORS

Introduction

13.1 Definitions Let H be a Hilbert space. By an *operator in* H we shall now mean a linear mapping T whose domain $\mathscr{D}(T)$ is a subspace of H and whose range $\mathscr{R}(T)$ lies in H.

It is not assumed that T is bounded or continuous. Of course, if T is continuous [relative to the norm topology that $\mathscr{D}(T)$ inherits from H] then T has a continuous extension to the closure of $\mathscr{D}(T)$, hence to H, since $\overline{\mathscr{D}(T)}$ is complemented in H. In that case, T is the restriction to $\mathscr{D}(T)$ of some member of $\mathscr{B}(H)$.

The *graph* $\mathscr{G}(T)$ of an operator T in H is the subspace of $H \times H$ that consists of the ordered pairs $\{x, Tx\}$, where x ranges over $\mathscr{D}(T)$. Obviously, S is an extension of T [that is, $\mathscr{D}(T) \subset \mathscr{D}(S)$ and $Sx = Tx$ for $x \in \mathscr{D}(T)$] if and only if $\mathscr{G}(T) \subset \mathscr{G}(S)$. This inclusion will often be written in the simpler form

$$T \subset S. \tag{1}$$

A *closed* operator in H is one whose graph is a closed subspace of $H \times H$. By the closed graph theorem, $T \in \mathscr{B}(H)$ if and only if $\mathscr{D}(T) = H$ and T is closed.

We wish to associate a Hilbert space adjoint T^* to T. Its domain $\mathscr{D}(T^*)$ is to consist of all $y \in H$ for which the linear functional

$$x \to (Tx, y) \tag{2}$$

is continuous on $\mathscr{D}(T)$. If $y \in \mathscr{D}(T^*)$, then the Hahn-Banach theorem extends the functional (2) to a continuous linear functional on H, and therefore there exists an element $T^*y \in H$ that satisfies

$$(3) \qquad (Tx, y) = (x, T^*y) \qquad [x \in \mathscr{D}(T)].$$

Obviously, T^*y will be *uniquely* determined by (3) if and only if $\mathscr{D}(T)$ is dense in H, that is, if and only if T is *densely defined.* The only operators T that will be given an adjoint T^* are therefore the densely defined ones. Routine verifications show then that T^* is also an operator in H, that is, that $\mathscr{D}(T^*)$ is a subspace of H and that T^* is linear.

Note that if $T \in \mathscr{B}(H)$, then the definition of T^* given here coincides with that given in Section 12.9. In particular, $\mathscr{D}(T^*) = H$ and $T^* \in \mathscr{B}(H)$.

Ordinary algebraic operations with unbounded operators must be handled with care; the domains have to be watched. Here are the natural definitions for the domains of sums and products:

$$(4) \qquad \mathscr{D}(S + T) = \mathscr{D}(S) \cap \mathscr{D}(T),$$

$$(5) \qquad \mathscr{D}(ST) = \{x \in \mathscr{D}(T): Tx \in \mathscr{D}(S)\}.$$

The usual associative laws

$$(6) \qquad (R + S) + T = R + (S + T), \qquad (RS)T = R(ST)$$

then hold. As regards the distributive laws, one of them, namely, $(R + S)T = RT + ST$, holds in its usual form, but the other one may only hold in the form

$$(7) \qquad T(R + S) \supset TR + TS,$$

since it can happen that $(R + S)x \in \mathscr{D}(T)$, even though one of Rx or Sx is not in $\mathscr{D}(T)$. Scalar multiplication is defined as follows: If $\alpha = 0$, then $\mathscr{D}(\alpha T) = H$ and $\alpha T = 0$. If $\alpha \neq 0$, then $\mathscr{D}(\alpha T) = \mathscr{D}(T)$ and $(\alpha T)x = \alpha(Tx)$ for $x \in \mathscr{D}(T)$.

13.2 Theorem *Suppose S, T, and ST are densely defined operators in H. Then*

$$(1) \qquad T^*S^* \subset (ST)^*.$$

If, in addition, $S \in \mathscr{B}(H)$, then

$$(2) \qquad T^*S^* = (ST)^*.$$

Note that (1) asserts that $(ST)^*$ is an extension of T^*S^*. The equality (2) implies that T^*S^* and $(ST)^*$ actually have the same domains.

PROOF. Suppose $x \in \mathscr{D}(ST)$ and $y \in \mathscr{D}(T^*S^*)$. Then

$$(3) \qquad (Tx, S^*y) = (x, T^*S^*y),$$

because $x \in \mathscr{D}(T)$ and $S^*y \in \mathscr{D}(T^*)$, and

(4) $$(STx, y) = (Tx, S^*y),$$

because $Tx \in \mathscr{D}(S)$ and $y \in \mathscr{D}(S^*)$. Hence

(5) $$(STx, y) = (x, T^*S^*y).$$

This proves (1).

Assume now that $S \in \mathscr{B}(H)$ and $y \in \mathscr{D}((ST)^*)$. Then $S^* \in \mathscr{B}(H)$, so that $\mathscr{D}(S^*) = H$, and

(6) $$(Tx, S^*y) = (STx, y) = (x, (ST)^*y)$$

for every $x \in \mathscr{D}(ST)$. Hence $S^*y \in \mathscr{D}(T^*)$, and therefore $y \in \mathscr{D}(T^*S^*)$. Now (2) follows from (1). ////

13.3 Definition An operator T in H is said to be *symmetric* if

(1) $$(Tx, y) = (x, Ty)$$

whenever $x \in \mathscr{D}(T)$ and $y \in \mathscr{D}(T)$. The densely defined symmetric operators are thus exactly those that satisfy

(2) $$T \subset T^*.$$

If $T = T^*$, then T is said to be *self-adjoint*.

These two properties evidently coincide when $T \in \mathscr{B}(H)$. In general, they do not.

Moreover, if $\mathscr{D}(T)$ is dense and $(Tx, y) = (x, Sy)$ for all $x \in \mathscr{D}(T)$ and $y \in \mathscr{D}(S)$, then $S \subset T^*$.

13.4 Example Let $H = L^2 = L^2([0, 1])$, relative to Lebesgue measure. We define operators T_1, T_2, and T_3 in L^2. Their domains are as follows:

$\mathscr{D}(T_1)$ consists of all absolutely continuous functions f on $[0, 1]$ with derivative $f' \in L^2$.

$$\mathscr{D}(T_2) = \mathscr{D}(T_1) \cap \{f: f(0) = f(1)\}.$$
$$\mathscr{D}(T_3) = \mathscr{D}(T_1) \cap \{f: f(0) = f(1) = 0\}.$$

These are dense in L^2. Define

(1) $$T_k f = if' \qquad \text{for } f \in \mathscr{D}(T_k),\ k = 1, 2, 3.$$

We claim that

(2) $$T_1^* = T_3, \qquad T_2^* = T_2, \qquad T_3^* = T_1.$$

Since $T_3 \subset T_2 \subset T_1$, it follows that T_2 is a self-adjoint extension of the symmetric (but not self-adjoint) operator T_3 and that the extension T_1 of T_2 is not symmetric.

Let us prove (2). Note that

$$(T_k f, g) = \int_0^1 (if')\bar{g} = \int_0^1 f\overline{(ig')} = (f, T_m g) \tag{3}$$

when $f \in \mathscr{D}(T_k)$, $g \in \mathscr{D}(T_m)$, and $m + k = 4$, since then $f(1)\bar{g}(1) = f(0)\bar{g}(0)$. It follows that $T_m \subset T_k^*$, or

$$T_1 \subset T_3^*, \qquad T_2 \subset T_2^*, \qquad T_3 \subset T_1^*. \tag{4}$$

Suppose now that $g \in \mathscr{D}(T_k^*)$ and $\phi = T_k^* g$. Put $\Phi(x) = \int_0^x \phi$. Then, for $f \in \mathscr{D}(T_k)$,

$$\int_0^1 if'\bar{g} = (T_k f, g) = (f, \phi) = f(1)\overline{\Phi(1)} - \int_0^1 f'\bar{\Phi}. \tag{5}$$

When $k = 1$ or 2, then $\mathscr{D}(T_k)$ contains nonzero constants, so that (5) implies $\Phi(1) = 0$. When $k = 3$, then $f(1) = 0$. It follows, in all cases, that

$$ig - \Phi \in \mathscr{R}(T_k)^\perp. \tag{6}$$

Since $\mathscr{R}(T_1) = L^2$, $ig = \Phi$ if $k = 1$, and since $\Phi(1) = 0$ in that case, $g \in \mathscr{D}(T_3)$. Thus $T_1^* \subset T_3$.

If $k = 2$ or 3, then $\mathscr{R}(T_k)$ consists of all $u \in L^2$ such that $\int_0^1 u = 0$. Thus

$$\mathscr{R}(T_2) = \mathscr{R}(T_3) = Y^\perp, \tag{7}$$

where Y is the one-dimensional subspace of L^2 that contains the constants. Hence (6) implies that $ig - \Phi$ is constant. Thus g is absolutely continuous and $g' \in L^2$, that is, $g \in \mathscr{D}(T_1)$. Thus $T_3^* \subset T_2$.

If $k = 2$, then $\Phi(1) = 0$, hence $g(0) = g(1)$, and $g \in \mathscr{D}(T_2)$. Thus $T_2^* \subset T_2$.

This completes the proof.

Before we turn to a more detailed study of the relations between symmetric operators and self-adjoint ones, we insert another example.

13.5 Example Let $H = L^2$, as in Example 13.4, define $Df = f'$ for $f \in \mathscr{D}(T_2)$, say (the exact domain is now not very important), and define $(Mf)(t) = tf(t)$. Then $(DM - MD)f = f$, or

$$DM - MD = I, \tag{1}$$

where I denotes the identity operator on the domain of D.

The identity operator appears thus as a commutator of two operators, of which only one is bounded. The question whether the identity is the commutator of two *bounded* operators on H arose in quantum mechanics. The answer is negative, not just in $\mathscr{B}(H)$, but in every Banach algebra.

13.6 Theorem *If A is a Banach algebra with unit element e, if $x \in A$ and $y \in A$, then*

$$xy - yx \neq e.$$

The following proof, due to Wielandt, does not even use the completeness of A.

PROOF. Assume $xy - yx = e$. Make the induction hypothesis

$$x^n y - yx^n = nx^{n-1} \neq 0, \tag{1}$$

which is assumed to hold for $n = 1$. If (1) holds for some positive integer n, then $x^n \neq 0$ and

$$\begin{aligned} x^{n+1}y - yx^{n+1} &= x^n(xy - yx) + (x^n y - yx^n)x \\ &= x^n e + nx^{n-1}x = (n+1)x^n, \end{aligned}$$

so that (1) holds with $n + 1$ in place of n. It follows that

$$n\|x^{n-1}\| = \|x^n y - yx^n\| \leq 2\|x^n\|\,\|y\| \leq 2\|x^{n-1}\|\,\|x\|\,\|y\|,$$

or $n \leq 2\|x\|\,\|y\|$, for every positive integer n. This is obviously impossible. ////

Graphs and Symmetric Operators

13.7 Graphs If H is a Hilbert space, then $H \times H$ can be made into a Hilbert space by defining the inner product of two elements $\{a, b\}$ and $\{c, d\}$ of $H \times H$ to be

$$(\{a, b\}, \{c, d\}) = (a, c) + (b, d), \tag{1}$$

where (a, c) denotes the inner product in H. We leave it as an exercise to verify that this satisfies all the properties listed in Section 12.1. In particular, the norm in $H \times H$ is given by

$$\|\{a, b\}\|^2 = \|a\|^2 + \|b\|^2. \tag{2}$$

Define

$$V\{a, b\} = \{-b, a\} \qquad (a \in H, b \in H). \tag{3}$$

Then V is a *unitary* operator on $H \times H$, which satisfies $V^2 = -I$. Thus $V^2M = M$ if M is any subspace of $H \times H$.

This operator yields a remarkable description of T^* in terms of T:

13.8 Theorem *If T is a densely defined operator in H, then*

$$\mathscr{G}(T^*) = [V\mathscr{G}(T)]^\perp, \tag{1}$$

the orthogonal complement of $V\mathscr{G}(T)$ in $H \times H$.

Note that once $\mathscr{G}(T^*)$ is known, so are $\mathscr{D}(T^*)$ and T^*.

PROOF. Each of the following four statements is clearly equivalent to the one that follows and/or precedes it.

$$\{y, z\} \in \mathscr{G}(T^*). \tag{2}$$

$$(Tx, y) = (x, z) \qquad \text{for every } x \in \mathscr{D}(T). \tag{3}$$

$$(\{-Tx, x\}, \{y, z\}) = 0 \qquad \text{for every } x \in \mathscr{D}(T). \tag{4}$$

$$\{y, z\} \in [V\mathscr{G}(T)]^\perp. \tag{5}$$

////

13.9 Theorem *If T is a densely defined operator in H, then T^* is a closed operator. In particular, self-adjoint operators are closed.*

PROOF. $M^\perp$ is closed, for every $M \subset H \times H$. Hence $\mathscr{G}(T^*)$ is closed in $H \times H$, by Theorem 13.8. ////

13.10 Theorem *If T is a densely defined closed operator in H, then*

$$H \times H = V\mathscr{G}(T) \oplus \mathscr{G}(T^*), \tag{1}$$

a direct sum of two orthogonal subspaces.

PROOF. If $\mathscr{G}(T)$ is closed, so is $V\mathscr{G}(T)$, since V is unitary, and therefore Theorem 13.8 implies that $V\mathscr{G}(T) = [\mathscr{G}(T^*)]^\perp$; see Theorem 12.4. ////

Corollary. *If $a \in H$ and $b \in H$, the system of equations*

$$-Tx + y = a$$

$$x + T^*y = b$$

has a unique solution with $x \in \mathscr{D}(T)$ and $y \in \mathscr{D}(T^*)$.

Our next theorem states some conditions under which a symmetric operator is self-joint.

13.11 Theorem *Suppose T is a densely defined operator in H, and T is symmetric.*

(*a*) *If $\mathscr{D}(T) = H$, then T is self-adjoint and $T \in \mathscr{B}(H)$.*
(*b*) *If T is self-adjoint and one-to-one, then $\mathscr{R}(T)$ is dense in H, and T^{-1} is self-adjoint.*
(*c*) *If $\mathscr{R}(T)$ is dense in H, then T is one-to-one.*
(*d*) *If $\mathscr{R}(T) = H$, then T is self-adjoint, and $T^{-1} \in \mathscr{B}(H)$.*

PROOF. (*a*) By assumption, $T \subset T^*$. If $\mathscr{D}(T) = H$, it is thus obvious that $T = T^*$. Hence T is closed (Theorem 13.9) and therefore continuous, by the closed graph theorem. (We could also refer to Theorem 5.1.)

(*b*) Suppose $y \perp \mathscr{R}(T)$. Then $x \to (Tx, y) = 0$ is continuous in $\mathscr{D}(T)$, hence $y \in \mathscr{D}(T^*) = \mathscr{D}(T)$, and $(x, Ty) = (Tx, y) = 0$ for all $x \in \mathscr{D}(T)$. Thus $Ty = 0$. Since T is assumed to be one-to-one, it follows that $y = 0$. This proves that $\mathscr{R}(T)$ is dense in H.

T^{-1} is therefore densely defined, with $\mathscr{D}(T^{-1}) = \mathscr{R}(T)$, and $(T^{-1})^*$ exists. The relations

$$\mathscr{G}(T^{-1}) = V\mathscr{G}(-T) \quad \text{and} \quad V\mathscr{G}(T^{-1}) = \mathscr{G}(-T) \tag{1}$$

are easily verified:

$$\{a, b\} \in \mathscr{G}(T^{-1}) \Leftrightarrow \{b, a\} \in \mathscr{G}(T) \Leftrightarrow \{b, -a\} \in \mathscr{G}(-T)$$
$$\Leftrightarrow \{a, b\} \in V\mathscr{G}(-T).$$

Being self-adjoint, T is closed (Theorem 13.9); hence $-T$ is closed, and hence T^{-1} is closed, by (1). Theorem 13.10 can now be applied to T^{-1} and to $-T$ and yields the orthogonal decompositions

$$H \times H = V\mathscr{G}(T^{-1}) \oplus \mathscr{G}((T^{-1})^*) \tag{2}$$

and

$$H \times H = V\mathscr{G}(-T) \oplus \mathscr{G}(-T) = \mathscr{G}(T^{-1}) \oplus V\mathscr{G}(T^{-1}). \tag{3}$$

Consequently,

$$\mathscr{G}((T^{-1})^*) = [V\mathscr{G}(T^{-1})]^{\perp} = \mathscr{G}(T^{-1}), \tag{4}$$

which shows that $(T^{-1})^* = T^{-1}$.

(*c*) Suppose $Tx = 0$. Then $(x, Ty) = (Tx, y) = 0$ for every $y \in \mathscr{D}(T)$. Thus $x \perp \mathscr{R}(T)$, and therefore $x = 0$.

(*d*) Since $\mathscr{R}(T) = H$, (*c*) implies that T is one-to-one, and $\mathscr{D}(T^{-1}) = H$. If $x \in H$ and $y \in H$, then $x = Tz$ and $y = Tw$, for some

$z \in \mathscr{D}(T)$ and $w \in \mathscr{D}(T)$, so that

$$(T^{-1}x, y) = (z, Tw) = (Tz, w) = (x, T^{-1}y).$$

Hence T^{-1} is symmetric, (*a*) implies that T^{-1} is self-adjoint (and bounded), and now it follows from (*b*) that $T = (T^{-1})^{-1}$ is also self-adjoint. ////

13.12 Theorem *If T is a densely defined closed operator in H, then $\mathscr{D}(T^*)$ is dense and $T^{**} = T$.*

PROOF. Since V is unitary, and $V^2 = -I$, Theorem 13.10 gives the orthogonal decomposition

$$H \times H = \mathscr{G}(T) \oplus V\mathscr{G}(T^*). \tag{1}$$

Suppose $z \perp \mathscr{D}(T^*)$. Then $(z, y) = 0$ and therefore

$$(\{0, z\}, \{-T^*y, y\}) = 0 \tag{2}$$

for all $y \in \mathscr{D}(T^*)$. Thus $\{0, z\} \in [V\mathscr{G}(T^*)]^\perp = \mathscr{G}(T)$, which implies that $z = T(0) = 0$. Consequently, $\mathscr{D}(T^*)$ is dense in H, and T^{**} is defined.

Another application of Theorem 13.10 gives therefore

$$H \times H = V\mathscr{G}(T^*) \oplus \mathscr{G}(T^{**}). \tag{3}$$

By (1) and (3),

$$\mathscr{G}(T^{**}) = [V\mathscr{G}(T^*)]^\perp = \mathscr{G}(T), \tag{4}$$

so that $T^{**} = T$. ////

We shall now see that operators of the form T^*T have interesting properties. In particular, $\mathscr{D}(T^*T)$ cannot be very small.

13.13 Theorem *Suppose T is a densely defined closed operator in H, and $Q = I + T^*T$.*

(*a*) *Under these assumptions, Q is a one-to-one mapping of*

$$\mathscr{D}(Q) = \mathscr{D}(T^*T) = \{x \in \mathscr{D}(T): Tx \in \mathscr{D}(T^*)\}$$

onto H, and there are operators $B \in \mathscr{B}(H)$, $C \in \mathscr{B}(H)$ that satisfy $\|B\| \le 1$, $\|C\| \le 1$, $C = TB$, and

$$B(I + T^*T) \subset (I + T^*T)B = I. \tag{1}$$

Also, $B \ge 0$, and T^*T is self-adjoint.

(*b*) *If T' is the restriction of T to $\mathscr{D}(T^*T)$, then $\mathscr{G}(T')$ is dense in $\mathscr{G}(T)$.*

Here, and in the sequel, the letter I denotes the identity operator with domain H.

PROOF. If $x \in \mathscr{D}(Q)$ then $Tx \in \mathscr{D}(T^*)$, so that

$$(x, x) + (Tx, Tx) = (x, x) + (x, T^*Tx) = (x, Qx). \tag{2}$$

Therefore $\|x\|^2 \le \|x\| \|Qx\|$, which shows that Q is one-to-one.

By Theorem 13.10 there corresponds to every $h \in H$ a unique vector $Bh \in \mathscr{D}(T)$ and a unique $Ch \in \mathscr{D}(T^*)$ such that

$$\{0, h\} = \{-TBh, Bh\} + \{Ch, T^*Ch\}. \tag{3}$$

It is clear that B and C are linear operators in H, with domain H. The two vectors on the right of (3) are orthogonal to each other (Theorem 13.10). The definition of the norm in $H \times H$ implies therefore that

$$\|h\|^2 \ge \|Bh\|^2 + \|Ch\|^2 \qquad (h \in H), \tag{4}$$

so that $\|B\| \le 1$ and $\|C\| \le 1$.

Consideration of the components in (3) shows that $C = TB$ and that

$$h = Bh + T^*Ch = Bh + T^*TBh = QBh \tag{5}$$

for every $h \in H$. Hence $QB = I$. In particular, B is a one-to-one mapping of H onto $\mathscr{D}(Q)$. If $y \in \mathscr{D}(Q)$, then $y = Bh$ for some $h \in H$, hence $Qy = QBh = h$, and $BQy = Bh = y$. Thus $BQ \subset I$, and (1) is proved.

If $h \in H$, then $h \in Qx$ for some $x \in \mathscr{D}(Q)$, so that

$$(Bh, h) = (BQx, Qx) = (x, Qx) \ge 0, \tag{6}$$

by (2). Thus $B \ge 0$, B is self-adjoint (Theorem 12.32), and now (*b*) of Theorem 13.11 shows that Q is self-adjoint, hence so is $T^*T = Q - I$.

This completes the proof of part (*a*).

Since T is a closed operator, $\mathscr{G}(T)$ is a closed subspace of $H \times H$; hence $\mathscr{G}(T)$ is a Hilbert space. Assume $\{z, Tz\} \in \mathscr{G}(T)$ is orthogonal to $\mathscr{G}(T')$. Then, for every $x \in \mathscr{D}(T^*T) = \mathscr{D}(Q)$,

$$\begin{aligned} 0 = (\{z, Tz\}, \{x, Tx\}) &= (z, x) + (Tz, Tx) = (z, x) + (z, T^*Tx) \\ &= (z, Qx). \end{aligned}$$

But $\mathscr{R}(Q) = H$. Hence $z = 0$. This proves (*b*). ////

13.14 Definition A symmetric operator T in H is said to be *maximally symmetric* if T has no proper symmetric extension, i.e., if the assumptions

$$T \subset S, \qquad S \text{ symmetric} \tag{1}$$

imply that $S = T$.

13.15 Theorem *Self-adjoint operators are maximally symmetric.*

PROOF. Suppose T is self-adjoint, S is symmetric (that is, $S \subset S^*$), and $T \subset S$. This inclusion implies obviously (by the very definition of the adjoint) that $S^* \subset T^*$. Hence

$$S \subset S^* \subset T^* = T \subset S,$$

which proves that $S = T$. ////

It should be noted that maximally symmetric operators need not be self-adjoint; see Example 13.21 and Exercise 10.

13.16 Theorem *If T is a symmetric operator in H (not necessarily densely defined), the following statements are true:*

(*a*) $\|Tx + ix\|^2 = \|x\|^2 + \|Tx\|^2 \qquad [x \in \mathscr{D}(T)]$.

(*b*) *T is a closed operator if and only if $\mathscr{R}(T + iI)$ is closed.*

(*c*) *$T + iI$ is one-to-one.*

(*d*) *If $\mathscr{R}(T + iI) = H$, then T is maximally symmetric.*

(*e*) *The preceding statements are also true if i is replaced by $-i$.*

PROOF. Statement (*a*) follows from the identity

$$\|Tx + ix\|^2 = \|x\|^2 + \|Tx\|^2 + (ix, Tx) + (Tx, ix),$$

combined with the symmetry of T. By (*a*),

$$(T + iI)x \leftrightarrow \{x, Tx\}$$

is an isometric one-to-one correspondence between the range of $T + iI$ and the graph of T. This proves (*b*). Next, (*c*) is also an immediate consequence of (*a*). If $\mathscr{R}(T + iI) = H$ and T_1 is a *proper* extension of T [that is, $\mathscr{D}(T)$ is a proper subset of $\mathscr{D}(T_1)$], then $T_1 + iI$ is a proper extension of $T + iI$ which cannot be one-to-one. By (*c*), T_1 is not symmetric. This proves (*d*).

It is clear that this proof is equally valid with $-i$ in place of i. ////

The Cayley Transform

13.17 Definition The mapping

$$t \to \frac{t - i}{t + i} \tag{1}$$

sets up a one-to-one correspondence between the real line and the unit circle (minus the point 1). The symbolic calculus studied in Chapter 12 shows therefore that every self-adjoint $T \in \mathscr{B}(H)$ gives rise to a unitary operator

(2) $$U = (T - iI)(T + iI)^{-1}$$

and that every unitary U whose spectrum does not contain the point 1 is obtained in this way.

This relation $T \leftrightarrow U$ will now be extended to a one-to-one correspondence between symmetric operators, on the one hand, and isometries, on the other.

Let T be a symmetric operator in H. Theorem 13.16 shows that

(3) $$\|Tx + ix\|^2 = \|x\|^2 + \|Tx\|^2 = \|Tx - ix\|^2 \qquad (x \in \mathscr{D}(T)).$$

Hence there is an isometry U, with

(4) $$\mathscr{D}(U) = \mathscr{R}(T + iI), \qquad \mathscr{R}(U) = \mathscr{R}(T - iI),$$

defined by

(5) $$U(Tx + ix) = Tx - ix \qquad (x \in \mathscr{D}(T)).$$

Since $(T + iI)^{-1}$ maps $\mathscr{D}(U)$ onto $\mathscr{D}(T)$, U can also be written in the form

(6) $$U = (T - iI)(T + iI)^{-1}.$$

This operator U is called the *Cayley transform of T*. Its main features are summarized in Theorem 13.19. It will lead to an easy proof of the spectral theorem for self-adjoint (not necessarily bounded) operators.

13.18 Lemma *Suppose U is an operator in H which is an isometry: $\|Ux\| = \|x\|$ for every $x \in \mathscr{D}(U)$.*

(*a*) *If $x \in \mathscr{D}(U)$ and $y \in \mathscr{D}(U)$, then $(Ux, Uy) = (x, y)$.*

(*b*) *If $\mathscr{R}(I - U)$ is dense in H, then $I - U$ is one-to-one.*

(*c*) *If any one of the three spaces $\mathscr{D}(U)$, $\mathscr{R}(U)$, and $\mathscr{G}(U)$ is closed, so are the other two.*

PROOF. Any of the identities listed in Exercise 2 of Chapter 12 proves (*a*). To prove (*b*), suppose $x \in \mathscr{D}(U)$ and $(I - U)x = 0$, that is, $x = Ux$. Then

$$(x, (I - U)y) = (x, y) - (x, Uy) = (Ux, Uy) - (x, Uy) = 0$$

for every $y \in \mathscr{D}(U)$. Thus $x \perp \mathscr{R}(I - U)$, so that $x = 0$ if $\mathscr{R}(I - U)$ is dense in H. The proof of (*c*) is a consequence of the relations

$$\|Ux - Uy\| = \|x - y\| = \frac{1}{\sqrt{2}} \|\{x, Ux\} - \{y, Uy\}\|,$$

which hold for all $x, y \in \mathscr{D}(U)$. ////

13.19 Theorem *Suppose U is the Cayley transform of a symmetric operator T in H. Then the following statements are true:*

(a) *U is closed if and only if T is closed.*

(b) *$\mathscr{R}(I - U) = \mathscr{D}(T)$, $I - U$ is one-to-one, and T can be reconstructed from U by the formula*

$$T = i(I + U)(I - U)^{-1}.$$

(The Cayley transforms of distinct symmetric operators are therefore distinct.)

(c) *U is unitary if and only if T is self-adjoint.*

Conversely, if V is an operator in H which is an isometry, and if $I - V$ is one-to-one, then V is the Cayley transform of a symmetric operator in H.

PROOF. By Theorem 13.16, T is closed if and only if $\mathscr{R}(T + iI)$ is closed. By Lemma 13.18, U is closed if and only if $\mathscr{D}(U)$ is closed. Since $\mathscr{D}(U) = \mathscr{R}(T + iI)$, by the definition of the Cayley transform, (a) is proved.

The one-to-one correspondence $x \leftrightarrow z$ between $\mathscr{D}(T)$ and $\mathscr{D}(U) = \mathscr{R}(T + iI)$, given by

$$z = Tx + ix, \qquad Uz = Tx - ix \tag{1}$$

can be rewritten in the form

$$(I - U)z = 2ix, \qquad (I + U)z = 2Tx. \tag{2}$$

This shows that $I - U$ is one-to-one, that $\mathscr{R}(I - U) = \mathscr{D}(T)$, so that $(I - U)^{-1}$ maps $\mathscr{D}(T)$ onto $\mathscr{D}(U)$, and that

$$2Tx = (I + U)z = (I + U)(I - U)^{-1}(2ix) \qquad [x \in \mathscr{D}(T)]. \tag{3}$$

This proves (b).

Assume now that T is self-adjoint. Then

$$\mathscr{R}(I + T^2) = H \tag{4}$$

by Theorem 13.13. Since

$$(T + iI)(T - iI) = I + T^2 = (T - iI)(T + iI) \tag{5}$$

[the three operators (5) have domain $\mathscr{D}(T^2)$], it follows from (4) that

$$\mathscr{D}(U) = \mathscr{R}(T + iI) = H \tag{6}$$

and

$$\mathscr{R}(U) = \mathscr{R}(T - iI) = H. \tag{7}$$

Since U is an isometry, (6) and (7) imply that U is unitary (Theorem 12.13).

To complete the proof of (*c*), assume that U is unitary. Then

$$[\mathscr{R}(I-U)]^{\perp} = \mathscr{N}(I-U) = \{0\}, \tag{8}$$

by (*b*) and the normality of $I - U$ (Theorem 12.12), so that $\mathscr{D}(T) = \mathscr{R}(I-U)$ dense in H. Thus T^* is defined, and $T \subset T^*$.

Fix $y \in \mathscr{D}(T^*)$. Since $\mathscr{R}(T + iI) = \mathscr{D}(U) = H$, there exists $y_0 \in \mathscr{D}(T)$ such that

$$(T^* + iI)y = (T + iI)y_0 = (T^* + iI)y_0. \tag{9}$$

The last equality holds because $T \subset T^*$. If $y_1 = y - y_0$, then $y_1 \in \mathscr{D}(T^*)$ and, for every $x \in \mathscr{D}(T)$,

$$((T - iI)x, y_1) = (x, (T^* + iI)y_1) = (x, 0) = 0. \tag{10}$$

Thus $y_1 \perp \mathscr{R}(T - iI) = \mathscr{R}(U) = H$, and so $y_1 = 0$, and $y = y_0 \in \mathscr{D}(T)$. Hence $T^* \subset T$, and (*c*) is proved.

Finally, let V be as in the statement of the converse. Then there is a one-to-one correspondence $z \leftrightarrow x$ between $\mathscr{D}(V)$ and $\mathscr{R}(I - V)$, given by

$$x = z - Vz. \tag{11}$$

Define S on $\mathscr{D}(S) = \mathscr{R}(I - V)$ by

$$Sx = i(z + Vz) \qquad \text{if } x = z - Vz. \tag{12}$$

If $x \in \mathscr{D}(S)$ and $y \in \mathscr{D}(S)$, then $x = z - Vz$ and $y = u - Vu$ for some $z \in \mathscr{D}(V)$ and $u \in \mathscr{D}(V)$. Since V is an isometry, it now follows from (*a*) of Lemma 13.18 that

$$\begin{aligned}(Sx, y) &= i(z + Vz, u - Vu) = i(Vz, u) - i(z, Vu) \\ &= (z - Vz, iu + iVu) = (x, Sy).\end{aligned} \tag{13}$$

Hence S is symmetric. Since (12) can be written in the form

$$2iVz = Sx - ix, \qquad 2iz = Sx + ix \qquad [z \in \mathscr{D}(V)], \tag{14}$$

we see that

$$V(Sx + ix) = Sx - ix \qquad [x \in \mathscr{D}(S)] \tag{15}$$

and that $\mathscr{D}(V) = \mathscr{R}(S + iI)$. Therefore V is the Cayley transform of S.

////

13.20 The deficiency indices If U_1 and U_2 are Cayley transforms of symmetric operators T_1 and T_2, it is clear that $T_1 \subset T_2$ if and only if $U_1 \subset U_2$. Problems about symmetric extensions of symmetric operators reduce therefore to (usually easier) problems about extensions of isometries.

Let us now consider a *closed* and *densely defined symmetric* operator T in H, with Cayley transform U. Then $\mathscr{R}(T + iI)$ and $\mathscr{R}(T - iI)$ are closed (see Theorem 13.16), and U is an isometry carrying the first onto the second. The dimensions of the orthogonal complements of these two spaces are called the *deficiency indices* of T. (The *dimension* of a Hilbert space is, by definition, the cardinality of any one of its orthonormal bases.)

Since $\mathscr{R}(I - U) = \mathscr{D}(T)$ is now assumed to be dense in H, *every* isometric extension U_1 of U has $\mathscr{R}(I - U_1)$ dense in H, so that $I - U_1$ is one-to-one (Lemma 13.18) and U_1 is the Cayley transform of a symmetric extension T_1 of T.

The following three statements are easy consequences of Theorem 13.19 and the preceding discussion; we still assume that T is closed, symmetric, and densely defined.

(*a*) *T is self-adjoint if and only if both its deficiency indices are* 0.

(*b*) *T is maximally symmetric if and only if at least one of its deficiency indices is* 0.

(*c*) *T has a self-adjoint extension if and only if its two deficiency indices are equal.*

The proofs of (*a*) and (*b*) are obvious. To see (*c*), use (*c*) of Theorem 13.19 and note that every unitary extension of U must be an isometry of $[\mathscr{R}(T + iI)]^{\perp}$ onto $[\mathscr{R}(T - iI)]^{\perp}$.

13.21 Example Let V be the right shift on ℓ^2. Then V is an isometry and $I - V$ is one-to-one (Chapter 12, Exercise 18), and so V is the Cayley transform of a symmetric operator T. Since $\mathscr{D}(V) = \ell^2$ and $\mathscr{R}(V)$ has codimension 1, the deficiency indices of T are 0 and 1.

This provides us with an example of a densely defined, maximally symmetric, closed operator T which is not self-adjoint.

Resolution of the Identity

13.22 Notation $\mathfrak{M}$ will now be a σ-algebra in a set Ω, H will be a Hilbert space, and $E: \mathfrak{M} \to \mathscr{B}(H)$ will be a resolution of the identity, with all the properties listed in Definition 12.17. Theorem 12.21 describes a symbolic calculus which associates to every $f \in L^{\infty}(E)$ an operator $\Psi(f) \in \mathscr{B}(H)$, by the formula

$$(1)\qquad (\Psi(f)x, y) = \int_\Omega f\,dE_{x,y} \qquad (x \in H, y \in H).$$

This will now be extended to unbounded measurable functions f (Theorem 13.24). We shall use the same notations as in Definition 12.17.

13.23 Lemma *Let $f\colon \Omega \to \mathcal{C}$ be measurable. Put*

$$(1)\qquad \mathscr{D}_f = \left\{x \in H: \int_\Omega |f|^2\,dE_{x,x} < \infty\right\}.$$

Then $\mathscr{D}_f$ is a dense subspace of H. If $x \in H$ and $y \in H$, then

$$(2)\qquad \int_\Omega |f|\,d|E_{x,y}| \le \|y\|\left\{\int_\Omega |f|^2\,dE_{x,x}\right\}^{1/2}.$$

If f is bounded and $v = \Psi(f)z$, then

$$(3)\qquad dE_{x,v} = \bar{f}\,dE_{x,z} \qquad (x \in H, z \in H).$$

PROOF. If $z = x + y$, and $\omega \in \mathfrak{M}$, then

$$\|E(\omega)z\|^2 \le (\|E(\omega)x\| + \|E(\omega)y\|)^2 \le 2\|E(\omega)x\|^2 + 2\|E(\omega)y\|^2$$

or

$$(4)\qquad E_{z,z}(\omega) \le 2E_{x,x}(\omega) + 2E_{y,y}(\omega).$$

It follows that $\mathscr{D}_f$ is closed under addition. Scalar multiplication is even easier. Thus $\mathscr{D}_f$ is a subspace of H.

For $n = 1, 2, 3, \ldots$, let ω_n be the subset of Ω in which $|f| < n$. If $x \in \mathscr{R}(E(\omega_n))$ then

$$(5)\qquad E(\omega)x = E(\omega)E(\omega_n)x = E(\omega \cap \omega_n)x$$

so that

$$(6)\qquad E_{x,x}(\omega) = E_{x,x}(\omega \cap \omega_n) \qquad (\omega \in \mathfrak{M}),$$

and therefore

$$(7)\qquad \int_\Omega |f|^2\,dE_{x,x} = \int_{\omega_n} |f|^2\,dE_{x,x} \le n^2\|x\|^2 < \infty.$$

Thus $\mathscr{R}(E(\omega_n)) \subset \mathscr{D}_f$. Since $\Omega = \bigcup_{n=1}^\infty \omega_n$, the countable additivity of $\omega \to E(\omega)y$ implies that $y = \lim E(\omega_n)y$ for every $y \in H$, so that y lies in the closure of $\mathscr{D}_f$. Hence $\mathscr{D}_f$ is dense.

If $x \in H$, $y \in H$, and f is a bounded measurable function on Ω, the Radon-Nikodym theorem [23] shows that there is a measurable function u on Ω, with $|u| = 1$, such that

$$uf\, dE_{x,y} = |f|\, d|E_{x,y}|. \tag{8}$$

Hence

$$\int_\Omega |f|\, d|E_{x,y}| = (\Psi(uf)x, y) \le \|\Psi(uf)x\|\,\|y\|. \tag{9}$$

By Theorem 12.21,

$$\|\Psi(uf)x\|^2 = \int_\Omega |uf|^2\, dE_{x,x} = \int_\Omega |f|^2\, dE_{x,x}. \tag{10}$$

Now (9) and (10) give (2) for bounded f. The general case follows from this.

Finally (3) holds because

$$\int_\Omega g\, dE_{x,v} = (\Psi(g)x, v) = (\Psi(g)x, \Psi(f)z)$$

$$= (\Psi(\bar{f})\Psi(g)x, z) = (\Psi(\bar{f}g)x, z) = \int_\Omega g\bar{f}\, dE_{x,z}$$

for every bounded measurable g, by Theorem 12.21. ////

13.24 Theorem *Let E be a resolution of the identity, on a set Ω.*

(a) *To every measurable $f\colon \Omega \to \mathcal{C}$ corresponds a densely defined closed operator $\Psi(f)$ in H, with domain $\mathscr{D}(\Psi(f)) = \mathscr{D}_f$, which is characterized by*

$$(\Psi(f)x, y) = \int_\Omega f\, dE_{x,y} \qquad (x \in \mathscr{D}_f,\ y \in H) \tag{1}$$

and which satisfies

$$\|\Psi(f)x\|^2 = \int_\Omega |f|^2\, dE_{x,x} \qquad (x \in \mathscr{D}_f). \tag{2}$$

(b) *The multiplication theorem holds in the following form: If f and g are measurable, then*

$$\Psi(f)\Psi(g) \subset \Psi(fg) \qquad \text{and} \qquad \mathscr{D}(\Psi(f)\Psi(g)) = \mathscr{D}_g \cap \mathscr{D}_{fg}. \tag{3}$$

Hence $\Psi(f)\Psi(g) = \Psi(fg)$ if and only if $\mathscr{D}_{fg} \in \mathscr{D}_g$.

(c) *For every measurable $f : \Omega \to \mathcal{C}$,*

$$\Psi(f)^* = \Psi(\bar{f}) \tag{4}$$

and

(5) $$\Psi(f)\Psi(f)^* = \Psi(|f|^2) = \Psi(f)^*\Psi(f).$$

PROOF. If $x \in \mathscr{D}_f$ then $y \to \int_\Omega f\,dE_{x,y}$ is a bounded conjugate-linear functional on H, whose norm is at most $(\int |f|^2\,dE_{x,x})^{1/2}$, by (2) of Lemma 13.23. It follows that there is a unique element $\Psi(f)x \in H$ that satisfies (1) for every $y \in H$ and that

(6) $$\|\Psi(f)x\|^2 \le \int_\Omega |f|^2\,dE_{x,x} \qquad (x \in \mathscr{D}_f).$$

The linearity of $\Psi(f)$ on $\mathscr{D}_f$ follows from (1), since $E_{x,y}$ is linear in x.

Associate with each f its *truncations* $f_n = f\phi_n$, where $\phi_n(p) = 1$ if $|f(p)| \le n$, $\phi_n(p) = 0$ if $|f(p)| > n$.

Then $\mathscr{D}_{f-f_n} = \mathscr{D}_f$, since each f_n is bounded, and therefore (6) shows, by the dominated convergence theorem, that

(7) $$\|\Psi(f)x - \Psi(f_n)x\|^2 \le \int_\Omega |f - f_n|^2\,dE_{x,x} \to 0 \qquad \text{as } n \to \infty,$$

for every $x \in \mathscr{D}_f$. Since f_n is bounded, (2) holds with f_n in place of f (Theorem 12.21). Hence (7) implies that (2) holds as stated.

This proves (*a*), except for the assertion that $\Psi(f)$ is closed. The latter follows from Theorem 13.9 if (4) (to be proved presently) is applied to $\bar{f}$ in place of f.

We turn to the proof of (*b*).

Assume first that f is bounded. Then $\mathscr{D}_{fg} \subset \mathscr{D}_g$. If $z \in H$ and $v = \Psi(\bar{f})z$, Equation (3) of Lemma 13.23 and Theorem 12.21 show that

$$\begin{aligned}(\Psi(f)\Psi(g)x, z) &= (\Psi(g)x, \Psi(\bar{f})z) = (\Psi(g)x, v)\\ &= \int_\Omega g\,dE_{x,v} = \int_\Omega fgE_{x,z} = (\Psi(fg)x, z).\end{aligned}$$

Hence

(8) $$\Psi(f)\Psi(g)x = \Psi(fg)x \qquad (x \in \mathscr{D}_g, f \in L^\infty).$$

If $y = \Psi(g)x$, it follows from (8) and (2) that

(9) $$\int_\Omega |f|^2\,dE_{y,y} = \int_\Omega |fg|^2\,dE_{x,x} \qquad (x \in \mathscr{D}_g, f \in L^\infty).$$

Now let f be arbitrary (possibly unbounded). Since (9) holds for all $f \in L^\infty$, it holds for all measurable f. Since $\mathscr{D}(\Psi(f)\Psi(g))$ consists of all $x \in \mathscr{D}_g$ such that $y \in \mathscr{D}_f$, and since (9) shows that $y \in \mathscr{D}_f$ if and only if $x \in \mathscr{D}_{fg}$, we see that

(10) $$\mathscr{D}(\Psi(f)\Psi(g)) = \mathscr{D}_g \cap \mathscr{D}_{fg}.$$

If $x \in \mathscr{D}_g \cap \mathscr{D}_{fg}$, if $y = \Psi(g)z$, and if the truncations f_n are defined as above, then $f_n \to f$ in $L^2(E_{y,y})$, $f_n g \to fg$ in $L^2(E_{x,x})$, and now (8) (with f_n in place of f) and (2) imply

$$\Psi(f)\Psi(g)x = \Psi(f)y = \lim_{n\to\infty} \Psi(f_n)y = \lim_{n\to\infty} \Psi(f_n g)x = \Psi(fg)x.$$

This proves (3) and hence (*b*).

Suppose now that $x \in \mathscr{D}_f$ and $y \in \mathscr{D}_{\bar{f}} = \mathscr{D}_f$. It follows from (7) and Theorem 12.21 that

$$(\Psi(f)x, y) = \lim_{n\to\infty} (\Psi(f_n)x, y) = \lim_{n\to\infty} (x, \Psi(\bar{f}_n)y) = (x, \Psi(\bar{f})y).$$

Thus $y \in \mathscr{D}(\Psi(f)^*)$, and

(11) $$\Psi(\bar{f}) \subset \Psi(f)^*$$

To pass from (11) to (4) we have to show that every $z \in \mathscr{D}(\Psi(f)^*)$ lies in $\mathscr{D}_f$. Fix z; put $v = \Psi(f)^*z$. Since $f_n = f\phi_n$, the multiplication theorem gives

(12) $$\Psi(f_n) = \Psi(f)\Psi(\phi_n).$$

Since $\Psi(\phi_n)$ is self-adjoint, we conclude from Theorems 13.2 and 12.21 that

$$\Psi(\phi_n)\Psi(f)^* \subset [\Psi(f)\Psi(\phi_n)]^* = \Psi(f_n)^* = \Psi(\bar{f}_n).$$

Hence

(13) $$\Psi(\phi_n)v = \Psi(\bar{f}_n)z \qquad (n = 1, 2, 3, \ldots).$$

Since $|\phi_n| \le 1$, (13) and (2) imply

(14) $$\int_\Omega |f_n|^2 \, dE_{z,z} = \int_\Omega |\phi_n|^2 \, dE_{v,v} \le E_{v,v}(\Omega)$$

for $n = 1, 2, 3, \ldots$. Hence $z \in \mathscr{D}_f$, and (4) is proved.

Finally, (5) follows from (4) by another application of the multiplication theorem, because $\mathscr{D}_{f\bar{f}} \subset \mathscr{D}_f$. ////

Remark. If g is bounded, then $\mathscr{D}_{fg} \subset \mathscr{D}_g$ (simply because $\mathscr{D}_g = H$) so that $\Psi(f)\Psi(g) = \Psi(fg)$. This was used in (12). It also shows, for bounded g, that

(15) $$\Psi(g)\Psi(f) \subset \Psi(f)\Psi(g),$$

because $\Psi(g)\Psi(f) \subset \Psi(gf) = \Psi(fg)$. If g is the characteristic function of a measurable set $\omega \subset \Omega$, (15) becomes

$$E(\omega)\Psi(f) \subset \Psi(f)E(\omega). \tag{16}$$

If $x \in \mathscr{D}_f \cap \mathscr{R}(E(\omega))$, it follows that

$$E(\omega)\Psi(f)x = \Psi(f)E(\omega)x = \Psi(f)x. \tag{17}$$

Thus $\Psi(f)$ maps $\mathscr{D}_f \cap \mathscr{R}(E(\omega))$ into $\mathscr{R}(E(\omega))$.

This should be compared with the discussion of invariant subspaces in Section 12.27.

Note also that, by analogy with (3),

$$\Psi(f) + \Psi(g) \subset \Psi(f + g). \tag{18}$$

Equality holds if and only if $\mathscr{D}_{f+g} = \mathscr{D}_f \cap \mathscr{D}_g$, which is true whenever at least one of f, g is bounded.

13.25 Theorem *In the situation of Theorem 13.24, $\mathscr{D}_f = H$ if and only if $f \in L^\infty(E)$.*

PROOF. Assume $\mathscr{D}_f = H$. Since $\Psi(f)$ is a closed operator, the closed graph theorem implies that $\Psi(f) \in \mathscr{B}(H)$. If $f_n = f\phi_n$ is a truncation of f, it follows from the multiplication theorem, combined with Theorem 12.21, that

$$\|f_n\|_\infty = \|\Psi(f_n)\| = \|\Psi(f)\Psi(\phi_n)\| \le \|\Psi(f)\|,$$

since $\|\Psi(\phi_n)\| = \|\phi_n\|_\infty \le 1$. Thus $\|f\|_\infty \le \|\Psi(f)\|$, and $f \in L^\infty(E)$. The converse is contained in Theorem 12.21. ////

13.26 Definition The *resolvent set* of a linear operator T in H is the set of all $\lambda \in \mathcal{C}$ such that $T - \lambda I$ is a one-to-one mapping of $\mathscr{D}(T)$ onto H whose inverse belongs to $\mathscr{B}(H)$.

In other words, $T - \lambda I$ should have an inverse $S \in \mathscr{B}(H)$, which satisfies

$$S(T - \lambda I) \subset (T - \lambda I)S = I.$$

For instance, Theorem 13.13 states that -1 lies in the resolvent set of T^*T if T is densely defined and closed.

The *spectrum* $\sigma(T)$ of T is the complement of the resolvent set of T, just as for bounded operators.

Some properties of $\sigma(T)$, for unbounded T, are described in Exercises 17 to 20.

For the next theorem, we refer to Section 12.20 for the definition of the essential range of a function, with respect to a given resolution of the identity.

13.27 Theorem *Suppose E is a resolution of the identity on a set Ω, $f: \Omega \to \mathcal{C}$ is measurable, and*

$$\omega_\alpha = \{p \in \Omega: f(p) = \alpha\} \qquad (\alpha \in \mathcal{C}).$$

(*a*) *If α is in the essential range of f and $E(\omega_\alpha) \neq 0$, then $\Psi(f) - \alpha I$ is not one-to-one.*

(*b*) *If α is in the essential range of f but $E(\omega_\alpha) = 0$, then $\Psi(f) - \alpha I$ is a one-to-one mapping of $\mathcal{D}_f$ onto a dense proper subspace of H, and there exist vectors $x_n \in H$, with $\|x_n\| = 1$, such that*

$$\lim_{n\to\infty} [\Psi(f)x_n - \alpha x_n] = 0.$$

(*c*) *$\sigma(\Psi(f))$ is the essential range of f.*

In the terminology used earlier for bounded operators, we may say that α lies in the *point spectrum* of $\Psi(f)$ in case (*a*) and in the *continuous spectrum* of $\Psi(f)$ in case (*b*). The conclusion of (*b*) is sometimes stated by saying that α is an *approximate eigenvalue of* $\Psi(f)$.

PROOF. We shall assume, without loss of generality, that $\alpha = 0$.

(*a*) If $E(\omega_0) \neq 0$, there exists $x_0 \in \mathcal{R}(E(\omega_0))$ with $\|x_0\| = 1$. Let ϕ_0 be the characteristic function of ω_0. Then $f\phi_0 = 0$, hence $\Psi(f)\Psi(\phi_0) = 0$, by the multiplication theorem. Since $\Psi(\phi_0) = E(\omega_0)$, it follows that

$$\Psi(f)x_0 = \Psi(f)E(\omega_0)x_0 = \Psi(f)\Psi(\phi_0)x_0 = 0.$$

(*b*) The hypothesis is now that $E(\omega_0) = 0$ but $E(\omega_n) \neq 0$ for $n = 1, 2, 3, \ldots$, where

$$\omega_n = \left\{p \in \Omega: |f(p)| < \frac{1}{n}\right\}.$$

Choose $x_n \in \mathcal{R}(E(\omega_n))$, $\|x_n\| = 1$; let ϕ_n be the characteristic functions of ω_n. The argument used in (*a*) leads to

$$\|\Psi(f)x_n\| = \|\Psi(f\phi_n)x_n\| \leq \|\Psi(f\phi_n)\| = \|f\phi_n\|_\infty \leq \frac{1}{n}.$$

Thus $\Psi(f)x_n \to 0$ although $\|x_n\| = 1$.

If $\Psi(f)x = 0$ for some $x \in \mathscr{D}_f$, then

$$\int_\Omega |f|^2 \, dE_{x,x} = \|\Psi(f)x\|^2 = 0.$$

Since $|f| > 0$ a.e. $[E_{x,x}]$, we must have $E_{x,x}(\Omega) = 0$. But $E_{x,x}(\Omega) = \|x\|^2$. Hence $\Psi(f)$ is one-to-one.

Likewise $\Psi(f)^* = \Psi(\bar{f})$ is one-to-one. If $y \perp \mathscr{R}(\Psi(f))$, then $x \to (\Psi(f)x, y) = 0$ is continuous in $\mathscr{D}_f$, hence $y \in \mathscr{D}(\Psi(f)^*)$, and

$$(x, \Psi(\bar{f})y) = (\Psi(f)x, y) = 0 \qquad (x \in \mathscr{D}_f).$$

Therefore, $\Psi(\bar{f})y = 0$, and $y = 0$. This proves that $\mathscr{R}(\Psi(f))$ is dense in H.

Since $\Psi(f)$ is closed, so is $\Psi(f)^{-1}$. If $\mathscr{R}(\Psi(f))$ filled H, the closed graph theorem would imply that $\Psi(f)^{-1} \in \mathscr{B}(H)$. But this is impossible, in view of the sequence $\{x_n\}$ constructed above.

Hence (*b*) is proved.

(*c*) It follows from (*a*) and (*b*) that the essential range of f is a subset of $\sigma(\Psi(f))$. To obtain the opposite inclusion, assume 0 is not in the essential range of f. Then $g = 1/f \in L^\infty(E)$, $fg = 1$, hence $\Psi(f)\Psi(g) = \Psi(1) = I$, which proves that $\mathscr{R}(\Psi(f)) = H$. Since $|f| > 0$, $\Psi(f)$ is one-to-one, as in the proof of (*b*). Therefore $\Psi(f)^{-1} \in \mathscr{B}(H)$, by the closed graph theorem.

This completes the proof. ////

The following theorem is sometimes called the *change of measure principle*.

13.28 Theorem *Suppose*

(*a*) *$\mathfrak{M}$ and $\mathfrak{M}'$ are σ-algebras in sets Ω and Ω',*
(*b*) *$E: \mathfrak{M} \to \mathscr{B}(H)$ is a resolution of the identity, and*
(*c*) *$\phi: \Omega \to \Omega'$ has the property that $\phi^{-1}(\omega') \in \mathfrak{M}$ for every $\omega' \in \mathfrak{M}'$.*

If $E'(\omega') = E(\phi^{-1}(\omega'))$, then $E': \mathfrak{M}' \to \mathscr{B}(H)$ is also a resolution of the identity, and

$$(1) \qquad \int_{\Omega'} f \, dE'_{x,y} = \int_\Omega (f \circ \phi) \, dE_{x,y}$$

for every $\mathfrak{M}'$-measurable $f: \Omega' \to \mathbb{C}$ for which either of these integrals exists.

PROOF. For characteristic functions f, (1) is just the definition of E'. Hence (1) holds for simple functions f. The general case follows from this. The proof that E' is a resolution of the identity is a matter of straightforward verifications and is omitted. ////

The Spectral Theorem

13.29 Normal operators A (not necessarily bounded) linear operator T in H is said to be *normal* if T is closed and densely defined and if

$$T^*T = TT^*.$$

Every $\Psi(f)$ that arises in Theorem 13.24 is normal; this is part of the statement of the theorem. We shall now see, just as in the bounded case discussed in Chapter 12, that all normal operators can be represented in this way, by means of resolutions of the identity on their spectra (Definition 13.26). For self-adjoint operators, this can be deduced very quickly from the unitary case, via the Cayley transform (Theorem 13.30). For normal operators in general, a different proof will be given in Theorem 13.33.

13.30 Theorem *To every self-adjoint operator A in H corresponds a unique resolution E of the identity, on the Borel subsets of the real line, such that*

$$(1)\qquad (Ax, y) = \int_{-\infty}^{\infty} t\, dE_{x,y}(t) \qquad (x \in \mathscr{D}(A),\ y \in H).$$

Moreover, E is concentrated on $\sigma(A) \subset (-\infty, \infty)$, in the sense that $E(\sigma(A)) = I$.

As before, this E will be called the *spectral decomposition of A.*

PROOF. Let U be the Cayley transform of A, let Ω be the unit circle with the point 1 removed, and let E' be the spectral decomposition of U (see Theorems 12.23 and 12.26). Since $I - U$ is one-to-one (Theorem 13.19), $E'(\{1\}) = 0$, by (*b*) of Theorem 12.29, and therefore

$$(2)\qquad (Ux, y) - \int_{\Omega} \lambda\, dE'_{x,y}(\lambda) \qquad (x \subset H,\ y \subset H).$$

Define

$$(3)\qquad f(\lambda) = \frac{i(1 + \lambda)}{1 - \lambda} \qquad (\lambda \in \Omega),$$

and define $\Psi(f)$ as in Theorem 13.24 with E' in place of E:

$$(4)\qquad (\Psi(f)x, y) = \int_{\Omega} f\, dE'_{x,y} \qquad (x \in \mathscr{D}_f,\ y \in H).$$

Since f is real-valued, $\Psi(f)$ is self-adjoint (Theorem 13.24), and since $f(\lambda)(1-\lambda) = i(1+\lambda)$, the multiplication theorem gives

$$\Psi(f)(I-U) = i(I+U). \tag{5}$$

In particular, (5) implies that $\mathscr{R}(I-U) \subset \mathscr{D}(\Psi(f))$. By Theorem 13.19,

$$A(I-U) = i(I+U), \tag{6}$$

and $\mathscr{D}(A) = \mathscr{R}(I-U) \subset \mathscr{D}(\Psi(f))$. Comparison of (5) and (6) shows now that $\Psi(f)$ is a self-adjoint extension of the self-adjoint operator A. By Theorem 13.15, $A = \Psi(f)$. Thus

$$(Ax, y) = \int_\Omega f\, dE'_{x,y} \qquad [x \in \mathscr{D}(A),\ y \in H]. \tag{7}$$

By (c) of Theorem 13.27, $\sigma(A)$ is the essential range of f. Thus $\sigma(A) \subset (-\infty, \infty)$. Note that f is one-to-one in Ω. If we define

$$E(f(\omega)) = E'(\omega) \tag{8}$$

for every Borel set $\omega \subset \Omega$, we obtain the desired resolution E which converts (7) to (1).

Just as (1) was derived from (2) by means of the Cayley transform, (2) can be derived from (1) by using the inverse of the Cayley transform. The uniqueness of the representation (2) (Theorem 12.23) leads therefore to the uniqueness of the resolution E that satisfies (1).

This completes the proof. ////

The machinery developed in Theorem 13.24 can now be applied to self-adjoint operators. The following theorem furnishes an example of this.

13.31 Theorem *Let A be a self-adjoint operator in H.*

(a) *$(Ax, x) \geq 0$ for every $x \in \mathscr{D}(A)$ (briefly: $A \geq 0$) if and only if $\sigma(A) \subset [0, \infty)$.*

(b) *If $A \geq 0$, there exists a unique self-adjoint $B \geq 0$ such that $B^2 = A$.*

PROOF. The proof of (a) is so similar to that of Theorem 12.32 that we omit it.

Assume $A \geq 0$, so that $\sigma(A) \subset [0, \infty)$, and

$$(Ax, y) = \int_0^\infty t\, dE_{x,x}(t) \qquad [x \in \mathscr{D}(A),\ y \in H], \tag{1}$$

where $\mathscr{D}(A) = \{x \in H: \int t^2\, dE_{x,y}(t) < \infty\}$; the domain of integration is $[0, \infty)$. Let $s(t)$ be the nonnegative square root of $t \geq 0$, and put

$B = \Psi(s)$; explicitly,

$$(Bx, y) = \int_0^\infty s(t)\, dE_{x,y}(t) \qquad (x \in \mathscr{D}_s,\ y \in H). \tag{2}$$

The multiplication theorem (*b*) of Theorem 13.24, with $f = g = s$, shows that $B^2 = A$. Since s is real, B is self-adjoint [(*c*) of Theorem 13.24], and since $s(t) \geq 0$, (2) with $x = y$, shows that $B \geq 0$.

To prove uniqueness, suppose C is self-adjoint, $C \geq 0$, $C^2 = A$, and E^C is its spectral decomposition:

$$(Cx, y) = \int_0^\infty s\, dE^C_{x,y}(s) \qquad (x \in \mathscr{D}(C),\ y \in H). \tag{3}$$

Apply Theorem 13.28 with $\Omega = [0, \infty)$, $\phi(s) = s^2$, $f(t) = t$, and

$$E'(\phi(\omega)) = E^C(\omega) \qquad \text{for } \omega \subset [0, \infty), \tag{4}$$

to obtain

$$(Ax, y) = (C^2x, y) = \int_0^\infty s^2\, dE^C_{x,y}(s) = \int_0^\infty t\, dE'_{x,y}(t). \tag{5}$$

By (1) and (5), the uniqueness statement in Theorem 13.30 shows that $E' = E$. By (4), E determines E^C, and hence C. ////

The following properties of normal operators will be used in the proof of the spectral theorem 13.33.

13.32 Theorem *If N is a normal operator in H, then*

(*a*) $\mathscr{D}(N) = \mathscr{D}(N^*)$,

(*b*) $\|Nx\| = \|N^*x\|$ *for every* $x \in \mathscr{D}(N)$, *and*

(*c*) *N is maximally normal.*

PROOF. If $y \in \mathscr{D}(N^*N) = \mathscr{D}(NN^*)$, then $(Ny, Ny) = (y, N^*Ny)$ because $Ny \in \mathscr{D}(N^*)$, and $(N^*y, N^*y) = (y, NN^*y)$ because $N^*y \in \mathscr{D}(N)$ and $N = N^{**}$ (Theorem 13.12). Since $N^*N = NN^*$, it follows that

$$\|Ny\| = \|N^*y\| \qquad \text{if } y \in \mathscr{D}(N^*N). \tag{1}$$

Now pick $x \in \mathscr{D}(N)$. Let N' be the restriction of N to $\mathscr{D}(N^*N)$. By Theorem 13.13, $\{x, Nx\}$ lies in the closure of the graph of N'. Hence there are vectors $y_i \in \mathscr{D}(N^*N)$ such that

$$\|y_i - x\| \to 0 \text{ as } i \to \infty \tag{2}$$

and

$$(3) \qquad \|Ny_i - Nx\| \to 0 \text{ as } i \to \infty.$$

By (1), $\|N^*y_i - N^*y_j\| = \|Ny_i - Ny_j\|$, so that (3) implies that $\{N^*y_i\}$ is a Cauchy sequence in H. Hence there exists $z \in H$ such that

$$(4) \qquad \|N^*y_i - z\| \to 0 \text{ as } i \to \infty.$$

Since N^* is a closed operator, (2) and (4) imply that $\{x, z\} \in \mathscr{G}(N^*)$.

From this we conclude first that $x \in \mathscr{D}(N^*)$, so that $\mathscr{D}(N) \subset \mathscr{D}(N^*)$, and secondly that

$$(5) \qquad \|N^*x\| = \|z\| = \lim \|N^*y_i\| = \lim \|Ny_i\| = \|Nx\|.$$

This proves (*b*) and half of (*a*). For the other half, note that N^* is also normal (since $N^{**} = N$), so that

$$(6) \qquad \mathscr{D}(N^*) \subset \mathscr{D}(N^{**}) = \mathscr{D}(N).$$

Finally, suppose M is normal and $N \subset M$. Then $M^* \subset N^*$, so that

$$(7) \qquad \mathscr{D}(M) = \mathscr{D}(M^*) \subset \mathscr{D}(N^*) = \mathscr{D}(N) \subset \mathscr{D}(M),$$

which gives $\mathscr{D}(M) = \mathscr{D}(N)$; hence $M = N$. ////

13.33 Theorem *Every normal operator N in H has a unique spectral decomposition E, which satisfies*

$$(1) \qquad (Nx, y) = \int_{\sigma(N)} \lambda \, dE_{x,y}(\lambda) \qquad (x \in \mathscr{D}(N), y \in H).$$

Moreover, $E(\omega)S = SE(\omega)$ for every Borel set to $\omega \subset \sigma(N)$ and for every $S \in \mathscr{B}(H)$ that commutes with N, in the sense that $SN \subset NS$.

It also follows from (1) and Theorem 13.24 that $E(\omega)N \subset NE(\omega)$.

PROOF. Our first objective is to find self-adjoint projections P_i, with pairwise orthogonal ranges, such that $P_i N \subset NP_i \in \mathscr{B}(H)$, NP_i is normal, and $x = \sum P_i x$ for every $x \in H$. The spectral theorem for bounded normal operators will then be applied to the operators NP_i, and this will lead to the desired result.

By Theorem 13.13, there exist $B \in \mathscr{B}(H)$ and $C \in \mathscr{B}(H)$ such that $B \geq 0$, $\|B\| \leq 1$, $C = NB$, and

$$(2) \qquad B(I + N^*N) \subset I = (I + N^*N)B.$$

Since $N^*N = NN^*$, (2) implies

$$BN = BN(I + N^*N)B = B(I + N^*N)NB \subset NB = C. \tag{3}$$

Consequently, $BC = B(NB) = (BN)B \subset CB$. Since B and C are bounded, it follows that $BC = CB$ and therefore that C commutes with every bounded Borel function of B. (See Section 12.24.)

Choose $\{t_i\}$ so that $1 = t_0 > t_1 > t_2 > \cdots$, $\lim t_i = 0$. Let p_i be the characteristic function of $(t_i, t_{i-1}]$, for $i = 1, 2, 3, \ldots$, and put $f_i(t) = p_i(t)/t$. Each f_i is bounded on $\sigma(B) \subset [0, 1]$. Let E^B be the spectral decomposition of B. The equality (2) shows that B is one-to-one, that is, 0 is not in the point spectrum of B. Hence $E^B(\{0\}) = 0$, and E^B is concentrated on $(0, 1]$.

Define

$$P_i = p_i(B) \qquad (i = 1, 2, 3, \ldots). \tag{4}$$

Since $p_i p_j = 0$ if $i \neq j$, the projections P_i have mutually orthogonal ranges. Since $\sum p_i$ is the characteristic function of $(0, 1]$, we have

$$\sum_{i=1}^{\infty} P_i x = E^B((0, 1])x = x \qquad (x \in H). \tag{5}$$

Since $p_i(t) = tf_i(t)$,

$$NP_i = NBf_i(B) = Cf_i(B) \in \mathscr{B}(H), \tag{6}$$

and $P_i N = f_i(B)BN \subset f_i(B)C$, by (3), so that

$$P_i N \subset NP_i. \tag{7}$$

By (6), $\mathscr{D}(NP_i) = H$, so that

$$\mathscr{R}(P_i) \subset \mathscr{D}(N) \qquad (i = 1, 2, 3, \ldots). \tag{8}$$

Hence, if $P_i x = x$, (7) implies $P_i Nx = NP_i x = Nx$. Thus N carries $\mathscr{R}(P_i)$ into $\mathscr{R}(P_i)$, or: *$\mathscr{R}(P_i)$ is an invariant subspace of N.*

Next, we wish to prove that each NP_i is normal. By (7) and Theorem 13.2,

$$(NP_i)^* \subset (P_i N)^* = N^* P_i. \tag{9}$$

But $NP_i \in \mathscr{B}(H)$, so that $(NP_i)^*$ has domain H. Hence

$$(NP_i)^* = N^* P_i, \tag{10}$$

and now Theorem 13.32 shows, by (8) and (10), that

$$\|NP_i x\| = \|N^* P_i x\| = \|(NP_i)^* x\| \qquad (x \in H). \tag{11}$$

By Theorem 12.12, (11) implies that NP_i is normal.

Hence (5), (6), and (7) show that our first objective has now been reached.

By Theorem 12.23, each NP_i has a spectral decomposition E^i, defined on the Borel subsets of $\mathcal{C}$.

Since N carries $\mathscr{R}(P_i)$ into $\mathscr{R}(P_i)$, P_i commutes with NP_i. Therefore P_i commutes with $E^i(\omega)$, for every Borel set $\omega \subset \mathcal{C}$, so that

$$(12) \qquad E^i(\omega)P_i x = P_i E^i(\omega)x \in \mathscr{R}(P_i) \qquad (x \in H,\ i = 1, 2, 3, \ldots).$$

Since these ranges are pairwise orthogonal, and since (5) implies

$$(13) \qquad \sum_{i=1}^{\infty} \|E^i(\omega)P_i x\|^2 \le \sum_{i=1}^{\infty} \|P_i x\|^2 = \|x\|^2,$$

the series $\sum E^i(\omega)P_i x$ converges, in the norm of H, and it makes sense to define

$$(14) \qquad E(\omega) = \sum_{i=1}^{\infty} E^i(\omega)P_i$$

for all Borel sets $\omega \subset \mathcal{C}$.

It is easy to check that E is a resolution of the identity. Hence there is a normal operator M, defined by

$$(15) \qquad (Mx, y) = \int \lambda \, dE_{x,y}(\lambda) \qquad (x \in \mathscr{D}(M),\ y \in H),$$

where the domain of integration is $\mathcal{C}$, and

$$(16) \qquad \mathscr{D}(M) = \left\{x \in H : \int |\lambda|^2 \, dE_{x,x}(\lambda) < \infty\right\}.$$

Our assertion (1) will not be proved by showing that $M = N$.

For any $x \in H$, (14) shows that

$$(17) \qquad E_{x,x}(\omega) = \|E(\omega)x\|^2 = \sum_{i=1}^{\infty} \|E^i(\omega)P_i x\|^2 = \sum_{i=1}^{\infty} E^i_{x_i,x_i}(\omega),$$

where $x_i = P_i x$. If $x \in \mathscr{D}(N)$, then $P_i Nx = NP_i x$, so that

$$(18) \qquad \sum_{i=1}^{\infty} \int |\lambda|^2 \, dE^i_{x_i,x_i}(\lambda) = \sum_{i=1}^{\infty} \|NP_i x_i\|^2 = \sum_{i=1}^{\infty} \|P_i Nx\|^2 = \|Nx\|^2.$$

It follows from (17) and (18) that the integral in (16) is finite for every $x \in \mathscr{D}(N)$. Hence

$$(19) \qquad \mathscr{D}(N) \subset \mathscr{D}(M).$$

If $x \in \mathscr{R}(P_i)$, then $x = P_i x$, and so $E(\omega)x = E^i(\omega)x$; thus $E_{x};$ $E^i_{x,y}$ for every $y \in H$. Hence

$$(Nx, y) = (NP_i x, y) = \int \lambda \, dE^i_{x,y}(\lambda) = \int \lambda \, dE_{x,y}(\lambda) = (Mx,$$

Consequently

$$P_i Nx = NP_i x = MP_i x \qquad [x \in \mathscr{D}(N),\, i = 1, 2, 3, \ldots]. \tag{20}$$

If $Q_i = P_1 + \cdots + P_i$, it follows that $Q_i Nx = MQ_i x$. Thus

$$\{Q_i x, Q_i Nx\} \in \mathscr{G}(M) \qquad [x \in \mathscr{D}(N),\, i = 1, 2, 3, \ldots]. \tag{21}$$

Since $\mathscr{G}(M)$ is closed, it follows from (5) and (21) that $\{x, Nx\} \in \mathscr{G}(M)$, that is, that $Nx = Mx$ for every $x \in \mathscr{D}(N)$. Thus $N \subset M$, by (19), and now the maximality of N (Theorem 13.32) implies $N = M$.

This gives the representation (1), with $\mathcal{C}$ in place of $\sigma(N)$. That E is actually concentrated on $\sigma(N)$ follows from (*c*) of Theorem 13.27.

To prove the uniqueness of E, consider the operator

$$T = N(I + \sqrt{N^*N})^{-1}, \tag{22}$$

where $\sqrt{N^*N}$ is the unique positive square root of N^*N. If (1) holds, it follows from Theorem 13.24 that

$$T = \int \phi \, dE, \tag{23}$$

where $\phi(\lambda) = \lambda/(1 + |\lambda|)$, so that $T \in \mathscr{B}(H)$, and since ϕ is one-to-one on $\mathcal{C}$, Theorem 13.28 implies that the spectral decomposition E^T of T satisfies

$$E(\omega) = E^T(\phi(\omega)) \tag{24}$$

for every Borel set $\omega \subset \mathcal{C}$. The uniqueness of E follows now from that of E^T (Theorem 12.23).

Finally, assume $S \in \mathscr{B}(H)$ and $SN \subset NS$. Put $Q = Q_n = E(\tilde{\omega})$, ›re $\tilde{\omega} = \{\lambda : |\lambda| < n\}$, and n is some positive integer. Then $\mathscr{B}(H)$ is normal and is given by

$$NQ = \int f \, dE,$$

$\tilde{\omega}$, $f(\lambda) = 0$ outside $\tilde{\omega}$. Theorem 13.28 implies that ›sition E' of NQ satisfies $E'(\omega) = E(f^{-1}(\omega))$, or

$$\cap\, \tilde{\omega}) = QE(\omega) \qquad \text{if } 0 \notin \omega,$$

$$\cup\, (\mathcal{C} - \tilde{\omega})) = E(\{0\}) + I - Q.$$

$$E'(\omega) \qquad \text{if } \omega \subset \tilde{\omega}.$$

$NQ = QNQ$, so that

$$QSNQ \subset QNSQ \subset (NQ)(QSQ).$$

Since $(QSQ)(NQ) \in \mathscr{B}(H)$, the inclusions in (28) are actually equalities. Now Theorem 12.23 implies that QSQ commutes with every $E'(\omega)$.

Consider a bounded ω, and take n so large that $\omega \subset \tilde{\omega}$. By (27)

$$QSE(\omega) = QSQE'(\omega) = E'(\omega)QSQ = E(\omega)SQ$$

so that

$$Q_n SE(\omega) = E(\omega)SQ_n \qquad (n = 1, 2, 3, \ldots). \tag{29}$$

It now follows from Proposition 12.18 that

$$SE(\omega) = E(\omega)S \tag{30}$$

if ω is bounded [let $n \to \infty$ in (29)], and hence also if ω is any Borel set in $\mathcal{C}$. ////

Semigroups of Operators

13.34 Definitions Let X be a Banach space, and suppose that to every $t \in [0, \infty)$ is associated an operator $Q(t) \in \mathscr{B}(X)$, in such a way that

(a) $Q(0) = I$,
(b) $Q(s + t) = Q(s)Q(t)$ for all $s \geq 0$ and $t \geq 0$, and
(c) $\lim_{t \to 0} \|Q(t)x - x\| = 0$ for every $x \in X$.

If (a) and (b) hold, $\{Q(t)\}$ is called a *semigroup* (or, more precisely, a *one-parameter semigroup*). Such semigroups have exponential representations, provided that the mapping $t \to Q(t)$ satisfies some continuity assumption. The one that is chosen here, namely (c), is easy to work with.

Motivated by the fact that every continuous complex function that satisfies $f(s + t) = f(s)f(t)$ has the form $f(t) = \exp(At)$, and that f is determined by the number $A = f'(0)$, we associate with $\{Q(t)\}$ the operators A_ε, by

$$A_\varepsilon x = \frac{1}{\varepsilon}[Q(\varepsilon)x - x] \qquad (x \in X, \varepsilon > 0), \tag{1}$$

and define

$$Ax = \lim_{\varepsilon \to 0} A_\varepsilon x \tag{2}$$

for all $x \in \mathscr{D}(A)$, that is, for all x for which the limit (2) exists in the norm topology of X.

It is clear that $\mathscr{D}(A)$ is a subspace of X and that A is thus a linear operator in X.

This operator, which is essentially $Q'(0)$, is called the *infinitesimal generator* of the semigroup $\{Q(t)\}$.

13.35 Theorem *If the semigroup $\{Q(t)\}$ satisfies the preceding hypotheses, then*

(a) *there are constants C, γ such that*

$$\|Q(t)\| \le Ce^{\gamma t} \qquad (0 \le t < \infty);$$

(b) *$t \to Q(t)x$ is a continuous map of $[0, \infty)$ into X, for every $x \in X$;*
(c) *$\mathscr{D}(A)$ is dense in X and A is closed;*
(d) *the differential equation*

$$\frac{d}{dt} Q(t)x = AQ(t)x = Q(t)Ax$$

holds for every $x \in \mathscr{D}(A)$;
(e) *for every $x \in X$,*

$$Q(t)x = \lim_{\varepsilon \to 0} (\exp(tA_\varepsilon))x,$$

the convergence being uniform on every compact subset of $[0, \infty)$; and
(f) *if $\lambda \in \mathbb{C}$ and $\operatorname{Re} \lambda > \gamma$, the integral*

$$R(\lambda)x = \int_0^\infty e^{-\lambda t} Q(t)x\, dt$$

defines an operator $R(\lambda) \in \mathscr{B}(X)$ [the so-called resolvent of $\{Q(t)\}$] whose range is $\mathscr{D}(A)$ and which inverts $\lambda I - A$.

It is remarkable that (e) holds for every $x \in X$, not just for $x \in \mathscr{D}(A)$. The limit in (e), as well as the one that is implicit in the derivative used in (d), is understood to refer to the norm topology of X. It follows from (f) that $\sigma(A)$ lies in the half plane $\{\lambda: \operatorname{Re} \lambda \le \gamma\}$.

PROOF. (a) If there were a sequence $t_n \to 0$ with $\|Q(t_n)\| \to \infty$, the Banach-Steinhaus theorem would imply the existence of an $x \in X$ for which $\{\|Q(t_n)x\|\}$ is unbounded, contrary to our assumption that

(1) $$\|Q(t)x - x\| \to 0 \qquad \text{as } t \to 0.$$

Hence there is a $\delta > 0$ and a $C < \infty$ such that $\|Q(t)\| \le C$ on $[0, \delta]$. If now $0 \le t < \omega$ and n is the positive integer satisfying $(n-1)\delta \le t < nd$, then $\|Q(t/n)\| \le C$, and the functional equation

(2) $$Q(s + t) = Q(s)Q(t)$$

leads to

$$\|Q(t)\| = \|Q(t/n)^n\| \leq C^n \leq C^{1+t/\delta}, \tag{3}$$

which proves (*a*), with $e^\gamma = C^{1/\delta}$.

(*b*) If $0 \leq s < t \leq T$, then (*a*) and (2) imply that

$$\begin{aligned}\|Q(t)x - Q(s)x\| &\leq \|Q(s)\| \cdot \|Q(t-s)x - x\| \\ &\leq Ce^{\gamma T}\|Q(t-s)x - x\|,\end{aligned}$$

which tends to 0 as $t - s \to 0$.

(*c*) Because of (*b*), the X-valued integrals

$$M_t x = \frac{1}{t}\int_0^t Q(s)x\, ds \qquad (x \in X,\ t > 0) \tag{4}$$

can be defined. In fact, $M_t \in \mathscr{B}(X)$ and $\|M_t\| \leq Ce^{\gamma t}$, by (*a*). We claim that

$$A_\varepsilon M_t x = A_t M_\varepsilon x \qquad (\varepsilon > 0,\ t > 0,\ x \in X). \tag{5}$$

To prove (5), insert the integrand $Q(s)x\, ds$ into

$$\int_\varepsilon^{t+\varepsilon} - \int_0^t = \int_t^{t+\varepsilon} - \int_0^\varepsilon.$$

By (2), the left side becomes

$$\begin{aligned}\int_0^t [Q(\varepsilon + s) - Q(s)]x\, ds &= [Q(\varepsilon) - I]\int_0^t Q(s)s\, ds \\ &= \varepsilon A_\varepsilon t M_t x.\end{aligned}$$

In the same way, the right side becomes $tA_t \varepsilon M_\varepsilon x$. This gives (5).

As $\varepsilon \to 0$, the right side of (5) converges to $A_t x$. Thus $M_t x \in \mathscr{D}(A)$, which proves that $\mathscr{D}(A)$ is dense in X, because $M_t x \to x$ as $t \to 0$. Moreover,

$$AM_t x = A_t x \qquad (x \in X). \tag{6}$$

To show that A is closed, suppose $x_n \in \mathscr{D}(A)$, $x_n \to x$, and $Ax_n \to y$. Since $Q(s)$ commutes with $Q(t)$, A_ε commutes with M_t, and therefore A commutes with M_t on $\mathscr{D}(A)$. Thus (6) gives

$$A_t x_n = AM_t x_n = M_t Ax_n.$$

Letting $n \to \infty$, we get

$$A_t x = M_t y. \tag{7}$$

As $t \to 0$, the right side of (7) converges, to y; hence the same is true of $A_t x$. This says that $x \in \mathscr{D}(A)$ and that $Ax = y$. The graph of A is therefore closed.

(*d*) Multiplying (6) by t gives

$$A \int_0^t Q(s)x \, ds = Q(t)x - x. \tag{8}$$

The integrand is continuous. Differentiation of the integral therefore proves (*d*), since $Q(t)Ax = AQ(t)x$ for $x \in \mathscr{D}(A)$. [Note that $Q(t)A_\varepsilon = A_\varepsilon Q(t)$.]

(*e*) We need an estimate for the norm of

$$\exp(tA_\varepsilon) = e^{-t/\varepsilon} \exp\left(\frac{t}{\varepsilon} Q(\varepsilon)\right)$$

$$= e^{-t/\varepsilon} \sum_{n=0}^{\infty} \frac{t^n Q(n\varepsilon)}{n!\,\varepsilon^n}$$

Replace the norm of this sum by the sum of the norms, apply the estimate (*a*), and sum the resulting series, to obtain

$$\|\exp(tA_\varepsilon)\| \le C \exp\left\{\frac{t}{\varepsilon}(e^{t\varepsilon\gamma} - 1)\right\} < C \exp(te^{\gamma t}) \tag{9}$$

for $0 < \varepsilon < 1$. Now define, for fixed $x \in X$,

$$\varphi(s) = \{\exp((t-s)A_\varepsilon)\}Q(s)x \qquad (0 \le s \le t). \tag{10}$$

If $x \in \mathscr{D}(A)$, it follows from (*d*) that

$$\varphi'(s) = \{\exp((t-s)A_\varepsilon)\}Q(s)(Ax - A_\varepsilon x). \tag{11}$$

Thus (*a*) and (9) show that there is a $K(t) < \infty$ such that

$$\|\varphi'(s)\| \le K(t)\|Ax - A_\varepsilon x\| \tag{12}$$

whenever $0 \le s \le t$, $0 < \varepsilon \le 1$, and $x \in \mathscr{D}(A)$.

Since $\varphi(t) = Q(t)x$ and $\varphi(0) = \{\exp(tA_\varepsilon)\}x$, (12) implies

$$\|Q(t)x - \{\exp(tA_\varepsilon)\}x\| \le tK(t)\|Ax - A_\varepsilon x\|, \tag{13}$$

for $x \in \mathscr{D}(A)$, $0 < \varepsilon \le 1$. This gives (*e*) for $x \in \mathscr{D}(A)$.

However, $\|Q(t) - \exp(tA_\varepsilon)\|$ is bounded on $0 \le t \le T$, $0 < \varepsilon \le 1$, by (*a*) and (9). These operators form therefore an equicontinuous family (Chapter 4, Exercise 3); it follows that their convergence on the dense set $\mathscr{D}(A)$ forces their convergence on all of X (Chapter 2, Exercise 14). This proves (*e*).

(*f*) It follows from (*a*) that

$$\|R(\lambda)\| \le C \int_0^\infty e^{(\gamma - \text{Re}\,\lambda)t} \, dt = \frac{C}{\text{Re}\,(\lambda - \gamma)} < \infty \tag{14}$$

if Re $\lambda > \gamma$. Thus $R(\lambda) \in \mathscr{B}(X)$. The definition of $R(\lambda)$ shows that

$$\varepsilon A_\varepsilon R(\lambda)x = \int_0^\infty e^{-\lambda t} Q(t+\varepsilon)x\, dt - \int_0^\infty e^{-\lambda t} Q(t)x\, dt.$$

If we replace t by $t - \varepsilon$ in the first integral, we are led to

$$A_\varepsilon R(\lambda)x = \frac{e^{\lambda\varepsilon} - 1}{\varepsilon} R(\lambda)x - \frac{1}{\varepsilon} e^{\lambda\varepsilon} \int_0^\varepsilon e^{-\lambda t} Q(t)x\, dt. \tag{15}$$

As $\varepsilon \to 0$, the right side of (15) converges to $\lambda R(\lambda)x - x$. This shows that $R(\lambda)x \in \mathscr{D}(A)$ and that

$$(\lambda I - A)R(\lambda)x = x \qquad (x \in X). \tag{16}$$

On the other hand, if $x \in \mathscr{D}(A)$, we can apply (*d*) to

$$R(\lambda)A_\varepsilon x = \int_0^\infty e^{-\lambda t} Q(t) A_\varepsilon x\, dt \tag{17}$$

and see that

$$R(\lambda)Ax = \int_0^\infty e^{-\lambda t} \frac{d}{dt} Q(t)x\, dt = -x + \lambda R(\lambda)x \tag{18}$$

by an integration by parts. Thus

$$R(\lambda)(\lambda I - A)x = x \qquad (x \in \mathscr{D}(A)). \tag{19}$$

In particular, $\mathscr{D}(A)$ lies in the range of $R(\lambda)$. This completes the proof. ////

It is now natural to ask whether the limit can be removed from the conclusion (*e*), that is, under what conditions the exponential representation $Q(t) = \exp(tA)$ is valid. Theorems 13.36 and 13.38 give answers to these questions.

13.36 Theorem *If $\{Q(t)\}$ is as in Theorem 13.35, then any of the following three conditions implies the other two:*

(*a*) $\mathscr{D}(A) = X$.

(*b*) $\lim_{\varepsilon \to 0} \|Q(\varepsilon) - I\| = 0$.

(*c*) *$A \in \mathscr{B}(X)$ and $Q(t) = e^{tA}$* $\quad (0 \le t < \infty)$.

PROOF. We shall use the same notations as in the proof of Theorem 13.35.

If (*a*) holds, the Banach-Steinhaus theorem implies that the norms of the operators A_ε are bounded, for all sufficiently small $\varepsilon > 0$. Since $Q(\varepsilon) - I = \varepsilon A_\varepsilon$, (*b*) follows from (*a*).

If (*b*) holds, then also $\|M_t - I\| \to 0$ as $t \to 0$. Fix $t > 0$, so small that M_t is invertible in $\mathscr{B}(X)$. Since $M_t A_\varepsilon = A_t M_\varepsilon$, we have

$$A_\varepsilon = (M_t)^{-1} A_t M_\varepsilon. \tag{1}$$

As $\varepsilon \to 0$, (1) shows first of all that $A_\varepsilon x$ converges, for every $x \in X$ [since $M_\varepsilon x \to x$ and $(M_t)^{-1} A_t \in \mathscr{B}(X)$], second that $A = (M_t)^{-1} A_t$, and third that

$$\|A_\varepsilon - A\| \le \|(M_t)^{-1} A_t\| \|M_\varepsilon - I\| \to 0 \qquad \text{as } \varepsilon \to 0. \tag{2}$$

The formula $Q(t) = \exp(tA)$ follows now from (*e*) of Theorem 13.35, since (2) implies that

$$\lim_{\varepsilon \to 0} \|\exp(tA_\varepsilon) - \exp(tA)\| = 0 \qquad (0 \le t < \infty). \tag{3}$$

Thus (*c*) follows from (*b*).

The implication (*c*) → (*a*) is trivial. ////

Infinitesimal generators have the following characterization.

13.37 The Hille-Yosida theorem *A densely defined operator A in a Banach space X is the infinitesimal generator of a semigroup $\{Q(t)\}$ as in Definition* 13.34 *if and only if there are constants C, γ so that*

$$\|(\lambda I - A)^{-m}\| \le C(\lambda - \gamma)^{-m} \tag{1}$$

for all $\lambda > \gamma$ and all positive integers m.

PROOF. If A is related to $\{Q(t)\}$ as in Theorem 13.35, we saw there that $(\lambda I - A)^{-1} = R(\lambda)$, for $\lambda > \gamma$, where

$$R(\lambda)x = \int_0^\infty e^{-\lambda t} Q(t)x\, dt \tag{2}$$

is the Laplace transform of $Q(t)x$. Hence $R(\lambda)^2 x$ is the transform of the convolution

$$\int_0^t Q(t-s)Q(s)x\, ds = tQ(t)x. \tag{3}$$

(The formalism is the same as for Fourier transforms.) Continuing in this way, we find that

$$R(\lambda)^m x = \frac{1}{(m-1)!} \int_0^\infty t^{m-1} e^{-\lambda t} Q(t)x\, dt \tag{4}$$

for $m = 1, 2, 3, \ldots$. Therefore, with C and γ as in (*a*) of Theorem 13.35,

$$\|R(\lambda)^m\| \le \frac{C}{(m-1)!}\int_0^\infty t^{m-1}e^{-(\lambda-\gamma)t}\,dt = C(\lambda-\gamma)^{-m}. \tag{5}$$

This proves the necessity of (1).

For the converse, set $S(\varepsilon) = (I - \varepsilon A)^{-1}$, so that (1) becomes

$$\|S(\varepsilon)^m\| \le C(1-\varepsilon\gamma)^{-m} \qquad (0 < \varepsilon < \varepsilon_0,\ m = 1, 2, 3, \ldots), \tag{6}$$

and the relations

$$(I-\varepsilon A)S(\varepsilon)x = x = S(\varepsilon)(I-\varepsilon A)x \tag{7}$$

hold, the first for all $x \in X$, the second for all $x \in \mathscr{D}(A)$.

If $x \in \mathscr{D}(A)$, then $x - S(\varepsilon)x = -\varepsilon S(\varepsilon)x$, so that

$$\lim_{\varepsilon\to 0} S(\varepsilon)x = x. \tag{8}$$

But since $\|S(\varepsilon)\| \le C(1-\varepsilon_0\gamma)^{-1}$, $\{S(\varepsilon): 0 < \varepsilon < \varepsilon_0\}$ is equicontinuous, and hence (8) holds for all $x \in X$.

Next we set

$$T(t, \varepsilon) = \exp\,(tAS(\varepsilon)) \tag{9}$$

and claim that

$$\|T(t,\varepsilon)\| \le C\exp\left\{\frac{\gamma t}{1-\varepsilon\gamma}\right\} \qquad (0 < \varepsilon < \varepsilon_0,\ t > 0). \tag{10}$$

Indeed, the relation $\varepsilon AS(\varepsilon) = S(\varepsilon) - I$ [see (7)] shows that

$$T(t,\varepsilon) = e^{-t/\varepsilon}\sum_{m=0}^{\infty}\frac{t^m}{m!\,\varepsilon^m}S(\varepsilon)^m. \tag{11}$$

Now (10) follows from (6) and (11).

For $x \in \mathscr{D}(A)$, (7) and (9) show that

$$\frac{d}{dt}\{T(t,\varepsilon)T(t,\delta)^{-1}x\} = T(t,\varepsilon)T(t,\delta)^{-1}(S(\varepsilon)-S(\delta))Ax.$$

If we integrate this and apply $T(t, \delta)$ to the result, we obtain

$$T(t,\varepsilon)x - T(t,\delta)x = \int_0^t T(u,\varepsilon)T(t-u,\delta)(S(\varepsilon)-S(\delta))Ax\,du. \tag{12}$$

If we use (8) with Ax in place of x, and refer to (10), we see that the right side of (12) converges to 0 when $\varepsilon \to 0$ and $\delta \to 0$. The limit

$$Q(t)x = \lim_{\varepsilon\to 0} T(t,\varepsilon)x \tag{13}$$

exists therefore for every $x \in \mathscr{D}(A)$, uniformly on every bounded subset of $[0, \infty)$. Moreover, (10) shows that $\|Q(t)\| \leq Ce^{\gamma t}$. By equicontinuity, and the assumption that $\mathscr{D}(A)$ is dense, we see now that (13) holds for all $x \in X$. Since $T(t, \varepsilon)$ is defined by (9), it follows that $\{Q(t)\}$ is a semigroup, as in Definition 13.34.

Let $\tilde{A}$ be the infinitesimal generator of $\{Q(t)\}$. Then, by (*f*) of Theorem 13.35,

$$(14) \qquad (\lambda I - \tilde{A})^{-1}x = \int_0^\infty e^{-\lambda t}Q(t)x\,dt \qquad (\lambda > \gamma).$$

On the other hand, $AS(\varepsilon)$ is the infinitesimal generator of $\{\exp(tAS(\varepsilon))\} = \{T(t, \varepsilon)\}$. Thus

$$(15) \qquad (\lambda I - AS(\varepsilon))^{-1}x = \int_0^\infty e^{-\lambda t}T(t, \varepsilon)x\,dt.$$

By (13) this becomes

$$(16) \qquad (\lambda I - A)^{-1}x = \int_0^\infty e^{-\lambda t}Q(t)x\,dt.$$

Comparison of (14) and (16) shows now that $\lambda I - A$ and $\lambda I - \tilde{A}$ have the same inverse for all sufficiently large λ, and this implies that $\tilde{A} = A$. ////

For our final theorem, we return to the Hilbert space setting.

13.38 Theorem *Assume that $\{Q(t): 0 \leq t < \infty\}$ is a semigroup of normal operators $Q(t) \in \mathscr{B}(H)$, which satisfies the continuity condition*

$$(1) \qquad \lim_{t \to 0} \|Q(t)x - x\| = 0 \qquad (x \in H).$$

The infinitesimal generator A of $\{Q(t)\}$ is then a normal operator in H, there is a $\gamma < \infty$ such that Re $\lambda \leq \gamma$ *for every $\lambda \in \sigma(A)$, and*

$$(2) \qquad Q(t) = e^{tA} \qquad (0 \leq t < \infty).$$

If each $Q(t)$ is unitary, then there is a self-adjoint operator S in H such that

$$(3) \qquad Q(t) = e^{itS} \qquad (0 \leq t < \infty).$$

This representation of unitary semigroups is a classical theorem of M. H. Stone.

Note: Although $\mathscr{D}(A)$ may be a *proper* subspace of H, the operators e^{tA} are defined in all of H and are bounded. To see this, let E^A be the spectral decomposition of A (Theorem 13.33). Since $|e^{t\lambda}| \leq e^{t\gamma}$ for all

$\lambda \in \sigma(A)$, the symbolic calculus described in Theorem 12.21 allows us to define bounded operators e^{tA} by

$$e^{tA} = \int_{\sigma(A)} e^{t\lambda}\, dE^A(\lambda) \qquad (0 \le t < \infty). \tag{4}$$

The theorem has an easy converse: If A is as in the conclusion, then (2) obviously defines a semigroup of normal operators, and (1) holds because

$$\|Q(t)x - x\|^2 = \int_{\sigma(A)} |e^{t\lambda} - 1|^2\, dE^A_{x,x}(\lambda) \to 0 \tag{5}$$

as $t \to 0$, by the dominated convergence theorem.

PROOF. Since each $Q(s)$ commutes with each $Q(t)$. Theorem 12.16 implies that $Q(s)$ and $Q(t)^*$ commute. The smallest closed subalgebra of $\mathscr{B}(H)$ that contains all $Q(t)$ and all $Q(t)^*$ is therefore normal. Let Δ be its maximal ideal space, and let E be the corresponding resolution of the identity, as in Theorem 12.22.

Let f_t and a_ε be the Gelfand transforms of $Q(t)$ and A_ε, respectively. Then

$$a_\varepsilon = \frac{f_\varepsilon - 1}{\varepsilon} \qquad (\varepsilon > 0), \tag{6}$$

and a simple computation gives

$$a_{2\varepsilon} - a_\varepsilon = \frac{\varepsilon}{2}(a_\varepsilon)^2, \tag{7}$$

since $f_{2\varepsilon} = (f_\varepsilon)^2$. Define

$$b(p) = \lim_{n\to\infty} a_{2^{-n}}(p) \tag{8}$$

for those $p \in \Delta$ at which this limit exists (as a complex number), and define $b(p) = 0$ at all other $p \in \Delta$. Then b is a complex Borel function on Δ. Put $B = \Psi(b)$, as in Theorem 13.24, with domain

$$\mathscr{D}(B) = \left\{x \in H : \int_\Delta |b|^2\, dE_{x,x} < \infty\right\}. \tag{9}$$

Then B is a normal operator in H.

We will show that $A = B$.

If $x \in \mathscr{D}(A)$ then $\|A_\varepsilon x\|$ is bounded, as $\varepsilon \to 0$. Hence there exists $C_x < \infty$ such that

$$\int_\Delta |a_\varepsilon|^2\, dE_{x,x} = \|A_\varepsilon x\|^2 \le C_x \qquad (0 < \varepsilon \le 1) \tag{10}$$

and therefore

$$\int_{\Delta} |a_{2\varepsilon} - a_{\varepsilon}| \, dE_{x,x} \le \frac{\varepsilon}{2} C_x \qquad (0 < \delta \le 1), \tag{11}$$

by (7). Take $\varepsilon = 2^{-n}$ ($n = 1, 2, 3, \ldots$) in (11) and add the resulting inequalities. It follows that

$$\sum_{n=1}^{\infty} |a_{2^{-n+1}} - a_{2^{-n}}| < \infty \qquad \text{a.e. } [E_{x,x}]. \tag{12}$$

The limit (8) exists therefore a.e. $[E_{x,x}]$, and now Fatou's lemma and (10) imply that

$$\int_{\Delta} |b|^2 \, dE_{x,x} \le C_x. \tag{13}$$

Consequently, $\mathscr{D}(A) \subset \mathscr{D}(B)$.

Part (*a*) of Theorem 13.35 shows that $\|\exp(A_\varepsilon)\| \le \gamma_1 < \infty$ for $0 < \varepsilon \le 1$, where γ_1 depends on $\{Q(t)\}$. Hence $|\exp a_\varepsilon(p)| \le \gamma_1$ for every $p \in \Delta$, since the Gelfand transform is an isometry on B^*-algebras. It now follows from (8) that $|\exp b(p)| \le \gamma_1$ for every $p \in \Delta$. Hence there exists $\gamma < \infty$ such that

$$\operatorname{Re} b(p) \le \gamma \qquad (p \in \Delta). \tag{14}$$

For every $x \in \mathscr{D}(A)$ and every $t \ge 0$,

$$\|\exp(tA_\varepsilon)x - \exp(tB)x\|^2 = \int_{\Delta} |\exp(ta_\varepsilon) - \exp(tb)|^2 \, dE_{x,x} \tag{15}$$

tends to 0 as $\varepsilon \to 0$ through the sequence $\{2^{-n}\}$, because the integrand is bounded by $4\gamma_1^{2t}$ and its limit is 0 a.e. $[E_{x,x}]$. Hence (*e*) of Theorem 13.35 implies that

$$Q(t)x = e^{tB}x \qquad [x \in \mathscr{D}(A)]. \tag{16}$$

However, e^{tb} is a bounded function on Δ, $e^{tB} \in \mathscr{B}(H)$, and since (16) shows that the continuous operators $Q(t)$ and e^{tB} coincide on the dense set $\mathscr{D}(A)$, we conclude that

$$Q(t) = e^{tB} \qquad (0 < t < \infty). \tag{17}$$

It follows from (17) that

$$A_\varepsilon x - Bx = \left(\frac{e^{\varepsilon B} - I}{\varepsilon} - B\right)x \tag{18}$$

so that

$$\|A_\varepsilon x - Bx\|^2 = \int_{\Delta} \left|\frac{e^{\varepsilon b} - 1}{\varepsilon} - b\right|^2 dE_{x,x}. \tag{19}$$

As $\varepsilon \to 0$, the integrand (19) tends to 0, at every point of Δ. Since $|(e^z - 1)/z|$ is bounded on every half-plane $\{z: \text{Re } z \le c\}$, and since the integrand (19) can be written in the form

$$\left| \frac{e^{\varepsilon b} - 1}{\varepsilon b} - 1 \right|^2 |b|^2,$$

it follows from (14) and the dominated convergence theorem that

$$\lim_{\varepsilon \to 0} \|A_\varepsilon x - Bx\|^2 = 0 \qquad \text{if } x \in \mathscr{D}(B). \tag{20}$$

This proves that $\mathscr{D}(B) \subset \mathscr{D}(A)$ and that $A = B$.

That the real part of $\sigma(A)$ is bounded above follows now from (14) and (c) of Theorem 13.27.

This completes the proof, except for the final statement about unitary semigroups. If each $Q(t)$ is unitary, then $|f_\varepsilon| = 1$, (6) shows that $\lim a_\varepsilon$ is pure imaginary at every point at which it exists, as $\varepsilon \to 0$, hence $b(p)$ is pure imaginary at every $p \in \Delta$, and if $S = -iB$ then (17) gives (3), and (c) of Theorem 13.24 shows that S is self-adjoint. ////

Exercises

Throughout this set of exercises, the letter H denotes a Hilbert space, unless the contrary is stated.

1. The associative law $(T_1 T_2)T_3 = T_1(T_2 T_3)$ has been used freely throughout this chapter. Prove it. Prove also that $T_1 \subset T_2$ implies $ST_1 \subset ST_2$ and $T_1 S \subset T_2 S$.
2. Let T be a densely defined operator in H. Prove that T has a closed extension if and only if $\mathscr{D}(T^*)$ is dense in H. In that case, prove that T^{**} is an extension of T.
3. By Theorem 13.8, $\mathscr{D}(T^*) = \{0\}$ for a densely defined operator T in H if and only if $\mathscr{G}(T)$ is dense in $H \times H$. Show that this can actually happen.

   Suggestion: Let $\{e_n: n = 1, 2, 3, \ldots\}$ be an orthonormal basis of H; let $\{x_n\}$ be a dense subset of H; define $Te_n = x_n$; and extend T linearly to $\mathscr{D}(T)$, the set of all *finite* linear combinations of the basis vectors e_n. Show that the graph of this T is dense in $H \times H$.
4. Suppose T is a densely defined, closed operator in H, and $T^*T \subset TT^*$. Does it follow that T is normal?
5. Suppose T is a densely defined operator in H, and $(Tx, x) = 0$ for every $x \in \mathscr{D}(T)$. Does it follow that $Tx = 0$ for every $x \in \mathscr{D}(T)$?
6. If T is an operator in H, define

   $$\mathscr{N}(T) = \{x \in \mathscr{D}(T): Tx = 0\}.$$

   If $\mathscr{D}(T)$ is dense, prove that

   $$\mathscr{N}(T^*) = \mathscr{R}(T)^\perp \cap \mathscr{D}(T^*).$$

If T is also closed, prove that

$$\mathcal{N}(T) = \mathcal{R}(T^*)^\perp \cap \mathcal{D}(T).$$

This generalizes Theorem 12.10.

7. Consider the following three boundary value problems. The differential equation is

$$f'' - f = g,$$

where $g \in L^2([0, 1])$ is given. The choices of boundary conditions are
   (i) $f(0) = f(1) = 0$.
   (ii) $f'(0) = f'(1) = 0$.
   (iii) $f(0) = f(1)$ and $f'(0) = f'(1)$.

   Show that each of these problems has a unique solution f such that f' is absolutely continuous and $f'' \in L^2([0, 1])$. *Hint:* Combine Example 13.4 with Theorem 13.13.

   Do this also by solving the problems explicitly.

8. (a) Prove the self-adjointness of the operator T in $L^2(R)$, defined by $Tf = if'$, with $\mathcal{D}(T)$ consisting of all absolutely continuous $f \in L^2$ such that $f' \in L^2$.

   Hint: You may need to know that $f(t) \to 0$ as $t \to \pm\infty$ for every $f \in \mathcal{D}(T)$. Prove this. Or prove more, namely, that every $f \in \mathcal{D}(T)$ is the Fourier transform of an L^1-function.

   (b) Fix $g \in L^2(R)$. Use Theorem 13.13 to prove that the equation

$$f'' - f = g$$

   has a unique absolutely continuous solution $f \in L^2$, which has $f' \in L^2$, $f'' \in L^2$, and f' absolutely continuous.

   Prove also, by direct calculation, that

$$f(x) = -\frac{1}{2}\int_{-\infty}^{x} e^{t-x} g(t)\, dt - \frac{1}{2}\int_{x}^{\infty} e^{x-t} g(t)\, dt.$$

   This solution can also be found by means of Fourier transforms.

9. Let H^2 be the space of all holomorphic functions $f(z) = \sum c_n z^n$ in the open unit disc that satisfy

$$\|f\|^2 = \sum_{n=0}^{\infty} |c_n|^2 < \infty.$$

   Show that H^2 is a Hilbert space which is isomorphic to ℓ^2 via the one-to-one correspondence $f \leftrightarrow \{c_n\}$.

   Define $V \in \mathcal{B}(H^2)$ by $(Vf)(z) = zf(z)$. Show that V is the Cayley transform of the symmetric operator T in H^2, given by

$$(Tf)(z) = i\,\frac{1+z}{1-z}\, f(z).$$

   Find the ranges of $T + iI$ and of $T - iI$; show that one is H^2 and one has codimension 1. (Compare with Example 13.21.)

10. With H^2 as in Exercise 9, define V now by

$$(Vf)(z) = zf(z^2).$$

Show that V is an isometry which is the Cayley transform of a closed symmetric operator T in H^2, whose deficiency indices are 0 and ∞.

11. Prove part (*c*) of Lemma 13.18.

12. (*a*) In the context of Theorem 13.24, how are the operators $\Psi(f+g)$ and $\Psi(f)+\Psi(g)$ related?

(*b*) If f and g are measurable and g is bounded, prove that $\Psi(g)$ maps $\mathscr{D}_f$ into $\mathscr{D}_f$.

(*c*) Prove that $\Psi(f) = \Psi(g)$ if and only if $f = g$ a.e. $[E]$, that is, if and only if

$$E(\{p\colon f(p) \neq g(p)\}) = 0.$$

13. Is the operator C that occurs in the proof of Theorem 13.33 normal?

14. Prove that every normal operator N in H, bounded or not, has a polar decomposition

$$N = UP = PU,$$

where U is unitary, P is self-adjoint, $P \geq 0$. Moreover, $\mathscr{D}(P) = \mathscr{D}(N)$.

15. Prove the following extension of Theorem 12.16: If $T \in \mathscr{B}(H)$, if M and N are normal operators in H, and if $TM \subset NT$, then also $TM^* \subset N^*T$.

16. Suppose T is a closed operator in H, $\mathscr{D}(T) = \mathscr{D}(T^*)$, and $\|Tx\| = \|T^*x\|$ for every $x \in \mathscr{D}(T)$. Prove that T is normal. *Hint:* Begin by proving that

$$(Tx, Ty) = (T^*x, T^*y) \qquad (x \in \mathscr{D}(T),\ y \in \mathscr{D}(T)).$$

17. Prove that the spectrum $\sigma(T)$ of any operator T in H is a closed subset of $\mathbb{C}$. (See Definition 13.26.) *Hint:* If $ST \subset TS = I$, and $S \in \mathscr{B}(H)$, then $S(I - \lambda S)^{-1}$ is a bounded inverse of $T - M$, for small $|\lambda|$.

18. Put $\phi(t) = \exp(-t^2)$. Define $S \in \mathscr{B}(L^2)$, where $L^2 = L^2(R)$, by

$$(Sf)(t) = \phi(t)f(t-1) \qquad (f \in L^2),$$

so that $(S^2f)(t) = \phi(t)\phi(t-1)f(t-2)$, etc. (Note that S is presented in its polar decomposition $S = PU$.)

Find S^*. Compute that

$$\|S^n\| = \exp\left\{-\frac{(n-1)n(n+1)}{12}\right\} \qquad (n = 1, 2, 3, \ldots).$$

Conclude that S is one-to-one, that $\mathscr{R}(S)$ is dense in L^2, and that $\sigma(S) = \{0\}$. Define T, with domain $\mathscr{D}(T) = \mathscr{R}(S)$, by

$$TSf = f \qquad (f \in L^2).$$

Prove that $\sigma(T)$ is empty.

19. Let T_1, T_2, T_3 be as in Example 13.4, put

$$\mathscr{D}(T_4) = \{f \in \mathscr{D}(T_1)\colon f(0) = 0\},$$

and define $T_4 f = if'$ for all $f \in \mathscr{D}(T_4)$.

Prove the following assertions:

(*a*) Every $\lambda \in \mathcal{C}$ is in the point spectrum of T_1.

(*b*) $\sigma(T_2)$ consists of the numbers $2\pi n$, where n runs through the integers; each of these is in the point spectrum of T_2.

(*c*) $\mathscr{R}(T_3 - \lambda I)$ has codimension 1 for every $\lambda \in \mathcal{C}$. Hence $\sigma(T_3) = \mathcal{C}$. The point spectrum of T_3 is empty.

(*d*) $\sigma(T_4)$ is empty.

Hint: Study the differential equation $if' - \lambda f = g$.

This illustrates how sensitive the spectrum of a differential operator is to its domain (in this case, to the boundary conditions that are imposed).

20. Show that every nonempty closed subset of $\mathcal{C}$ is the spectrum of some normal operator in H (if $\dim H = \infty$).

21. Define $Q(t) \in \mathscr{B}(L^2)$, where $L^2 = L^2(R)$, by

$$(Q(t)f)(s) = f(s + t).$$

(*a*) Prove that each $Q(t)$ is unitary.

(*b*) Prove that $\{Q(t)\}$ satisfies the conditions stated in Definition 13.34.

(*c*) If A is the infinitesimal generator of $\{Q(t)\}$, prove that $f \in \mathscr{D}(A)$ if and only if $\int |y\hat{f}(y)|^2 \, dy < \infty$ (where $\hat{f}$ is the Fourier transform of f) and that $Af = f'$ for all $f \in \mathscr{D}(A)$.

(*d*) Prove that $\sigma(A)$ is the imaginary axis. More precisely, show that $A - \lambda I$ is one-to-one for every $\lambda \in \mathcal{C}$, that λ lies in the resolvent set of A if and only if λ is not pure imaginary, and that the range of $A - \lambda I$ is a proper dense subspace of L^2 if λ is pure imaginary.

Hint: $g \in \mathscr{R}(A - \lambda I)$ if and only if $g \in L^2$ and also $\hat{g}(y)/(iy - \lambda)$ is in L^2.

22. If $f \in H^2$ (see Exercise 9) and $f(z) = \sum c_n z^n$, define

$$[Q(t)f](z) = \sum_{n=0}^{\infty} (n + 1)^{-t} c_n z^n \qquad (0 \le t < \infty).$$

Show that each $Q(t)$ is self-adjoint (and positive). Find the infinitesimal generator A of the semigroup $\{Q(t)\}$. Is A self-adjoint? Show that A has pure point spectrum, at the points $\log 1, \log \frac{1}{2}, \log \frac{1}{3}, \ldots$.

23. For $f \in L^2(R)$, $x \in R$, $0 < y < \infty$, define

$$[Q(y)f](x) = \frac{1}{\pi} \int_{-\infty}^{\infty} \frac{y}{(x - \xi)^2 + y^2} f(\xi) \, d\xi,$$

and put $Q(0)f = f$. Show that $\{Q(y): 0 \le y < \infty\}$ satisfies the conditions imposed in Definition 13.34 and that $\|Q(y)\| = 1$ for all y.

[The integral represents a harmonic function in the upper half-plane, with boundary values f. The semigroup property of $\{Q(y)\}$ can be deduced from this, as well as from a look at the Fourier transforms of the functions $Q(y)f$.]

Find the domain of the infinitesimal generator A of $\{Q(y)\}$, and prove that

$$Af = -Hf',$$

where H is the Hilbert transform (Chapter 7, Exercise 24).

Prove that $-A$ is positive and self-adjoint.

24. Show that every isometric operator in H has a closed isometric extension.

25. On the other hand, show that some symmetric operators in H have no closed symmetric extension by completing the following outline.

Let $\{e_1, e_2, e_3, \ldots\}$ be an orthonormal basis of H. Let X be the set of all finite sums $\sum \alpha_i e_i$, subject to the condition $\sum \alpha_i = 0$. Prove that X is a dense subspace of H. Define $U \in \mathscr{B}(H)$ by

$$U\left(\sum_1^\infty \alpha_i e_i\right) = \alpha_1 e_1 - \sum_2^\infty \alpha_i e_i,$$

and let V be the restriction of U to X. Then V is an isometry, with $\mathscr{D}(V) = X$, and $I - V$ is one-to-one on X. Hence V is the Cayley transform of a symmetric operator T. Any closed symmetric extension of T would correspond to a closed isometric extension V_1 of V, with $I - V_1$ one-to-one. But $\mathscr{D}(V)$ is dense in H; hence V has only one closed isometric extension, namely U, and $I - U$ is not one-to-one.

APPENDIX A

COMPACTNESS AND CONTINUITY

A1 Partially ordered sets A set $\mathscr{P}$ is said to be *partially ordered* by a binary relation $\leq$ if:

(*i*) $a \leq b$ and $b \leq c$ implies $a \leq c$.
(*ii*) $a \leq a$ for every $a \in \mathscr{P}$,
(*iii*) $a \leq b$ and $b \leq a$ implies $a = b$.

A subset $\mathscr{Q}$ of a partially ordered set $\mathscr{P}$ is said to be *totally ordered* if every pair $a, b \in \mathscr{Q}$ satisfies either $a \leq b$ or $b \leq a$.

Hausdorff's maximality theorem states:

Every nonempty partially ordered set $\mathscr{P}$ contains a totally ordered subset $\mathscr{Q}$ which is maximal with respect to the property of being totally ordered.

A proof (using the axiom of choice) may be found in [23]. Explicit applications of the theorem occur in the proofs of the Hahn-Banach theorem, of the Krein-Milman theorem, and of the theorem that every proper ideal in a commutative ring with unit lies in a maximal ideal. It will now be applied once more (A2) to prepare the way to an easy proof of the Tychonoff theorem.

A2 Subbases A collection $\mathscr{S}$ of open subsets of a topological space X is said to be a *subbase* for the topology τ of X if the collection of all finite intersections of members of $\mathscr{S}$ forms a base for τ. (See Section 1.5.) Any subcollection of $\mathscr{S}$ whose union is X will be called an $\mathscr{S}$-cover of X. By definition, X is compact provided that every open cover of X has a finite subcover. It is enough to verify this property for $\mathscr{S}$-covers:

Alexander's subbase theorem. *If $\mathscr{S}$ is a subbase for the topology of a space X, and if every $\mathscr{S}$-cover of X has a finite subcover, then X is compact.*

PROOF. Assume X is not compact. We will deduce from this that X has an $\mathscr{S}$-cover $\tilde{\Gamma}$ without finite subcover.

Let $\mathscr{P}$ be the collection of all open covers of X that have no finite subcover. By assumption, $\mathscr{P} \neq \varnothing$. Partially order $\mathscr{P}$ by inclusion, let Ω be a maximal totally ordered subcollection of $\mathscr{P}$, and let Γ be the union of all members of Ω. Then

(*a*) Γ is an open cover of X,
(*b*) Γ has no finite subcover, but
(*c*) $\Gamma \cup \{V\}$ has a finite subcover, for every open $V \notin \Gamma$.

Of these, (*a*) is obvious. Since Ω is totally ordered, any finite subfamily of Γ lies in some member of Ω, hence cannot cover X; this gives (*b*), and (*c*) follows from the maximality of Ω.

Put $\tilde{\Gamma} = \Gamma \cap \mathscr{S}$. Since $\tilde{\Gamma} \subset \Gamma$, (*b*) implies that $\tilde{\Gamma}$ has no finite subcover. To complete the proof, we show that $\tilde{\Gamma}$ covers X.

If not, some $x \in X$ is not covered by $\tilde{\Gamma}$. By (*a*), $x \in W$ for some $W \in \Gamma$. Since $\mathscr{S}$ is a subbase, there are sets $V_1, \ldots, V_n \in \mathscr{S}$ such that $x \in \bigcap V_i \subset W$. Since x is not covered by $\tilde{\Gamma}$, no V_i belongs to Γ. Hence (*c*) implies that there are sets $Y_1, \ldots, Y_n$, each a finite union of members of Γ, such that $X = V_i \cup Y_i$ for $1 \leq i \leq n$. Hence

$$X = Y_1 \cup \cdots \cup Y_n \cup \bigcap_{i=1}^{n} V_i \subset Y_1 \cup \cdots \cup Y_n \cup W,$$

which contradicts (*b*). ////

A3 Tychonoff's theorem *If X is the cartesian product of any nonempty collection of compact spaces X_α, then X is compact.*

PROOF. If $\pi_\alpha(x)$ denotes the X_α-coordinate of a point $x \in X$, then, by definition, the topology of X is the weakest one that makes each $\pi_\alpha\colon X \to X_\alpha$ continuous; see Section 3.8. Let $\mathscr{S}_\alpha$ be the collection of all sets $\pi_\alpha^{-1}(V_\alpha)$, where V_α is any open subset of X_α. If $\mathscr{S}$ is the union of all $\mathscr{S}_\alpha$, it follows that $\mathscr{S}$ is a subbase for the topology of X.

Suppose Γ is an $\mathscr{S}$-cover of X. Put $\Gamma_\alpha = \Gamma \cap \mathscr{S}_\alpha$. Assume (to get a contradiction) that no Γ_α covers X. Then there corresponds to each α a point $x_\alpha \in X_\alpha$ such that Γ_α covers no point of the set $\pi_\alpha^{-1}(x_\alpha)$, and if $x \in X$ is chosen so that $\pi_\alpha(x) = x_\alpha$, then x is not covered by Γ. But Γ is a cover of X.

Hence at least one Γ_α covers X. Since X_α is compact, some finite subcollection of Γ_α covers X. Since $\Gamma_\alpha \subset \Gamma$, Γ has a finite subcover, and now Alexander's theorem implies that X is compact. ////

A4 Theorem *If K is a closed subset of a complete metric space X, then the following three properties are equivalent:*

(*a*) *K is compact.*
(*b*) *Every infinite subset of K has a limit point in K.*
(*c*) *K is totally bounded.*

Recall that (*c*) means that K can be covered by finitely many balls of radius ε, for every $\varepsilon > 0$.

PROOF. Assume (*a*). If $E \subset K$ is infinite and no point of K is a limit point of E, there is an open cover $\{V_\alpha\}$ of K such that each V_α contains at most one point of E. Therefore $\{V_\alpha\}$ has no finite subcover, a contradiction. Thus (*a*) implies (*b*).

Assume (*b*), fix $\varepsilon > 0$, and let d be the metric of X. Pick $x_1 \in K$. Suppose $x_1, \ldots, x_n$ are chosen in K so that $d(x_i, x_j) \geq \varepsilon$ if $i \neq j$. If possible, choose $x_{n+1} \in K$ so that $d(x_i, x_{n+1}) \geq \varepsilon$ for $1 \leq i \leq n$. This process must stop after a finite number of steps, because of (*b*). The ε-balls centered at $x_1, \ldots, x_n$ then cover K. Thus (*b*) implies (*c*).

Assume (*c*), let Γ be an open cover of K, and suppose (to reach a contradiction) that no finite subcollection of Γ covers K. By (*c*), K is a union of finitely many closed sets of diameter ≤ 1. One of these, say K_1, cannot be covered by finitely many members of Γ. Do the same with K_1 in place of K, and continue. The result is a sequence of closed sets K_i such that

(*i*) $K \supset K_1 \supset K_2 \supseteq \cdots$,
(*ii*) diam $K_n \leq 1/n$, and
(*iii*) no K_n can be covered by finitely many members of Γ.

Choose $x_n \in K_n$. By (*i*) and (*ii*), $\{x_n\}$ is a Cauchy sequence which (since X is complete and each K_n is closed) converges to a point

$x \in \bigcap K_n$. Hence $x \in V$ for some $V \in \Gamma$. By (*ii*), $K_n \subset V$ when n is sufficiently large. This contradicts (*iii*). Thus (*c*) implies (*a*). ////

Note that the completeness of X was used only in going from (*c*) to (*a*). In fact, (*a*) and (*b*) are equivalent in any metric space.

A5 Ascoli's theorem *Suppose X is a compact space, $C(X)$ is the sup-normed Banach space of all continuous complex functions on X, and $\Phi \subset C(X)$ is pointwise bounded and equicontinuous. More explicitly,*

(*a*) $\sup \{|f(x)| : f \in \Phi\} < \infty$ *for every* $x \in X$, *and*
(*b*) *if* $\varepsilon > 0$, *every* $x \in X$ *has a neighborhood* V *such that* $|f(y) - f(x)| < \varepsilon$ *for all* $y \in V$ *and for all* $f \in \Phi$.

Then Φ is totally bounded in $C(X)$.

Corollary. *Since $C(X)$ is complete, the closure of Φ is compact, and every sequence in Φ contains a uniformly convergent subsequence.*

PROOF. Fix $\varepsilon > 0$. Since X is compact, (*b*) shows that there are points $x_1, \ldots, x_n \in X$, with neighborhoods $V_1, \ldots, V_n$, such that $X = \bigcup V_i$ and such that

(1) $$|f(x) - f(x_i)| < \varepsilon \qquad (f \in \Phi,\ x \in V_i,\ 1 \le i \le n).$$

If (*a*) is applied to $x_1, \ldots, x_n$ in place of x, it follows from (1) that Φ is uniformly bounded:

(2) $$\sup \{|f(x)| : x \in X, f \in \Phi\} = M < \infty.$$

Put $D = \{\lambda \in \mathbb{C} : |\lambda| \le M\}$, and associate to each $f \in \Phi$ a point $p(f) \in D^n \subset \mathbb{C}$, by setting

(3) $$p(f) = (f(x_1), \ldots, f(x_n)).$$

Since D^n is a finite union of sets of diameter $< \varepsilon$. there exist $f_1, \ldots, f_m \in \Phi$ such that every $p(f)$ lies within ε of some $p(f_k)$.

If $f \in \Phi$, there exists k, $1 \le k \le m$, such that

(4) $$|f(x_i) - f_k(x_i)| < \varepsilon \qquad (1 \le i \le n).$$

Every $x \in X$ lies in some V_i, and for this i

(5) $$|f(x) - f(x_i)| < \varepsilon \quad \text{and} \quad |f_k(x) - f_k(x_i)| < \varepsilon.$$

Thus $|f(x) - f_k(x)| < 3\varepsilon$ for every $x \in X$.

The 3ε-balls centered at $f_1, \ldots, f_k$ therefore cover Φ. Since ε was arbitrary, Φ is totally bounded. ////

A6 Sequential continuity If X and Y are Hausdorff spaces and if f maps X into Y, then f is said to be *sequentially continuous* provided that $\lim_{n\to\infty} f(x_n) = f(x)$ for every sequence $\{x_n\}$ in X that satisfies $\lim_{n\to\infty} x_n = x$.

Theorem

(a) *If $f: X \to Y$ is continuous, then f is sequentially continuous.*

(b) *If $f: X \to Y$ is sequentially continuous, and if every point of X has a countable local base (in particular, if X is metrizable), then f is continuous.*

PROOF. (*a*) Suppose $x_n \to x$ in X, V is a neighborhood of $f(x)$ in Y, and $U = f^{-1}(V)$. Since f is continuous, U is a neighborhood of x, and therefore $x_n \in U$ for all but finitely many n. For these n, $f(x_n) \in V$. Thus $f(x_n) \to f(x)$ as $n \to \infty$.

(*b*) Fix $x \in X$, let $\{U_n\}$ be a countable local base for the topology of X at x, and assume that f is not continuous at x. Then there is a neighborhood V of $f(x)$ in Y such that $f^{-1}(V)$ is not a neighborhood of x. Hence there is a sequence x_n, such that $x_n \in U_n$, $x_n \to x$ as $n \to \infty$, and $x_n \notin f^{-1}(V)$. Thus $f(x_n) \notin V$, so that f is not sequentially continuous. ////

A7 Totally disconnected compact spaces A topological space X is said to be *totally disconnected* if none of its connected subsets contains more than one point.

A set $E \subset X$ is said to be *connected* if there exists no pair of open sets V_1, V_2 such that

$$E \subset V_1 \cup V_2, \qquad E \cap V_1 \neq \emptyset, \qquad E \cap V_2 \neq \emptyset,$$

but $E \cap V_1 \cap V_2 = \emptyset$.

Theorem. *Suppose $K \subset V \subset X$, where X is a compact Hausdorff space, V is open, and K is a component of X. Then there is a compact open set A such that $K \subset A \subset V$.*

Corollary. *If X is a totally disconnected compact Hausdorff space, then the compact open subsets of X form a base for its topology.*

PROOF. Let Γ be the collection of all compact open subsets of X that contain K. Since $X \in \Gamma$, $\Gamma \neq \emptyset$. Let H be the intersection of all members of Γ.

Suppose $H \subset W$, where W is open. The complements of the members of Γ form an open cover of the compact complement of W.

Since Γ is closed under finite intersections, it follows that $A \subset W$ for some $A \in \Gamma$.

We claim that H is connected. To see this, assume $H = H_0 \cup H_1$, where H_0 and H_1 are disjoint compact sets. Since $K \subset H$ and K is connected, K lies in one of these. Say $K \subset H_0$. By Urysohn's lemma, there are disjoint open sets W_0, W_1 such that $H_0 \subset W_0$, $H_1 \subset W_1$, and the preceding paragraph shows that some $A \in \Gamma$ satisfies $A \subset W_0 \cup W_1$. Put $A_0 = A \cap W_0$. Then $K \subset A_0$, A_0 is open, and A_0 is compact, because $A \cap W_0 = A \cap \bar{W}_0$. Thus $A_0 \in \Gamma$. Since $H \subset A_0$, it follows that $H_1 = \varnothing$.

Thus H is connected. Since $K \subset H$ and K is a component, we see that $K = H$. The preceding argument, with K and V in place of H and W, shows that $A \subset V$ for some $A \in \Gamma$. ////

APPENDIX B

NOTES AND COMMENTS

The abstract tendency in analysis which developed into what is now known as *functional analysis* began at the turn of the century with the work of Volterra, Fredholm, Hilbert, Fréchet, and F. Riesz, to mention only some of the principal figures. They studied integral equations, eigenvalue problems, orthogonal expansions, and linear operations in general. It is of course no accident that the Lebesgue integral was born in the same period.

The normed space axioms appear in F. Riesz' work on compact operators in $C([a, b])$ (*Acta Math.*, vol. 41, pp. 71–98, 1918), but the first abstract treatment of the subject is in Banach's 1920 thesis (*Fundam. Math.*, vol. 3, pp. 133–181, 1922). His book [2], published in 1932, was tremendously influential. It contains what is still the basic theory of Banach spaces, but with some omissions which, from our vantage point, seem curious.

One of these is the complete absence of complex scalars, in spite of Wiener's observation (*Fundam. Math.*, vol. 4, pp. 136–143, 1923) that the axioms can be formulated just as well over $\mathcal{C}$, and, more importantly, that a theory of Banach-space-valued holomorphic functions can then be developed whose basic features are very similar to the classical complex-valued case. Very little (if anything) was done with this until 1938. (See the notes for Chapter 3 in this appendix.)

Even more puzzling, in retrospect, is Banach's treatment of weak convergence—surely one of his most important contributions to the subject.

In spite of the vigorous development of topology in the twenties, and in spite of von Neumann's explicit description of weak neighborhoods in a Hilbert space and in operator algebras (*Math. Ann.*, vol. 102, pp. 370–427, 1930; see p. 379), Banach deals only with weakly convergent *sequences*. Since the adjunction of all limits of weakly convergent subsequences of a set need not lead to a weakly sequentially closed set (see Exercise 9, Chapter 3), he is forced into complicated notations such as transfinite closures, but he never uses the much simpler and more satisfactory concept of *weak topologies.*[1]

Occasionally, unnecessary separability assumptions are made in [2]. This is also true of von Neumann's axiomatization of Hilbert space (*Math. Ann.*, vol. 102, pp. 49–131, 1930), where separability is included among the defining properties. In this fundamental paper on unbounded operators, he establishes the spectral theorem for them, thus generalizing what Hilbert had done for the bounded ones more than 20 years earlier. Another basic contribution to operator theory was M. H. Stone's 1932 book [28].

Although continuous functions obviously play an important role in Banach's book, he considers only their vector space structure. They are never multiplied. But multiplication was not neglected for very long. In his work on the tauberian theorem (*Ann. Math.*, vol. 33, pp. 1–100, 1932) Wiener stated and used the fact that the Banach space of absolutely convergent Fourier series satisfies the multiplicative inequality $\|xy\| \leq \|x\| \|y\|$. M. H. Stone's generalization of the Weierstrass approximation theorem (*Trans. Amer. Math. Soc.*, vol. 41, pp. 375–481, 1937; especially pp. 453–481) is undoubtedly the best-known instance of the explicit use of the ring structure of spaces of continuous functions. Von Neumann's interest in operator theory, which stemmed from quantum mechanics, led him to a systematic study of operator algebras. M. Nagumo (*Jap. J. Math.*, vol. 13, pp. 61–80, 1936) initiated the abstract study of normed rings. But what really got this subject off the ground was Gelfand's discovery of the important role played by the maximal ideals of a commutative algebra (*Mat. Sbornik N. S.*, vol. 9, pp. 3–24, 1941) and his construction of what is now known as the Gelfand transform.

Before the middle forties, the interest of functional analysts was focused almost exclusively on *normed* spaces. The first major paper on the general theory of locally convex spaces is that of J. Dieudonné and L. Schwartz in *Ann. Inst. Fourier* (*Grenoble*), vol. 1, pp. 61–101, 1949. One of

---

[1] Banach is obviously one of the major heroes of this story. The preceding remarks are not intended to be in any way derogatory (as some readers of the first edition thought) or to belittle the importance and originality of his work. Their intent is merely to contrast our present mathematical environment with what it was then.

its principal motivations was Schwartz' construction of the theory of distributions [26]. (The first version of this book appeared in 1950.) Just as Banach and Gelfand had predecessors, so did Schwartz. As Bochner points out in his review of Schwartz' book (*Bull. Amer. Math. Soc.*, vol. 58, pp. 78–85, 1952), the idea of "generalized functions" goes back at least as far as Riemann. It was applied in Bochner's *Vorlesungen über Fouriersche Integrale* (Leipzig, 1932), a book that played a very important role in the development of harmonic analysis. Sobolev's work also predates Schwartz. But it was Schwartz who built all this into a smoothly operating very general structure that turned out to have many applications, especially to partial differential equations.

The following expository articles describe some of the history of our subject in greater detail.

Bonsall, F. F.: "A Survey of Banach Algebra Theory," *Bull. London Math. Soc.*, vol. 2, pp. 257–274, 1970.

Hildebrandt, T. H.: "Integration in Abstract Spaces," *Bull. Amer. Math. Soc.*, vol. 59, pp. 111–139, 1953.

Horváth, J.: "An Introduction to Distributions," *Amer. Math. Monthly*, vol. 77, pp. 227–240, 1970.

Lorch, E. R.: "The Structure of Normed Abelian Rings," *Bull. Amer. Math. Soc.*, vol. 50, pp. 447–463, 1944.

Taylor, A. E.: "Notes on the History and Uses of Analyticity in Operator Theory," *Amer. Math. Monthly*, vol. 78, pp. 331–342, 1971.

Trèves, F.: "Applications of Distributions to PDE Theory," *Amer. Math. Monthly*, vol. 77, pp. 241–248, 1970.

Volume 1 of the series *Studies in Mathematics* (published by the Mathematical Association of America, 1962, edited by R. C. Buck) contains articles by

Goffman, C.: "Preliminaries to Functional Analysis"

Lorch, E. R.: "The Spectral Theorem"

McShane, E. J.: "A Theory of Limits".

Stone, M. H.: "A Generalized Weierstrass Approximation Theorem"

There are two special issues of *Bull. Amer. Math. Soc.*: One (May 1958) is devoted to the work of John von Neumann; the other (January 1966) to that of Norbert Wiener.

The origins of functional analysis are well described in Dieudonné's book [36].

We now give detailed references to some items in the text.

Chapter 1

For the general theory of topological vector spaces, see [5], [14], [15], [31], [32].

Section 1.8 (*e*). In Banach's definition of an F-space, he postulated only the *separate* continuity of scalar multiplication and proved that joint continuity was a consequence. See [4], pp. 51–53, for a proof based on Baire's theorem. Another proof (due to S. Kakutani) does not require completeness of X but uses Lebesgue measure in the scalar field; see [33], pp. 31–32.

Theorem 1.24. This metrization theorem was first proved (in the more general context of topological groups) by G. Birkhoff (*Compositio Math.*, vol. 3, pp. 427–430, 1936) and by S. Kakutani (*Proc. Imp. Acad. Tokyo*, vol. 12, pp. 128–142, 1936). Part (*d*) of the theorem is perhaps new.

Section 1.33. The Minkowski functional of a convex set is sometimes called its *support function*.

Theorem 1.39 is due to A. Kolmogorov (*Studia Math.*, vol. 5, pp. 29–33, 1934). It may well be the first theorem about locally convex spaces.

Section 1.46. The construction of the function g by repeated averaging may be found on pp. 80–84 of S. Mandelbrojt's 1942 Rice Institute Pamphlet "Analytic Functions and Classes of Infinitely Differentiable Functions," where it is credited to H. E. Bray.

Section 1.47. Of particular interest among the F-spaces that are not locally convex but have enough continuous linear functionals to separate points are certain subspaces of L^p, the H^p-spaces (with $0 < p < 1$). For a detailed study of these, see the paper by P. L. Duren, B. W. Romberg, and A. L. Shields in *J. Reine Angew. Math.*, vol. 238, pp. 32–60, 1969, and those by Duren and Shields in *Trans. Amer. Math. Soc.*, vol. 141, pp. 255–262, 1969, and in *Pac. J. Math.*, vol. 32, pp. 69–78, 1970, as well as [40].

Chapter 2

Basically, all results of this chapter are in [2].

Exercise 11. Charles Horowitz constructed a bilinear map from $R^3 \times R^3$ *onto* R^4 which is not open at (0, 0), in *Proc. Amer. Math. Soc.*, vol. 53, pp. 293–294, 1975. P. J. Cohen (*J. Func. Anal.*, vol. 16, pp. 235–239, 1974) had earlier constructed a much more complicated example, mapping $\ell^1 \times \ell^1$ onto ℓ^1.

Exercise 13. A *barrel* is a closed, convex, balanced, absorbing set. A space is *barreled* if every barrel contains a neighborhood of 0. Exercise 13 asserts: Topological vector spaces of the second category are barreled. There exist barreled spaces of the first category, and certain versions of the Banach-Steinhaus theorem are valid for them. See [14], p. 104; also [15].

Barreled spaces with the Heine-Borel property are often called *Montel spaces*; see Sec. 1.45.

Chapter 3

Theorem 3.2 is in [2]. Its complex version, Theorem 3.3, was proved by H. F. Bohnenblust and A. Sobczyk, *Bull. Amer. Math. Soc.*, vol. 44, pp. 91–93, 1938, and by G. A. Soukhomlinoff, *Mat. Sbornik*, vol. 3, pp. 353–358, 1938. The latter also considered quaternion scalars. In *Proc. Amer. Math. Soc.*, vol. 50, pp. 322–327, 1975, J. A. R. Holbrook presents a proof in which real scalars are not treated separately, and he includes a simplified version of Nachbin's work on Hahn-Banach extensions of linear transformations (in place of linear functionals); see *Trans. Amer. Math. Soc.*, vol. 68, pp. 28–46, 1950.

Theorem 3.6. For a partial converse, see J. H. Shapiro, *Duke Math. J.*, vol. 37, pp. 639–645, 1970.

Theorem 3.15. See L. Alaoglu, *Ann. Math.*, vol. 41, pp. 252–267, 1940. For separable Banach spaces, the theorem is in [2], p. 123.

Theorem 3.18. A shorter proof, based on seminorms, may be found on p. 223 of [32].

Section 3.22. Compact convex sets with no extreme point exist in some F-spaces. See [40].

Theorem 3.23 was proved, for weak*-compact convex subsets of the dual of a Banach space, by M. Krein and D. Milman, in *Studia Math.*, vol. 9, pp. 133–1940.

Theorem 3.25 appeared in *Dokl. Akad. Nauk SSSR*, vol. 57, pp. 119–122, 1947.

The history of vector-valued integration is described by T. H. Hildebrandt in *Bull. Amer. Math. Soc.*, vol. 59, pp. 111–139, 1953. The "weak" integral of Definition 3.26 was developed by B. J. Pettis, *Trans. Amer. Math. Soc.*, vol. 44, pp. 277–304, 1938.

The history of vector-valued holomorphic functions is described by A. E. Taylor in *Amer. Math. Monthly*, vol. 78, pp. 331–342, 1971.

Theorem 3.31. That weakly holomorphic functions (with values in a complex Banach space) are strongly holomorphic was proved by N. Dunford in *Trans. Amer. Math. Soc.*, vol. 44, pp. 304–356, 1938.

Theorem 3.32 was used by A. E. Taylor to prove that the spectrum of every bounded linear operator on a complex Banach space is nonempty (*Bull. Amer. Math. Soc.*, vol. 44, pp. 70–74, 1938). Since every Banach algebra A is isomorphic to a subalgebra of $\mathscr{B}(A)$ (see the proof of Theorem 10.2), Taylor's result contains (*a*) of Theorem 10.13.

Exercise 9 is due to von Neumann, *Math. Ann.*, vol. 102, pp. 370–427, 1930; see p. 380.

Exercise 10 is patterned after a construction in the appendix of [2].

Exercise 25. If K is also separable and metric, then such a μ exists even on E, rather than on $\bar{E}$. This is Choquet's theorem. See [20]. For a recent paper on this, see R. D. Bourgin, *Trans. Amer. Math. Soc.*, vol. 154, pp. 323–340, 1971.

Exercise 28 (*c*). This is the easy part of the Eberlein-Šmulian theorem. See [4], pp. 430–433 and p. 466. Another characterization of weak compactness has been given by R. C. James, *Trans. Amer. Math. Soc.*, vol. 113, pp. 129–140, 1964: A weakly closed set S in a Banach space X is weakly compact if and only if every $x^* \in X^*$ attains its supremum on S.

Exercise 33. See [14], p. 133.

Chapter 4

A large part of this chapter is in [2].

Compact operators used to be called *completely continuous*. As defined by Hilbert (in ℓ^2) this means that weakly convergent sequences are mapped to strongly convergent ones. The presently used definition was given by F. Riesz (*Acta Math.*, vol. 41, pp. 71–98, 1918). In reflexive spaces, the two definitions coincide (Exercise 18).

Section 4.5. R. C. James has constructed a *nonreflexive* Banach space X which is isometrically isomorphic with X^{**} (*Proc. Natl. Acad. Sci. USA*, vol. 37, pp. 174–177, 1951).

Theorems 4.19 and 4.25 were proved by J. Schauder (*Studia Math.*, vol. 2, pp. 183–196, 1930). For generalizations to arbitrary topological vector spaces, see J. H. Williamson, *J. London Math. Soc.*, vol. 29, pp. 149–156, 1954; also [5], chap. 9.

Exercise 13. It was a problem of long-standing whether every compact operator in every separable Banach space can be approximated (in the operator norm) by operators with finite-dimensional ranges. The first counterexample was constructed by P. Enflo, in *Acta Math.*, vol. 130, pp. 309–317, 1973. (This also gave a negative solution to the so-called *basis problem.*) Ramifications of this approximation problem are discussed in [41].

Exercise 15. These operators are usually called *Hilbert-Schmidt operators.* See [4], chap. XI.

Exercise 17. Operators of this type are discussed by A. Brown, P. R. Halmos, and A. L. Shields in *Acta Sci. Math. Szeged.*, vol. 26, pp. 125–137, 1965.

Exercise 19. This "max-min duality" was exploited by W. W. Rogosinski and H. S. Shapiro to obtain very detailed information about certain extremum problems for holomorphic functions. See *Acta Math.*, vol. 90, pp. 287–318, 1953.

Exercise 21. This was proved by M. Krein and V. Šmulian in *Ann. Math.*, vol. 41, pp. 556–583, 1940. See also [4], pp. 427–429.

Chapter 5

Theorem 5.1. For a more general version, see R. E. Edwards, *J. London Math. Soc.*, vol. 32, pp. 499–501, 1957.

Theorem 5.2 is due to A. Grothendieck, *Can. J. Math.*, vol. 6, pp. 158–160, 1954. His proof is less elementary than the one given here.

Theorem 5.3. For more on trigonometric series with gaps, see *J. Math. Mech.*, vol. 9, pp. 203–228, 1960; also, sec. 5.7 of [24], J. P. Kahane's article in *Bull. Amer. Math. Soc.*, vol. 70, pp. 199–213, 1964, and [42].

Theorem 5.5 was first proved by A. Liapounoff, *Bull. Acad. Sci. USSR*, vol. 4, pp. 465–478, 1940. The proof of the text is due to J. Lindenstrauss, *J. Math. Mech.*, vol. 15, pp. 971–972, 1966. J. J. Uhl (*Proc. Amer. Math. Soc.*, vol. 23, pp. 158–163, 1969) generalized the theorem to measures whose values lie in a reflexive Banach space or in a separable dual space.

Theorem 5.7. The idea to use Krein-Milman to prove Stone-Weierstrass is due to L. de Branges, *Proc. Amer. Math. Soc.*, vol. 10, pp. 822–824, 1959. E. Bishop's generalization is in *Pac. J. Math.*, vol. 11, pp. 777–783, 1961. The proof given here is that of I. Glicksberg, *Trans. Amer. Math. Soc.*, vol. 105, pp. 415–435, 1962. C. Hamburger pointed out to me that one does not need to assume that A contains the constants. A very elementary approach to Bishop's theorem was found by Mao Chao-Lin, *C. R. Acad. Sci. Paris*, vol. 301, pp. 349–350, 1985.

Theorem 5.9. Bishop proved this in *Proc. Amer. Math. Soc.*, vol. 13, pp. 140–143, 1962. For the special case of the disc algebra, see *Proc. Amer. Math. Soc.*, vol. 7, pp. 808–811, 1956, and L. Carleson's paper in *Math Z.*, vol. 66, pp. 447–451, 1957. Other applications occur in Chapter 6 of [25] and in Chapter 10 of [45]. See also [29].

Theorem 5.10. The proof follows that of M. Heins, *Ann. Math.*, vol. 52, pp. 568–573, 1950, where the same method is applied to a large class of interpolation problems.

Theorem 5.11 was proved by S. Kakutani in *Proc. Imp. Acad. Tokyo*, vol. 14, pp. 242–245, 1938. The proof given here was communicated to me by Isaac Namioka and is due to F. Hahn, *Math. Systems Theory*, vol. 1, pp. 55–57, 1968. The lemma avoids the use of nets and subnets at the end of the proof.

Theorem 5.14. This simple construction of the Haar measure of a compact group is essentially that of von Neumann (*Compositio Math.*, vol. 1, pp. 106–114, 1934). His is even more elementary and self-contained, though a little longer, since he uses no fixed point theorem. (In *Trans. Amer. Math. Soc.*, vol. 36, pp. 445–492, 1934, he uses the same method to construct mean values of almost periodic functions.) If compactness is replaced by local compactness, the construction of Haar measure becomes more difficult. See [18], [11], [16].

Theorem 5.18 was proved (for Banach spaces) in *Proc. Amer. Math. Soc.*, vol. 13, pp. 429–432, 1962. For further results on uncomplemented subspaces, see H. P. Rosenthal's 1966 AMS Memoir *Projections onto Translation-Invariant Subspaces of $L^p(G)$* and his paper in *Acta Math.*, vol. 124, pp. 205–248, 1970. There are also positive results. For example c_0 is complemented in any separable Banach space which contains it (isomorphically) as a closed subspace. A very short proof of this theorem of A. Sobczyk was obtained by W. A. Veech in *Proc. Amer. Math. Soc.*, vol. 28, pp. 627–628, 1971.

Section 5.19. That H^1 is uncomplemented in L^1 was first proved by D. J. Newman, *Proc. Amer. Math. Soc.*, vol. 12, pp. 98–99, 1961. The proof given here is in *Proc. Amer. Math. Soc.*, vol. 13, pp. 429–432, 1962.

Theorem 5.21. F. F. Bonsall's paper in *Quart. J. Math. Oxford*, vol. 37, pp. 129–136, 1986, contains this and further applications of Theorem 5.22.

Theorems 5.23, 5.28. The history of these fixed point theorems is described on pp. 470–471 of [4]. A proof of Brouwer's theorem that is both elementary and simple may be found on pp. 38–40 of *Dimension Theory* by Hurewicz and Wallman, Princeton University Press, Princeton, N.J., 1948.

Chapter 6

The standard reference is, of course, [26]. See also [5], [8], [27], [31]. [13] contains a very concise introduction to the subject.

Definition 6.3. $\mathscr{D}(\Omega)$ is here topologized as the *inductive limit* of the Fréchet spaces $\mathscr{D}_K(\Omega)$. See [15], pp. 217–225, for a systematic discussion of this notion in an abstract setting.

Chapter 7

For those aspects of Fourier analysis that are related to distributions, we refer to [26] and [13]. The group-theoretic aspects of the subject are discussed in [11] and [24]. The standard work on Fourier series is [34].

Theorem 7.4. The intimate relation between Fourier transforms and differentiation is no accident; Fourier series were invented, in the eighteenth century, as tools to solve differential equations.

Theorem 7.5 is sometimes called the Riemann-Lebesgue lemma.

Theorem 7.9 was originally proved by M. Plancherel in *Rend. Palermo.*, vol. 30, pp. 289–335, 1910.

Theorems 7.22 and 7.23. These proofs are as in [13] but contain more details.

Theorem 7.25 is due to S. L. Sobolev, *Mat. Sbornik*, vol. 4, pp. 471–497, 1938.

Exercise 16. This is taken from L. Schwartz' first counterexample to the *spectral synthesis problem* (*C. R. Acad. Sci. Paris*, vol. 227, pp. 424–426, 1948). For further information on this problem, see C. S. Herz (*Trans. Amer. Math. Soc.*, vol. 94, pp. 181–232, 1960) and chap. 7 of [24].

Exercise 17. See C. S. Herz, *Ann. Math.*, vol. 68, pp. 709–712, 1958.

Chapter 8

General references: [1], [13], [27], [30].

The existence of fundamental solutions (Theorem 8.5) was established independently by L. Ehrenpreis (*Amer. J. Math.*, vol. 76, pp. 883–903, 1954) and by B. Malgrange in his thesis (*Ann. Inst. Fourier*, vol. 6, pp. 271–355, 1955–1956). Lemma 8.3 is Malgrange's. He proves it for Fourier transforms f of test functions. He integrates over a ball where we have used a torus. As far as applications are concerned, this makes hardly any difference. The point is to get some useful majorization of f by fP, that is, to have division by P under control. Ehrenpreis solved this division problem in a different way and went on to solve more general division problems of this type. See [13] and [30] for further references and more detailed results.

It is essential in Theorem 8.5 that the coefficients of the differential operator under consideration be constant. This follows from an equation constructed by H. Lewy (*Ann. Math.*, vol. 66, pp. 155–158, 1957), which has C' coefficients but no solution. Hörmander ([13], chap. VI) has investigated this nonexistence phenomenon very completely.

Section 8.8. Many other types of Sobolev spaces have been studied. See [13], chap. II.

Theorem 8.12. See K. O. Friedrichs, *Comm. Pure Appl. Math.*, vol. 6, pp. 299–325, 1953, and P. D. Lax, *Comm. Pure Appl. Math.*, vol. 8, pp. 615–633, 1955. Lax treats the periodic case first, via Fourier series, and then uses the bootstrap proposition to obtain the general case. He does not assume that the highest-order terms are constant. See also [4], pp. 1703–1708.

Exercise 10. G is the so-called "Green's function" of $P(D)$.

Exercise 16. This is a theorem about zero sets of homogeneous polynomials (with complex coefficients) in R^n. See [1], p. 46.

Chapter 9

Section 9.1. See A. Tauber, *Monatsh. Math.*, vol. 8, pp. 273–277, 1897, and J. E. Littlewood, *Proc. London Math. Soc.*, vol. 9, pp. 434–448, 1910.

Theorem 9.3. The use of distributions in this proof is as in J. Korevaar's paper in *Proc. Amer. Math. Soc.*, vol. 16, pp. 353–355, 1965.

Theorem 9.4 to Theorem 9.7. N. Wiener, *Ann. Math.*, vol. 33, pp. 1–100, 1932, and H. R. Pitt, *Proc. London Math. Soc.*, vol. 44, pp. 243–288, 1938. Later proofs gave various generalizations; see [24], p. 159, for further references. See also A. Beurling, *Acta Math.*, vol. 77, pp. 127–136, 1945.

Section 9.9. The prime number theorem was first proved, independently, by J. Hadamard (*Bull. Soc. Math. France*, vol. 24, pp. 199–220, 1896) and by Ch. J. de la Vallée-Poussin (*Ann. Soc. Sci. Bruxelles*, vol. 20, pp. 183–256, 1896). Both used complex variable methods. Wiener gave the first tauberian proof, as an application of his general theorem. "Elementary" proofs were found in 1949 by A. Selberg and by P. Erdös. For a simpler elementary proof, see N. Levinson, *Amer. Math. Monthly*, vol. 76, pp. 225–245, 1969. The complex variable proofs still give the best error estimates; see W. J. Le Veque, *Topics in Number Theory*, vol. II, p. 251, Addison-Wesley Publishing Company, Reading, Mass., 1956.

Theorem 9.12. A. E. Ingham, *J. London Math. Soc.*, vol. 20, pp. 171–180, 1945.

The material on the renewal equation is from S. Karlin, *Pac. J. Math.*, vol. 5, pp. 229–257, 1955, where references to earlier work may be found. Nonlinear versions of the renewal equation are discussed by J. Chover and P. Ney in *J. d'Analyse Math.*, vol. 21, pp. 381–413, 1968; see also B. Henry, *Duke Math. J.*, vol. 36, pp. 547–558, 1969.

Exercise 7. This approximation problem is much less delicate in L^2. See [23], sec. 9.16.

Chapter 10

General references: [7], [12], [16], [19], [21]. In [16] and [21], a great deal of basic theory is developed without assuming the presence of a unit. [21] contains some material about real algebras.

Gelfand's paper (*Mat. Sbornik*, vol. 9, pp. 3–24, 1941) contains Theorems 10.2, 10.13, and 10.14, some symbolic calculus, and Theorem 11.9. For Fourier transforms of measures, the spectral radius formula (*b*) of Theorem 10.13 had been obtained earlier by A. Beurling (*Proc. IX Congrès de Math. Scandinaves, Helsingfors*, pp. 345–366, 1938). See also the note to Theorem 3.32.

Theorem 10.9. The commutative case was obtained independently by A. M. Gleason (*J. Anal. Math.*, vol. 19, pp. 171–172, 1967) and by J. P. Kahane and W. Zelazko (*Studia Math.*, vol. 29, pp. 339–343, 1968). W. Zelazko (*Studia Math.*, vol. 30, pp. 83–85, 1968) removed the commutativity hypothesis. The proof given in the text contains some simplifications. See also Theorem 1.4.4 of [3], and J. A. Siddiqi, *Can. Math. Bull.*, vol. 13, pp.

219–220, 1970. M. Roitman and Y. Sternfeld (*Trans. Amer. Math. Soc.*, vol. 267, pp. 111–124, 1981) found a more algebraic proof, which uses no function theory. Related results concerning ideals of finite codimension were found by N. V. Rao (*J. Func. Anal.*, vol. 82, pp. 237–258, 1989).

Theorem 10.19. H. A. Seid (*Amer. Math. Monthly*, vol. 77, pp. 282–283, 1970) obtains the same conclusions, without assuming that A has a unit, if $M = 1$.

Theorem 10.20 says that $\sigma(x)$ is an upper semicontinuous function of x. An example of Kakutani ([21], p. 282) shows that $\sigma(z)$ is not, in general, a continuous function of x. See also Exercise 20.

Section 10.21. The terms *operational calculus* or *functional calculus* are also frequently used. [12] contains a very thorough treatment of the symbolic calculus in Banach algebras.

Theorem 10.34 (*d*) is due to E. R. Lorch (*Trans. Amer. Math. Soc.*, vol. 52, pp. 238–248, 1942).

Theorem 10.35. Lomonosov's proof was published in *Func. Anal. and Appl.*, vol. 7, pp. 55–56, 1973. Even for a single operator it is much simpler and more far-reaching than anything that was known before. A. J. Michaels gave an account of Hilden's contribution in *Adv. in Math.*, vol. 25, pp. 56–58, 1977.

As regards earlier work, N. Aronszajn and K. T. Smith (*Ann. Math.*, vol. 60, pp. 345–350, 1954) proved that every compact operator on a Banach space has a proper invariant subspace. A. R. Bernstein and A. Robinson (*Pac. J. Math.*, vol. 16, pp. 421–431, 1966) obtained the same conclusion for bounded operators T on a Hilbert space that have $p(T)$ compact for some polynomial p. Their proof uses nonstandard analysis; P. R. Halmos converted it into one that uses only classical concepts (*Pac. J. Math.*, vol. 16, pp. 433–437, 1966).

Since some operators, even on a Hilbert space, commute with no compact one (Exercise 26), Lomonosov's theorem does not settle the invariant subspace problem. In fact, operators without invariant subspaces have been found in certain nonreflexive Banach spaces (P. Enflo, *Acta Math.*, vol. 158, pp. 213–313, 1987), and even in ℓ^1 and c_0 (C. J. Read, *Proc. London Math. Soc.*, vol. 53, pp. 583–607, 1989). See also Section 12.27.

Exercise 22. This is one of the simplest cases of the Arens-Royden theorem for commutative Banach algebras. It relates the group G/G_1 to the topological structure of the maximal ideal space of A. See H. L. Royden's article in *Bull. Amer. Math. Soc.*, vol. 69, pp. 281–298, 1963, that by R. Arens in F. T. Birtel, ed., *Function Algebras*, pp. 164–168, Scott, Foresman and Company, Glenview, Ill., 1966, and [6] and [29].

Exercise 23. For the precise structure of G/G_1 in this case, see J. L. Taylor, *Acta Math.*, vol. 126, pp. 195–225, 1971.

Exercise 24. See C. Le Page, *C. R. Acad. Sci. Paris*, vol. 265, pp. A235–A237, 1967.

Exercise 26. The invariant subspaces of this shift operator are completely known. This is Beurling's theorem (*Acta Math.*, vol. 81, pp. 239–255, 1949). Helson and Lowdenslager (*Acta Math.*, vol. 99, pp. 165–202, 1958) used different methods and extended Beurling's theorem to other settings.

Chapter 11

Theorem 11.7. The case $n = 1$ was proved in elementary fashion by P. J. Cohen in *Proc. Amer. Math. Soc.*, vol. 12, pp. 159–163, 1961. For $n > 1$, the proof of the text seems to be the only one that is known.

Theorem 11.9. When A has no unit, then Δ is locally compact (but not compact) and $\hat{A} \subset C_0(\Delta)$; the origin of A^* is then in the closure of Δ. See [16], pp. 52–53.

Theorem 11.10 is what has been called an "automatic continuity" theorem. (Theorems 10.7 and 11.31 are other examples.) This is a concept which brings classical analysis into contact with axiomatic set theory. For example, "Kaplansky's problem" is the following: Is it true, for every compact Hausdorff space X and every Banach algebra A, that every homomorphism from $C(X)$ into A is continuous? The work of Dales, Esterle, Solovay, and Woodin has shown that this question is undecidable in ZFC (Zermelo-Frankel set theory plus the axiom of choice). See [38] for details.

Example 11.13 (*d*) shows why there are very close relations between commutative Banach algebras, on the one hand, and holomorphic functions of several complex variables on the other. This topic is not at all pursued in the present book. Very good, up-to-date accounts of it may be found in the books by Browder [3]. Gamelin [6], Stout [29], and Wermer [47]. A symbolic calculus for functions of several Banach algebra elements can be developed. See R. Arens and A. P. Calderon, *Ann. Math.*, vol. 62, pp. 204–216, 1955, and J. L. Taylor, *Acta Math.*, vol. 125, pp. 1–38, 1970.

Example 11.13 (*e*) shows why certain parts of Fourier analysis may be derived easily from the theory of Banach algebras. This is done in [16] and [24].

Theorem 11.18 was proved by Gelfand and Naimark in *Mat. Sbornik*, vol. 12, pp. 197–213, 1943. In the same paper they also proved that every B^*-algebra A (commutative or not) is isometrically *-isomorphic to an algebra of bounded operators on some Hilbert space (Theorem 12.41), *if* $e + x^*x$ is invertible for every $x \in A$. That this additional hypothesis is redundant was proved 15 years later by I. Kaplansky [(*f*) of Theorem 11.28]. See [21], p. 248, for references to the rather tangled history of this theorem. B. J. Glickfeld (*Ill. J. Math.*, vol. 10, pp. 547–556, 1966) has shown that A is a B^*-algebra if $\|\exp(ix)\| = 1$ for every hermitian $x \in A$.

Theorem 11.20. The idea to pass from A to A/R, in order to prove the theorem without assuming the involution to be continuous, is due to J. W. M. Ford (*J. London Math. Soc.*, vol. 42, pp. 521–522, 1967).

Theorem 11.23. See R. S. Foguel, *Ark. Mat.*, vol. 3, pp. 449–461, 1957.

Theorem 11.25. See P. Civin and B. Yood, *Pac. J. Math.*, vol. 9, pp. 415–436, 1959; especially p. 420. Also [21], p. 182.

Theorem 11.28. A recent treatment of these matters was given by V. Pták, *Bull. London Math. Soc.*, vol. 2, pp. 327–334, 1970. Also, see the note to Theorem 11.18.

Theorem 11.31. See [19], [21]. H. F. Bohnenblust and S. Karlin (*Ann. Math.*, vol. 62, pp. 217–229, 1955) have found relations between positive functionals, on the one hand, and the geometry of the unit ball of a Banach algebra on the other.

Theorem 11.32. See [7]. Also [16], p. 97, and [21], p. 230.

Theorem 11.33 is in [20], for continuous involutions.

Exercise 13. Part (*g*) contradicts the second half of corollary (4.5.3) in [21]. It also affects Theorem (4.8.16) of [21].

Exercise 14. This was first proved by S. Bochner (*Math. Ann.*, vol. 108, pp. 378–410, 1933; especially p. 407), using essentially the same machinery that we used in Theorem 7.7. See [24] for a somewhat different proof. The proof that is suggested here shows that the presence or absence of a unit element makes a difference in studying positive functionals. See [16], p. 96, and [21], p. 219.

Chapter 12

General references: [4], [9], [10], [17], [22].

Theorem 12.16. B. Fuglede proved the case $M = N$ in *Proc. Natl. Acad. Sci. USA*, vol. 36, pp. 35–40, 1950, including the unbounded case (Chapter 13, Exercise 15). His proof used the spectral theorem and was extended to the case $M \neq N$ by C. R. Putnam (*Amer. J. Math.*, vol. 73, pp. 357–362, 1951), who also obtained Theorem 12.36. The short proof of the text is due to M. Rosenblum, *J. London Math. Soc.*, vol. 33, pp. 376–377, 1958.

Theorem 12.22. The extension process that is used here to go from continuous functions to bounded ones is as in [16], pp. 93–94.

Theorem 12.23. See [4], pp. 926–936, for historical remarks about the spectral theorem. See also P. R. Halmos' article in *Amer. Math. Monthly*, vol. 70, pp. 241–247, 1963, for a different description of the spectral theorem.

Theorem 12.38 was proved by P. R. Halmos, G. Lumer, and J. Schäffer, in *Proc. Amer. Math. Soc.*, vol. 4, pp. 142–149, 1953. D. Deckard and C. Pearcy (*Acta Sci., Math. Szeged.*, vol. 28, pp. 1–7, 1967) went further and proved that the range of the exponential function is neither open nor closed in the group of invertible operators. Their paper contains several references to intermediate results.

Theorem 12.39. See [21], p. 227.

Theorem 12.41. Closed *-subalgebras of $\mathscr{B}(H)$ are called *C^*-algebras.* Before Theorem 12.41 was known (see the note to Theorem 11.18), B^*-algebras were studied separately, but now the term *B^*-algebra* is no longer used much.

Theorems 12.43, 12.44. Several types of ergodic theorems are discussed in [4] and [43].

Exercise 2 is very familiar if $N = 4$.

Exercise 18. The relation between shift operators and the invariant subspace problem is discussed by P. R. Halmos in *J. Reine Angew. Math.*, vol. 208, pp. 102–112, 1961.

Exercise 27. See P. Civin and B. Yood, *Pac. J. Math.*, vol. 9, pp. 415–436, 1959, for many results about involutions.

Exercise 32. Part (*c*) implies that every uniformly convex Banach space is reflexive. See Exercise 1 of Chapter 4 and the note to Exercise 28 of Chapter 3. All L^p-spaces (with $1 < p < \infty$) are uniformly convex. See J. A. Clarkson, *Trans. Amer. Math. Soc.*, vol. 40, pp. 396–414, 1936, or [15], pp. 355–359.

Chapter 13

General references: [4], [12], [22].

Theorem 13.6 was first proved by A. Wintner, *Phys. Rev.*, vol. 71, pp. 738–739, 1947. The more algebraic proof of the text is H. Wielandt's, *Math. Ann.*, vol. 121, p. 21, 1949. It was generalized by D. C. Kleinecke (*Proc. Amer. Math. Soc.*, vol. 8, pp. 535–536, 1957), to yield the following theorem about *derivations:* If D is a continuous linear operator in a Banach algebra A such that $D(xy) = xDy + (Dx)y$ for all $x, y \in A$, then the spectral radius of Dx is 0 for every x that commutes with Dx. This was also proved by Shirokov (*Uspehi*, vol. 11, no. 4, pp. 167–168, 1956) and, in commutative Banach algebras, by Singer and Wermer (*Math. Ann.*, vol. 129, pp. 260–264, 1955). See p. 20 of I. Kaplansky's article "Functional Analysis" in *Some Aspects of Analysis and Probability*, John Wiley & Sons, New York, 1958.

A. Brown and C. Pearcy (*Ann. Math.*, vol. 82, pp. 112–127, 1965) have proved, for separable H, that an operator $T \in \mathscr{B}(H)$ is a commutator if and only if T is *not* of the form $M + C$, where $\lambda \neq 0$ and C is compact. See also C. Schneeberger, *Proc. Amer. Math. Soc.*, vol. 28, pp. 464–472, 1971.

The Cayley transform, its relation to deficiency indices, and the proof of Theorem 13.30 are in von Neumann's paper in *Math. Ann.*, vol. 102, pp. 49–131, 1929–1930, and so is the spectral theorem for normal unbounded operators. The material on graphs is in his paper in *Ann. Math.*, vol. 33, pp. 294–310, 1932. Our proof of Theorem 13.33 is like that of F. Riesz and E. R. Lorch, *Trans. Amer. Math. Soc.*, vol. 39, pp. 331–340, 1936. See also [4], chap. XII.

Definition 13.34. The continuity condition we impose can be weakened: if (*a*) and (*b*) hold, and if $Q(t)x \to x$ *weakly*, as $t \to 0$, for every $x \in X$, then (*c*) holds. See [33], pp. 233–234. The proof uses more from the theory of vector-valued integration than the present book contains.

Theorems 13.35–13.37 are proved in [4], [12], [22], [33], and [46].

Theorem 13.38. This is due to M. H. Stone, *Ann. Math.*, vol. 33, pp. 643–648, 1932; see also B. Sz. Nagy, *Math. Ann.* vol. 112, pp. 286–296, 1936.

Exercise 25 was pointed out to me by Sheldon Axler. It corrects an error made on p. 341 of the first edition of this book.

Appendix A

Section A2. J. W. Alexander, *Proc. Natl. Acad. Sci. USA*, vol. 25, pp. 296–298, 1939.

Section A3. A. Tychonoff proved this for cartesian products of intervals (*Math. Ann.*, vol. 102, pp. 544–561, 1930) and used it to construct what is now known as the Čech (or Stone-Čech) compactification of a completely regular space. E. Čech (*Ann. Math.*, vol. 38, pp. 823–844, 1937; especially p. 830) proved the general case of the theorem and studied properties of the compactification. Thus it appears that Čech proved the Tychonoff theorem, whereas Tychonoff found the Čech compactification—a good illustration of the historical reliability of mathematical nomenclature.

BIBLIOGRAPHY

1. AGMON, S.: *Lectures on Elliptic Boundary Value Problems*, D. Van Nostrand Company, Princeton, N.J., 1965.
2. BANACH, S.: *Théorie des Opérations linéaires*, Monografje Matematyczne, vol. 1, Warsaw, 1932.
3. BROWDER, A.: *Introduction to Function Algebras*, W. A. Benjamin, New York, 1969.
4. DUNFORD, N., and J. T. SCHWARTZ: *Linear Operators*, Interscience Publishers, a division of John Wiley & Sons, New York, pt. I, 1958; pt. II, 1963; pt. III, 1971.
5. EDWARDS, R. E.: *Functional Analysis*, Holt, Rinehart and Winston, New York, 1965.
6. GAMELIN, T. W.: *Uniform Algebras*, Prentice-Hall, Englewood Cliffs, N.J., 1969.
7. GELFAND, I. M., D. RAIKOV, and G. E. SHILOV: *Commutative Normed Rings*, Chelsea Publishing Company, New York, 1964. (Russian original, 1960.)
8. GELFAND, I. M., and G. E. SHILOV: *Generalized Functions*, Academic Press, New York, 1964. (Russian original, 1958.)
9. HALMOS, P. R.: *Introduction to Hilbert Space and the Theory of Spectral Multiplicity*, Chelsea Publishing Company, New York, 1951.
10. HALMOS, P. R.: *A Hilbert Space Problem Book*, D. Van Nostrand Company, Princeton, N.J., 1967.
11. HEWITT, E., and K. A. ROSS: *Abstract Harmonic Analysis*, Springer-Verlag OHG, Berlin, vol. 1, 1963; vol. 2, 1970.
12. HILLE, E., and R. S. PHILLIPS, *Functional Analysis and Semigroups*, Amer. Math. Soc. Colloquium Publ. 31, Providence, R.I., 1957.
13. HÖRMANDER, L.: *Linear Partial Differential Operators*, Springer-Verlag OHG, Berlin, 1963.
14. KELLEY, J. L., and I. NAMIOKA: *Linear Topological Spaces*, D. Van Nostrand Company, Princeton, N.J., 1963.
15. KÖTHE, G.: *Topological Vector Spaces*, Springer-Verlag, New York, vol. 1, 1969; vol. 2, 1979.
16. LOOMIS, L. H.: *An Introduction to Abstract Harmonic Analysis*, D. Van Nostrand Company, Princeton, N.J., 1953.

17. LORCH, E. R.: *Spectral Theory*, Oxford University Press, New York, 1962.
18. NACHBIN, L.: *The Haar Integral*, D. Van Nostrand Company, Princeton, N.J., 1965.
19. NAIMARK, M. A.: *Normed Rings*, Erven P. Noordhoff, Groningen, Netherlands, 1960. (Original Russian edition, 1955.)
20. PHELPS, R. R.: *Lectures on Choquet's Theorem*, D. Van Nostrand Company, Princeton, N.J., 1966.
21. RICKART, C. E.: *General Theory of Banach Algebras*, D. Van Nostrand Company, Princeton, N.J., 1960.
22. RIESZ, F., and B. SZ.-NAGY: *Functional Analysis*, Frederick Ungar Publishing Company, New York, 1955.
23. RUDIN, W.: *Real and Complex Analysis*, 3d ed., McGraw-Hill Book Company, New York, 1987.
24. RUDIN, W.: *Fourier Analysis on Groups*, Interscience Publishers, a division of John Wiley & Sons, New York, 1962.
25. RUDIN, W.: *Function Theory in Polydiscs*, W. A. Benjamin, New York, 1969.
26. SCHWARTZ, L.: *Théorie des distributions*, Hermann & Cie, Paris, 1966.
27. SHILOV, G. E.: *Generalized Functions and Partial Differential Equations*, Gordon and Breach, Science Publishers, New York, 1968. (Russian original, 1965.)
28. STONE, M. H.: *Linear Transformations in Hilbert Space and Their Applications to Analysis*, Amer. Math. Soc. Colloquium Publ. 15, New York, 1932.
29. STOUT, E. L.: *The Theory of Uniform Algebras*, Bogden and Quigley, Tarrytown, N.Y., 1971.
30. TRÈVES, F.: *Linear Partial Differential Equations with Constant Coefficients*, Gordon and Breach, Science Publishers, New York, 1966.
31. TRÈVES, F.: *Topological Vector Spaces, Distributions, and Kernels*, Academic Press, New York, 1967.
32. WILANSKY, A.: *Functional Analysis*, Blaisdell, New York, 1964.
33. YOSIDA, K.: *Functional Analysis*, Springer-Verlag, New York, 1968.
34. ZYGMUND, A.: *Trigonometric Series*, 2d ed., Cambridge University Press, New York, 1959.

Supplementary Bibliography

35. DALES, H. G., and W. H. WOODIN: *An Introduction to Independence for Analysts*, *London Math. Soc. Lecture Notes*, vol. 115, Cambridge University Press, Cambridge, 1987.
36. DIEUDONNÉ, J. A.: *History of Functional Analysis*, North Holland, Amsterdam, 1981.
37. DIXMIER, J.: *C*-algebras*, North Holland, Amsterdam, 1977.
38. DOUGLAS, R. G.: *Banach Algebra Techniques in Operator Theory*, Academic Press, New York, 1972.
39. HELSON, H.: *Lectures on Invariant Subspaces*, Academic Press, New York, 1964.
40. KALTON, N. J., N. T. PECK, and J. W. ROBERTS: *An F-space Sampler*, *London Math. Soc. Lecture Notes*, vol. 89, Cambridge University Press, Cambridge, 1984.
41. LINDENSTRAUSS, J., and L. TZAFRIRI: *Classical Banach Spaces*, Springer-Verlag, Berlin, vol. 1, 1977, vol. 2, 1979.
42. LOPÉZ, J. M., and K. A. ROSS: *Sidon Sets*, Marcel Dekker, New York, 1975.
43. PETERSEN, K.: *Ergodic Theory*, Cambridge University Press, Cambridge, 1983.
44. RADJAVI, H., and P. ROSENTHAL: *Invariant Subspaces*, Springer-Verlag, New York, 1973.
45. RUDIN, W.: *Function Theory in the Unit Ball of* $\mathbb{C}^n$, Springer-Verlag, New York, 1980.
46. TRÈVES, F.: *Basic Linear Partial Differential Equations*, Academic Press, New York, 1975.
47. WERMER, J.: *Banach Algebras and Several Complex Variables*, 2d ed., Springer-Verlag, New York, 1976.

LIST OF SPECIAL SYMBOLS

The numbers that follow the symbols indicate the sections where their meanings are explained.

Spaces

$C(\Omega)$	1.3	$\mathscr{B}(X, Y)$	4.1
$H(\Omega)$	1.3	$\mathscr{B}(X)$	4.1
C_K^∞	1.3	X^{**}	4.5
$\mathscr{N}(\Lambda)$	1.16	$M^\perp$	4.6, 12.4
R^n	1.19	${}^\perp N$	4.6
$\mathbb{C}^n$	1.19	$\mathscr{N}(T)$	4.11
X/N	1.40	$\mathscr{R}(T)$	4.11
$\mathbb{L}$	1.43	H^1	5.19
$\mathscr{D}_K$	1.46	$\mathscr{D}$	6.1
$C^\infty(\Omega)$	1.46	$\mathscr{D}(\Omega)$	6.2
Lip α	Exercise 22, Chapter 1	$\mathscr{D}'(\Omega)$	6.7
ℓ^p	Exercise 5, Chapter 2	$\mathscr{S}_n$	7.3
X^*	3.1	$C_0(R^n)$	7.5
X_w	3.11	$\mathscr{S}'_n$	7.11
ℓ^∞	Exercise 4, Chapter 3	$C^{(p)}(\Omega)$	7.24

Operators

Number Theoretic Functions and Symbols

Other Symbols

Symbol	Section	Meaning
$\|\phi\|_N$	6.2	norm in $\mathscr{D}(\Omega)$
$x \cdot y$	6.10	scalar product
$\lvert x \rvert$	6.10	length of vector
x^α	6.10	monomial
S_Λ	6.24	support
$\check{u}$	6.29	$\check{u}(x) = u(-x)$
$u * v$	6.29, 6.34, 6.37, 7.1	convolution
m_n	7.1	Lebesgue measure on R^n
e_t	7.1	character
$\hat{f}(t)$	7.1	Fourier transform
e_z	7.20	exponential
rB	7.22	ball of radius r
E	8.1	fundamental solution
σ_n	8.2	Haar measure on T^n
μ_s	8.8	measure related to H^s
$Z(Y)$	9.3	zero set
μ_a, μ_s	9.14	Lebesgue decomposition of μ
e	10.1	unit element
$G(A)$	10.10	group of invertible elements
$\sigma(x)$	10.10	spectrum
$\rho(x)$	10.10	spectral radius
A_Ω	10.26	members of A with spectrum in Ω
$\tilde{f}$	10.26	A-valued holomorphic functions
Δ	11.5	maximal ideal space
U^n	11.7	polydisc
$\hat{x}$	11.8	Gelfand transform
$\Gamma(S)$	11.21	centralizer
(x, y)	12.1	inner product
$\perp$	12.1	orthogonality relation
E	12.17	resolution of identity
$E_{x,y}$	12.17	spectral measure
$T \subset S$	13.1	inclusion of operators

INDEX